World Sustainability Series

Series Editor

Walter Leal Filho, European School of Sustainability Science and Research, Research and Transfer Centre "Sustainable Development and Climate Change Management", Hamburg University of Applied Sciences, Hamburg, Germany

Due to its scope and nature, sustainable development is a matter which is very interdisciplinary, and draws from knowledge and inputs from the social sciences and environmental sciences on the one hand, but also from physical sciences and arts on the other. As such, there is a perceived need to foster integrative approaches, whereby the combination of inputs from various fields may contribute to a better understanding of what sustainability is, and means to people. But despite the need for and the relevance of integrative approaches towards sustainable development, there is a paucity of literature which address matters related to sustainability in an integrated way.

Notes on the quality assurance and peer review of this publication

Prior to publication, the works published in this book are initially assessed and reviewed by an in-house editor. If suitable for publication, manuscripts are sent for further review, which includes a combined effort by the editorial board and appointed subject experts, who provide independent peer-review. The feedback obtained in this way was communicated to authors, and with manuscripts checked upon return before finally accepted. The peer-reviewed nature of the books in the "World Sustainability Series" means that contributions to them have, over many years, been officially accepted for tenure and promotion purposes.

Walter Leal Filho · Thais Dibbern ·
Salvador Ruiz de Maya ·
María-del-Carmen Alarcón-del-Amo ·
Longinos Marin Rives
Editors

The Contribution of Universities Towards Education for Sustainable Development

Springer

Editors
Walter Leal Filho
Faculty of Life Sciences
Hamburg University of Applied Sciences
Hamburg, Germany

Salvador Ruiz de Maya
Department of Marketing
University of Murcia
Murcia, Spain

Longinos Marin Rives
Department of Marketing
University of Murcia
Murcia, Spain

Thais Dibbern
Department of Science and Technology Policy
University of Campinas (UNICAMP)
Campinas, Brazil

María-del-Carmen Alarcón-del-Amo
Department of Marketing
University of Murcia
Murcia, Spain

ISSN 2199-7373 ISSN 2199-7381 (electronic)
World Sustainability Series
ISBN 978-3-031-49855-8 ISBN 978-3-031-49853-4 (eBook)
https://doi.org/10.1007/978-3-031-49853-4

This Springer imprint is published by the registered company Springer Nature Switzerland AG
The registered company address is: Gewerbestrasse 11, 6330 Cham, Switzerland

Paper in this product is recyclable.

Contents

Are the UN SDGs Being Implemented in Geography Courses at Spanish Universities?

Carlos Martínez-Hernández and Carmen Mínguez

Abstract In the present territorial context (socioenvironmental global changes and inequalities), UN has conformed a framework of sustainability goals (SDGs) through 2030 Agenda, which can be transferred to citizens by education. This way, Geography can be considered "the science of sustainability", so geography education must be adapted to the new framework. Is it like that for higher education? To what extent? In this work, we aim to assess the implementation of the SDGs framework in Geography Degrees in Spain according to teachers and students. We present the results of a validated survey to 319 participants. Data analysis with descriptive and inferential techniques have been performed. As key findings, all SDGs are significantly less present than participants think they should be and only SGDs 11 and 13 are highly punctuated (urbanism and climatology, traditional research themes in geography). In general, teachers better know the potentiality of geography for a sustainable education, and they usually evaluate significatively higher. We can conclude that there is a need to "sustain" the curriculum and even that the adaptation to this framework could improve the working opportunities of geographers.

Keywords Geography education · Global change · 2030 agenda · University degree · Education for sustainable development

C. Martínez-Hernández (✉)
Faculty of Education, Complutense University of Madrid, 1, 28040 Madrid, Spain
e-mail: cmartinezhernandez@ucm.es

C. Mínguez
Faculty of Geography and History, Complutense University of Madrid, 28040 Madrid, Spain
e-mail: mcmingue@ucm.es

W. Leal Filho et al. (eds.), *The Contribution of Universities Towards Education for Sustainable Development*, World Sustainability Series,
https://doi.org/10.1007/978-3-031-49853-4_1

1 Introduction

The global present moment is strongly marked by human action, by the volume of inhabitants and the use we make of resources. This leads to a continuous degradation of natural resources, with major environmental and social effects at the global level (Kerr et al. 2007; Crutzen 2021).

Therefore, a new era based on sustainable development and social welfare has become a need for many academics (Holmberg and Sandbrook 2019; Fletcher et al. 2019; Hummels and Argyrou 2021). This discourse has also been adopted by international organizations, such as the United Nations (UN), which designed a series of global goals: 8 "Millennium Development Goals" (2000–2015) and 17 "Sustainable Development Goals" (SDGs) later, as part of the 2030 Agenda for Sustainable Development (United Nations 2017).

The SDGs are part of a globally shared plan to ensure present and future prosperity. This plan aims to end poverty through a series of strategies to improve the health and education of the population. To this end, it is considered essential that the strategies are aimed at reducing inequalities and promoting economic growth, taking into account two key aspects: climate change and the preservation of the natural environment (mainly forests and oceans).

The involvement of governments, companies and civil society is required to achieve the UN goals. UNESCO, as the responsible organization for education in the UN goals, has designed strategies for implementing the SDGs in education (Murga-Menoyo 2021). Universities are optimal institutions for addressing education for sustainable development (Lim et al. 2022), as they can bring research and teaching of the Agenda 2030 together, improving a global citizenship among their diverse actors (Faham et al. 2017). However, to get an efficacy transfer to global sustainability, higher education institutions must adapt this framework all around the world (Leal Filho et al. 2023). Some authors highlight the importance of facing the obstacle of faculty motivation (Brahm and Kühner 2019). A common proposal is implanting transdisciplinary programs (Jia et al. 2019) and a sustainability-based curricula (Žalėnienė and Pereira 2021), so that any future graduates can obtain an eco-social literacy basement (Wals 2009), in line with education for sustainable development. Nevertheless, there is yet uncertainty on how much universities are having success (Leal Filho et al. 2023) and research on specific countries, universities or degrees is of increasing interest.

The discipline of Geography allows a holistic understanding of global sustainability, bringing environmental and human aspects together, so it can highly contribute to the generation of knowledge and skills to face future challenges (Meadows 2020). This fact represents a relevant opportunity for geographic education at all academic levels, especially in universities, to train specialists in the study of territorial problems as well as in the search for solutions for a "glocal" sustainable citizenship (Sklad et al. 2016).

Training in sustainability would improve the skills and ethical background of geography students and then facilitate their incorporation into the labor market. We hypothesize that the 2030 Agenda is not yet actually present in Geography degrees as it could potentially be. Therefore, this research aims to evaluate the degrees of Geography in Spain from the perspective of both students and teachers in order to provide their perception on how the SDGs are implemented in the teaching programs. So, it could be concluded how Geography courses contribute in higher education towards education for sustainable development.

2 Sustainable Development Goals in Geography Higher Education

Geography, as a science based on knowledge of spatial phenomena on a planetary scale, must be adaptative and anticipate changes, becoming relevant to society (Harden 2012; Moss et al. 2013). Research has been updated to non-conformist themes, and it sometimes provides an alternative, but there is not a full transfer to teaching, wasting the power of knowledge (Castree 2015a). This transfer would bring the train of the future professionals, through theoretical content and certain skills, methodologies and competencies to ensure anticipation (Meadows 2020). This represents a significant opportunity cost for degrees (Harden 2012) and becomes one of the main challenges of geographic education.

The approach to a culture of sustainability in the curricula helps to improve the quality of university studies (Collado et al. 2022) and its image (Žalėnienė and Pereira 2021). The International Year of Global Understanding (IYGU) in 2016 was a milestone to learn about other educational practices and reinforce the role of geography as 'the science for sustainability', thanks to the holistic vision of geography on territorial aspects (Meadows 2020).

As recognized in the International Charter for Geographic Education (Commission on Geographical Education 1992), geography provides students with knowledge about natural and socioeconomic systems. This helps to understand the interaction of all their elements and/or the relationships between systems, from a same-level approach between natural and human sciences. However, in order to get geography relevant in the teaching of sustainable development, contents must be considered in a systemic environmental way, reflectively integrating social and environmental aspects and analyzing them together as part of the same problem (Haubrich et al. 2007).

Learning Geography provides knowledge as well as skills, attitudes and values which are necessary to implement the sustainable development of the future world (Commission on Geographical Education 1992). In this way, geographic education can specifically contribute to the acquisition of vital competencies around sustainability: (i) ecosocial competencies, to calculate risks, evaluate alternatives, and become interested in geographic phenomena from different perspectives; (ii) critical competencies, to reflect on cause-effect relationships, consequences and contextualize local experiences in global phenomena; (iii) action competencies, to acquire sensitivity and become aware of the impact of their own behavior and that which is generated by society of which they are a part; and (iv) critical thinking competencies, to access accurate information and detect misinformation. Geographic education provides competencies that can allow students to make decisions in accordance with the principles of sustainability, combining social and environmental points of view.

The new Spanish educational curriculum already includes the potential didactic contribution of geography in the teaching of sustainable development. It states some specific ecosocial literacy competences to be developed in Geography and similar subjects. For primary education, it is stated that students must: "Identify the causes and consequences of human intervention in the environment, from the social, economic, cultural, technological and environmental points of view, to improve the ability to face problems, seek solutions and act individually and cooperatively in its resolution, and to put into practice sustainable lifestyles consistent with respect, care and protection of people and the planet" (BOE 52 2022, p. 24,416). Also, the curriculum of the Geography subject in compulsory secondary education includes a specific competence to educate in sustainable development. It is related to human intervention in the environment, as in primary education: "Understanding the interconnected and inter- and eco-dependent nature of the human activities, through the identification and analysis of relevant ecosocial problems, to promote habits and attitudes ethically committed to the achievement of sustainable ways of life" (BOE 76 2022, p. 41,649). In non-compulsory secondary education, the Geography subject is limited to the context of Spain, and the competence required for the education in sustainable development is adapted to this circumstance: "Recognize current and future ecosocial challenges in Spain, debating from the geographical perspective on the messages received through unofficial, official and informal channels, and developing critical thinking, to transform unsustainable consumption patterns and adopt formal healthy lifestyles" (BOE 82 2022, p. 46,214).

As Haubrich points out, "to analyse a problem is not sufficient in modern geography education. To think about alternative solutions and responses has to be discussed, if education shall qualify students for their participation for example in local agenda activities or generally in politics" (Haubrich 2007, p. 33). However, beyond research on the contribution of geography to the education for sustainable development in a theoretical way (Meadows 2020) or in specific courses or contexts (Bagoly-Simó and Kriewaldt 2022), there has been scarcely in-depth discussion on the roots of geography training studies, moreover in Spain, to check if they guarantee the competencies required by an ethical framework of sustainability and the current labor market itself, within a context of crisis and global change.

In Spain, geographic education has a long tradition as a university degree. It was a specialty from XIX Century and an independent degree after a regulation change in 1990. It was based on fundamental subjects of Physical Geography, Human Geography and Regional Geography, from spatial planning and management of natural resources approach. Between 2001 and 2011 it was adapted to the European Higher Education Area (EHEA), following the model of the Spanish 4 + 1 degree (Tulla 2010; De Miguel and De Lázaro 2016).

Since EHEA regulation, the degrees have been occasionally modified, but not adapted to the new framework of the major global challenges, which are largely of a socio-environmental nature and becoming increasingly powerful, so they might require a thematic, philosophical and methodological adaptation of geography higher education (Castree 2015a).

Within the context of global changes, a significant part of the career opportunities for geography students are related to the 2030 Agenda, and higher education is a key element to achieving this (Yli-Panula et al. 2019). Geographers can work for the private or administrative sector, applying sustainability competencies to interdisciplinary spatial thinking and becoming more competitive and ethical for the labor market (Teichert et al. 2021), which is traditionally difficult to access (De Cos and Reques 2010).

3 Materials and Methods

3.1 Data Collection Tool and Participants

An online questionnaire (https://arcg.is/1HLDe10), designed in ARCGIS-Survey, was used as a data collection instrument for students and teachers of undergraduate degrees in Geography. The questionnaire was structured into nine sections, in order to systematize the opinion of the respondents according to the theoretical framework of the SDGs and their presence in university geographic education.

This questionnaire underwent a validation process through an expert judgment and a subsequent pilot test:

- Adequacy (3.8 ∓ 0.5 out of 4) and pertinence (3.9 ∓ 0.3 out of 4) of all the questionnaire items were excellent. The Cronbach's Alpha reliability test yielded a very high statistical consistency ($\alpha = 0.944$) and allowed the judgement to be considered valid. The experts made some minor observations on the improvement in the wording of some items, and they were considered.
- The responses of the pilot test were subjected to the Cronbach's Alpha reliability test, with an excellent result ($\alpha = 0.943$). Respondents were also asked to rate the appropriateness of a series of items on the completion of the questionnaire, resulting in an average of 3.4 ∓ 0.9 out of 4, which serves to positively culminate the questionnaire validation process.

The final version of the questionnaire was disseminated among all the universities in Spain where Geography degrees are taught ($n = 26$). Responses could be submitted for a period of two months. The submission period closed with 319 valid responses. It is a representative sample, with an unknown universe that tends to infinity, a confidence interval of 95% and a margin of error of 5.5%, according to the sampling formula of Bartlett et al. (2001).

The research data were classified according to the sample categorical variables of gender (67.4% female, 31.7% male), age (36.4% < 26, 21.0% 26–40, 35.1% 41–60,7.5% > 60), university (100% of Spanish universities), student or teacher condition (52.7% teachers, 47.3% students), teaching areas (11.6% Physical G., 19.7% Human G., 19.4% Regional G., 1.9% Other), student status (37.3% active, 10.0% graduates), student seniority (4.4% pre-Bologna plan, 42.9% post), and postgraduate students (7.8%).

3.2 Research Variables and Statistical Analysis

The responses to the questionnaire enabled a series of research variables to be configured. In this research 36 were used, associated with the nominal variables for classifying the sample: 17 for the importance of each SDG, another 17 for their actual presence, and 2 for the opinion about the 2030 Agenda. The SDGs variables are ordinal quantitative, with a range of values from 1 to 4, and the rest are qualitative, with values of text.

The statistical treatment of the research was carried out using SPSS v25 (IBM) and Excel (Microsoft Office 365). Descriptive statistics were calculated for all the research variables. The open opinion variables were subjected to a reclassification and descriptive statistics for frequencies of each type of opinion.

Inferential statistics were carried out to determine the significant differences between the scores of the research variables and the different categories of the classification variables. The nominal categories were reclassified into ordinal groups and statistical tests were performed according to the number of independent groups and nonparametric quantitative dependent variables: Mann–Whitney U significance test for 2 independent samples and Kruskal–Wallis test for K independent samples. Previously, the Kolmogorov–Smirnov normality test ($n > 50$) was applied ($p = 0.000$).

Finally, a longitudinal study on the SDGs score between an initial moment of potentiality or expectation and a final moment of the actual situation was also carried out, in order to determine if the differences were statistically significant. For this purpose, the Wilcoxon nonparametric significance test for two related samples was used.

The results of all the inferential statistical tests and the normality test have been compiled in Appendix I. The most important information is referred to and shown in the Results section.

4 Results

At least 75% of respondents consider that all the SDGs of the 2030 Agenda should have a high or very high academic presence in geography courses of Spanish universities (Table 1). They specially highlight SDGs 11 (Sustainable cities and communities)

Table 1 Presence of every SDG (expectative versus reality) in Geography Spanish higher education in a scale 1–4 (low–high)

SDG	Expectative					Reality				
	Average	SD	Q1	Mean	Q3	Average	SD	Q1	Mean	Q3
1*	3.2	0.9	3	4	4	2.2	1.0	1	3	3
2*	3.2	0.9	3	4	4	2.0	1.0	1	3	3
3*	3.2	0.9	3	4	4	2.3	1.0	1	3	3
4*	3.3	0.9	3	4	4	2.5	1.0	2	3	3
5*	3.2	0.9	3	4	4	2.6	1.1	2	4	4
6*	3.6	0.6	3	4	4	2.7	0.9	2	3	3
7*	3.6	0.7	3	4	4	2.7	1.0	2	3	3
8*	3.3	0.8	3	4	4	2.5	1.0	2	3	3
9*	3.3	0.8	3	4	4	2.5	1.0	2	3	3
10*	3.5	0.7	3	4	4	2.7	1.0	2	4	4
11*	3.7	0.5	4	4	4	3.1	0.9	3	4	4
12*	3.4	0.7	3	4	4	2.6	1.0	2	3	3
13*	3.8	0.5	4	4	4	3.1	0.9	3	4	4
14*	3.2	0.9	3	4	4	2.0	1.1	1	3	3
15*	3.6	0.7	3	4	4	2.8	1.0	2	4	4
16*	3.1	0.9	3	4	4	2.2	1.0	1	3	3
17*	3.2	0.9	3	4	4	2.2	1.0	1	3	3
All*	3.4	0.8	3	4	4	2.5	1.0	2	3	3

* Significative differences (p-value < 0.05)

and 13 (Climate action). On the other hand, when assessing the actual presence of the SDGs according to their experience, the scores considerably decrease in all of them. Only the SDGs 11 and 13 maintain a majority perception of a high actual presence (mean of 3.2). The worst rated are SDG 2 (Zero hunger) and 14 (Life below water), with an average of around 2.0.

The differences between the level considered adequate and the level of actual presence are around 1 point on average and are always significative. The largest decreases, around 1.2 points on average, coincides with the SDGs with less expectation: 2 (Zero hunger) and 14 (Life below water). The smallest decreases, although statistically significant, are around 0.6 points on average and correspond to SDG 5 (Gender equality) and 11 (Sustainable cities and communities). The last one obtained the highest value in both its expectation and its actual final presence.

The punctuation of the expected and experienced SDGs usually presents statistically significant differences between teachers and students (Fig. 1). In general, teachers give higher scores in expectation. On the contrary, the actual presence of SDGs is usually significantly higher punctuated by students.

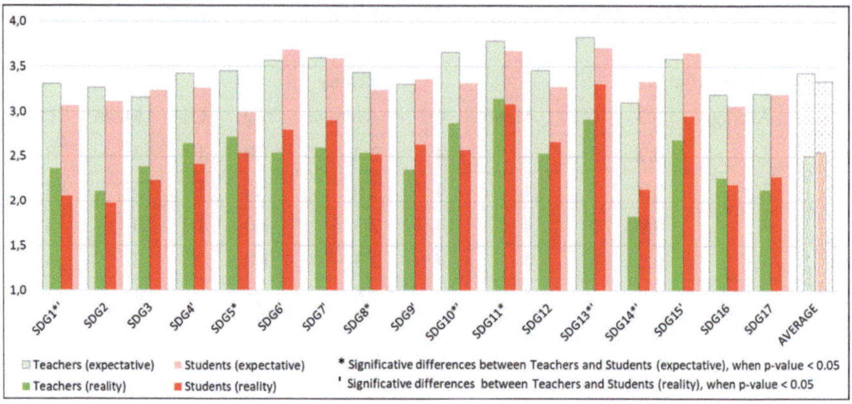

Fig. 1 Statistical differences between the average willing presence of every SDG in Geography Spanish higher education (expectative) and the real perceived one (reality) regarding teachers and students in a scale 1–4 (low–high)

Statistically significant differences can be found in several SDGs regarding the age of the respondents, coinciding with the differences between teachers and students (see Appendix I). There is a clear pattern: younger ages as students versus older ages as teachers. For gender, it is noteworthy that women give significantly higher scores in SDGs 4 and 5, both in expectation and reality.

There are also significant differences according to some particularities of teachers and students. For example, the SDGs which are more linked to humanistic research (like 1, 3, 8 or 10) are less valued by Physical Geography teachers, who give higher scores to others, such as 14 and 15, which are more environmentalists. Among the students, those who are still in training always give higher scores than former graduates. Nonetheless, the year of commencement of studies does not generate statistically significant differences.

Forty percent of the respondents expressed their opinion on the adequacy of including the 2030 Agenda in a curricular way, from both an academic and professional point of view. Among them, 34% demand a higher presence of practical contents related to the SDGs, like more case studies and field work in the context of the 2030 Agenda, or the use of GIT at the service of the SDGs. Next, 28.1% claim for an improvement in conceptual contents on sustainability, adapted to current social responsibility topics such as global change, the circular economy, food security, geopolitical conflicts or social participation systems. A greater importance of attitudinal contents is only called for by 8%, in relation to transversal ethical background of sustainability and critical thinking. However, the opinion of most of the respondents (40%) is directed towards a better mixture of the previous kind of contents, through the "sustainability" of the curriculum, that is, the inclusion of competences in sustainability, the creation of a specific subject, the teaching of seminars and workshops, service-learning experiences and greater training and experimentation in development cooperation.

On the other hand, when expressing opinion on the importance of the 2030 Agenda itself and its relationship with Geography, up to 23% of the respondents make a negative assessment, considering it is insufficient and late, as well as in line with neoliberal policies. However, 34% recognize it as a very important milestone and with great potential in Geography, and 53% detail in what way: sustainability of the curriculum and teacher training, thematic updating to global change and more practical content.

5 Discussion

Geography is widely considered a basic discipline in the study of sustainability (Žalėnienė and Pereira 2021; Maude 2022; Olcina Cantos 2022), which is based on the linkage and knowledge of many of the factors involved and manifested in research, specifically in Spain. However, this conviction is not transmitted in teaching, and according to our results, a significantly greater academic presence of the SDGs in university geographic education would be desirable. It is also necessary as it is essential in the training of future professionals and because it would improve the image that geography projects to society (De Cos and Reques 2010).

The discipline might need an update at different levels: the end of the division between physical and human geography, which weakens the capacity for a global conception of the world and an openness to collaboration with other disciplines (Goudie 2017); current narratives that recognize the reality and changes of the moment, while generating alternatives (Harden 2020; Castree 2015a, b); and a discourse of its own, which helps society to perceive the discipline as useful and associated with sustainability (Castree et al. 2022).

However, these aspects, which are already visible in the research agenda, are still limited in teaching (Mínguez et al. 2021). In the Spanish case, we can observe that the SDGs are addressed unequally. Only two, related to urban planning and climate, have a high and very high presence both in expectation and reality. They are hot topics in geography, with a long academic and research trajectory. Although a greater presence of SGDs would be always desirable, there are some SDGs which great relation to geography research that achieve only a medium presence, such as SDG 16 (peace, justice and institutions). Our results can be a call to solve this neglect.

As the general situation in other studies on the perception of teachers and students (Mínguez et al. 2021), teachers usually give higher scores. Due to their research experience, they might know the potential of SDGs better than students, but this also brings them to be more critical when valuing their actual presence. Among teachers, significant differences appear according to the traditional division between social and environmental geographic researchers (De Miguel and Sebastián-López 2022), with different conceptions regarding the human or nature root of the SDGs. Among students, those who have already graduated score worse than active students, maybe due to discouragement in the face of the difficult and poor-quality access to the labor market (Colegio de Geógrafos 2018).

The situation might be generalized and structural since the university and the beginning study year do not influence the scores. Age seems to influence in a similar way to the student–teacher binomial. Gender generates significant differences in relation to the SDGs on education and gender equality, which are better scored by women, in expectation and reality. This might reflect the systemic privileged position of men (Heras-Sevilla et al. 2021), who can be less aware of education for equality and coeducation.

The qualitative opinions bring this idea out: the SDGs are not theoretically and practically much present in geography courses and it would be desirable to manage a greater integration of contents, because of ethical and professional reasons. Implicitly, they are demanding a "sustainabilitization" of the curriculum, that is, the development of competencies in sustainability (Leal Filho et al. 2023): critical analysis, systemic reflection, collaborative decision-making and a sense of generational responsibility (Murga-Menoyo 2021). Sustainability competencies can easily be channeled into geographic education (Rodríguez-Domenech 2022), as respondents outline: in theoretical themes on global change and other practical ones such as fieldwork, case studies and GIT. Sustainability programming could also include workshops and even the creation of a new subject, as also suggested, in line with syllabus on Education for Sustainable Development or for Global Citizenship, which have already been introduced in educative programs all over the world (Sklad et al. 2016).

Finally, it seems to be consensus on the importance of the 2030 Agenda and a structural incorporation of sustainability education in the degrees. Most countries are immersed in a new development framework (Leal Filho et al. 2023), defined by the UN in sustainability terms (MDGs first, SDGs later, and a sure update of goals in the future). Within this framework, as Murga-Menoyo (2021) states, the acquisition of competencies in sustainability is an urgent pedagogical task, so education must be adapted. Geography degrees offer great potential for transmitting knowledge on the SDGs (Pretorius 2018), and teachers and students clearly verify this idea according to our results.

Although the SDGs present some limitations, defined by some respondents as a reflection of new forms of absorbent governmentality and neoliberalism policies, it must be recognized that they have contributed to lifting more than one billion people out of extreme poverty (Liverman 2018). Indeed, in space terms, they are a solid model for territorial management, what requires a high level of professional involvement of geographers. Therefore, an important opportunity cost for the degrees is detected (Harden 2012). In the context of the new era of the global change, the SDGs can structure the adaption of geography higher education to the labor market and the consolidated framework of social responsibility on sustainability. Regarding our results, despite the need has been recognized, the implementation in geography courses at Spanish universities is still pending.

6 Conclusions

According to the results, the UN SDGs are not fully implemented in geography courses at Spanish universities, although teachers and students claim for a higher presence. The learning of geography as the science of sustainability would provide updated knowledge, skills, methodologies and competencies, and an ethical background.

Teachers are usually more significantly critical when demanding the implementation of the Agenda 2030. The most valued SDGs, by teachers and students, are 11 and 13, which are related to two important topics in Geography: cites and climate. Nevertheless, other hot topics are poorly assessed while highly desired. An example is observed for leading geographic areas like GIT. The open opinions can be summarized as a request to "sustain" the curriculum. Indeed, the situation might be structural in Spain as there are not significant differences between universities.

Although the study sample is significant and it is national-scaled, the research could be expanded to other countries to determine the global extension of the results. There is an important limitation in methodology that could be solved in future detailed works: results are drawn from subjective perspectives, so they could be complemented by the study of syllabus sheets for all the geography courses. For future detailed works, the inclusion of information on kind of subjects and other specific academic issues would be also noteworthy.

In times of global change, universities are immersed in training for a sustainable development. If Geography pretends to be a science of sustainability, future geographers should be trained in professional competences to overcome crisis situations and in ethical responsibility. This research demonstrates that students and teachers sadly think that geography courses in Spanish universities do not fully manage the UN SDGs, but they are strongly convinced that they could do. Curricula should become more "sustainable", and it should be easy for the teaching of geography as it is a natural particularity of geography as a discipline.

Appendix I. Statistical Tests: Significant Difference (*) When p-value < 0.05

I-a. Normality test: Kolmogorov–Smirnov (Lilliefors meaning correction)

Variables	Statistics		
	K-S	fg	Sig
SDG1_expect	0.276	318	0.000*
SDG2_expect	0.276	318	0.000*
SDG3_expect	0.268	318	0.000*
SDG4_expect	0.341	318	0.000*

(continued)

(continued)

Variables	Statistics		
	K-S	fg	Sig
SDG5_expect	0.314	318	0.000*
SDG6_expect	0.420	318	0.000*
SDG7_expect	0.413	318	0.000*
SDG8_expect	0.301	318	0.000*
SDG9_expect	0.308	318	0.000*
SDG10_expect	0.381	318	0.000*
SDG11_expect	0.467	318	0.000*
SDG12_expect	0.313	318	0.000*
SDG13_expect	0.483	318	0.000*
SDG14_expect	0.275	318	0.000*
SDG15_expect	0.418	318	0.000*
SDG16_expect	0.244	318	0.000*
SDG17_expect	0.260	318	0.000*
SDG1_real	0.227	318	0.000*
SDG2_real	0.236	318	0.000*
SDG3_real	0.191	318	0.000*
SDG4_real	0.213	318	0.000*
SDG5_real	0.210	318	0.000*
SDG6_real	0.214	318	0.000*
SDG7_real	0.224	318	0.000*
SDG8_real	0.209	318	0.000*
SDG9_real	0.203	318	0.000*
SDG10_real	0.189	318	0.000*
SDG11_real	0.248	318	0.000*
SDG12_real	0.201	318	0.000*
SDG13_real	0.240	318	0.000*
SDG14_real	0.276	318	0.000*
SDG15_real	0.220	318	0.000*
SDG16_real	0.238	318	0.000*
SDG17_real	0.227	318	0.000*

I-b. Significance statistical test for the independent variable "Degree relation" (Groups: 1 = student; 2 = teacher): Mann–Whitney U

Nonparametric dependent variables	U	Statistics		
		W	Z	Sig
SDG1_expect	10,697.000	22,173.000	−2.598	0.009*
SDG2_expect	11,518.000	22,994.000	−1.524	0.128
SDG3_expect	12,151.500	26,347.500	−0.696	0.486
SDG4_expect	11,326.500	22,802.500	−1.848	0.065
SDG5_expect	9298.000	20,774.000	−4.498	0.000*
SDG6_expect	11,496.000	25,692.000	−1.789	0.074
SDG7_expect	12,283.000	23,759.000	−0.599	0.549
SDG8_expect	10,916.000	22,392.000	−2.368	0.018*
SDG9_expect	12,131.500	26,327.500	−0.738	0.460
SDG10_expect	9633.000	21,109.000	−4.318	0.000*
SDG11_expect	11,336.000	22,812.000	−2.253	0.024*
SDG12_expect	11,458.000	22,934.000	−1.651	0.099
SDG13_expect	11,497.000	22,973.000	−2.126	0.034*
SDG14_expect	10,770.500	24,966.500	−2.516	0.012*
SDG15_expect	12,054.000	26,250.000	−0.955	0.340
SDG16_expect	11,876.000	23,352.000	−1.050	0.294
SDG17_expect	12,494.000	26,690.000	−0.249	0.803
SDG1_real	10,737.000	22,213.000	−2.384	0.017*
SDG2_real	11,884.000	23,360.000	−0.932	0.351
SDG3_real	11,568.500	23,044.500	−1.321	0.187
SDG4_real	11,000.000	22,476.000	−2.039	0.041*
SDG5_real	11,311.000	22,787.000	−1.642	0.101
SDG6_real	10,843.500	24,871.500	−2.259	0.024*
SDG7_real	10,516.500	24,544.500	−2.674	0.008*
SDG8_real	12,317.500	23,793.500	−0.371	0.711
SDG9_real	10,719.000	24,747.000	−2.397	0.017*
SDG10_real	10,226.500	21,702.500	−3.029	0.002*
SDG11_real	11,643.500	23,119.500	−1.255	0.209
SDG12_real	11,799.500	25,827.500	−1.030	0.303
SDG13_real	9950.000	23,978.000	−3.466	0.001*
SDG14_real	10,498.500	24,526.500	−2.743	0.006*
SDG15_real	10,977.000	25,005.000	−2.079	0.038*
SDG16_real	12,245.500	23,721.500	−0.464	0.643
SDG17_real	11,460.000	25,488.000	−1.465	0.143

I-c. Significance statistical test for the independent variable "Genre" (Groups: 1 = female; 2 = male): Mann–Whitney U

Nonparametric dependent variables	U	Statistics		
		W	Z	Sig
SDG1_expect	10,719.500	33,939.500	−0.196	0.845
SDG2_expect	9697.000	32,917.000	−1.648	0.099
SDG3_expect	9877.000	33,097.000	−1.393	0.164
SDG4_expect	8976.000	32,196.000	−2.780	0.005*
SDG5_expect	7826.500	31,046.500	−4.376	0.000*
SDG6_expect	10,798.500	34,018.500	−0.097	0.923
SDG7_expect	10,599.000	33,819.000	−0.420	0.675
SDG8_expect	9849.500	33,069.500	−1.467	0.142
SDG9_expect	10,394.500	33,614.500	−0.672	0.502
SDG10_expect	9983.500	33,203.500	−1.343	0.179
SDG11_expect	10,341.000	33,561.000	−0.939	0.348
SDG12_expect	9456.500	32,676.500	−2.048	0.041*
SDG13_expect	10,473.500	33,693.500	−0.749	0.454
SDG14_expect	9488.500	32,708.500	−1.955	0.051
SDG15_expect	10,174.500	33,394.500	−1.124	0.261
SDG16_expect	9871.000	33,091.000	−1.394	0.163
SDG17_expect	10,296.500	33,516.500	−0.799	0.424
SDG1_real	9693.000	14,844.000	−1.540	0.124
SDG2_real	10,127.000	15,278.000	−0.949	0.343
SDG3_real	10,803.500	15,954.500	−0.005	0.996
SDG4_real	9330.500	32,335.500	−2.032	0.042*
SDG5_real	9193.500	32,198.500	−2.217	0.027*
SDG6_real	9443.500	14,594.500	−1.894	0.058
SDG7_real	10,181.000	15,332.000	−0.868	0.385
SDG8_real	10,708.500	15,859.500	−0.136	0.892
SDG9_real	10,476.000	15,627.000	−0.456	0.649
SDG10_real	10,000.500	15,151.500	−1.113	0.266
SDG11_real	9938.500	32,943.500	−1.225	0.220
SDG12_real	10,678.500	33,683.500	−0.177	0.859
SDG13_real	10,649.000	15,800.000	−0.224	0.823
SDG14_real	10,571.500	33,576.500	−0.332	0.740
SDG15_real	10,043.000	33,048.000	−1.056	0.291
SDG16_real	10,222.000	15,373.000	−0.811	0.417
SDG17_real	10,738.500	15,889.500	−0.095	0.924

I-d. Significance statistical test for the independent variable "Student status" (Groups: 1 = active; 2 = old): Mann–Whitney U

Nonparametric dependent variables	U	Statistics		
		W	Z	Sig
SDG1_expect	1857.500	2385.500	−0.225	0.822
SDG2_expect	1599.500	2127.500	−1.478	0.139
SDG3_expect	1738.000	2266.000	−0.819	0.413
SDG4_expect	1483.000	2011.000	−2.100	0.036*
SDG5_expect	1750.000	8890.000	−0.739	0.460
SDG6_expect	1841.000	8981.000	−0.376	0.707
SDG7_expect	1852.000	8992.000	−0.286	0.775
SDG8_expect	1714.000	2242.000	−0.941	0.347
SDG9_expect	1891.500	2419.500	−0.063	0.950
SDG10_expect	1496.500	8636.500	−2.033	0.042*
SDG11_expect	1602.000	8742.000	−1.749	0.080
SDG12_expect	1659.500	2187.500	−1.212	0.225
SDG13_expect	1693.000	8833.000	−1.296	0.195
SDG14_expect	1528.500	2056.500	−1.897	0.058
SDG15_expect	1870.500	9010.500	−0.195	0.845
SDG16_expect	1804.500	2332.500	−0.482	0.630
SDG17_expect	1824.500	2352.500	−0.390	0.696
SDG1_real	1460.500	1988.500	−2.158	0.031*
SDG2_real	1273.500	1801.500	−3.057	0.002*
SDG3_real	1463.500	1991.500	−2.095	0.036*
SDG4_real	1121.000	1649.000	−3.710	0.000*
SDG5_real	1361.000	1889.000	−2.571	0.010*
SDG6_real	1766.500	2294.500	−0.662	0.508
SDG7_real	1653.000	2181.000	−1.209	0.227
SDG8_real	1189.500	1717.500	−3.438	0.001*
SDG9_real	1492.000	2020.000	−1.970	0.049*
SDG10_real	1624.000	2152.000	−1.348	0.178
SDG11_real	1810.000	8950.000	−0.459	0.646
SDG12_real	1636.000	2164.000	−1.283	0.199
SDG13_real	1797.500	2325.500	−0.533	0.594
SDG14_real	1507.500	2035.500	−1.887	0.059
SDG15_real	1742.500	2270.500	−0.778	0.437
SDG16_real	1469.500	1997.500	−2.088	0.037*
SDG17_real	1163.500	1691.500	−3.544	0.000*

I-e. Significance statistical test for the independent variable "Age" (Groups: 1 = <25; 2 = 25–40; 3 = 41–60; 4 = >60): Kruskall-Wallis

Nonparametric dependent variables	Statistics		
	H	fg	Sig
SDG1_expect	9.823	3	0.020*
SDG2_expect	6.428	3	0.093
SDG3_expect	1.797	3	0.616
SDG4_expect	6.874	3	0.076
SDG5_expect	19.047	3	0.000*
SDG6_expect	0.916	3	0.822
SDG7_expect	1.456	3	0.692
SDG8_expect	6.855	3	0.077
SDG9_expect	1.888	3	0.596
SDG10_expect	18.311	3	0.000*
SDG11_expect	8.870	3	0.031*
SDG12_expect	5.185	3	0.159
SDG13_expect	7.297	3	0.063
SDG14_expect	10.227	3	0.017*
SDG15_expect	0.289	3	0.962
SDG16_expect	2.885	3	0.410
SDG17_expect	3.688	3	0.297
SDG1_real	4.194	3	0.241
SDG2_real	4.503	3	0.212
SDG3_real	4.833	3	0.184
SDG4_real	2.873	3	0.412
SDG5_real	1.840	3	0.606
SDG6_real	6.131	3	0.105
SDG7_real	12.389	3	0.006*
SDG8_real	2.204	3	0.531
SDG9_real	7.293	3	0.063
SDG10_real	3.470	3	0.325
SDG11_real	2.572	3	0.462
SDG12_real	5.061	3	0.167
SDG13_real	9.887	3	0.020*
SDG14_real	7.867	3	0.049*
SDG15_real	5.521	3	0.137
SDG16_real	0.612	3	0.894
SDG17_real	5.527	3	0.137

I-f. Significance statistical test for the independent variable "Teaching areas" (Groups: 1 = Regional; 2 = Physical; 3 = Human; 4 = Others): Kruskall-Wallis.

Nonparametric dependent variables	Statistics		
	H	fg	Sig
SDG1_expect	9.306	3	0.025*
SDG2_expect	3.734	3	0.292
SDG3_expect	4.898	3	0.179
SDG4_expect	9.042	3	0.029*
SDG5_expect	6.108	3	0.106
SDG6_expect	2.541	3	0.468
SDG7_expect	2.016	3	0.569
SDG8_expect	3.183	3	0.364
SDG9_expect	4.237	3	0.237
SDG10_expect	14.867	3	0.002*
SDG11_expect	8.921	3	0.030*
SDG12_expect	5.102	3	0.164
SDG13_expect	2.032	3	0.566
SDG14_expect	2.962	3	0.397
SDG15_expect	5.597	3	0.133
SDG16_expect	5.885	3	0.117
SDG17_expect	3.754	3	0.289
SDG1_real	24.317	3	0.000*
SDG2_real	6.499	3	0.090
SDG3_real	12.325	3	0.006*
SDG4_real	6.117	3	0.106
SDG5_real	10.817	3	0.013*
SDG6_real	3.075	3	0.380
SDG7_real	3.146	3	0.370
SDG8_real	24.155	3	0.000*
SDG9_real	8.258	3	0.041*
SDG10_real	39.893	3	0.000*
SDG11_real	24.002	3	0.000*
SDG12_real	10.187	3	0.017*
SDG13_real	13.976	3	0.003*
SDG14_real	7.991	3	0.046*
SDG15_real	16.055	3	0.001*
SDG16_real	19.466	3	0.000*
SDG17_real	8.116	3	0.044*

I-g. Significance statistical test for the independent variable "Beginning year" (Groups: one group per year 1987–2021): Kruskall-Wallis

Nonparametric dependent variables	Statistics		
	H	fg	Sig
SDG1_expect	12.274	18	0.833
SDG2_expect	14.986	18	0.663
SDG3_expect	15.123	18	0.654
SDG4_expect	26.004	18	0.100
SDG5_expect	21.920	18	0.236
SDG6_expect	21.107	18	0.274
SDG7_expect	16.497	18	0.558
SDG8_expect	15.729	18	0.611
SDG9_expect	15.974	18	0.594
SDG10_expect	18.144	18	0.446
SDG11_expect	23.391	18	0.176
SDG12_expect	15.596	18	0.621
SDG13_expect	13.742	18	0.746
SDG14_expect	16.475	18	0.559
SDG15_expect	10.300	18	0.922
SDG16_expect	20.290	18	0.317
SDG17_expect	21.521	18	0.254
SDG1_real	16.402	18	0.565
SDG2_real	22.241	18	0.221
SDG3_real	25.942	18	0.101
SDG4_real	18.653	18	0.413
SDG5_real	16.568	18	0.553
SDG6_real	13.903	18	0.735
SDG7_real	15.034	18	0.660
SDG8_real	18.821	18	0.403
SDG9_real	15.043	18	0.659
SDG10_real	9.507	18	0.947
SDG11_real	19.904	18	0.338
SDG12_real	19.041	18	0.389
SDG13_real	17.241	18	0.507
SDG14_real	24.852	18	0.129
SDG15_real	17.110	18	0.516
SDG16_real	20.654	18	0.297
SDG17_real	19.941	18	0.336

I-h. Significance statistical test for the associated groups: "Expectative"—"Reality" (longitudinal study): Wilcoxon signed rank test.

Nonparametric dependent variables	Statistics	
	Z	Sig
SDG1	−12.023	0.000*
SDG2	−12.72	0.000*
SDG3	−11.121	0.000*
SDG4	−10.978	0.000*
SDG5	−8.159	0.000*
SDG6	−12.417	0.000*
SDG7	−11.589	0.000*
SDG8	−10.873	0.000*
SDG9	−11.221	0.000*
SDG10	−10.584	0.000*
SDG11	−10.115	0.000*
SDG12	−10.782	0.000*
SDG13	−10.213	0.000*
SDG14	−12.694	0.000*
SDG15	−11.398	0.000*
SDG16	−11.35	0.000*
SDG17	−12.048	0.000*

References

Bagoly-Simó P, Kriewaldt J (2022) Future geography teachers for the planet. Powerful (disciplinary) knowledge and education for sustainable development (ESD) in initial teacher education (ITE). Int Res Geograp Environ Educ. in press. https://doi.org/10.1080/10382046.2023.2158622

Bartlett JE, Kotrlik JW, Higgins CC (2001) Organizational research: determining appropriate sample size in survey research. Inf Technol Learn Perform J 19(1):43–50

Brahm T, Kühner P (2019) Faculty motivation as a key for integrating sustainable development in higher education curricula. Int J Innov Sustain Dev 13(2):181–197. https://doi.org/10.1504/IJISD.2019.098985

BOE 52 (2022) Real Decreto 157/2022, de 1 de marzo, por el que se establecen la ordenación y las enseñanzas mínimas de la Educación Primaria, Ministerio de Educación del Gobierno de España, Madrid [Royal Decree 157/2022, of March 1, establishing the organization and minimum teaching requirements for Primary Education, Ministry of Education of the Government of Spain, Madrid]. https://www.boe.es/boe/dias/2022/03/02/pdfs/BOE-A-2022-3296.pdf

BOE 76 (2022) Real Decreto 217/2022, de 29 de marzo, por el que se establece la ordenación y las enseñanzas mínimas de la Educación Secundaria Obligatoria, Ministerio de Educación del Gobierno de España, Madrid [Royal Decree 217/2022, of March 29, establishing the organization and minimum education requirements for Compulsory Secondary Education, Ministry of

Education of the Government of Spain, Madrid]. https://www.boe.es/boe/dias/2022/03/30/pdfs/BOE-A-2022-4975.pdf
BOE 82 (2022) Real Decreto 243/2022, de 5 de abril, por el que se establecen la ordenación y las enseñanzas mínimas del Bachillerato, Ministerio de Educación del Gobierno de España, Madrid [Royal Decree 243/2022, of April 5, 2002, establishing the organization and the minimum teaching requirements of the Baccalaureate, Ministry of Education of the Government of Spain, Madrid]. https://www.boe.es/boe/dias/2022/04/06/pdfs/BOE-A-2022-5521.pdf
Castree N (2015a) Geography and global change science: relationships necessary, absent and possible. Geogr Res 53(1):1–15. https://doi.org/10.1111/1745-5871.12100
Castree N (2015b) Changing the Anthropo(s)cene: geographers, global environmental change and the politics of knowledge. Dialogues Hum Geogr 5(3):301–316. https://doi.org/10.1177/2043820615613216
Castree N, Leszczynski A, Stallins JA, Schwanen T, Patel Z (2022) Reconstituting geography for the 21st century. Environ Plan 1(1):3–6
Colegio de, Geógrafos (2018) IV Informe Perfiles Profesionales de la Geografía [IV Geography Professional Profiles Report]. Colegio de Geógrafos, Madrid
Collado S, Moreno JD, Martín-Albo J (2022) Innovation for environmental sustainability: longitudinal effects of an education for sustainable development intervention on university students' pro-environmentalism. Int J Sustain High Educ 23(6):1277–1293. https://doi.org/10.1108/IJSHE-07-2021-0315
Commission on Geographical Education (1992) International charter on geographical education. International Geographical Union, Washington. https://www.igu-cge.org/wp-content/uploads/2018/02/1.-English.pdf
Crutzen PJ (2021) We live in the Anthropocene, so will our grandchildren. Her Russ Acad Sci 91(1):13–16. https://doi.org/10.1134/S1019331621010020
De Cos O, Reques P (2010) "Espacio Europeo de Educación Superior y Geografía: la importancia de la formación en competencias y la empleabilidad" [European Higher Education Area and Geography: the importance of skills training and employability]. Boletín AGE 52:295–312
De Miguel R, De Lázaro ML (2016) Educating geographers in Spain. Geography teaching renewal by implementing the European Higher Education Area. J Geogr High Educ 40(2):267–283. https://doi.org/10.1080/03098265.2016.1139556
De Miguel González R, Sebastián-López M (2022) Education on sustainable development goals: geographical perspectives for gender equality in sustainable cities and communities. Sustainability 14(7):4042. https://doi.org/10.3390/su14074042
Faham E, Rezvanfar A, Movahed SH, Rajabi M (2017) Using system dynamics to develop education for sustainable development in higher education with the emphasis on the sustainability competencies of students. Technol Forecast Soc Chang 123:307–326. https://doi.org/10.1016/j.techfore.2016.03.023
Fletcher R, Murray Mas I, Blanco-Romero A, Blázquez-Salom M (2019) Tourism and degrowth: an emerging agenda for research and praxis. J Sustain Tour 27(12):1745–1763. https://doi.org/10.1080/09669582.2019.1679822
Goudie AS (2017) The integration of human and physical geography revisited. Can Geogr/le Geographe Canadien 61(1):19–27. https://doi.org/10.1111/cag.12315
Harden C (2012) Framing and reframing questions of human-environment interactions. Ann Assoc Am Geogr 102:737–747. https://doi.org/10.1080/00045608.2012.678035
Harden C (2020) The dance of sustainability: a call to engage geographers in local-and global-scale research. In: Sarmiento F (ed) The elgar companion to geography, transdisciplinarity and sustainability. Edward Elgar Publishing, Florida, pp 79–92
Haubrich H, Reinfried S, Schleicher Y (2007) Lucerne declaration on geographical education for sustainable development. International Geographical Union, Lucerne. https://www.iau-hesd.net/sites/default/files/documents/lucernedeclaration.pdf

Heras-Sevilla D, Ortega-Sánchez D, Rubia-Avi M (2021) Coeducation and citizenship: a study on initial teacher training in sexual equality and diversity. Sustainability 13(9):5233. https://doi.org/10.3390/su13095233

Holmberg J, Sandbrook R (2019) "Sustainable development: what is to be done?". In Holmberg J (ed) Policies for a small planet. Routledge, London, pp 19–38

Hummels H, Argyrou A (2021) Planetary demands: redefining sustainable development and sustainable entrepreneurship. J Clean Prod 278:123804. https://doi.org/10.1016/j.jclepro.2020.123804

Jia Q, Wang Y, Fengting L (2019) Establishing transdisciplinary minor programme as a way to embed sustainable development into higher education system: case by Tongji University, China. Int J Sustain High Educ 20(1):157–169. https://doi.org/10.1108/IJSHE-05-2018-0095

Kerr JT, Kharouba HM, Currie DJ (2007) The macroecological contribution to global change solutions. Science 316(5831):1581–1584. https://doi.org/10.1126/science.1133267

Leal Filho WL, Salvia AL, Eustachio JHPP (2023) An overview of the engagement of higher education institutions in the implementation of the UN sustainable development goals. J Clean Prod 386:135694. https://doi.org/10.1016/j.jclepro.2022.135694

Lim CK, Haufiku MS, Tan KL, Farid M, Ng TF (2022) Systematic review of education sustainable development in higher education institutions. Sustainability 14(20):13241. https://doi.org/10.3390/su142013241

Liverman DM (2018) Geographic perspectives on development goals: constructive engagements and critical perspectives on the MDGs and the SDGs. Dialogues Hum Geogr 8(2):168–185

Maude A (2022) Using geography's conceptual ways of thinking to teach about sustainable development. Int Res Geogr Environ Educ in Press. https://doi.org/10.1080/10382046.2022.2079407

Meadows ME (2020) Geography education for sustainable development. Geogr Sustain 1(1):88–92. https://doi.org/10.1016/j.geosus.2020.02.001

Mínguez C, Martínez-Hernández C, Yubero C (2021) Higher education and the sustainable tourism pedagogy: are tourism students ready to lead change in the post pandemic era? J Hosp Leis Sport Tour Educ 29:100329. https://doi.org/10.1016/j.jhlste.2021.100329

Moss R et al (2013) Hell and high water: practice-relevant adaptation science. Science 342:696–698

Murga-Menoyo MA (2021) Education in the anthropocene. Between Possibilism versus Utopia. Revista Interuniversitaria, vol 33, 2nd edn, pp 107–128. https://doi.org/10.14201/teri.25375

Olcina Cantos J (2022) Geography and environmental issues. In: Kolosov V, García-Álvarez J, Heffernan M, Schelhaas B (eds) A geographical century. Springer, Luxembourg, pp 211–223

Pretorius RW (2018) Towards an integrated disciplinary narrative and an enhanced role for geography in education for sustainability: reflections on south African Higher Education. In: Leal Leal Leal Filho W, Consorte-McCrea A (eds) Sustainability and Humanities: linking social values, theology and spirituality towards sustainability (World Sustainability Series), Springer, Cham, pp 291–311. https://doi.org/10.1007/978-3-319-95336-6_16

Rodríguez-Domenech MA (2022) "El papel de la Geografía en la enseñanza de los Objetivos de Desarrollo Sostenible. Aprendizaje a través de competencias geográficas" ["The role of geography in teaching the Sustainable Development Goals. Learning through geographic competencies"]. In: Olcina Cantos J, Morote Seguido AF (eds) La enseñanza de la geografía en el siglo XXI. Retos, recursos y propuestas docentes ante los nuevos desafíos globales [Teaching geography in the 21st century. Challenges, resources and teaching proposals in the face of new global challenges]. Universidad de Alicante, Alicante, pp 63–76

Sklad M, Friedman J, Park E, Oomen B (2016) 'Going glocal': a qualitative and quantitative analysis of global citizenship education at a Dutch liberal arts and sciences college. High Educ 72:323–340. https://doi.org/10.1007/s10734-015-9959-6

Teichert C, Liefner I, Otto A (2021) How wide is the gap? Comparing geography graduates' labor market success with that of peers from business and computer science. J Geogr High Educ 1–29. https://doi.org/10.1080/03098265.2021.1960490

Tulla A (2010) Los nuevos planes de estudio de los títulos de Grado en Geografía adaptados al modelo del Espacio Europeo de Educación Superior (EEES) [The new curricula for undergraduate degrees in Geography adapted to the model of the European Higher Education Area (EHEA)]. Estudios Geográficos 268:319–338

United Nations (2017) The sustainable development goals report. United Nations, New York

Wals A (2009) Review of contexts and structures for education for sustainable development. UNESCO, Paris. https://unesdoc.unesco.org/ark:/48223/pf0000184944

Yli-Panula E, Jeronen E, Lemmetty P (2019) Teaching and learning methods in geography promoting sustainability. Educ Sci 10(1):5. https://doi.org/10.3390/educsci10010005

Žalėnienė I, Pereira P (2021) Higher education for sustainability: a global perspective. Geogr Sustain 2(2):99–106. https://doi.org/10.1016/j.geosus.2021.05.001

Contributing to Food Security and Sovereignty: Promoting Sustainable Production, Using as an Example the Standardization of the Conditions of a Low Energy Consumption Process for the Dehydration of Strawberry Flakes in México

Abraham Juárez Ma. del Rosario, Ramírez Moreles Lidia Abigail, Joel Everardo Valtierra Olivares, Everardo Mares Mares, Carreño Aguilera Gilberto, and Baltazar Vera Juan Carlos

Abstract The problems caused by bad eating habits are one of the leading health problems in North America, which is why great efforts have been made to improve the quality of people's food through food products that provide a high nutritional value. The strawberry is a fruit with excellent nutritional properties and has been used to complement different foods. However, this fruit has a short postharvest life, which generates significant losses; thus, there are shortages in some seasons of the year. Therefore, developing sustainable methodologies that increase this product's shelf life is vital. In this work, the standardization of infrared-assisted dehydration

A. J. Ma. del Rosario
División de Ciencias de la Vida, Departamento de Alimentos, Universidad de Guanajuato, Campus Irapuato-Salamanca, Ex Hacienda "El Copal", Km 9 carretera Irapuato-Silao, A.P. 311, C.P. 36500 Irapuato, Gto, México

R. M. L. Abigail
Tonic World Center S.A de C.V., Irapuato, Gto, México

J. E. V. Olivares · B. V. J. Carlos (✉)
División de Ingenierías, Departamento de Ingeniería en Minas, Metalurgia y Geología, Universidad de Guanajuato, Campus Guanajuato, Exhda San Matías S/N, Col San Matías, C.P 36020 Guanajuato, Gto, México
e-mail: jc.baltazarvera@ugto.mx

E. M. Mares
Instituto Tecnológico Superior de Guanajuato, Ingeniería en Industrias Alimentarias, Carr. Guanajuato Puentecillas km 10.5 Predio El Carmen, Guanajuato, México

C. A. Gilberto
División de Ingenierías, Departamento de Ingeniería Geomática e Hidráulica, Universidad de Guanajuato, Campus Guanajuato, Juarez # 77 Col. Centro, Guanajuato C.P 36000, Gto, México

W. Leal Filho et al. (eds.), *The Contribution of Universities Towards Education for Sustainable Development*, World Sustainability Series,
https://doi.org/10.1007/978-3-031-49853-4_2

of strawberry flakes was carried out, varying the temperature and time of the process to obtain a more durable product that preserves its nutritional properties. The results show that the best process conditions are for a time of 1:30 h. at a temperature of 80 °C, which allows obtaining a dehydrated product with a considerably longer shelf life, improving the storage possibilities and availability of this fruit, contributing to thus with food security and sovereignty, facilitating the trade of this product.

Keywords Strawberry · Flakes · Postharvest life · Sustainable process

1 Introduction

To comprehensively address the various problems facing humanity, as of 2015, the United Nations adopted the 2030 Agenda for Sustainable Development, which focuses on a shared plan for peace and prosperity for people and the planet, now and in the future. To achieve the goals outlined, 17 Sustainable Development Goals (SDGs) were established through strategies that improve health and education, reduce inequality, and stimulate economic growth [Department of Economic and Social Affairs, 2022]. According to this, there are two SDGs related to aspects of human nutrition; SDG 2 is called "Zero Hunger," which consists of ending hunger, achieving food security, improving nutrition, and promoting sustainable agriculture [Department of Economic and Social Affairs (Goal 2), 2022] and SDG 3 called "Health and well-being," which consists of guaranteeing a healthy life and promoting well-being for all at all ages [Department of Economic and Social Affairs (Goal 3), 2022]. This work deals with the study of technology that allows increasing the lifetime of perishable food (strawberries), which contributes to the commercialization and access of this product to the population contributing to the food security referred to in SDG 2. Similarly, strawberry has high nutritional and nutraceutical properties, which helps individuals have a balanced diet by having greater access to these. The products contribute to what is sought by SDG 3.

The current panorama related to eating habits may have changed concerning the time before the pandemic caused by COVID-19; this can be observed in the decrease in activities that involve eating away from home and in the increase in food preparation at home. This condition resulted from the security measures implemented during the pandemic and the precautions people took to avoid infection with the COVID-19 virus (Parekh and Deirlein 2020). In the same way, these new circumstances have led to a gradual increase in the self-preparation of food with characteristics that people consider healthier compared to the traditional offer of street food. These changes have been reported by observational studies, which indicate that, for example, in the case of the United States, half of the adults that make up the analyzed sample have made changes to their diet, most of which have been in the sense of improvement. (Cosgrove and Wharton 2021; Miller et al. 2021). Similar results were reported in the case of Canada, which revealed an improvement in diet quality between April and May 2020 compared to the time before the pandemic (Lamarche et al. 2021).

However, there is evidence that the pandemic significantly affected the food security of some sectors of the population, such as people with low resources, those who experience conditions of depression and anxiety, or those who, for various reasons, are socially vulnerable (Picchioni et al. 2021; Fitzpatrick et al. 2021). In this sense, in some regions of the world, actions have been developed to mitigate the adverse effects of the pandemic on the food security of their populations (USDA Economic Research Service 2021). However, it is necessary to analyze the effects according to the socioeconomic and health reality of each region of the world and thus be able to have a more realistic picture of the current state of food security in each area (Murphy et al. 2021; Benson et al. 2021).

In the case of Latin America and the Caribbean, the conditions that significantly affect food security and are related to poverty are diverse and complex. Nevertheless, one factor that has been observed to have a predominant influence is climate change and its effects on the various ecosystems of the region. This aspect impacts the production because of the continuous wear and tear of land use; the low stability of food production; inadequate and discontinuous availability of essential food products; accessibility to food decreased due to the socioeconomic conditions of the communities, and low food consumption of good nutritional and biological quality (Kopittke et al. 2019; Fernández et al. 2018; Vargas et al. 2018).

Since a significant sector of the population of these regions depends economically on traditional agricultural and livestock production, the effects mentioned above represent a severe risk to the food security of their communities and the population to which these products are aimed (Abeldaño et al. 2021). Therefore, it is vital to identify specific risks to production and generate comprehensive and sustainable solutions that allow adequate production of food products suitable for the population in general. In the case of Mexico, the production of agricultural and livestock products for human consumption greatly exceeded the requirements of the country's population between the 1940s and 1960s; likewise, due to the exponential increase in the population in this region, the energy crisis, and the slowdown in the production of farm products that occurred in the 1970s, caused this behavior to be reversed. The country was no longer self-sufficient in food supplies (Luiselli 2017). For this reason, the Mexican government implemented policies focused on stimulating field production, mainly on subsidies and support for farmers and livestock producers. However, it was not until 1994, with the signing of the North America Free Trade Agreement (NAFTA) among the United States, Mexico, and Canada, that considerable progress was achieved in modernization and technology transfer to increase and improve agricultural and livestock production in this region.

Food safety is directly related to consuming nutrient-rich foods, including fruits and vegetables. In this sense, the consumption of this type of essential food to maintain a healthy diet and avoid chronic diseases due to poor nutrition has decreased thanks to factors such as the cultural change in food consumption, increasing the consumption of foods considered unhealthy, the increase in costs and decrease in the availability of various vegetable products among others (Miranda et al. 2021). Therefore, the socioeconomically most vulnerable sectors are those with the most significant difficulty in acquiring adequate food for a quality diet, substituting them with products of low nutritional value and, to some extent, harmful to health. Therefore, in various regions of the world, some alternatives have been developed to help mitigate this food safety problem, such as more significant financial incentives to stimulate the production of healthy, high-quality vegetables and animal products, promotion of direct interaction between producers and consumers in order to reduce the costs represented by intermediaries, development of particular crops of products from the region in the case of areas where conditions allow it (Drisdelle et al. 2020). These actions, together with the policies to raise awareness about the importance of good nutrition, have taken a significant rise in recent times in practically all regions of the world. They have made it possible to advance in the fight against the lack of food security and poverty in all its forms. This scenario has been strengthened with a significant scientific advance in the nutritional and, in some cases, nutraceutical properties of various products such as vegetables. It also generates a solid perception of the importance of these products for a healthy diet that allows a better life quality. That is why developing research projects that contribute to the best production and marketing of these is of the utmost importance to improve nutrition in all regions of the planet.

The strawberry (Fragaria x ananassa) is a fruit of high added value because it is considered a functional food due to its biological activities and benefits for human health (Chaves et al. 2017). The main pharmacological characteristics of this fruit are related to its majority phytochemical composition, which includes anthocyanins, which have anticancer and antioxidant properties (Mozafari et al. 2019; Glensk et al. 2021). To carry out an adequate fruit dehydration process, technologies have been developed that allow this process to be made more efficient, such as simple or assisted osmotic dehydration (Tylewicks et al. 2019; Jiang et al. 2021), microwave drying at vacuum (de Bruijn and Borquez 2014; Ishibashi et al. 2022), hot air drying (Meng et al. 2022; Wen et al. 2022), among others. These methods present advantages and areas of opportunity in terms of energy consumption and conservation of the physicochemical and biological properties of the product, for which reason continuing with the development of technology and methods for the dehydration process of these products is significantly essential.

The use of infrared radiation in food processing has gained significant interest due to the inherent advantage over heating with hot air since it is produced volumetrically, which allows shortening of process times, intense and uniform heating, energy efficiency, and high-quality dry products (Jeevarathinam et al. 2021; Geng et al. 2022). Similarly, studies have reported that using infrared radiation in combination with hot air can reduce the latter's flow, reducing the consumption of services for this process (Al-Hilphy et al. 2020). In this work, the dehydration of strawberry flakes by infrared is studied to know the effect of this process on the product's physicochemical properties.

2 Methodology

Strawberry fruits of the Camino Real variety were used, which were washed with plenty of water; then 0.5 cm thick strawberry slices were made, and 50 g of these slices were placed in the infrared tray. The infrared was set at a temperature of 60 °C at a power of 200 W for treatment 1 (T1). A temperature of 80 °C at the same power for treatment 2 (T2) was compared with a standard test (T3), and a scale was placed under the infrared to calculate the weight loss every 15 min rotating the tray every 10 min.

For color evaluation, a Hunter Lab colorimeter was used under the L*, a* and b* scale; in this sense, the determination was made in the region of the equatorial diameter; Total Soluble Solids (TSS) were measured with the help of a Handheld digital refractometer and expressed as °Brix; pH was measured with a HANNA equipment; moisture was calculated using a DRAWELL model LIO-300 drying oven in which 2 g of the sample were placed in crucibles at constant weight for a period of 5 h using the algorithm:

$$\theta = \frac{m_{wet} - m_{dry}}{\rho_w . V_b}$$

where m_{wet} and m_{dry} are the masses of the sample before and after oven drying; ρ_w is the density of water and V_b is the volume of the sample before drying. For the statistical analysis, a multilevel factorial design was carried out with two factors, one of 5 level and the other of two levels, with three replicates.

3 Results

3.1 Physicochemical Properties

Table 1 shows the results obtained for the physicochemical tests for the 2 treatments and the standard based on the dehydration.

The exposure of the strawberry to high temperatures causes its physicochemical properties to be affected because the heat causes it to lose its attractive red color, turning it into a burnt red; the Brix degrees and the pH were also affected in T3 by the effect of heat, since, in this sample, the sugars of the strawberry suffered a deterioration generating a bitter taste. Likewise, the strawberry is a mostly water fruit, so both were significantly affected by this parameter. However, it is worth mentioning that T3 was more affected by the high temperature and the short time of exposure to heat; they did not allow it to be lost.

Table 1 Physicochemical properties for the two treatments and the standard based on the dehydrated

Physicochemical properties	T1	T2	T3
Colorimetry	l = 25.94 a = 24.61 b = 12.65	l = 25.94 a = 12.65 b = 24.61	l = 36.26 a = 22.00 b = 18.60
°Brix	7.40	7.40	5.00
pH	4.00	4.00	5.00
Humidity (%)	83.65	83.55	4.75
Final weight (g)	50.00	50.00	3.50
Performance (With respect to the weight of dehydrated strawberry flakes in %	100	100	7.00

3.2 *Strawberry Weight Loss During Dehydration*

Figures 1 and 2 show the weight losses during the standardization of the dehydrated strawberry during T2 and T3, respectively, in which it can be observed that this parameter was higher in T3 compared to T2; likewise, it is shown that at two hours of exposure, to infrared in both treatments shows a higher rate of weight loss, with 20 g being obtained in T2 from 50 g of fresh flakes, compared to 5 g in T3.

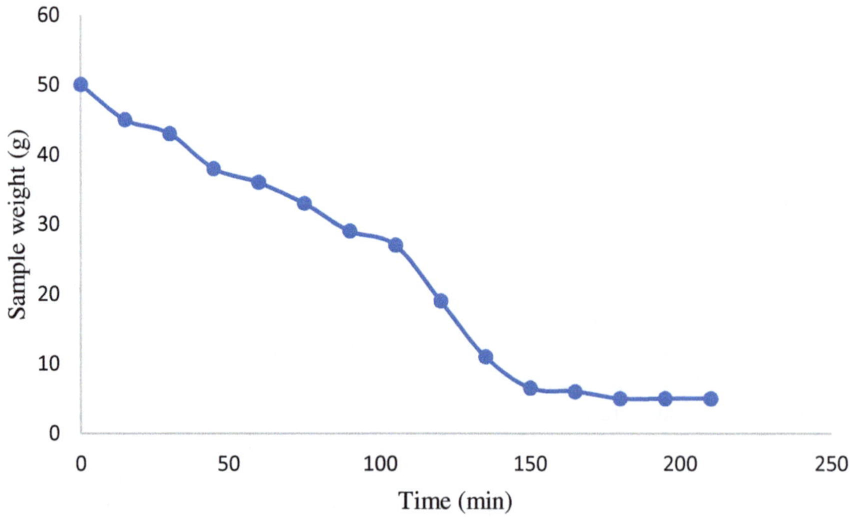

Fig. 1 Weight loss during standardization of dried strawberries during T2

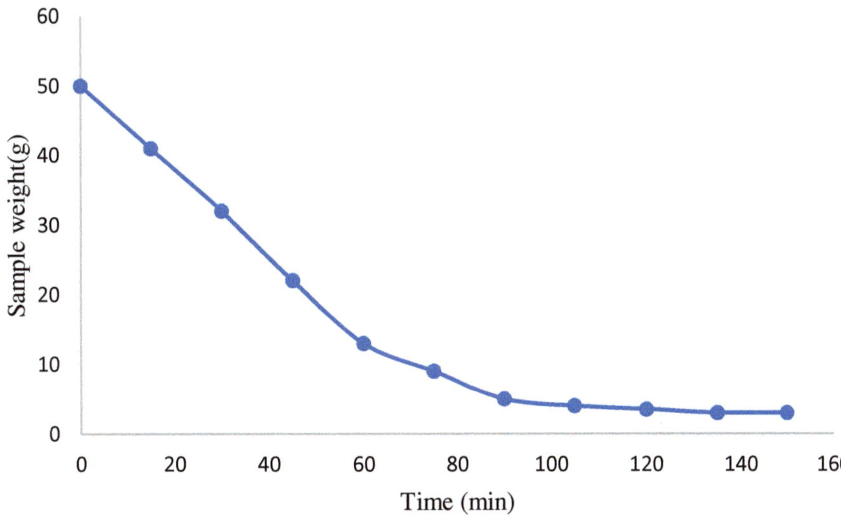

Fig. 2 Weight loss during standardization of dried strawberries during T3

3.3 *Statistical Analysis*

Table 2 shows the analysis of variance for the case study, in which the time and temperature factors are significant. Also, the time–temperature interaction is essential, as seen in the Pareto diagram of standardized effects (Fig. 3), where the most critical variable is temperature, then time, and finally, the interaction of both.

Figure 4 shows the interaction of time and temperature; in the case of a temperature of 80 °C and a time of 90 min, it can be observed that an optimal weight was reached approximately; this is because, after more time, little weight is lost, it is also observed that the same result is reached with a temperature of 60 °C but with a time of 150 min.

Table 2 Analysis of variance of the case study

	GL	SC	MC	F	P
Model	11	6054.72	550.43	1077.63	0.000
Blocks	2	1.81	0.91	1.77	0.198
Linear	5	5614.29	1122.86	2198.33	0.000
Time	4	3834.05	958.51	1876.58	0.000
Temperature	1	1780.24	1780.24	3485.35	0.000
2-term interactions	4	438.61	109.65	214.68	0.000
Time–Temperature	4	438.61	109.65	214.68	0.000
Error	18	9.19	0.51		
Total	29	6063.91			

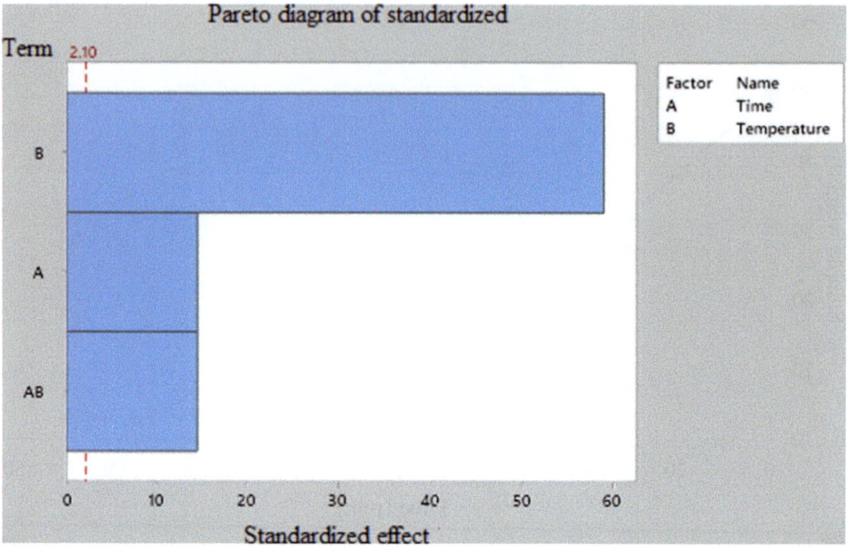

Fig. 3 Pareto diagram of standardized effects

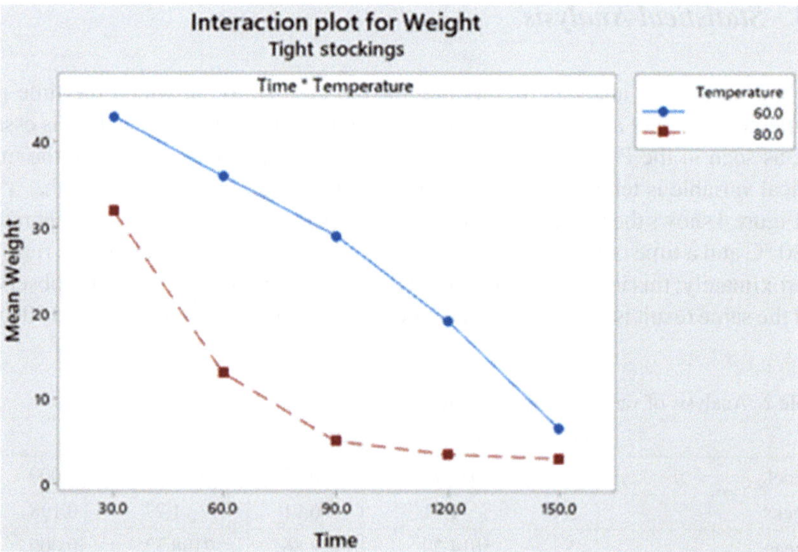

Fig. 4 Interaction of time and temperature

4 Conclusions

In the present study, standardization of infrared-assisted dehydration of strawberry flakes was accomplished by varying the temperature and time of the process. This experiment enabled the development of a more durable product that retains its nutritional properties. The obtained results demonstrate that a duration of 90 min and a temperature of 80 °C achieve dehydration of 5 g, while a temperature of 60 °C and 150 min of dehydration results in 6.5 g of weight loss. The time differential of 60 min may prove significant for pilot tests or production, as both time and electrical energy consumption can affect project profitability.

With this methodology, it was possible to obtain a dehydrated product with a considerably longer shelf life, improving the storage possibilities and availability of this fruit, thus contributing to food security and sovereignty, facilitating the trade of this product, which contributes to what has been proposed in SDGs 2 and 3 of the 2030 Agenda for Sustainable Development. In this sense, to achieve that this type of project represents a considerable impact, it is vital to carry out future work that allows to know its feasibility and later achieve its implementation at an industrial level, which will allow reaching the main objective, which is to achieve the accessibility of this product to consumers.

Finally, the development of this type of process contributes to the improvement of a healthy diet, making products of high nutritional value available to all sectors of the population, thus contributing to food security in the various regions of the world.

Acknowledgements The authors thank the University of Guanajuato for its support in carrying out this research project.

References

Abeldaño R, Lima G, González A (2021) Impact of slow-onset events related to climate change on food security in Latin America and the Caribbean. Curr Opin Environ Sustain 50:215–224. https://doi.org/10.1016/j.cosust.2021.04.011

Al-Hilphy A, Al-Shatty S, Al-Mtury A, Gavahian M (2020) Infraredassisted oil extraction for valorization of carp viscera: effects of process parameters, mathematical modeling, and process optimization. LWT 129:109541. https://doi.org/10.1016/j.lwt.2020.109541

Benson T, Murphy B, McCloat A, Mooney E, Dean M, Lavelle F (2021) From the pandemic to the pan: the impact of COVID-19 on parental inclusion of children in cooking activities: a cross-continental survey. Public Health Nutr 25:36–42. https://doi.org/10.1017/S1368980021001932

Chaves V, Calvete E, Reginatto F (2017) Quality properties and antioxidant activity of seven strawberry (*Fragaria x ananassa* duch) cultivars. Sci Hortic 225:293–298. https://doi.org/10.1016/j.scienta.2017.07.013

Cosgrove K, Wharton C (2021) Predictors of COVID-19-related perceived improvements in dietary health: results from a US cross-sectional study. Nutrients 13(6):2097. https://doi.org/10.3390/nu13062097

de Bruijn J, Bórquez R (2014) Quality retention in strawberries dried by emerging dehydration methods. Food Res Int 63:42–48. https://6342-48.10.1016/j.foodres.2014.03.029

Departamento de Asuntos Económicos y Sociales, Desarrollo Sostenible (2022). https://sdgs.un.org/es/goals

Departamento de Asuntos Económicos y Sociales, Desarrollo Sostenible (2022) Objetivo 2. https://sdgs.un.org/es/goals/goal2

Departamento de Asuntos Económicos y Sociales, Desarrollo Sostenible (2022) Objetivo 3. https://sdgs.un.org/es/goals/goal3

Drisdelle C, Kestens Y, Hamelin A, Mercille G (2020) Disparities in access to healthy diets: how food security and food shopping behaviors relate to fruit and vegetable intake. J Acad Nutr Diet 120(11):1847–1858. https://doi.org/10.1016/j.jand.2020.03.020

Fernández F, Blanco M, Ponce R, Vásquez-Lavín F, Roco L (2018) Implications of climate change for semi-arid dualistic agriculture: a case study in Central Chile. Reg Environ Change 19:89–100. https://doi.org/10.5751/ES-02348-130141

Fitzpatrick K, Harris C, Drawve G, Willis D (2021) Assessing food insecurity among US adults during the COVID-19 pandemic. J Hunger Environ Nutr 16:1–18. https://doi.org/10.1080/19320248.2020.1830221

Geng Z, Torki M, Kaveh M, Beigi M, Yang X (2022) Characteristics and multi-objective optimization of carrot dehydration in a hybrid infrared /hot air dryer. LWT 172:114229. https://doi.org/10.1016/j.lwt.2022.114229

Glensk M, Dudek M, Rybacka A, Wlodarczyk M, Fecka I (2021) Triterpenoids from strawberry *Fragaria* × *ananassa* Duch. cultivar Senga Sengana leaves. Indus Crops Products 169, 113668. https://doi.org/10.1016/j.indcrop.2021.113668

Ishibashi R, Numata T, Tanigawa H, Tsuruta T (2022) *In-situ* measurements of drying and shrinkage characteristics during microwave vacuum drying of radish and potato. J Food Eng 323:110988. https://doi.org/10.1016/j.jfoodeng.2022.110988

Jeevarathinam G, Pandiselvam R, Pandiarajan T, Preetha P, Balakrishnan M, Thirupathi V, Kothakota A (2021) Infrared assisted hot air dryer for turmeric slices: effect on drying rate and quality parameters. LWT 144:111258. https://doi.org/10.1016/j.lwt.2021.111258

Jiang J, Zhang M, Devahasting S, Yu D (2021) Effect of ultrasound-assisted osmotic dehydration pretreatments on drying and quality characteristics of pulsed fluidized bed microwave freeze-dried strawberries. LWT 145:111300. https://doi.org/10.1016/j.lwt.2021.111300

Kopittke P, Menzies N, Wang P, McKenana B, Lombi E (2019) Soil and the intensification of agriculture for global food security. Environ Int 132:105078. https://doi.org/10.1016/j.envint.2019.105078

Lamarche B, Brassard D, Lapointe A, Laramée C, Kearney M, Cote M, Belanger-Gravel A, Desroches S, Lemieux S, Plante C (2021) Changes in diet quality and food security among adults during the COVID-19-related early lockdown: results from NutriQuébec. Am J Clin Nutr 113(4):984–992. https://doi.org/10.1093/ajcn/nqaa363

Luiselli C (2017) Agricultura y alimentación en México. Evolución, desempeño y perspectivas. Universidad Nacional Autónoma de México. 1. http://pued.unam.mx/opencms/publicaciones/22/22.html

Meng Z, Cui X, Liu Y, Du C, Hu R, Wang S, Wang F (2022) Effect of electrohydrodynamics on hot air drying characteristics of fruits and vegetables. Sustain Energy Technol Assess 53:102716. https://doi.org/10.1016/j.seta.2022.102716

Miller S, Bruine W, Livings M, Wilson J, Weber K, Frazzini A, Babboni M, De la Haye K (2021) Self-reported dietary changes among Los Angeles County adults during the COVID-19 pandemic. Appetite 166:105586. https://doi.org/10.1016/j.appet.2021.105586

Miranda C, Ward R, Newman T, Berney S, Slagel N, Bussey-Jones J, Schmidt S, Lee J, Webb-Girard A (2021) Food security and clinical outcomes of the 2017 georgia fruit and vegetable prescription program. J Nutr Educ Behav Adv Res Pract Policy 53(9):770–778. https://www.jneb.org/article/S1499-4046(21)00689-8/fulltext

Mozfari A, Ghaderi N, Havas F, Dedejani F (2019) Comparative investigation of structural relationships among morphophysiological and biochemical properties of strawberry (*Fragaria* × *ananassa* Duch.) under drought and salinity stresses: a study based on in vitro culture. Scientia Horticulturae 256:108601. https://doi.org/10.1016/j.scienta.2019.108601

Murphy B, Benson T, McCloat A, Mooney E, Elliott C, Dean M, Lavelle F (2021) Changes in consumers' food practices during the COVID-19 lockdown, implications for diet quality and the food system: a cross-continental comparison. Nutrientes 13(1):20. https://doi.org/10.3390/nu13010020

Parekh N, Deierlein A (2020) Health behaviours during the coronavirus disease 2019 pandemic: implications for obesity. Public Health Nutr 17:3121–3125. https://doi.org/10.1017/S1368980020003031

Picchioni F, Goulao L, Roberfroid D (2021) The impact of COVID-19 on diet quality, food security and nutrition in low and middle income countries: a systematic review of the evidence. Clin Nutr 41(12):2955–2964. https://doi.org/10.1016/j.clnu.2021.08.015

Tylewicz U, Tappi S, Genovese J, Mozzon M, Rocculi P (2019) Metabolic response of organic strawberries and kiwifruit subjected to PEF assisted-osmotic dehydration. Innov Food Sci Emerg Technol 56:102190. https://doi.org/10.1016/j.ifset.2019.102190

USDA Economic Research Service (2021) Household food security in the United States in 2020. Economic Research Report No. (ERR-298). https://www.ers.usda.gov/publications/pub-details/?pubid=102075

Vargas R, Cabrera M, Cicowiez M, Escobar P, Hernández V, Vabrera J, Guzmán V (2018) Climate risk and food availability in Guatemala. Environ Dev Econ 23(5):558–579. https://doi.org/10.1017/S1355770X18000335

Wen X, Li W, Li W, Chen W, Zhang Z, Wu D, Yang Y (2022) Quality characteristics and non-volatile taste formation mechanism of *Lentinula edodes* during hot air drying. Food Chem 393:133378. https://doi.org/10.1016/j.foodchem.2022.133378

Towards Sustainable Mining Operations: Optimizing Thermal and Environmental Conditions for Underground Workers Using CFD Analysis

Juan Carlos Baltazar Vera, Yolanda Guadalupe Zamarripa Perez, Carolina de Jesús Rodríguez Rodríguez, Gilberto Carreño Aguilera, Adrian Zamorategui Molina, Lucía Guadalupe Alvarado Montalvo, Roberto Ontiveros Ibarra, Ma. del Rosario Abraham Juárez, Ricardo Solís Rodríguez, and Joel Everardo Valtierra Olivares

Abstract The work carried out to extract minerals in underground mines represents an activity that must be highly controlled to ensure adequate safety conditions for workers. One of the most important aspects to consider is the environmental conditions since the assurance of temperature and humidity values within the acceptable ranges for work is vital for developing the activities. Accordingly, providing the environment with a ventilation system that allows the parameters to be controlled to achieve the appropriate conditions is necessary. In this research, the study of adequate time to ventilate an underground mine development work was carried out, taking as a parameter the temperature and humidity present in it and the effect of this system in the reduction of said parameters; to achieve this objective, the calculations of the work dimensions, equipment to be used and required air flows were made according to current regulations. After that, a simulation was carried

J. C. B. Vera · Y. G. Z. Perez · C. de Jesús Rodríguez Rodríguez · L. G. A. Montalvo · R. O. Ibarra · R. S. Rodríguez · J. E. V. Olivares (✉)
División de Ingenierías, Departamento de Ingeniería en Minas, Metalurgia y Geología, Universidad de Guanajuato, Campus Guanajuato, Exhda San Matías S/N, Col San Matías, C.P 36020 Guanajuato, Gto, México
e-mail: jc.baltazarvera@ugto.mx

G. C. Aguilera
División de Ingenierías, Departamento de Ingeniería Geomática E Hidráulica, Universidad de Guanajuato, Campus Guanajuato, Juarez # 77 Col. Centro, C.P 36000 Guanajuato, Gto, México

A. Z. Molina
División de Ingenierías, Departamento de Ingeniería Civil y Ambiental, Universidad de Guanajuato, Campus Guanajuato, Juarez # 77 Col. Centro, C.P 36000 Guanajuato, Gto, México

Ma. del Rosario Abraham Juárez
División de Ciencias de la Vida, Departamento de Alimentos, Universidad de Guanajuato, Campus Irapuato-Salamanca, Ex Hacienda "El Copal", Km 9 carretera Irapuato-Silao, A.P. 311, C.P 36500 Irapuato, Gto, México

W. Leal Filho et al. (eds.), *The Contribution of Universities Towards Education for Sustainable Development*, World Sustainability Series,
https://doi.org/10.1007/978-3-031-49853-4_3

out using the CFD FLUENT software to determine the time necessary for the best thermo-environmental conditions according to the standards.

Keywords Underground mines · Thermo-environmental conditions · CFD FLUENT · Ventilation system · Sustainable mining industry

1 Introduction

In 2015, the member countries of the United Nations adopted the 2030 Agenda for Sustainable Development, which consists of a shared plan for peace and prosperity for people and the planet, now and in the future. To achieve the goals set, 17 Sustainable Development Goals (SDGs) were established, requiring strategies that improve health and education, reduce inequality, and stimulate economic growth while addressing climate change and preserving natural environments (Departamento de Asuntos Económicos y Sociales 2022). According to this, the mining industry in Mexico has been concerned with developing increasingly sustainable and sustainable projects. In this sense, this document addresses an issue widely related to SDG 3 (Departamento de Asuntos Económicos y Sociales (Objective 3) 2022) since it analyzes the adequate ventilation time to maintain thermo-environmental conditions that allow workers to maintain good health. In the same way, it contributes to SDG 9 (Departamento de Asuntos Económicos y Sociales (Sector industrial),2022) by providing an overview of alternatives for efficient energy use in the ventilation process of underground works.

Mining is one of the oldest economic activities of humanity and has played a significant role in developing various industrial sectors and producing goods and services (Rhatigan 2020; Tetreault 2020). However, the chain of processes related to the extraction and benefits of minerals represents a considerable environmental impact on the water resource and the degradation of the soil in the environment in which it develops, which must be adequately treated or else severe damage to the environment and serious effects to the health of the population living in these areas can be generated (Dibattista et al. 2023). Among the most reported environmental and health effects due to inadequate treatment of mining waste and poor work practices is the severe damage to the soil and water resources necessary for agriculture, as well as the increased incidence of skin diseases, malaria, physical injuries, and fatal accidents both in workers and in the general population (Quarm et al. 2022).

Sustainable processes in the mining sector focus on economic, environmental, and social aspects (Upadhyay 2022). Regarding environmental topics, the industry strives for sustainable use of water, land, and atmospheric resources (Brodny and Tutuk 2022) to maintain a balance with the environment. Mining research and efforts have focused on implementing comprehensive management systems, including effective management of water resources, sustainable processes, and personnel training (Ofosu and Sarpong 2022).

Some mining companies have carried out rehabilitation activities of forest resources in mining areas. Upon return to competent forestry administration, have allowed for a balance between the activity and the protection of the environment. This approach is part of the vision called 'Mining for Sustainable Development' (Yildiz 2020), which proposes the development of the mining sector sustainably, improving the living conditions of the population in the area and preserving its natural resources.

Underground mining is necessary to extract minerals from the earth's crust's rocky massifs at considerable depths (Pomasoncco et al. 2022). In order to work in such an environment, it is crucial to maintain thermo-environmental conditions and adequate air quality to ensure workers' excellent health and equipment functionality. Ventilation systems are implemented to allow a constant flow of fresh air (Nikolakis et al. 2021; Brodny and Tutak 2021). These systems work on the principle of pressure differences to generate airflow, and controlling the difference ensures adequate flow for the tasks (Ars et al. 2021). Fans, particularly axial fans, are selected to provide adequate fresh air requirements, as they allow better flow control and have a more extensive range of operation than other equipment (Quevedo 2013).

To ensure the safety of workers in underground mining, regulations, and laws have been established to govern work operations and services in this industry. Adequate ventilation is crucial for worker safety, and Official Mexican Standards have been set to specify the necessary ventilation requirements for underground mines. These standards detail the characteristics that a ventilation system in an underground mine should have, ensuring that workers are protected and safe while performing their duties (Secretaria de Trabajo y Previsión Social 2012).

2 Methodology

The model of fluent k-e standard (Wilcox 1994) was used to simulate the ventilation system. Regarding the elaboration of the mesh of the work front, a construction method of upper-lower geometry was used in GAMBIT (Kendouci et al. 2013), using mesh units of the mixed type with tetrahedral and hexahedral cells with a total of 300,000 units (see Figs. 1 and 2) using the following mesh dimensions:

- Dimensions of 7 m × 7 m with a length of 150 m.
- The ventilation sleeve is formed by a concentric pipe of 0.5 m and a length of 120 m, located in the upper right.
- The extraction pipe has a diameter of 1 m with a length of 75 m, located in the center of the mine.
- The combustion generator has a diameter of 0.08 m with a height of 1 m.

In order to calculate the amount of air required, the following specifications were considered: Use of 1 power scoop of 278.94 HP and a low profile power truck of 409.01 HP. It was considered that the mining work is at 2012 m above sea level, and a group of 8 people worked in the development work. An extractor-injector

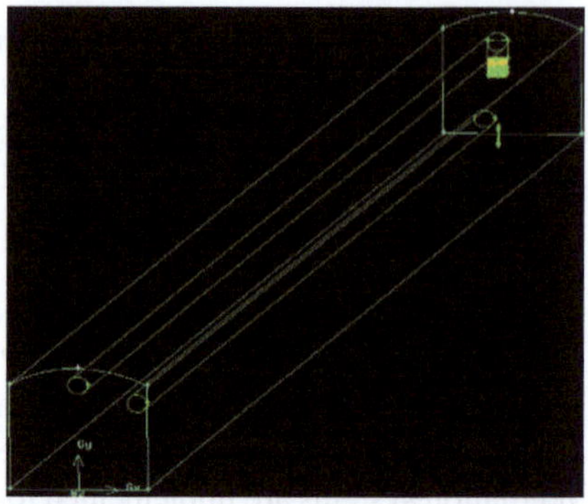

Fig. 1 Scheme of the work mine (Vizguerra et al. 2020)

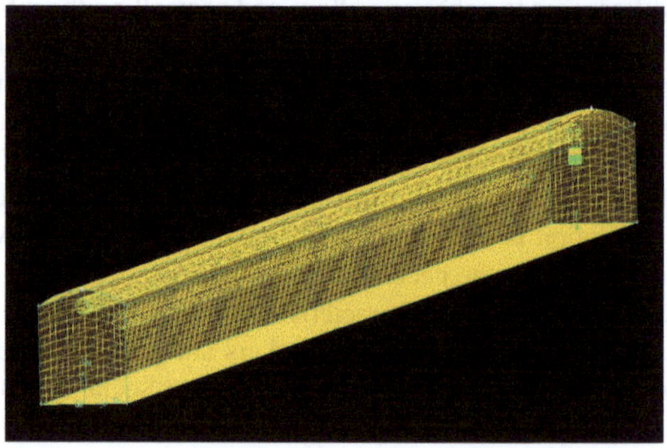

Fig. 2 Mine work mesh (Vizguerra et al. 2020)

ATC axial tube injector with injected aluminum propeller and glass fiber-reinforced polypropylene with seven blades at an angle of 45° was used. The properties of air mixture and emission gases, as well as the boundary conditions necessary to perform the simulation, are shown below (Tables 1 and 2).

The simulation was carried out by varying the ventilation process time until the appropriate parameters regarding temperature and humidity were obtained, considering the combustion reaction generated in the engines of the leading work teams. These reactions are shown below:

$$C_{10}H_{22} + 10.5O_2 \rightarrow 10CO + 11\,H_2O \tag{1}$$

$$2\,CO + 2\,H_2O + O_2 \rightarrow 2\,CO_2 + 2\,H_2O \tag{2}$$

Once the simulations have been carried out, varying the process time, profiles and contour schemes are generated to elucidate the differences in the desired parameters between each one.

Table 1 Properties of the air mixture and emission gases in underground mine (Vizguerra et al. 2020)

Border condition	Type of condition	Condition values	
		Velocity (m/s)	Temperature (K)
Mine entrance	Output pressure	–	298
Extractor inlet	Output pressure	15	300
Fresh air	Mass flow input	16.82	298
Burner air	Mass flow input	–	300
Combustion air	Mass flow input	–	315
Right wall	Wall	0	300
Left wall	Wall	0	300
Ceiling	Wall	0	300
Front of work	Wall	0	310

Table 2 Boundary conditions for the simulation in the development work of the underground mine (Vizguerra et al. 2020)

Border condition	Type of condition	Condition values
Interior default	Inside	Inside
Lighting_200_5_Interior_superior	Inside	Inside
Lighting_200_5_Interior_superior	Wall	Stationary wall
Upper shaft_3_ of the intermediate shaft	Wall	Stationary wall
Wall	Wall	Stationary wall
Output pressure	Output pressure	177,608.933 pa
Tank	Mass flow input	16.82 m/s

3 Results

3.1 Temperature Analysis for the Ventilation Process in the Case Study

Figure 3 shows the temperature contour of the underground work without implementing ventilation, in which this parameter ranges between 30 and 36 °C, with the highest temperatures being recorded in the area where there is machinery and personnel, which generates high-stress conditions and, therefore, unsafe and unsuitable for workers and equipment.

Figure 4 shows the behavior of the temperature as the ventilation process progresses; in this graph, as the time of this process progresses, the parameter of interest decreases, having values below that indicated in NOM-015-STPS- 2001 from a time of 80 s, in this sense it is observed that the temperature stabilizes at a value of 28 °C at 1300 s. The temperature contour (Fig. 5) corroborates the decrease

Fig. 3 Temperature contour in underground work without implementing the ventilation system

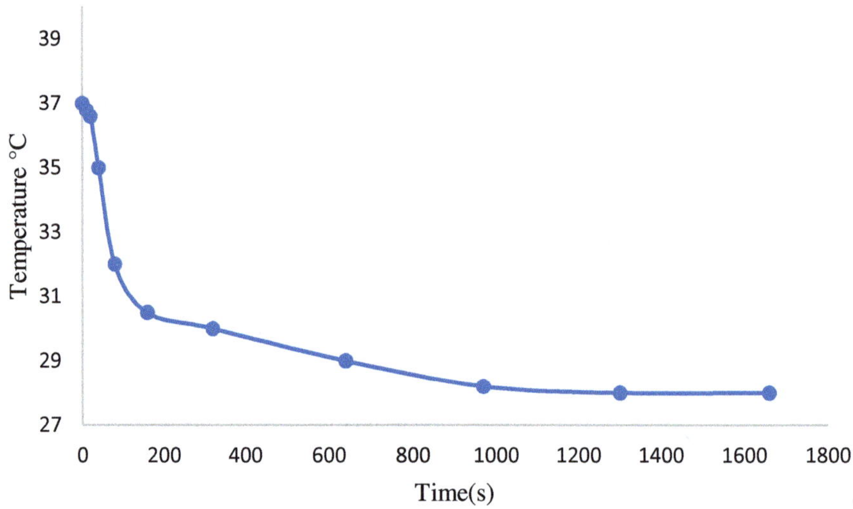

Fig. 4 Temperature behavior in the case study versus time of ventilation process

of this parameter in the underground work and a more homogeneous distribution of it after the ventilation process.

3.2 Relative Humidity Analysis for the Ventilation Process in the Case Study

Figure 6 shows the contour of relative humidity of the underground work without implementing ventilation; the graph indicates that this parameter ranges between 50 and 90 °C, registering the highest humidity in the entrance area to work, decreasing as

Fig. 5 Temperature contour after the ventilation process in the case study

Fig. 6 Relative humidity contour in underground work without implementing the ventilation system

it advances. The work area is consistent because the highest temperature is reported in this area, displacing the humidity in colder places.

Figure 7 shows the behavior of relative humidity as the ventilation process progresses. In this graph, the parameter of interest decreases as the time of this process progresses. This process happens when evacuating stale air, the vacuum generated is occupied by fresh air, which maintains a considerably lower humidity. Concerning stale air, in this sense, the graph shows that a substantial decrease in this parameter is achieved at 160 s, maintaining constant values from this value. The preceding can be corroborated with the relative humidity contour (Fig. 8), which shows the displacement of stale air with higher humidity towards the exhaust area. Therefore, humidity homogeneity is achieved in the work area.

4 Conclusions

With the application of the standard k-epsilon model in the simulation of ventilation at different process times to analyze the change in the thermo-environmental conditions (temperature and humidity) that intervene daily in work carried out in the different works of development in the underground mine, results were obtained within the permissible range established by the NOM-015-STPS-2001 standard.

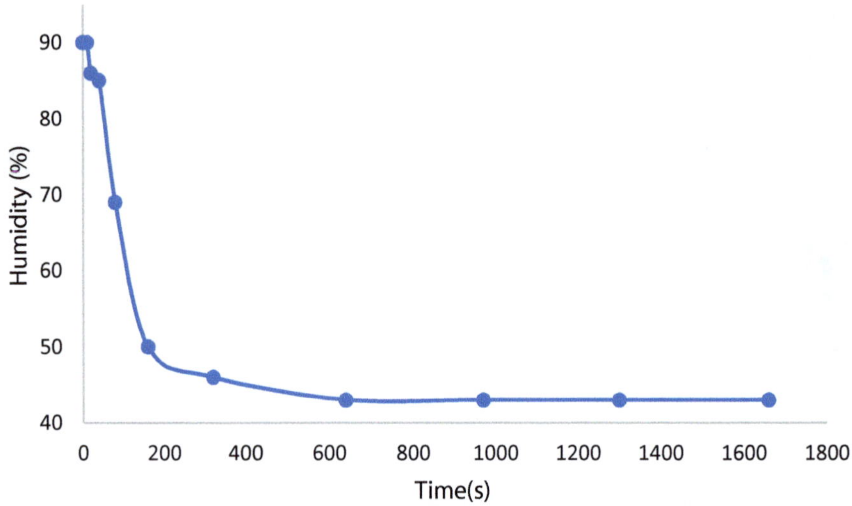

Fig. 7 Behavior of the relative humidity in the case study versus time of ventilation process

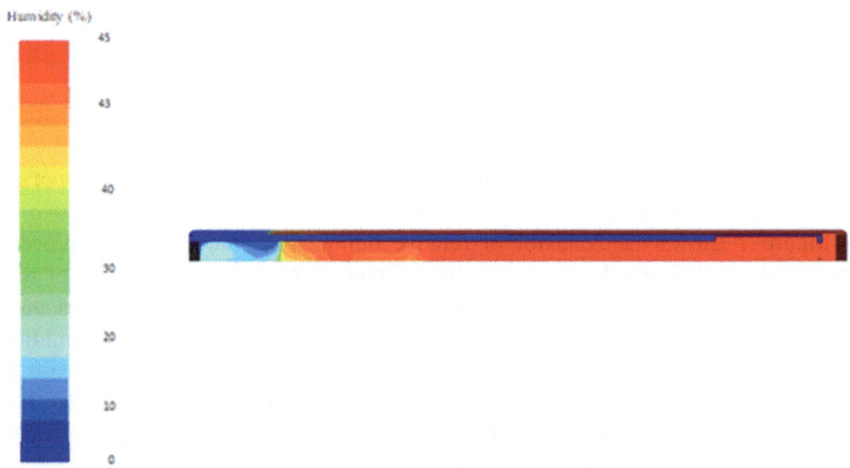

Fig. 8 Relative humidity contour after the ventilation process in the case study

These results show that this process represents an option to ventilate a determined area in underground mines.

In general, this work provides an acceptable option in the study of sustainable ventilation systems because, from these simulations, there is the possibility of generating implementation alternatives, considering critical parameters which contribute to the reduction of energy resources as well as the reduction of gas emissions in underground mines and the mitigation of inadequate thermo-environmental conditions. This scenario generates conditions to improve the health of workers and contributes to comfort by providing an adequate environment where these types of activities take place. These practices help to achieve the goals set out in SDGs 3 and 9 contemplated in the 2030 Agenda for Sustainable Development. Finally, the results show that the temperature within the range indicated in NOM-015-STPS-2001 is achieved at 1300 s having a humidity of 43%.

This work represents a study on the adequate time required to improve the thermo-environmental conditions in the work areas in underground works. In this sense, the analysis was developed considering only the area occupied by the equipment and workers. However, to obtain a more realistic overview, the network of air flows in the entire underground complex should be considered. Thus, future work will focus on the effect of the air flows of the system when the flow in the work areas is optimized. According to what was expressed in the previous paragraphs, this work contributes to the development of sustainable mining processes. Likewise, it is possible to reflect on how this type of study in the various areas of the mining-metallurgical sector and its environment represents an attractive path for the sustainable evolution of this activity.

Acknowledgements The authors thank the University of Guanajuato for its support in carrying out this research project.

References

Ars C, Gauthier J, Florsch N (2021) Underground ancient mine work ventilation modelling. J Archaeol Sci Rep 37.https://doi.org/10.1016/j.jasrep.2021.102805

Brodny J, Tutak M (2021) Applying computational fluid dynamics in research on ventilation safety during underground hard coal mining: a systematic literature review. Process Saf Environ Prot 151:373–400. https://doi.org/10.1016/j.psep.2021.05.029

Brodny J, Tutak M (2022) Challenges of the polish coal mining industry on its way to innovative and sustainable development. J Clean Prod 375:134061. https://doi.org/10.1016/j.jclepro.2022.134061

Departamento de Asuntos Económicos y Sociales, Desarrollo Sostenible [Sustainable Development] (2022). https://sdgs.un.org/es/goals

Departamento de Asuntos Económicos y Sociales, Desarrollo Sostenible [Sustainable Development] (2022) Objetivo 3 [Goal 3]. https://sdgs.un.org/es/goals/goal3

Departamento de Asuntos Económicos y Sociales, Desarrollo Sostenible [Sustainable Development] (2022) Sector industrial [Industrial Sector]. https://sdgs.un.org/es/topics/industry

Dibattista I, Camara A, Molderez I, Benassai E, Palozza F (2023) Socio-environmental impact of mining activities in Guinea: the case of bauxite extraction in the region of Boké. J Clean Prod 387:135720. https://doi.org/10.1016/j.jclepro.2022.135720

Kendouci M, Kharroubi B, Khelfaoui R, Bendida A, Dennai B, Maazouzi A (2013) Simulation of water filtration in porous zone based on Darcy's law. Proced Energy 36:163–168. https://doi.org/10.1016/j.egypro.2013.07.019

Nikolakis N, Kantaris G, Bourmpoucha K, Alexopoulos K (2021) A cyber-physical system approach for enabling ventilation on-demand in an underground mining site. Proced CIRP 97:487–490. https://doi.org/10.1016/j.procir.2020.05.265

Ofosu G, Sarpong D (2022) Beyond the doom: Sustainable water management practices of small-scale mining operations. Resour Policy 77:102649. https://doi.org/10.1016/j.resourpol.2022.102649

Pomasoncco A, Trujillo C, Arauzo L, Raymundo C, Quispe G, Dominguez F (2022) Pre-split blasting design to reduce costs and improve safety in underground mining. Energy Rep 8:1208–1225. https://doi.org/10.1016/j.egyr.2022.07.109

Quarm J, Anning A, Fei-Baffoe B, Siaw V, Amuah E (2022) Perception of the environmental, socio-economic and health impacts of artisanal gold mining in the Amansie West District, Ghana. Environ Chall 9:100653. https://doi.org/10.1016/j.envc.2022.100653

Quevedo C (2013) Sistema de ventilación de diez kilometros del túnel de conducción de la Central Hidroeléctrica Huanza [Ten-kilometer ventilation system for the Huanza Hydroelectric Power Plant's conduction tunnel]. Thesis, Universidad Nacional Mayor De San Marcos 1:28–29

Rhatigan J (2020) Mining meaning: telling spatial histories of the Britannia Mine. J Hist Geogr 67:36–47. https://doi.org/10.1016/j.jhg.2019.10.010

Secretaría de Trabajo y Previsión Social. (2001) NORMA Oficial Mexicana NOM-015-STPS-2001, Condiciones térmicas elevadas o abatidas-Condiciones de seguridad e higiene. Estados Unidos Mexicanos [Mexican Official Standard NOM-015-STPS-2001, Elevated or lowered thermal conditions-Safety and hygiene conditions. United Mexican States]. https://www.dof.gob.mx/nota_detalle.php?codigo=728016&fecha=14/06/2002#gsc.tab=0

Secretaría de Trabajo y Previsión Social. (2012) NORMA Oficial Mexicana NOM-023-STPS-2012, Minas subterráneas y minas a cielo abierto - Condiciones de seguridad y salud en el trabajo. Estados Unidos Mexicanos [Mexican Official Standard NOM-023-STPS-2012, Underground mines and open pit mines - Occupational safety and health conditions. United Mexican States]. https://dof.gob.mx/nota_detalle.php?codigo=5272056&fecha=11/10/2012#gsc.tab=0

Tetreault D (2020) The new extractivism in Mexico: rent redistribution and resistance to mining and petroleum activities. World Dev 126:104714. https://doi.org/10.1016/j.worlddev.2019.104714

Upadhyay S (2022) Drivers for sustainable mining waste management–a mixed-method study on the Indian Mining Industry. Resour Policy 79:102904. https://doi.org/10.1016/j.resourpol.2022.102904

Vizguerra P, López R, Baltazar J, Valtierra J, Ontiveros I, Rodríguez C, García J, Carreño G, Aguilera A (2020) Towards More sustainable extractive industries: study of simulation of efficient ventilation systems in the emission reduction of gases for the development of mine works. International Business, trade and institutional sustainability. World Sustain Ser 1:591–613. https://doi.org/10.1007/978-3-030-26759-9_34

Wilcox D (1994) Turbulence modeling for CFD, vol 1. DCW Industries, Inc., USA. pp 1–17. https://cfd.spbstu.ru/agarbaruk/doc/2006_Wilcox_Turbulence-modeling-for-CFD.pdf

Yildiz T (2020) Evaluation of forestland use in mining operation activities in Turkey in terms of sustainable natural resources. Land Use Policy 96:104638. https://doi.org/10.1016/j.landusepol.2020.104638

Education for Sustainable Development in Higher Education Institutions

Gemma Tejedor, Fermín Sánchez-Carracedo, and Jordi Segalàs

Abstract This paper presents a tool to facilitate the embedding of sustainability in engineering degree curricula. The tool is the Engineering Sustainability Competency Map of the EDINSOST2-SDG project. The map has been presented in previous publications, but this paper presents the most simplified version and the methodology followed to carry out the simplification. The map contains 29 learning outcomes related to sustainability that should be developed in all engineering degrees. The relationship of each of the learning outcomes with the Sustainable Development Goals is presented. In particular, the relationship with the targets defined by the United Nations in 2015 and with the learning objectives defined by UNESCO in 2017.

Keywords Sustainability · Education for sustainable development · Sustainable development goals (SDG) · UNESCO learning objectives · EDINSOST · Competency map

1 Introduction

In the context of the European Higher Education Area, an increasing number of universities are incorporating professional skills into their curricula (Ferreras-Garcia et al. 2021; Forcael et al. 2022). Students acquire the technical competencies of the degree and complete their training through the professional competencies. These competencies refer to transversal skills and experiences applicable to a wide variety

G. Tejedor (✉) · F. Sánchez-Carracedo · J. Segalàs
University Research Institute for Sustainability Science and Technology, Universitat Politècnica de Catalunya, Jordi Girona 1-3, 08034 Barcelona, Spain
e-mail: gemma.tejedor@upc.edu

F. Sánchez-Carracedo
e-mail: fermin.sanchez@upc.edu

J. Segalàs
e-mail: jordi.segalas@upc.edu

W. Leal Filho et al. (eds.), *The Contribution of Universities Towards Education for Sustainable Development*, World Sustainability Series,
https://doi.org/10.1007/978-3-031-49853-4_4

of jobs, situations, and tasks in most areas of knowledge (Gardelliano 2002). Some examples of professional competencies are teamwork, communication and interpersonal skills, ethics, social commitment, and sustainability. Several initiatives have helped define the real-world career needs of engineering students. Accreditation systems introduce professional competencies in the American (ABET 2021), Canadian (CEAB 2021) or European (ENAEE 2021) engineering education curricula. The CDIO (Conceive, Design, Implement, and Operate) project, promoted by the MIT Department of Aeronautics and Astronautics, and extended to many other universities around the world, is also aimed at fostering personal and interpersonal skills in students (Berggren et al. 2003). The University of Deusto in the Basque Country led the Tuning project (Gonzalez and Wagennar 2008), initially aimed at Europe and later extended to Latin America, with the aim of standardizing engineering curricula in terms of professional competencies. The Tuning project defines 30 professional competencies classified in three categories (instrumental, interpersonal, and systemic). It currently coordinates, together with the University of Groningen, the Tuning Academy (http://tuningacademy.org/), with projects in more than 58 countries in America, Africa, Asia, and Europe.

Competency-based education continues to be a challenge today (Sánchez-Carracedo et al. 2018a; Leicht et al. 2018). Each professional competency can be deployed throughout the entire curriculum by means of different strategies, beyond its occasional introduction in certain subjects. The development of a competency must be included, in a planned way, in the different subjects that make up the competency itinerary. Coordination between the subjects is essential to establish an itinerary for each competency, to which the different subjects contribute according to their characteristics, such as the level they occupy in the degree, type (compulsory or optional), degree of specialization, etc. In this way, the subjects of the competency itinerary cover the expected learning outcomes in different depths and forms throughout the entire curriculum.

Figure 1 shows an example of a competency itinerary. The red squares represent the subjects that are integrated into a competency itinerary of a given competency in a 4-year curriculum. The squares correspond to the subjects of a degree curriculum structured in 8 semesters. The red squares represent the subjects that are integrated into the competency itinerary. Students reach the different domain levels of the competency as they progress through the curriculum. To ensure that the students acquire the competency, planning must be careful, and the subjects must be well coordinated.

In general, engineering professors tend to have little training in professional competencies, and more specifically in competencies in sustainability, ethics, or social commitment. This makes it more difficult for them to integrate learning objectives for sustainability in their subjects (Busquets et al. 2021; Sánchez-Carracedo et al. 2018c).

To facilitate efficient and systematic embedding of sustainability competencies in engineering curricula, the EDINSOST projects propose a methodology based on the use of competency maps (Sánchez-Carracedo et al. 2018c). From 2016 to December 2022, two projects will have been carried out, EDINSOST1 and EDINSOST2-SDG,

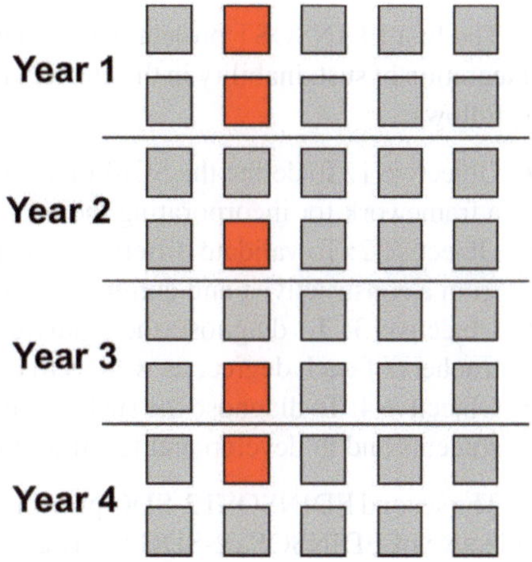

Fig. 1 Example of a competency itinerary throughout a 4-year curriculum

lasting three and four years, respectively, and coordinated by the University Research Institute for Sustainability Science and Technology of the Universitat Politècnica de Catalunya-BarcelonaTech. Both projects have the objective of proposing a methodology for embedding and evaluating sustainability in the curricula of Bachelor's and Master's degrees at Spanish universities. Fifty-nine researchers from eleven Spanish universities have participated in both projects. The scope of application of the EDINSOST projects covers thirteen degrees in the areas of knowledge of Education, Engineering and Business Administration and Management.

The EDINSOST projects have developed several tools that enable the following questions to be answered in each of the degrees analyzed:

(a) What learning outcomes should graduates have acquired on completion of their university studies?
(b) To what extent is sustainability present in the curriculum?
(c) To what extent do students perceive that they have acquired the learning outcomes related to sustainability foreseen in the curriculum?

To answer the first question, the Sustainability Competency Map (SCM) of each degree has been developed (Sánchez-Carracedo et al. 2021). This map contains the set of learning outcomes related to sustainability that must be achieved by the students of the degree. These learning outcomes are easily extrapolated to other degrees in other areas of knowledge, given the transversal nature of the sustainability competency.

The second question is answered through the degree's Sustainability Presence Map (Sánchez-Carracedo et al. 2019). This map identifies how much each of the learning outcomes of the SCM has been developed in the degree. To that end, all the subjects of the competency itinerary in sustainability are analyzed.

Finally, to answer the third question, a questionnaire has been developed for the students in order to find out what they really perceive that they have learnt (Sánchez-Carracedo et al. 2018b).

The first EDINSOST project was an analytical project, focused on exploring the integration of sustainability in the curricula of the degrees studied. Its objectives were as follows:

- Objective 1: To define the SCM of each of the participating degrees and establish a framework for incorporating the map into the degree in a holistic way
- Objective 2: To validate different didactic strategies for addressing sustainability from a constructivist and community pedagogical approach
- Objective 3: To diagnose the status of the sustainability training needs of the teachers of each degree, as well as to develop and test training proposals; and
- Objective 4: To diagnose the sustainability competency level of current university students and to develop and test training proposals.

The second EDINSOST2-SDG project has been an intervention project. The main objective of EDINSOST2-SDG has been to introduce the Sustainable Development Goals (SDG) in the tools developed by EDINSOST1 and apply the results to the degrees participating in the project.

The objectives of EDINSOST2-SDG were the following:

- Objective 1: Identify the SDGs in the EDINSOST' SCM
- Objective 2: Improve the learning outcomes of sustainability and SDGs in current degrees
- Objective 3: Faculty training in Education for Sustainability and Sustainable Development Goals
- Objective 4: Analyze the learning in Sustainability and SDGs of students during their training at the university.

In 2015, the UN approved the 2030 Agenda (UN 2015) on Sustainable Development, an opportunity for countries and their societies to embark on a new path to improve everyone's life, leaving no one behind. The Agenda has 17 Sustainable Development Goals and 169 targets. UNESCO (Rieckmann 2017) defined 15 learning objectives in the cognitive, socio-emotional, and behavioral fields for each SDG, with a total of 255 learning objectives. The SCM has been explicitly related to the targets and learning objectives of the 17 SDGs.

The SCM is a key element for the design of the curriculum to enable the different learning outcomes to be distributed among the subjects of the sustainability itinerary. It also makes it possible to determine what sustainability training teachers should receive. In this paper we present the SCM that contains the learning outcomes common to all engineering degrees, developed by the EDINSOST2-SDG project.

2 Methodology

In EDINSOST2-SDG, a SCM was designed for each degree based on the map developed in EDINSOST1. The SCM design and validation process can be consulted in Sánchez-Carracedo et al. (2021). The SCM of the EDINSOST2-SDG project is a matrix designed on the basis of the four sustainability competencies defined by CRUE (2012).

- C1: Critical contextualization of knowledge by establishing interrelations with social, economic, environmental, local and/or global problems.
- C2: Sustainable use of resources and prevention of negative impacts on the natural and social environment.
- C3: Participation in community processes that promote sustainability.
- C4: Application of ethical principles related to the values of sustainability in personal and professional behavior.

Each competency is more precisely defined by one or more Competency Units (CU), which are described using the three dimensions of sustainability (environmental, social, and economic) plus a holistic dimension as a starting point. Whenever possible, only the holistic dimension has been used to make the map as simple as possible. Competencies C1, C3 and C4 have a single CU that is defined holistically. The C2 competency has four CUs, one for each dimension. Each CU is described in the form of learning outcomes using the simplified Miller pyramid as a taxonomy (Miller 1990). The taxonomy has three domain levels: Know, know-how and demonstrate + do (the original Miller pyramid has four levels, since the 'demonstrate' and 'do' levels are treated independently). Each CU has one or more learning outcomes at each level of the taxonomy. In total, the SCM of the EDINSOST2-SDG project contains 53 learning outcomes common to all engineering degrees. Table 1 shows a schematic of this map.

To relate the learning outcomes of the SCM with the SDGs, a table was created in which the learning outcomes of the map are related to the targets of the SDGs (UN 2015) and the learning objectives of UNESCO (Rieckmann 2017). Details on the process of creating and validating this table can be found in Sánchez-Carracedo et al. (2021).

Table 1 Scheme of the Engineering SCM of the EDINSOST2-SDG project. Each of the map cells can contain one or more learning outcomes (LOu)

SCM of Engineering degrees			Domain levels of the simplified Miller pyramid		
Competency	**Dimension**	**Competency Unit**	**Know**	**Know How**	**Demo+do**
C1	Holistic	CU 1.1			
C2	*Environmental*	CU 2.1			
	Social	CU 2.2			
	Economic	CU 2.3			
	Holistic	CU 2.4			
C3	Holistic	CU 3.1			
C4	Holistic	CU 4.1			

Learning Outcomes

In 2021, the Universitat Politècnica de Catalunya - BarcelonaTech (UPC) decided to use the EDINSOST model to incorporate Sustainability in the curricula of all its degrees. After analyzing the tools designed by the EDINSOST2-SDG project (Sánchez-Carracedo et al. 2021), the UPC asked the project researchers to simplify the tools so that they would be easier to use by teachers and management teams unfamiliar with sustainability. The EDINSOST team then started a process to reduce the number of learning outcomes of the Engineering SCM. For this purpose, the objective was defined to reduce the number of learning outcomes of each cell of the map, so that each CU corresponded, whenever possible, to a single learning outcome for each level of the taxonomy.

To achieve this goal and validate the process, the following steps were performed:

1. The EDINSOST team was divided into two groups, one to carry out the first approximation (working group-WG) and the other to validate the result (validation group-VG)
2. The WG analyzed the learning outcomes for each cell and grouped those that were most closely related and could be consistently written as a single learning outcome. The new learning outcomes were initially written by keeping the original wording. From this wording, a simplification process was undertaken with the aim of obtaining a complete but shorter wording. This process gave rise to an SCM in which each cell had one or, in some cases, two learning outcomes (there is only one cell that has three)
3. The VG analyzed the new learning outcomes and compared them with the initial ones to ensure that all the information described by the original learning outcomes was included in the new learning outcomes. Some deficiencies were detected in this regard, and some of the original learning outcomes were considered to be insufficiently represented by the new learning outcomes
4. The WG collected the suggestions of the VG and drafted a new proposal containing a total of 29 learning outcomes, and
5. The VG validated that the original learning outcomes were correctly represented in the new proposal.

Once the new simplified SCM was obtained, the tables relating the SCM learning outcomes with the UNESCO learning objectives and the UN targets UN were updated. To this end, each learning outcome of the simplified SCM was assigned to the learning objectives and targets that had been previously assigned to the learning outcomes of the original SCM involved in the simplified learning outcome.

3 Results

Before making the tables relating the SDGs to the SCM, we analyzed which of the UNESCO learning objectives (Rieckmann 2017) and the UN targets (UN 2015) were related to engineering degrees. Table 2 presents the results for the UNESCO learning objectives, and Table 3 for the UN targets.

Table 2 Relationship of the 255 UNESCO learning objectives with engineering. (In red those that are not related; in yellow those that are related to a specific degree; in green those that are related to all or almost all engineering degrees) Lob—Learning Objectives; Cx-Cognitive; Sx—Social Emotional; Bx—Behavioral

	SDG																
LOb	1	2	3	4	5	6	7	8	9	10	11	12	13	14	15	16	17
C1																	
C2																	
C3																	
C4																	
C5																	
S1																	
S2																	
S3																	
S4																	
S5																	
B1																	
B2																	
B3																	
B4																	
B5																	

Table 3 Relationship of the 169 UN targets with engineering (in red those that are not related, in yellow those that are related to a specific degree, in green those that are related to all or almost all engineering degrees)

SDG																
1	2	3	4	5	6	7	8	9	10	11	12	13	14	15	16	17
1.1	2.1	3.1	4.1	5.1	6.1	7.1	8.1	9.1	10.1	11.1	12.1	13.1	14.1	15.1	16.1	17.1
1.2	2.2	3.2	4.2	5.2	6.2	7.2	8.2	9.2	10.2	11.2	12.2	13.2	14.2	15.2	16.2	17.2
1.3	2.3	3.3	4.3	5.3	6.3	7.3	8.3	9.3	10.3	11.3	12.3	13.3	14.3	15.3	16.3	17.3
1.4	2.4	3.4	4.4	5.4	6.4	7.a	8.4	9.4	10.4	11.4	12.4	13.a	14.4	15.4	16.4	17.4
1.5	2.5	3.5	4.5	5.5	6.5	7.b	8.5	9.5	10.5	11.5	12.5	13.b	14.5	15.5	16.5	17.5
1.a	2.a	3.6	4.6	5.6	6.6		8.6	9.a	10.6	11.6	12.6		14.6	15.6	16.6	17.6
1.b	2.b	3.7	4.7	5.a	6.a		8.7	9.b	10.7	11.7	12.7		14.7	15.7	16.7	17.7
	2.c	3.8	4.a	5.b	6.b		8.8	9.c	10.a	11.a	12.8		14.a	15.8	16.8	17.8
		3.9	4.b	5.c			8.9		10.b	11.b	12.a		14.b	15..9	16.9	17.9
		3.a	4.c				8.1		10.c	11.c	12.b		14.c	15.a	16.1	17.10
		3.b					8.a				12.c			15.b	16.a	17.11
		3.c					8.b							15.c	16.b	17.12
		3.d														17.13
																17.14
																17.5
																17.16
																17.17
																17.18
																17.19

The next step was to relate the learning outcomes of the SCM with the learning objectives of UNESCO and the targets of the SDGs that were related to all engineering degrees (those marked in green in Tables 2 and 3). Details of this assignment can be found in Sánchez-Carracedo et al. (2021). The learning objectives and the targets marked in yellow were not considered, because the Engineering SCM is a map common to all engineering degrees. Each engineering degree must add its own learning outcomes and relate them to some of the learning objectives and targets marked in yellow.

As detailed in the Methodology Section, a simplification of the Engineering SCM was carried out to make it easier to manage by teachers and management teams unfamiliar with sustainability. Thus, in the simplified Engineering SCM only 29 learning outcomes remained. Table 4 shows the outline of the map with the final wording of the CUs and the coding of the learning outcomes, and Table 5 contains the precise definition of each learning outcome.

Finally, the relationship of the Engineering SCM learning outcomes (Table 5) with the UNESCO learning objectives (LOb) (Table 2) and the UN targets (Table 3) has been studied. Table 6 shows this relationship. Tables 2, 3 and 4 show that the SCM explicitly contemplates the acquisition of the necessary learning outcomes to achieve compliance with the SDG.

The relevance of each SDG in the SCM has also been analyzed by counting the number of times that learning objectives and targets of each SDG are related to some SCM learning outcome. Figure 2 shows that the most represented SDGs are SDG4: Quality Education, SDG9: Industry, innovation and infrastructure, and SDG12: Responsible consumption and production. These SDGs are closely related to engineering and sustainability education. Note that the SDG2: Zero hunger is not explicitly contemplated in the SCM, since the SCM is common to all engineering degrees and not all engineering degrees develop this SDG. However, learning outcomes related to this SDG should be included in specific degrees such as Agricultural Engineering.

4 Conclusions

This paper has presented the latest version of the Engineering Sustainability Competency Map of the EDINSOST2-SDG project. The map contains 29 learning outcomes related to sustainability that should be developed in all engineering degrees. The learning outcomes of the map are related to the SDGs, so that by introducing the learning outcomes in different subjects of the curriculum, these subjects also develop different SDGs, which is an urgent need in today's society. The most represented SDGs on the map are SDG4: Quality Education, SDG9: Industry, innovation and infrastructure, and SDG12: Responsible consumption and production. These SDGs are closely related to all engineering degrees and sustainability education.

Table 4 Main scheme of Engineering SCM

Engineering SCM					
Competency	Dimension	Competency unit	Learning domain levels		
			Level 1 know	Level 2 know how	Level 3 demonstrate + do
C1. Critical contextualization of knowledge establishing interrelations with social, economic, and environmental, local and/or global problems	Holistic	CU1.HO. Learners have a historical and contemporary perspective, and understand the systemic nature of environmental, social, and economic problems, as well as their interrelationships and future challenges, both locally and globally	C1.HO.1.1	C1.HO.2.1	C1.HO.3.1
C2. Sustainable use of resources and prevention of negative impacts on the natural and social environment	Environmental	CU2.EN. Learners are able to detect and analyze the environmental impact of their professional activity and to propose sustainable solutions	C2.EN.1.1 C2.EN.1.2	C2.EN.2.1	C2.EN.3.1
	Social	CU2.SO. Learners are able to detect and analyze the social impact of their professional activity and to propose sustainable solutions	C2.SO.1.1 C2.SO.1.2	C2.SO.2.1 C2.SO.2.2	C2.SO.3.1
	Economic	CU2.EC. Learners are able to manage the material, financial and human resources of the projects in their professional field with sustainability criteria to ensure their economic viability	C2.EC.1.1	C2.EC.2.1	C2.EC.3.1
	Holistic	CU2.HO. Learners are able to detect and analyze the environmental, social, and economic impact of their professional activity and to propose, design, organize and carry out sustainable actions	C2.HO.1.1 C2.HO.1.2 C2.HO.1.3	C2.HO.2.1 C2.HO.2.2	C2.HO.3.1

(continued)

Table 4 (continued)

Engineering SCM					
Competency	Dimension	Competency unit	Learning domain levels		
			Level 1 know	Level 2 know how	Level 3 demonstrate + do
C3. Participation in community processes that promote sustainability	Holistic	CU3.HO. Learners are capable of participating in inclusive reflection and decision-making processes with a global citizenship perspective, and of working from their professional field on interdisciplinary and transdisciplinary projects that guide society towards sustainable transitions	C3.HO.1.1 C3.HO.1.2	C3.HO.2.1	C3.HO.3.1
C.4 Application of ethical principles related to the values of sustainability in personal and professional behavior	Holistic	CU4.HO. Learners act in accordance with ethical and deontological principles related to the values of sustainability	C4.HO.1.1 C4.HO.1.2	C4.HO.2.1	C4.HO.3.1

Table 5 Precise definition of the learning outcomes of Engineering SCM

C1.HO.1.1	Learners know the main causes, consequences and proposed solutions to sustainability problems (social, economic and/or environmental), both local and global, especially in their professional field. For example, Sustainable Development Goals from Agenda 2030, IPCC reports
C1.HO.2.1	Learners reflect critically about sustainability in their professional field
C1.HO.3.1	Learners are able to relate a sustainability problem of a product or service in their professional field with the methods and strategies to face them
C2.EN.1.1	Learners know metrics (or tools) to measure the environmental impact of products and services related to their professional field (for example, environmental footprint, pollutant emissions, resource/energy consumption, biodiversity impact, waste generation, Directive 2014/95/UE for non-financial reporting)
C2.EN.1.2	Learners know strategies and/or technologies for reduction, reuse and recycling of natural resources and waste, related to products and services in their professional field
C2.EN.2.1	Learners know how to use appropriate metrics (or tools) to measure the environmental impact of products and services related to their professional field, throughout their life cycle (extraction, production, use and end-of-life)
C2.EN.3.1	Learners take into account environmental criteria in projects related to their professional field and include indicators to estimate/measure environmental impact
C2.SO.1.1	Learners know metrics (or tools) to measure and describe the social impact of products and services related to their professional field (for example, social life cycle assessment, ISO 26000, Directive 2014/95/UE for non-financial reporting)
C2.SO.1.2	Learners know the basic concepts about health, security and social justice related to their professional field (for example, ergonomics, accessibility, user experience, equity, diversity, common good, transparency, human rights, gender perspective, needs of the most vulnerable groups, discrimination, dignity, anti-corruption)
C2.SO.2.1	Learners know how to use appropriate metrics (or tools) to measure the social impact of products and services related to their professional field
C2.SO.2.2	Learners understand the direct and indirect consequences for security, health and social justice of products and services related to their professional field
C2.SO.3.1	Learners take into account security, health and social justice criteria in their projects and actions and include indicators to measure social impact
C2.EC.1.1	Learners know the basic concepts of resource management, applicable to the management of projects in their professional field and methods (or tools) to estimate their economic viability (for example, fixed and variable costs, amortizations, budgets, Gantt diagrams, externalities analysis, CANVAS analysis, SWOT analysis, business plan, strategic plan, cost–benefit)
C2.EC.2.1	Learners understand the economic viability plan of a project in their professional field and identify the economic consequences it will have on society
C2.EC.3.1	C2.EC.3.1. Learners are able to plan a project in their professional field, design an economic viability plan and follow up the economic management throughout its useful life
C2.HO.1.1	Learners know the strategic role of their profession on sustainability and the direct and indirect consequences of the use of products and services related to their professional field on society, the economy and the environment

(continued)

Table 5 (continued)

C2.HO.1.2	Learners know different economic approaches that promote sustainable development (for example, circular economy, economy of the common good, social economy, ecological economy)
C2.HO.1.3	Learners know the roles, rights and duties of the different stakeholders (professionals, companies, legislation, clients, consumers, etc.) on the production and consumption of products and services related to their professional field
C2.HO.2.1	Learners know how to analyze the different alternatives of a product or service in their professional field to decide which is the most sustainable
C2.HO.2.2	Learners know how to apply different sustainability approaches on production, consumption (responsible consumption) and recycling
C2.HO.3.1	Learners are able to bring new ideas and solutions to a project in their professional field to make it more sustainable, to propose sustainable projects, to follow up and dismantle appropriately and to select which indicators will be used to measure sustainability
C3.HO.1.1	Learners know the main economic and environmental stakeholders related to their professional field
C3.HO.1.2	Learners know techniques and/or tools to promote, in processes and projects in their professional field, their collaboration, the consideration of needs and expectations (information processes, consultation, participation, integration) and cooperation among them (scenario building techniques, co-creation of knowledge, etc.)
C3.HO.2.1	Learners know how to collaborate with the different stakeholders involved to a project in their professional field, to identify their needs and expectations and to assess the implications they may have on the sustainability of the project
C3.HO.3.1	Learners are able to use techniques and/or tools to promote collaboration and cooperation in interdisciplinary and transdisciplinary contexts of their professional field, participating in processes of reflection and decision making as agents of change towards sustainable transitions
C4.HO.1.1	Learners know the code of ethics of their profession, the main ethical issues and the laws and regulations related to sustainability
C4.HO.1.2	Learners know the concepts of social commitment and corporate social responsibility, their possibilities and limitations
C4.HO.2.1	Learners are capable of identifying and critically assessing the implications of ethical and deontological principles related to the values of sustainability in their professional field, and of critically assessing the responsible action of companies
C4.HO.3.1	Learners are capable of exercising their profession, and of actively participating in responsible action in the entities in which they develop their profession, taking into account ethical principles related to the values of sustainability (for example, equality, justice, precautionary principle, prevention of damage, responsibility towards present and future generations, protection and restoration of a healthy environment, social, economic and environmental human rights)

Table 6 Specific relationship of the learning outcomes (LOu) of the SCM with the learning objectives (LOb) and targets (T) of the SDGs

SCM LOu	SDG																																	
	1		)		;		B		/		F		0		=		,		10		11		12		13		14		15		16		17	
	>(b	T	LOb	T	LOb	T	LOb	T	LOb	T	LOb	T	LOb	T	LOb	T	LOb	T	LOb	T	LOb	T	LOb	T	LOb	T	LOb	T	LOb	T	LOb	T	LOb	T
C1.HO.1.1	C3					:".	C4-C5	4.4-B"0	C5	5.1-/"/ 5.b		6.3-6.4	C1	7.1-7.3	C1	="B	C4	9.2-9.4 9.b		10.2	C1		C2	12.2-12.6 12.8	C3	13.3		14.1		15.1		16.10	C2	
C1.HO.2.1						:".		4.4-B"0				6.3-6.4					:)				C1			12.6		13.3				15.1		16.10		
C1.HO.3.1						:".		4.4-B"0				6.3-6.4	:)	0")		="B	C4-B1	9.2-9.4 9.b	::					12.6		13.3				15.1		16.5		
C2.EN.1.1						:".		4.4-B"0				6.3-6.4		0")		="B		9.2-9.4 9.a						12.2-12.4 12.5	C3	13.3		14.1		15.1			C5	
C2.EN.1.2						:".		4.4-B"0				6.3-6.4	C3	0")		="B	C1	9.2-9.4					C2	12.2-12.4 12.5				14.1		15.1				
C2.EN.2.1						:".		4.4-B"0			C4	6.3-6.4	C3	0")		="B	B2-C1 C2	9.2-9.4					C2	12.2-12.4 12.5		13.3		14.1		15.1				17.7
C2.EN.3.1						:".		4.4-B"0				6.3-6.4	:)	0")		="B	:)	9.2-9.4 9.b			:/		B1	12.2-12.4 12.5	B1	13.3		14.1		15.1			SE3	
C2.SO.1.1						3.6-:".		4.4-B"0		5.1-/"/ 5.b						=")		.")	C2	10.2													C5	
C2.SO.1.2					C1	3.6-:".		4.4-B"0								=")	C5	.")					C2											
C2.SO.2.1	C4				C1	:".	C3	4.4-B"0	C5							=")	C2-C3	.")	B1-C1 C4				C2								SE4			
C2.SO.2.2						3.6-:".		4.4-B"0	C5	5.1-/"/ 5.b						=")	C1-C2	.")	C2-SE1 SE5	10.2											C4-C5 SE4			
C2.SO.3.1	B1-:/					3.6-:".		4.4-B"0	:)	5.1-/"/ 5.b					B1-B5	=")		.")	B2-B4	10.2			B1										SE3	
C2.EC.1.1								4.4-B"0						0")		=")		9.2-9.4																
C2.EC.2.1								4.4-B"0						0")		8.2-="B		9.2-9.4						12.2										

(continued)

Table 6 (continued)

C2.EC.3.1								4.4- B"0						0")	:B	8.2- ="B		,"B																
C2.HO.1. 1						;".		4.4- B"0		5.b	C4	6.3-6.4	C3			="B	C1-C3	9.4-9.b				11.a		12.4- 12.5 12.6- 12.8				14.1				16.10	C4	
C2.HO.1.)						;".		4.4- B"0						0")	C1-B1 B3-B5		C5	9.2-9.4				11.a		12.2- 12.6									C5	
C2.HO.1. ;						;".		4.4- B"0						0")		="B	C5	9.2-9.b				11.a	C3	12.4- 12.6										
C2.HO.2. 1						;".		4.4- B"0				6.3-6.4		0")		="B	C2- C3-C4 C5-B2	9.2-9.4					C5-B4	12.2- 12.4 12.5- 12.6				14.1		15.1				
C2.HO.2.)						;".		4.4- B"0				6.3-6.4				="B		9.2-9.4					C4-C5	12.2- 12.5 12.6				14.1						
C2.HO.3. 1	:/					;".		4.4- B"0				6.3-6.4	:)	0")	B1- B3-B5	="B	B1-B2 C3	9.2-9.4	:B	10.2	B1-B4	11.a	B1-B3	12.2- 12.4 12.5- 12.6	:/	13.3		14.1		15.1			B1	
C3.HO.1.)								4.4- B"0		5.1- /"/ 5.b										10.2		11.a		12.8								16.10		
C3.HO.1. 1								4.4- B"0		5.1- /"/ 5.b										10.2				12.8								16.10	C4	
C3.HO.2. 1								4.4- B"0		5.1- /"/							:)			10.2		11.a		12.3- 12.8								16.10		
C3.HO.3. 1								4.4- B"0		5.1- /"/ 5.b			SE1				:/	9.b		10.2	B2-B4 SE1- SE2	11.a	SE2	12.6- 12.8	SE3					15.1		16.10	B1	
C4.HO.1. 1						;".		4.4- B"0		5.1- /"/ 5.b		6.3-6.4						9.4-9.b	C5	10.2				12.6- 12.4						15.1				
C4.HO.1.)								4.4- B"0		5.1- /"/ 5.b										10.2												16.10		
C4.HO.2. 1	SE5 -B1					;".		4.4- B"0		5.1- /"/ 5.b		6.3-6.4					B2-B5	,"B		10.2			C5- SE5	12.6	B1- :/							16.10	B1-B5	
C4.HO.3. 1	B1- :/					3.6- ;".		4.4- B"0		5.1- /"/ 5.b		6.3-6.4					B1-B3 :/	9.4-9.b		10.2	B1-B4		B1-B4 SE5	12.4- 12.6	:/					15.1		16.10	SE3- B1 :/	

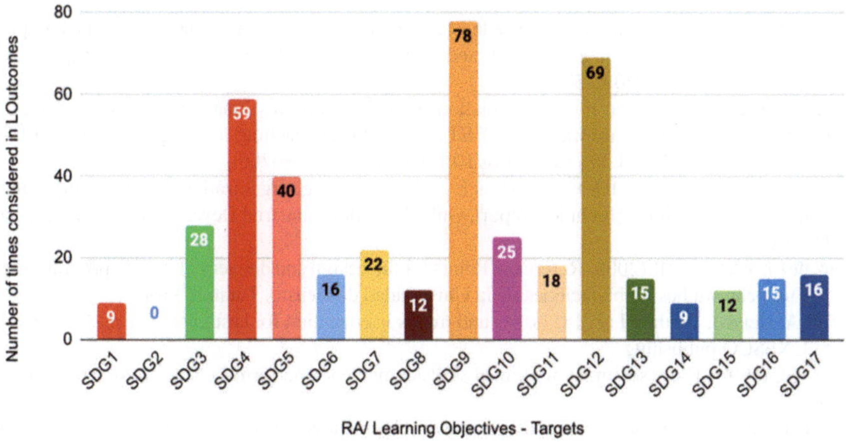

Fig. 2 Relevance of the SDGs in the Engineering SCM

Acknowledgements This work has been conducted out by all the members of the engineering group of the EDINSOST2-SDG project. In addition to the authors, the following researchers have participated (alphabetical order): Gorka Bueno, Pere Busquets, Sara Camacho, Joan Climent, Victor G. Galofré, Irene Josa, Boris Lazarrini, David López, Carme Martín, Rafael Miñano, Estíbaliz Sáez de Cámara, Bàrbara Sureda and Eva Vidal

References

ABET (2021) Criteria for accrediting engineering programs. Accreditation Board for Engineering and Technology. https://www.abet.org/accreditation/accreditation-criteria/criteria-for-accrediting-engineering-programs-2022-2023/

Berggren KF, Brodeur D, Crawley EF, Ingemarsson I, Litant WT, Malmqvist J, Östlund S (2003) CDIO: an international initiative for reforming engineering education. World Trans Eng Technol Educ 2(1):49–52

Busquets P, Segalas J, Gomera A, Antúnez M, Ruiz-Morales J, Albareda-Tiana S, Miñano R (2021) Sustainability education in the spanish higher education system: faculty practice, concerns and needs. Sustainability 13(15):1–14. https://doi.org/10.3390/su13158389

CEAB (Canadian Engineering Accreditation Board) (2021) Accreditation criteria and procedures. https://engineerscanada.ca/sites/default/files/2021-11/2021%20Accreditation%20Criteria%20Book%20Word.pdf

CRUE (2012) Directrices para la Introducción de la Sostenibilidad en el Curriculum. Madrid, Spain: Grupo de Trabajo de Calidad Ambiental y Desarrollo Sostenible de la Conferencia de Rectores de las Universidades Españolas [Guidelines for the Introduction of Sustainability in the Curriculum. Madrid, Spain: Working Group on Environmental Quality and Sustainable Development of the Conference of Rectors of Spanish Universities]

ENAEE (European Network for Accreditation of Engineering Education) (2021) EUR-ACE framework standards. https://www.enaee.eu/eur-ace-system/standards-and-guidelines/#standards-and-guidelines-for-accreditation-ofengineering-programmes

Ferreras-Garcia R, Sales-Zaguirre J, Serradell-López E (2021) Developing entrepreneurial competencies in higher education: a structural model approach. Educ Train 63(5):720–743. https://doi.org/10.1108/ET-09-2020-0257
Forcael E, Garcés G, Orozco F (2022) Relationship between professional competencies required by engineering students according to ABET and CDIO and teaching-learning techniques. IEEE Trans Educ 65(1):46–55. https://doi.org/10.1109/TE.2021.3086766
Gardelliano S (2002) UNIDO competencies. Strengthening organizational core values and managerial capabilities. Technical report, United Nations Industrial Development Organization UNIDO
Gonzalez J, Wagenaar R (2008) Reference Points for the design and delivery of degree programmes in European studies. Publicaciones de la Universidad de Deusto, Tunning Project
Leicht A, Heiss J, Byun WJ (2018) Issues and trends in education for sustainable development vol 5. UNESCO publishing
Miller GE (1990) The assessment of clinical skills/competence/performance. Acad Med 65(9):63–67
Rieckmann M (2017) Education for sustainable development goals: learning objectives. UNESCO publishing. https://unesdoc.unesco.org/ark:/48223/pf0000247444
Sánchez-Carracedo F, Soler A, Martín C, López D, Ageno A, Cabré J, García J, Aranda J, Gibert K (2018a) Competency maps: an effective model to integrate professional competencies across a STEM curriculum. J Sci Educ Technol 27(5):448–468. https://rdcu.be/NbGx
Sánchez-Carracedo F, Álvarez MJ, Barrón A, Caballero D, López E, Muñoz JM, Lugo-Muñoz M, Sureda B, Vidal E, Vidal S (2018b) Elaboración de un Cuestionario para Evaluar el nivel de Sostenibilidad de los Estudiantes de Grados en Ingeniería. In Actas de las XXIV Jornadas Sobre la Enseñanza Universitaria de la Informática [Elaboration of a Questionnaire to Evaluate the Level of Sustainability of Undergraduate Engineering Students. In Proceedings of the XXIV Conference on University Teaching of Computer Science]. Asociación de Enseñantes Universitarios de la Informática Barcelona, Spain 141–148. https://upcommons.upc.edu/bitstream/handle/2117/12173
Sánchez-Carracedo F, Segalàs J, Vidal E, Martín C, Climent J, López D, Cabré J (2018c) Improving engineering educators' sustainability competencies by using competency maps: the EDINSOST project. Int J Eng Educ 34:1527–1537. http://hdl.handle.net/2117/121887
Sánchez-Carracedo F, Moreno-Pino FM, Sureda B, Antúnez M, Gutiérrez IA (2019) Methodology to analyze the presence of sustainability in engineering curricula. Case of study: ten Spanish engineering degree curricula. Sustainability 11:4553. https://doi.org/10.3390/su11174553
Sánchez-Carracedo F, López D, Martín C, Vidal E, Cabré J, Climent J (2020) The sustainability matrix: a tool for integrating and assessing sustainability in the bachelor and master theses of engineering degrees. Sustainability 12:5755. https://doi.org/10.3390/su12145755
Sánchez-Carracedo F, Segalas J, Bueno G, Busquets P, Climent J, Galofré VG, Lazzarini B, López D, Martín C, Miñano R, Sáez de Cámara E, Sureda B, Tejedor G, Vidal E (2021) Tools for embedding and assessing sustainable development goals in engineering education. Sustainability 13(21):12154. https://doi.org/10.3390/su132112154
UN (2015) Transforming our world: the 2030 agenda for sustainable development. https://undocs.org/en/A/RES/70/1

Fieldwork Resource Pack as a Tool in the Teaching of Chemistry and Education for Sustainability in Secondary Schools

Nadine Sciortino and Mark Mifsud

Abstract Recent studies have shown a decline in the number of students that choose to study Chemistry, at Secondary level. A number of researchers (Gröger, Teaching chemistry in near-natural learning environments to address sustainability, 2013; Höper and Köller, Int J Math Sci Technol Educ 6(2), 2018; Höper et al., Chem Educ Res Pract 23:361–372, 2022) believe that since the environment can provide a context for learning, relocating lessons from the classroom and the laboratory into nature, can increase student motivation and improve sustainability competences in students. A mixed-methods approach was used to investigate teachers' and students' perceptions on the pedagogical effectiveness of fieldwork in Chemistry at Secondary level and if such an inquiry-based learning activity offers benefits, that cannot be achieved via laboratory experiments. The challenges that teachers face when organizing fieldwork were also investigated. The Fieldwork Pack, compiled as part of this study, was evaluated both for its usefulness in increasing effective learning and also in its approach to sustainability. Results indicate that both teachers and students endorse the idea of fieldwork in Chemistry, however unlike students, the majority of teachers did not agree that it should be compulsory. Respondents believe that working in-situ during field excursions allows students to achieve transferable skills, which cannot be attained through laboratory experiments. Furthermore, Chemistry fieldwork plays an important role in educating the future generations about the importance of environmental sustainability. Despite the benefits, fieldwork is time-consuming and difficult to fit in the curriculum, which discourages teachers from conducting fieldwork. In response to these findings, a number of recommendations are drawn up to further enhance the effectiveness of fieldwork in teaching Chemistry and to infuse Education for Sustainable Development (ESD) principles in Chemistry education.

N. Sciortino (✉)
Institute for Education, St. Michael Foundation for Education, Europe, San Ġwann, Malta
e-mail: nad.sciortino@gmail.com

M. Mifsud
CEER, University of Malta, Europe, Msida, Malta
e-mail: mark.c.mifsud@um.edu.mt

W. Leal Filho et al. (eds.), *The Contribution of Universities Towards Education for Sustainable Development*, World Sustainability Series,
https://doi.org/10.1007/978-3-031-49853-4_5

Keywords Chemistry · Fieldwork · Sustainability · Mixed-methods · Fieldwork Pack

1 Introduction

The following paper is divided into four sections as follows:

1. Introduction: focuses on the literature review about Chemistry Education in Malta and the importance of fieldwork in teaching Chemistry and Education for Sustainable Development as well as identifies the research gap.
2. Research design and Methodology: describes the mixed methods approach used and the compilation of data to produce the Fieldwork handbook.
3. Results and discussion: focuses on an interpretation of the data collected from the research study in relation to the respondents' views about using fieldwork as a tool to teach Chemistry and Education for Sustainable Development as well as feedback on the resources in the Fieldwork Handbook.
4. Conclusions and recommendations: gives a brief summary of the main outcomes from this research study and identifies any recommendations for future studies.

Chemistry Education in Malta

Chemistry is a subject that brings to light the variety of different materials, their properties and uses in everyday life. This subject also focuses on how chemical reactions can lead to the production of new materials that can be used to improve our quality of life and reduce the impacts of pollutants in our environment (Tundo et al. 2000). Chemistry topics are tackled at different stages in the Maltese education system, as described in Fig. 1.

The Maltese education system is currently experiencing a radical restructuring, shifting from a content-based method of teaching and learning to an outcomes-based

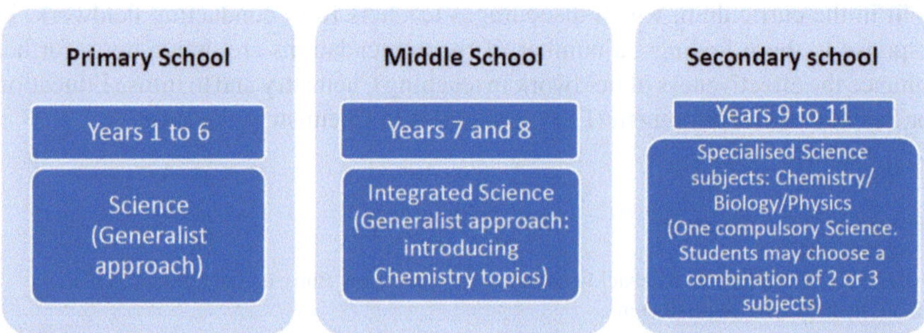

Fig. 1 Approaches of tackling Chemistry topics in schools. *Note* Approaches of tackling Chemistry topics in schools taken from 'Fieldwork as a tool in the teaching of Chemistry in Secondary Schools'. Own work

method (Schembri 2020). The development of a Learning Outcomes Framework (LOF), has been established to support the National Curriculum Framework (NCF) and the main aim is to give schools the flexibility to develop their own learning programmes and modify them according to the needs and abilities of their student cohorts (Schembri 2020). It also aims to reduce the focus on theory recall and shift importance to the development of essential 21st Century skills, thus creating a conducive learning environment to meet the needs of all learners (Attard Tonna and Bugeja 2016).

Up till now, Chemistry content has always been presented in an abstract and fragmented way, which focused mainly on rote learning and memorisation of theory (Farrugia et al. 2017). Furthermore, laboratory experiments have mainly been presented as cookbook-type experiments, which focus on proving theory (Mizzi 2019). The new LOF syllabus of Chemistry, which will be introduced from Years 9 to 11, addresses the holistic development of students and proposes that Chemistry can be taught in a way that instils curiosity, and motivates the students (Farrugia et al. 2017). This inquiry-based learning approach can help students develop higher order thinking and reasoning skills, which are essential to further learning (Goteti et al. 2012). The revised O-Level Chemistry syllabus thus allows more time for an investigative approach to teaching Chemistry, giving students an opportunity to understand its applications in everyday life (Farrugia et al. 2017). Practical work, which provokes thinking and discussion, has become an integral part of the new syllabus (Farrugia et al. 2017). One of the changes to the practical aspect of Chemistry is the inclusion of fieldwork, as part of the MATSEC coursework.

Defining Fieldwork

Howarth and Slingsby (2006, p.99) define fieldwork as "the direct observation of the environment and practical scientific enquiry outside the classroom". Fieldwork is regarded by teachers as a learning technique that makes it easier for students to understand concepts by experiencing the environment first-hand (Mercieca 2011). Furthermore, fieldwork gives the students the opportunity to acquire scientific knowledge and skills which cannot be taught in class or in the laboratory (Behrendt and Franklin 2014).

Benefits of Fieldwork

Literature shows that engaging students in fieldwork has many benefits such as improvement in academic achievement, enhanced critical thinking and problem-solving skills as well as increased co-operative skills and motivation to learn (Rahmawati and Koul 2016). The benefits which students acquire through fieldwork are explained in more detail in Table 1.

Purpose of Fieldwork in Teaching Chemistry and Education for Sustainable Development

Despite the work of chemists having a profound effect on the environment and our quality of life, society is sceptical about 'Chemistry', as it is often associated with toxic or hazardous chemicals which lead to accidents and catastrophes (Fenech

Table 1 Benefits which students acquire through fieldwork

i. Conceptual development and understanding	Fuller et al. (2006), explains that fieldwork helps to bridge the gap between the content that is learnt in the classroom and its application in the real-world, giving students a conceptual understanding of the theories learnt. Such experiential learning experiences provide students with a unique learning environment giving them the opportunity to observe and learn through discovery, which cannot be achieved in class, in the lab or through computer simulations (Tanner 2003)
ii. Improves academic achievement and enhances scientific skills	Different teaching and learning styles are used during fieldwork, helping increase student confidence in transferable skills such as observational skills, data collection and interpretation as well as report writing, thus helping them become all-round learners (Maw et al. 2011). Furthermore, research shows that learning outdoors leads to improved academic achievement in standardised tests, enhanced critical thinking and problem-solving skills (Taş and Gülen 2019)
iii. Increases awareness and appreciation of the environment	Research shows that when students are exposed to outdoor learning they develop a genuine appreciation and respect for the environment (Kenney et al. 2003). By incorporating topics in the curriculum that relate to green Chemistry, such as pollution, climate change and the use and misuse of natural resources, it increases the students' consciousness about the current environmental challenges (Fenech Caruana 2018). This creates citizens who become effective advocates about protecting the environment leading to an inclination to address these problems, both on a local and global scale (Barrows et al. 2016). Fieldwork, thus plays an important role in educating future generations about the need for a green society and the importance of environmental sustainability (Boca and Saraçlı 2019)
iv. Empowers students to foster social and personal development	Research also shows that the pedagogical techniques used in fieldwork ensure that students take responsibility for their learning (Hall et al. 2002). Furthermore, the challenges that they encounter during fieldwork can help students improve their self-confidence, persevere and build resilience (Hall et al. 2002). The application of scientific techniques while working on site, are powerful tools that empower students to foster personal and social development (Larsen et al. 2016). Fieldwork can help students develop and improve important 21st Century skills such as teamwork and leadership skills, communication and collaborative skills (Hall et al. 2002)

Caruana 2018). In fact, in today's world, food and cosmetics are often marketed as being "free of chemicals" or "without artificial additives", since society perceives natural products as being healthier (Krischer et al. 2016). Researchers believe that this could be a reason why students perceive Chemistry as "evil" and why it is the most unpopular science subject in schools (Krischer et al. 2016).

Statistics show that student interest in science declines during compulsory education (Christidou 2011), with Chemistry being the least studied science subject at both secondary and post-secondary level, in many countries, including Malta (Fenech Caruana 2018; Musumeci and Pirotta 2018). A local study by Farrugia et al. (2017) showed that the number of students sitting for the Chemistry O-Level, between 2004 and 2015, dropped by 15%. A number of factors have been found to hinder student motivation to study Chemistry. Many students perceive it as a difficult subject, because of its abstract nature (Fenech Caruana 2018). Furthermore, the fact that Chemistry is often reduced to boring theories, which are rushed in the way they are taught, allows little time for application of content, discussion and creativity. Although laboratory work makes the experience more memorable, they do not see the applications of Chemistry in their everyday lives (Sultana 2011). Höper et al. (2022) state that Chemistry education in schools still focuses on the academic side rather than on engaging learners with phenomena linked to their everyday life. All these factors contribute to a decrease in student interest in studying Chemistry (Fenech Caruana 2018).

Researchers believe that since nature is perceived in a positive way, relocating Chemistry lessons from the classroom and the laboratory into nature, can increase the motivation for learning (Gröger 2013) and help students remember the content in more detail (Rickinson et al. 2004). Chemistry can be experienced through nature-related phenomena and thus field trips in nature can be used to help change the students' attitudes towards studying Chemistry. Local science educators can use fieldwork to enable students to relate the concepts of chemistry to their everyday lives and to understand how chemical processes can be rendered less toxic, less hazardous and non-polluting (Fenech Caruana 2018).

The Situation Abroad—Chemistry Fieldwork Studies

Fieldwork studies in Biology, Environmental Science and Geography are very common, however outdoor teaching experiences in Chemistry are a rare occurrence (Maciejowska 2007). In fact, when it comes to Chemistry education, context-based learning and problem- solving investigations are often utilised within a classroom or laboratory context, with little research done on the importance of these techniques outside the classroom (Borrows 2019). Outdoor experiences in Chemistry, are usually related to visiting Science centres, factories and University/industrial labs. However, research shows that such learning experiences are often disassociated from context-based learning environments and lack transferable value (Thorburn and Allison 2010). Thus, outdoor learning experiences can help to address the issue levelled at Chemistry in terms of relevance to everyday life, since the environment can provide a context for learning the basics of Chemistry (Höper et al. 2022).

The Local Situation Regarding Fieldwork

Local fieldwork has formed an integral part of the curriculum at all stages of education up to tertiary level and has also been introduced at Primary school level (Pace 1979). While it has been made clear to teachers that fieldwork sessions play a crucial role in science pedagogy, its implementation has always been left up to the discretion of the teacher. It is only in recent years that fieldwork in Biology became a compulsory part of the MATSEC coursework (MATSEC 2010). While one may argue that the revised Chemistry syllabus puts more emphasis on the importance of Environmental Chemistry and the importance of sustainability with the introduction of fieldwork, this is not deemed as a compulsory requirement of the coursework (MATSEC 2024).

Christidou (2011), states that Science education at schools must put more emphasis on those aspects of Chemistry, that are more relevant to students' lives. Thus, fieldwork studies in Chemistry, can help students break away from their passive learning habits, giving them a chance to experience the life of a field scientist (Caruana and Treeby 2014). Such inquiry-based learning activities, can motivate students who have difficulty with the abstract nature of the subject, as it provides innovative ways of relating the material covered in theory to a more hands-on approach. Thus, the implementation of fieldwork in Chemistry might be a small step towards improving the students' interest and motivation to study and learn Chemistry (Gröger 2013).

Inquiry-Based Learning Through Fieldwork

Although the curriculum provides students with the foundations of Chemistry and gives them the opportunity to be actively involved in laboratory experiments, students have little or no hands-on experience of the work conducted by field scientists (Holstermann et al. 2010). New research about learning suggests that students learn best when the material is presented in a way that engages their interest and curiosity, thus encouraging them to discover and construct their own understanding of scientific concepts (Ansberry and Morgan 2005). Fieldwork encourages teachers to deviate from traditional teaching methods and implement experiential learning through hands-on activities (Slingsby 2006). Despite the acknowledged benefits of fieldwork, recent studies in fieldwork provision, both locally and abroad, have been in connection mainly with Biology and Geography, but there has been no research, conducted locally, on the importance of fieldwork in Chemistry. Furthermore, despite the inclusion of fieldwork in the LOF syllabus, no resources have yet been developed to guide educators in planning and conducting fieldwork. Farrugia et al. (2017), states that teachers need training, guidance and support when it comes to implementing the changes in the LOF syllabus. Teachers have also stated the need for a resource pack to help with planning and carrying out fieldwork (Aquilina 2010).

Research Gap

Despite the numerous research studies related to fieldwork, no local studies yet exist to determine the importance of fieldwork in Chemistry. The only publications traced in the University of Malta and Institute for Education libraries as well as higher institutions, regarding the purpose of fieldwork, were related to Biology, Geography and Integrated Science. This study thus aimed to fill a knowledge gap, by initiating and encouraging further research to explore the importance of fieldwork as a pedagogical tool in teaching Chemistry and Education for Sustainable development and its importance to help make the subject content more relevant to students, through an inquiry-based learning approach.

2 Research Design and Methodology

The study aimed to shed light on the current local situation and identify the main trends regarding fieldwork in teaching Chemistry and Education for Sustainable Development and to determine the challenges that teachers encounter when scheduling fieldwork. The study was also used to build a Fieldwork Pack made up of a teachers' and student handbook, which included resources linked to different activities that can be carried out before, during and after fieldwork to consolidate student learning and understanding. A mixed-methods approach was used to render the data collected from the study more creditable, since it was used to crosscheck the findings obtained from different research tools to determine if the findings could be confirmed or disputed (Carter et al. 2014).

Student and teacher questionnaires were used to collect quantitative data while group interviews were used to collect qualitative data. Questionnaires were distributed to secondary school Chemistry teachers from all sectors and to Year 10 and Year 11 Chemistry students from participating schools. In addition, two group interviews with teachers who volunteered to participate, were also carried out to enhance the questionnaire data collected. Any issues worth pursuing from the questionnaires were followed up through the group interviews (Cohen et al. 2018). This study was carried out during the scholastic year 2021–2022. A pilot study was conducted during November 2021, before the actual study.

All secondary schools around Malta were asked to participate in this study by distributing the Fieldwork Pack and questionnaire to all Chemistry teachers, to avoid researcher and participant bias. Stratified random sampling of students was used by choosing the Year 10 and Year 11 student cohorts and giving the participating schools a free choice to present it to students from each cohort. This addressed the issue of less homogeneity of the student population and ensured a more accurate representation of the Chemistry cohort (Pooja 2019).

Teacher and student questionnaires addressed demographic data as well as the effectiveness of fieldwork in teaching and learning Chemistry. The importance of introducing fieldwork in Chemistry, the benefits it offers which cannot be attained through laboratory experiments and if fieldwork should be compulsory, were also addressed. Feedback about the resources provided as part of this study and open-ended questions to determine how the teacher and student handbooks can be improved, were also included. Student questionnaires also targeted the relevance of chemistry in the students' academic and daily lives and multiple-response questions to determine which fieldwork activities students find enjoyable and how the experience can be made more rewarding. Teacher questionnaires also addressed the problems that teachers face when organising a fieldwork. Group interviews with teachers also focused on these points.

Data collected from the participants was analysed both quantitatively and qualitatively. A Kolmogorov–Smirnov test was carried out to determine if the data collected in the teacher and student questionnaires had a normal distribution. Any data which showed a normal distribution was analysed using One-Way ANOVA analysis, while data which did not show normal distribution was analysed using the Friedman and Kruskal Wallis tests. The Pearson's Chi-Squared test was also used to determine if variables such as teaching experience and school sector affected the number of fieldworks organised by teachers, and their opinion about the inclusion of fieldwork in Chemistry. Audio recordings from group interviews were transcribed and thematic analysis was carried out using the six steps described by Braun and Clarke (2006).

As part of this study, the manual 'Chemistry Fieldwork Pack' was tailor-made with activities and resources to guide teachers and students, intending on carrying out fieldwork in Chemistry, at O-level, since to date, there are no published works or resources for Chemistry teachers. Reference was made to local studies, even though related to Biology fieldwork, and to the LOF Syllabus, to identify the activities that should be included in the Fieldwork Pack. Research shows that connecting classroom activities during the preparation phase and follow-up work, with the activities carried out during the fieldwork, will ensure students accommodate understanding of new knowledge faster (Behrendt and Franklin 2014). Thus, outdoor activities should take place over three stages: preparation phase, fieldwork activity and follow-up activities (Remmen and Frøyland 2015). For this reason the teacher and student handbooks were divided into three sections: 'Pre-fieldwork activities', 'Fieldwork activities' and 'Post-fieldwork activities'. Feedback from teacher and student questionnaires was used to improve the resources prepared as part of this study. The research process timeline is summarised in Fig. 2.

Limitations of Study

The main limitation of the study was that due to the interruptions, caused by the COVID-19 pandemic, which affected education significantly between March, 2020 and September, 2021. Most schools refused to participate in the study to give teachers more time to catch-up with the syllabus. Furthermore, the student cohorts involved in the study, had limited to no experience of laboratory experiments or fieldwork, at the time of the study. Additionally, the student questionnaires were distributed online and answered at home, thus not all questionnaires were returned.

This research study was thus considered as exploratory in nature, which identified a number of important trends. However, further studies need to be carried out to more accurately identify the pedagogical importance of fieldwork in Chemistry and how it can be used to teach students about the importance of Green Chemistry and improve sustainability competences in students.

3 Results and Discussion

The concepts of Green Chemistry and Education for Sustainable Development started being promoted at various levels of education, including the O-level syllabus, with the recent inclusion of Chemistry fieldwork as part of the coursework. This exploratory study highlighted a number of important trends, which will be discussed in detail below:

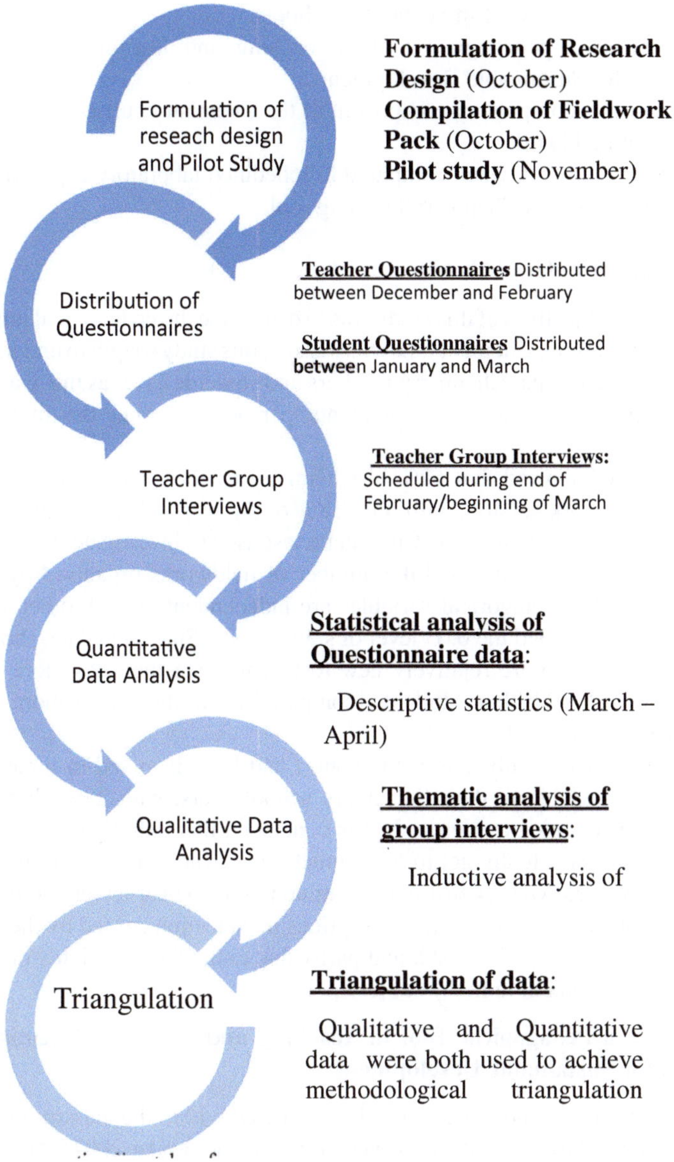

Fig. 2 Depicts the research. *Note* Research process timeline taken from 'Fieldwork as a tool in the teaching of Chemistry in Secondary Schools'. Own work

1. Importance of fieldwork in Secondary schools
2. Fieldwork as a pedagogical tool in teaching and learning Chemistry and Education for Sustainable Development
3. Fieldwork as a compulsory component of the Chemistry curriculum
4. Benefits of Fieldwork
5. Benefits of fieldwork when compared to scheduled laboratory experiments
6. Fieldwork Pack as a Chemistry teaching tool

Fieldwork in Secondary Schools

One of the main objectives of this study was to determine the pedagogical importance of fieldwork in learning Chemistry. For this reason, this study sought to understand the fieldwork experience that Chemistry teachers and students have, as this was deemed a factor that might influence the respondents' perspectives about the importance of fieldwork.

Data collected showed that 62% of Chemistry teachers have been involved in at least one fieldwork session, with only 38% of the respondents having organised fieldwork with their students. The Chi-square test used to investigate the association between teaching experience and the number of fieldworks organised by teachers showed that the two categorical variables are independent of each other, since the p-value (0.153) exceeded the 0.05 level of significance. Statistical data, showed that teachers (71.2%) who are relatively new to teaching are the ones more likely to conduct fieldwork with their students, compared to teachers with more teaching experience (70.0%).

Out of 86 students, only 1% of the cohort had been involved in three or more fieldwork sessions, during their Secondary school years, with 28% claiming that they have no fieldwork experience. It was concluded that although teachers understand "that students actually get to learn much more when they're out in the field" (Interviewee 3), fieldwork is still not a common pedagogical technique used. This could be attributed partly due to the disruptions in education caused by the COVID-19 pandemic, between 2020–2022, and partly due to the fact that fieldwork is not a compulsory requirement in many subjects.

Fieldwork as a Pedagogical Tool in Teaching and Learning Chemistry and Education for Sustainable Development

Chemistry lessons in school are typically conducted within the classroom or in the laboratory and fieldwork studies are very rare (Maciejowska 2007). Nonetheless, fieldwork has been included as part of the MATSEC coursework in the new LOF syllabus for Chemistry. However, fieldwork is not a compulsory requirement for attaining a good grade in Chemistry, since no marks are reserved specifically for it. With this in mind, another objective of this study was to understand the respondents' opinion about the inclusion of fieldwork in teaching Chemistry.

The study found that most teachers (67%) reacted favourably to the introduction of fieldwork in Chemistry, since "*Chemistry must be more tangible than theoretical*" (*Interviewee 2*). The students were equally supportive, with 65% agreeing that it should be incorporated in the O-Level coursework. Furthermore, 51% of the students agreed that relevant practical work through fieldwork is required when studying Chemistry as it can help them to familiarise themselves better with the work that is carried out by scientists (84%) and prepare them for a future career in the scientific field (77%). Most respondents agreed that fieldwork can help to bridge the gap between the abstract nature of the subject and its application in industry and everyday life. Fieldwork in Chemistry can also help students develop a higher degree of environmental awareness as well as develop higher order thinking skills and a critical mind to express themselves more confidently on the role of Sustainable Chemistry and its impacts on society.

Fieldwork as a Compulsory Component of the Chemistry Curriculum

Despite the large number of respondents in favour of Chemistry fieldwork, only 24% of teachers agreed that it should be a compulsory part of the curriculum. The Chi-square test indicated that there were no statistical differences in opinion between teachers from different sectors, since the p-value (0.07) exceeded the 0.05 level of significance. Statistical data showed that a large percentage of Independent school teachers (50%) were in favour of compulsory fieldwork while a larger percentage of Church school teachers (52.9%) were against. Majority of state school teachers (70%) were unsure. Educators argued that a number of obstacles such as lack of resources, adequate training sessions and an overburdened curriculum deters teachers from including a fieldwork in their pedagogy.

This was in contrast with the students' views, where 51% were in favour of compulsory fieldwork. The majority of students (59%) agreed that visual experiences outside the classroom are more engaging leading to a better understanding of the theory learnt in class. One student explained that making it compulsory will ensure that all schools conduct fieldwork as otherwise teachers may opt out, which may put students at a disadvantage.

Benefits of Fieldwork

i. Fieldwork provides students with an opportunity to learn in a different environment, which cannot be replicated in a laboratory setting

Despite the acknowledged benefits provided by learning through laboratory experimentation, the general consensus among teachers is that fieldwork provides students with an opportunity to analyse different environments, which cannot be replicated in a laboratory setting. Statistical analysis of the data collected gave this statement a mean rating score of 8.27 as shown in Fig. 3, and one-way ANOVA analysis showed that there is no statistical significance of the mean rating score provided to this statement, irrespective of the teachers' fieldwork experience. This statement was listed as one of the top three reasons for conducting fieldwork by 73% of the teachers, with 30% ranking it as the primary reason. Both teachers and students, claimed that fieldwork allows students to gain valuable skills, which cannot be acquired through laboratory work.

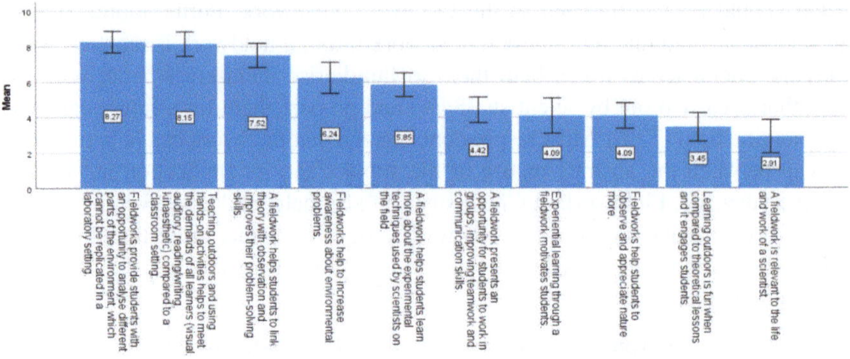

Fig. 3 Error bar graph displaying the mean rating score assigned to statements related to benefits of fieldwork. Note Error bar graph taken from 'Fieldwork as a tool in the teaching of Chemistry in Secondary Schools'. Own work

This implies that teachers and students perceive this educational activity as an important part of learning Chemistry, that can complement the material learned in class and in the laboratory. Research shows that since students carry out on-field tasks in an environment that is not uniform when compared to the predictable laboratory setting, it can lead to transformative learning (Scott et al. 2012). Furthermore, the direct experience achieved through fieldwork enables deeper learning when compared to classroom or laboratory lessons (Scott et al. 2012). Students also value such learning experiences because they are exposed to different field techniques conducted by scientists, helping them to experience Chemistry in real-world situations (Behrendt and Franklin 2014).

ii. Fieldwork helps students to link theory with observation and provides a context to learning and its applications in real life

The group interviews identified fieldwork as a means of providing context to learning, helping to bridge the gap between the abstract nature of Chemistry and its applications in real life. Interviewee 5 claimed that fieldwork can help students learn more "*because they're seeing the relevance of what they're doing in class, and actually link it to what's around us.*" This was corroborated by 29% of the students who listed it as the second beneficial reason for fieldwork and 55% of teachers who selected it as the third most beneficial reason. Statistical analysis of the data collected gave this statement a mean rating score of 7.52 (Fig. 3) and one-way ANOVA analysis showed that there is no statistical significance of the mean rating score provided to this statement, irrespective of fieldwork experience.

Both teachers (24%) and students (59%), agreed that visual learning experiences through fieldwork, can help students engage more with the subject, since they can put into practice the theories learnt in class and link them with real-world phenomena, fostering deeper learning. Local studies show that relocating lessons from the classroom and laboratory into nature helps students to develop connections between the theories learnt in class and what they experience, making it easier for learners to understand (Aquilina 2010; Mercieca 2011). Thus, formal education through fieldwork is seen as an innovative pedagogical approach that creates an alternative learning environment, complementing classroom learning (Larsen et al. 2016). Nature provides great pedagogical value since it offers students the chance to sharpen their skills and reinforce learning, through observation and hands-on activities, making education more holistic (Quay and Seaman 2013). Fieldwork can thus help to reduce the trend which has been observed in Malta in recent years, where the number of students sitting for the Chemistry O-Level has declined by 15% (Farrugia et al. 2017).

iii. Fieldwork helps students to appreciate nature and increases awareness about environmental problems

Teachers gave this statement a mean rating score of 6.24 (Fig. 3), while one-way ANOVA analysis showed that there was no statistical significance of the mean rating score provided to this statement, irrespective of fieldwork experience. Respondents also stated that fieldwork gives students a chance to engage with the natural environment. In fact, 70% of the students agreed that fieldwork increases awareness and care for the environment around them. This was also reflected in the quantitative data, where teachers listed this as the fourth benefit of conducting fieldwork.

Unfortunately, fieldwork provision at schools is decreasing at a time when outdoor education is extremely important to help students understand the impact of humans on the environment and the need to respect and protect nature (Mercieca 2011). This could be attributed to the fact that teachers feel that students view fieldwork as a fun outing with friends and therefore, it loses its educational value. Professor Lord May (2002), was quoted as saying that "*It is in the field that science becomes alive and where acting locally becomes thinking globally. Our young people are being let down if their science education does not include field experience*". In today's world, education is critical for promoting sustainable development to ensure that the world's natural resources are available for future generations to enjoy (Casinader and Kidman 2018). Education plays a key role in shaping the students' values to become socially responsible citizens for a greener society. Fenech Caruana (2018) explained that including the concept of Education for Sustainable Development in future Chemistry syllabi helps students to think more critically and to evaluate better the sustainability of chemical processes and products. While Green Chemistry does not feature as yet in any science curriculum in Malta, until post-secondary level, the inclusion of fieldwork in the O-Level Chemistry syllabus shows that Education for Sustainable development is gaining more recognition as an important area of Chemistry education.

Benefits of Fieldwork that Cannot be Attained Through Scheduled Laboratory Experiments

Considering that the respondents were aware of the benefits afforded by fieldwork, as explained above, another objective of this study was to understand if the respondents perceived any benefits that cannot be attained through laboratory experiments. Most teachers (64%) claimed that fieldwork offers advantages over laboratory work. This was supported by 52% of the students who believed that lab experiments are not enough to learn all the necessary skills when studying Chemistry.

Both teachers (49%) and students (21%) viewed working in-situ during field excursions as "*the biggest advantage of fieldwork, over just laboratory [...] experiments in class*" *(Interviewee 4)*. Popov (2015), states that when students work in outdoor environments, the phenomena they are supposed to learn about are not directly visible to the learner in the same way they are in the laboratory. Thus, fieldwork can provide benefits that cannot be attained through experiments, since it involves investigative work in a changing natural setting, rather than the manipulation of a factor in a controlled lab setting (Aziz 2017).

Respondents also explained that fieldwork allows students to gain experience in sampling and handling of equipment while acquiring data collection, observational and sampling skills, which cannot be attained through experiments. Additionally, fieldwork helps students understand the link that exists between working in-situ to collect samples and lab analysis, while carrying out experiments which cannot be replicated in a lab.

Finally, data showed that fieldwork can give academically challenged students who struggle in the classroom, a chance to excel in practical tasks carried out on field. Hutson et al. (2011), explain that fieldwork gives these 'at risk' students, a unique opportunity to create new connections with the material studied, leading to a conceptual understanding of the theories learnt in class. The increased interest in the subject stimulates curiosity which empowers students, leading to improved academic performances in class (Behrendt and Franklin 2014) and higher levels of enthusiasm towards learning (Claiborne et al. 2020).

Fieldwork Pack as a Tool for Teaching Chemistry and Education for Sustainable Development

Lack of resources has always been a major barrier to fieldwork. Thus, a Fieldwork Pack, consisting of a Teacher and Student Handbook, with a number of activities designed to be carried out before, during and after the fieldwork, was developed as part of this study. This was distributed to teachers and students to determine its usefulness as a pedagogical tool.

Most teachers (72%) affirmed that the Fieldwork Pack is very useful as a teaching tool for Chemistry and they would recommend it to their colleagues. With regards to the information included, 97% of teacher respondents agreed that it aids educators when planning fieldwork and clearly outlines how each activity can be devised effectively. Furthermore, 82% of teachers agreed that each section has a proper outline of what students need to do. A teacher also explained that the Student Handbook contains a lot of fieldwork activities that can be tailored to the student cohort. Interviewee 1 described the resources as "*really, really helpful*" while interviewee 5 stated that the Fieldwork Pack is "*really well-structured and it's very good for us and very good for them.*"

Students were also asked to rate the usefulness of the Student Handbook, with 69% of the respondents rating it highly. One student explained that the student handbook is 'well guided', while another explained that it is 'filled with good activities' and that it is 'well- thought out'. The lowest rating score given by 4% of the students was a 3, however these respondents, did not comment on how to improve the handbook.

The Friedman and Kruskal Wallis tests of the data collected from teachers' and students' questionnaires showed that there was no statistical difference in ratings provided to statements about the design and resources included in the Fieldwork Pack, irrespective of the school sector and year group. This shows that teachers and students were in agreement about the usefulness of the Fieldwork Pack as a pedagogical tool.

Education for Sustainable Development Resources in Fieldwork Pack

It is important to note that only 53% of student respondents felt that what they learn in the Chemistry curriculum is relevant to their lives. Thus, a number of activities were designed to promote environmental literacy by making use of the knowledge learned in class to render the syllabus content more enjoyable, while raising awareness about the impacts of human activities on the local environment and the importance of sustainability (Pavia 2008). Combining analytical with environmental chemistry also helps students to see the interdisciplinary nature of Chemistry in the local context and its relevance to their lives (Mandler et al. 2012).

When students were asked which of the suggested activities they will find enjoyable, the most preferred activity was the 'Microplastic content in sand activity' (33%), followed by qualitative analysis of the water samples (24%), the acid/alkali reactions during soil sampling (22%) and qualitative analysis of the ionic composition of soil and salt samples (21%). Activities linking the on-site use of commercial test kits to analyse the nitrate content of water samples to determine pollution levels as well as a suggestion to assign an activity where students act as detectives to determine the presence of ions, in different natural samples, were also included. Such activities allow students to experience the analytical procedures carried out by field scientists (Höper and Köller 2018). Such problem-based learning activities, which give students the chance to devise their own methods to isolate and identify the ionic composition of the samples collected, can help to improve students' understanding of qualitative analysis (Hicks and Bevsek 2012) while helping them determine the impact of pollution on the environment (Dalu et al. 2020). Following the students' suggestions, a number of activities were included in the Fieldwork Pack, as summarised in Table 2.

These results show that students at an early age are interested and sensitive to environmental issues. Research shows that there is the need to integrate experiments related to global issues, within the school curriculum, to help students understand the environmental impacts of pollution (Dalu et al. 2020) and thus instil an element of social responsibility, which can help to achieve sustainability.

Table 2 Fieldwork activities included in Fieldwork Pack, following students' suggestions

Suggestion	Action taken
Activities related to pollution	Activities to measure the Air Quality Index and particulate matter in the fieldwork area studied as well as activities to determine pollutants in water/sea and soil samples, were included
Titrations	A simplified version of a complexometric titration that can be used to determine the level of hardness in water samples collected was also included. Interviewees claimed that titrations are a difficult concept for students to comprehend, since they "*seem to be stand-alone*" in the syllabus (*Interviewee 5*). Thus, it was concluded that conducting a titration on the water samples collected, can help students understand the applications of this technique in various areas of Chemistry, including determination of sea water quality (Li and Liu 2018)

4 Conclusions and Recommendations

Most teachers and students endorsed the idea of fieldwork in Chemistry and viewed it as an important pedagogical tool which helps students to engage more with the subject since it allows them to see its application in a real-world context. However, unlike students, the majority of teachers did not agree that it should become a compulsory component of the syllabus. The main findings of the study are summarised in Fig. 4.

Respondents also reacted favourably to the concepts of Education for Sustainable Development included as part of the fieldwork activities. Teaching Chemistry through fieldwork can show students the true value of Chemistry in society and how it has

Main findings from study

Most teachers and students viewed Chemistry fieldwork as an important pedagogical tool which helps students understand the link that exists between what they learn in class and its application in real-life.

Respondents believe that working in-situ during field excursions allows students to develop transferable skills, which cannot be attained in a laboratory.

Respondents agreed that fieldwork gives students the opportunity to analyse different parts of the environment which cannot be replicated in a laboratory setting.

Fieldwork helps students to engage more with the subject and to understand the content better since it allows them to see its application in a real-world context.

Education is critical for promoting sustainable development. Thus, the introduction of fieldwork in the Chemistry curriculum can help students evaluate better the sustainability of chemical processes and their environmental impact.

Fig. 4 Summarises main findings from study

the potential to make a difference to our quality of life as it can help to reduce environmental degradation since it targets prevention of pollution.

Since this study is the first of its kind locally, more research is needed to determine the pedagogical importance of fieldwork in Chemistry. The need for support and training to encourage teachers to conduct fieldwork was clearly indicated and hopefully this study will trigger the necessary action to be taken by the authorities concerned. Furthermore, findings from this study show that education is critical for promoting sustainable development and thus the concept of Green Chemistry should be given more priority in future school curricula. While the current curriculum introduces secondary school students to environmental chemistry concepts such as pollution and toxicity of chemicals, it lacks any reference to 'sustainability'. This study shows that concepts of Green Chemistry can be taught more effectively through fieldwork. Such hands-on experiences can help students think more critically about the environmental impact of Chemistry and evaluate better the sustainability of chemical processes and the impact they have on their lives and the environment.

References

Ansberry KR, Morgan E (2005) Picture-perfect science lessons: using children's books to guide inquiry. (2nd Edition). Arlington: NSTA Press

Attard Tonna M, Bugeja G (2016) 'A reflection on the learning outcomes framework project [Commentary]. Malta Rev Educ Res 10(1):169–176

Aquilina V (2010) The role of Biology fieldwork in postsecondary institutions in Malta. Unpublished B.Ed. (Hons.) dissertation, University of Malta, Malta

Aziz H (2017) Comparison between field research and controlled laboratory research. Arch Clinical Biomed Res 1:101–104

Barrows CW, Murphy-Mariscal ML, Hernandez RR (2016) At a crossroads: the nature of natural History in the twenty-first century. BioScience 66:592–599

Boca G, Saraçli S (2019) Environmental education and student's perception, for sustainability. Sustainability 11:1553.https://doi.org/10.3390/su11061553

Braun V, Clarke V (2006) Using thematic analysis in psychology. Qual Res Psychol 3:77–101.https://doi.org/10.1191/1478088706qp063oa

Behrendt M, Franklin T (2014) A review of research on school field trips and their value in education. Int J Environ Sci Educ 9(3):235–245. https://doi.org/10.12973/ijese.2014.213a

Borrows P (2019) Chemistry doesn't just happen in test tubes. School Sci Rev 372:33–40

Caruana M, Treeby R (2014) Fieldwork at Xrobb l-Ghagin nature park: a teacher's manual for SEC level biology. Unpublished B.Ed. (Hons.) dissertation, University of Malta, Malta

Carter N, Bryant-Lukosius D, DiCenso A, Blythe J, Neville AJ (2014) The use of triangulation in qualitative research. Oncology Nursing Forum 41(5):545–7.https://doi.org/10.1188/14.ONF.545-547

Casinader N, Kidman G (2018) Fieldwork, sustainability, and environmental education: the centrality of geographical inquiry. Australian J Environ Educ 34(1):17.https://doi.org/10.1017/aee.2018.12

Claiborne L, Morrell J, Bandy J, Bruff D, Smith G, Fedesco H (2020) Teaching outside the classroom. Vanderbilt University Centre for Teaching. Retrieved [20–04–22] from https://cft.vanderbilt.edu/guides-sub-pages/teaching-outside-the-classroom/

Christidou V (2011) Interest, attitudes and images related to science: combining students' voices with the voices of school science, teachers, and popular science. Int J Environ Sci Educ 6(2):141–159

Cohen L, Manion L, Morrison K (2018) Research methods in education (8th Edition) Routledge: New York

Dalu M, Cuthbert R, Muhali H, Chari L, Manyani A, Masunungure C, Dalu T (2020). Is awareness on plastic pollution being raised in schools? understanding perceptions of primary and secondary school educators. Sustainability 12:6775.https://doi.org/10.3390/su12176775

Farrugia J, Mizzi D, Zahra R, Zarb D (2017) Changing the chemistry curriculum at secondary level : opportunities and challenges. New Perspectives in Science Education

Fenech Caruana M (2018) Attracting more students to opt for chemistry at post-Secondary level : potential barriers for students to take up the 'Chemistry Challenge'. Malta, 322–342

Fuller IAN, Edmondson S, France D, Higgitt D, Ratinen I (2006) International perspectives on the effectiveness of geography fieldwork for learning. J Geogr Higher Educ 30(1):89–101

Goteti M, Kantamreddi V, Goteti L (2012) Promoting higher order thinking skills using inquiry-based learning. Europ J Eng Educ 37:1–7.https://doi.org/10.1080/03043797.2012.661701

Gröger M (2013) Teaching chemistry in near-natural learning environments to address sustainability. Paper presented at the ESERA 2013 Conference, Strand 9 Environmental, health and outdoor science education, Nicosia

Hall T, Healey M, Harrison M (2002) Fieldwork and disabled students: discourses of exclusion and inclusion. Trans Inst British Geogr 27(2):213–231

Hicks RW, Bevsek HM (2012) Utilizing problem-based learning in qualitative analysis lab experiments. J Chem Educ 89(2):254–257.https://doi.org/10.1021/ED1001202

Holstermann N, Grube D, Bögeholz S (2010) Hands-on activities and their influence on students' interest. Res Sci Educ 40:743–757. https://doi.org/10.1007/s11165-009-9142-0

Höper J, Köller H (2018) Outdoor chemistry in teacher education—a case study about finding carbohydrates in nature. LUMAT: Int J Math Sci Technol Educ 6(2). https://doi.org/10.31129/LUMAT.6.2.314

Höper J, Jegstad KM, Remmen KB (2022) Student teachers' problem-based investigations of chemical phenomena in the nearby outdoor environment. Chem Educ Res Pract 23:361–372https://doi.org/10.1039/D1RP00127B

Howarth S, Slingsby D (2006) Biology Fieldwork in school grounds: a model of good practice in teaching science. School Sci Rev 87(320):99–105

Hutson T, Cooper S, Talbert T (2011). Describing connections between science content and future careers: implementing Texas curriculum for rural at-risk high school students using purposefully-designed field trips. Rural Educator 33:37–47

Kenney J, Militana H, Donohue M (2003) Helping teachers to use their school's backyard as an outdoor classroom: a report on the watershed learning center program. J Environ Educ 35:18-26.https://doi.org/10.1080/00958960309600591

Krischer D, Spitzer P, Gröger M (2016) Chemistry is toxic, nature is idyllic"—investigation of pupils' attitudes. J Health Environ Educ 8:7–13

Larsen C, Walsh C, Almond N, Myers C (2016) The "real value" of field trips in the early weeks of higher education: the student perspective. Educ Studies, 1–12.https://doi.org/10.1080/03055698.2016.1245604

Li D, Liu S (2018) Water quality monitoring and management. (1st Edition). Academic Press

Maciejowska I (2007) Chemical Education outside the classroom—Step by Step. Gamtamokslinis ugdymas 3(20)

Mandler D, Mamlok-Naaman R, Blonder R, Yayon M, Hofstein A (2012) High-school chemistry teaching through environmentally oriented curricula. Chem Educ Res Pract 13(2):80–92https://doi.org/10.1039/C1RP90071D

MATSEC (2010) Secondary education certification examination: biology syllabus. University of Malta

MATSEC (2024) Secondary education certificate examination: chemistry syllabus. University of Malta
Mizzi D (2019) Supporting science teachers teaching outside specialism: teachers' views of a professional development programme. Europ J Teacher Educ 44(5):706 725. https://doi.org/10.1080/02619768.2020.1793951
Maw SJ, Mauchline AL, Park JR (2011) biological fieldwork provision in higher education. Biosci Educ 17(1):1–14
Mercieca A (2011) An appraisal of the scope of biology fieldwork in Gozo. Unpublished B.Ed. (Hons.) dissertation, University of Malta, Malta
Musumeci M, Pirotta D (2018) Choice in the science subjects: trends in Malta. Conference: international conference new perspectives in science education Edition 7, Florence
Pace F (1979) Fieldwork activities for Secondary level, Form I and II. A Personal Note to all Heads of schools and to Science teachers. (Unpublished)
Pavia S (2008) An environmental resource pack for Dwejra, Gozo. Unpublished B.Ed. (Hons.) dissertation, University of Malta, Malta
Pooja B (2019) Types of sampling in research. J Pract Cardiovascular Sci 5:157https://doi.org/10.4103/jpcs.jpcs_62_19
Popov O (2015) Outdoor science in teacher education. Contemporary approaches to activity theory: interdisciplinary perspectives on human behavior, 128–142
Quay J, Seaman J (2013) John dewey and education outdoors: making sense of the 'educational situation' through more than a century of progressive reforms.https://doi.org/10.1007/978-94-6209-215-0
Rahmawati Y, Koul R (2016) Fieldwork, co-teaching and co-generative dialogue in lower secondary school environmental science. Educ Res 26(1):147–164
Remmen K, Frøyland M (2015). Supporting student learning processes during preparation, fieldwork and follow-up work: Examples from upper secondary school in Norway. Nordic Studies Sci Educ, 11
Rickinson M. Dillon J, Teamey K, Morris M, Choi MY, Sanders D, Benefield P (2004) A review of research on outdoor learning, Preston Montford, Shropshire: Field Studies Council
Schembri H (2020) Chapter 2: how is the learning outcomes framework responding to an internationalised school culture in primary schools in Malta? Malta J Educ (MJE) Internationalisation Multiculturalism Maltese Educ Soc 1(1)
Scott G, Goulder R, Wheeler P, Scott L, Tobin M, Marsham S (2012) The value of fieldwork in life and environmental sciences in the context of higher education: a case study in learning about biodiversity. J Sci Educ Technol 21:11–21. https://doi.org/10.1007/s10956-010-9276-x
Slingsby D (2006) The future of school science lies outdoors. J Biological Educ 40(2):51–52. https://doi.org/10.1080/00219266.2006.9656013
Sultana A (2011) An investigation of the attitudes of secondary level students in Gozo (Malta) towards Science Education. Unpublished B.Ed. (Hons.) dissertation, University of Malta, Malta
Taş E, Gülen S (2019) Analysis of the influence of outdoor education activities on seventh grade students. Participatory Educ Res 6:122–143
Tanner CK (2003) Into the woods, wetlands, and prairies. Educ Leadership 58(7):64–66
Thorburn M, Allison P (2010) Are we ready to go outdoors now? The prospects for outdoor education during a period of curriculum renewal in Scotland. Curriculum J 21:97–108. https://doi.org/10.1080/09585170903560824
Tundo P, Anastas P, Black D, Breen J, Collins T, Memoli S, Miyamoto J, Polyakoff M, Tumas W (2000) Synthetic pathways and processes in green chemistry. Introductory overview. Pure Appl Chem 72:1207–1228. https://doi.org/10.1351/pac200072071207

Sustainability in Higher Education Institutions and the Brazilian Environmental Agenda in Public Administration

Laís Viera Trevisan, Janaina Mazutti, Thais Dibbern, and Walter Leal Filho

Abstract The sustainability topic is being increasingly discussed both in the academy and in the campuses' daily life. Against a worldwide scenario of socio-environmental catastrophes, pandemic, wars, and climate change, this discussion becomes even more urgent. From this perspective, public institutions should be models for the reduction of negative socio-environmental impacts, especially the organisations which work with knowledge, such as higher education institutions (HEIs). Therefore, this study aims to analyse the Brazilian Environmental Agenda in Public Administration (*Agenda Ambiental na Administração Pública*—A3P), within the scope of public HEIs. Brazil is one of the richest countries in biodiversity and, therefore, responsible for its preservation. In this study, the A3P is highlighted as a key tool in the implementation of organisational initiatives in HEIs towards sustainable development. Nevertheless, the involvement and adherence of HEIs to this Agenda is a challenge to the traditional university models and requires a commitment from the entire university community.

Keyword Sustainability · Environmental and social responsibility · Higher education institutions · Environmental agenda in public administration

L. Viera Trevisan (✉)
Hamburg University of Applied Sciences (HAW/Hamburg/Germany) and Federal University of Rio Grande do Sul (UFRGS/Brazil), Alegre, Brazil
e-mail: lais.trevisan@ufrgs.br

J. Mazutti
University of Passo Fundo (UPF/Brazil), Rio Grande do Sul, Brazil
e-mail: janainamazutti@gmail.com

T. Dibbern
University of Campinas (UNICAMP/Brazil), Campinas, Brazil
e-mail: dibbern.thais@gmail.com

W. Leal Filho
Manchester Metropolitan University (MMU/Manchester/UK) and Hamburg University of Applied Sciences (HAW/Hamburg/Germany), Hamburg, Germany
e-mail: walter.leal2@haw-hamburg.de

W. Leal Filho et al. (eds.), *The Contribution of Universities Towards Education for Sustainable Development*, World Sustainability Series,
https://doi.org/10.1007/978-3-031-49853-4_6

1 Introduction

Based on the demands of contemporary society, concerned with current forms of production and consumption, the sustainability topic has been receiving attention on the agenda of governments, international organisations, higher education institutions (HEIs), and other organisations.

In the last decades, some global initiatives towards sustainability that can be mentioned are: (1) the United Nations (UN) Conference on the Human Environment, held in Stockholm/Sweden, in 1972; (2) the UN Conference on Environment and Development, in Rio de Janeiro/Brazil, in 1992, in which the Agenda 21 was created; (3) The Rio + 20, in 2012; and (4) the 2030 Agenda for Sustainable Development, which brought together, in September 2015, representatives of the 193 UN member states in New York/United States to endorse the 17 Sustainable Development Goals (SDGs) for the next 15 years.

In this regard, Cavalcante (2012) notes the importance that both the business sector and the public administration have to the creation of environmental and social responsibility policies and actions, which can help increase citizens' awareness of sustainability, as well as behavioural changes and ethical and responsible practices.

The Brazilian government, seeking to stimulate and promote sustainable practices among public entities, instituted, in 1999, the Brazilian Environmental Agenda in Public Administration (A3P). Although it is not mandatory, this Agenda has been highly widespread and recommended by the Brazilian Environmental Ministry (*Ministério do Meio Ambiente*—MMA) (Kruger et al. 2011). According to data from the MMA (2023), the A3P has a total of 1,084 partner institutions. As part of public administration, Public HEIs are also encouraged to join the A3P. However, a few HEIs are part of it.

Several authors have been discussing the role of HEIs in sustainable development, seeking to build concepts, identify their contributions, and propose strategies to promote sustainable practices (Arruda and Rodrigues 2018; Cavalcante 2012; Santos et al. 2017) and, more specifically, in public HEIs (Almeida et al. 2020; Da Rosa et al. 2015; Figueiredo et al. 2019; Gazzoni et al. 2018; Lanzarin et al. 2018; Zeitoune et al. 2019).

In this regard, considering HEIs should research and develop sustainable practices in their institutional spaces, seeking to eliminate waste and reduce the consumption of natural resources (Tauchen and Brandli 2006), in this study, it is argued that environmental management systems in universities campuses, through the adherence to A3P guidelines can serve as models of sustainable management for society. Thus, this study aims to analyse the Brazilian Environmental Agenda in Public Administration, within the scope of public HEIs, as a possible tool for the implementation of management practices and models to ensure sustainability on campuses.

2 Sustainability in Higher Education Institutions

The Brundtland Commission report, presented at the UN meeting in 1987, entitled 'Our Common Future' (World Commission on Environment and Development—WCED 1987), agreed on the urgent need for sustainability. From this, according to Baron (2001), companies began to be pressured by stakeholders to engage in environmental and social responsibility. Therefore, companies' managers had to be able to determine how their organisations can become more socially responsible, ecologically sustainable, and economically competitive. This increasing pressure on the social, environmental, and economic performance of a wide variety of stakeholders raises important research questions in many fields of business administration.

Research for sustainable development, also known as the Science of Sustainability, produces answers about transformative processes. This implies that complex concepts require equally complex frameworks, considering temporal and spatial dimensions, as well as the stakeholders involved. Therefore, emerging complexity leads to a social process of change towards a desired quality of life and pluralistic approaches that involve diverse actors, at various levels, creating an integrated and shared vision of concerns and solutions, and solving trade-offs iteratively (Baker-Shelley et al. 2017; Zeijl Rozema et al. 2008). Daly (2006, p. 39) defines sustainability 'as a way of affirming the value of longevity and intergenerational justice, recognizing morality and finitude', emphasizing that young people in contemporary society have an interest in the future state of the world that remains for them.

Considering contemporary issues, such as globalisation, climate change, biodiversity loss, social inequality and a global cross-border governance gap, the potential that HEIs have to prepare young people for such a future is imperative. This mission embodies principles of intergenerational justice and socio-ecological balance. For Rockstrom et al. (2009), universities must provide a response to the current crisis, through education, research, and social transformation of modes of production and consumption. Although sustainability has different meanings for each individual or organisation (Velazquez et al. 2006), this study assumes the definition of a sustainable university as one that:

> [...] addresses, involves and promotes, at a regional or global level, the minimization of negative environmental, economic, social and health effects generated in the use of resources in order to fulfil the functions of teaching, research, dissemination, partnership and stewardship to help society transition to sustainable lifestyles (Velazquez et al. 2006, p. 812).

This definition is based on social, economic, and environmental aspects, that is, on the tripod of sustainability (triple bottom line), a triple result of sustainable development, as defined in the Brundtland Report (Elkington 1999; WCED 1987).

The great challenges of sustainable development go far beyond the experiences, skills, and resources of any individual or organisation (Kania and Kramer 2011). Therefore, HEIs should play a role as cross-sectoral actors and facilitators of transformational changes. This would be an opportunity to prepare the entire campus community to make decisions towards an increasingly complex, dynamic, and uncertain future (Baker-Shelley et al. 2017; Lanzarin et al. 2018; Yarime and Tanaka 2012).

However, Viegas et al. (2015) highlight the huge challenge of managing an organisation, at the same time, to be competitive, innovative, and sustainable. This requires reflection on their values, behaviours, and attitudes.

A study conducted by Lozano et al. (2015) in 70 HEIs observed that although there is some progress towards integrated and holistic thinking, in general, the implementation of sustainability in HEIs remains compartmentalised, according to the functionalist logic and Cartesian models. However, HEIs are characterised by offering courses whose contents fundamentally suggest working with 'relevant knowledge', which implies 'contextualising and integrating knowledge' (Morin 2000). In this sense, Morin et al. (2002) point out the need to think about HEIs under the paradigm of complexity, seeking to reform their structures, while providing a reform of thought.

An important strategy for a sustainable culture is to make employees aware that environmental conservation also contributes to the expansion of their levels of health, well-being, and quality of life. Thus, it is imporant to incorporate into environmental educational programs the fact that the adoption of sustainable practices directly implies the survival of humanity and the planet (Andrade 2002; Nascimento et al. 2015). Furthermore, according to Viegas et al. (2015), other examples are to use the analysis of organisational practices and visible symbolic elements, such as the apparent behaviour of people, the ways in which resources are used, the information that is transmitted, the type of knowledge disseminated, organisational rituals, relationships with the environment and, in short, the culture of the organisation. Therefore, becoming 'eco-efficient' implies reflecting on 'non-renewable resources' and, based on that, seeking alternatives to transforming attitudes and habits (Nascimento et al. 2015, p. 494).

According to Cavalcante (2012), the implementation of each environmental and social practice in public institutions is categorical for the development of the country. The author assumes that the measures will be reflected in the country's budget, since transformations arising from the behavioural sphere avoid losses, thus reducing costs. Therefore, universities should research and develop sustainable practices in their institutional spaces, in which the elimination of waste and the reduction of consumption of natural resources must prevail, necessarily implying a change in behaviour (Tauchen and Brandli 2006; Zeitoune et al. 2019). Tauchen and Brandli (2006) emphasize five challenges for implementing sustainability in HEIs: (1) dominant sustainability challenges in the region; (2) financial and independence structure of the institution; (3) institutional organisation; (4) democratic dimension of the process; and (5) interaction and communication with society. Gazzoni et al. (2018) note that there should be an interaction between HEIs and society and, in agreement with Fouto (2002), believe that they are related through four levels of intervention: (1) education of decision-makers for a sustainable future; (2) investigation of solutions, paradigms and values that serve a sustainable society; (3) operation of university campuses as models and practical examples of sustainability at the local scale; and (4) coordination and communication between the previous levels and between them and society. Thus, it is important to understand this interaction through the lens of Environmental and Social Responsibility and University Social Responsibility.

3 Environmental and Social Responsibility and University Social Responsibility

Although Environmental and Social Responsibility are dimensions of corporate sustainability, generally, they have been practiced by companies in a disintegrated way, according to functionalist logic. More recently, efforts have been made by some organisations to bring these actions together within the same area, under the title of 'Sustainability'. The concept of Corporate Social Responsibility has evolved to encompass environmental issues. However, several organisations and researchers use the term 'Environmental and Social Responsibility' as a way of integrating these practices (Cohen et al. 2017).

Corporate Social Responsibility efforts have a significant impact on the social and political environment in which they operate (Detomasi 2008). Hrdlicka and Kruglianskas (2010) identified that, during the 2008 global financial crisis, companies that demonstrated commitment to sustainability and used their principles in business strategy had better results compared to their competitors that did not. For Scherer et al. (2009), the political-social engagement of corporations has become a widespread phenomenon. Despite criticism, a growing number of business leaders seem to take on responsibilities that go beyond financial and legal requirements. Undoubtedly, some of these initiatives can be part of public relations strategies and serve the purpose of creating a positive corporate image. However, it is evident that the presence of any monitoring system is an element of organisational transformation.

These systemic effects over time on the organisational community respond to calls for greater political and economic responsibility towards society (Soderbaum 2008), consequently, culminating in Environmental and Social Responsibility practices. There are several definitions for this term, but basically, they all have the same focus, which considers political justice and impartiality, practices that lead to equity, reciprocity, and impartiality (Treviño and Nelson 2021; Becker et al. 2010). In Brazil, Environmental and Social Responsibility practices gained greater visibility after the UN Conference on Environment and Development held in Rio de Janeiro, the ECO-92, which consolidated the concern for environmental issues. Moreover, the Agenda 21 was created as a sustainable development guideline that reconciles environmental protection, social justice, and economic efficiency (Santos et al. 2017).

In 2015, the UN established the 2030 Agenda and its 17 Sustainable Development Goals (SDGs) to be achieved by the 193 signatory member states. The SDGs address a range of social and environmental issues to be pursued globally, including education, health, social protection, employment opportunities, climate change, social inequality, and environmental protection, among other topics (Sustainable Development Solutions Network 2020).

In this regard, HEIs have a key role concerning environmental and social responsibility as they have broad competence around the creation and dissemination of knowledge, are responsible for training environmentally responsible citizens, develop research focused on the topic, manage the inflow and output of natural capital, contribute to national and global innovation and for the transformation of neighbouring communities and regions, as well as have an impact on economic development and social well-being, and can promote solutions through models, research, and partnerships (Beynaghi et al. 2016; Trencher et al. 2014; Zilahy and Huisingh 2009).

In HEIs, social responsibility is usually linked to university outreach, considering its relationship with society. However, the concept of social responsibility should not be focused only on specific outreach actions, given that the educational process is systemic and inseparable (Nunes et al. 2017). In this regard, University Social Responsibility emerges as a tool for organisational change applied to the university reality, and approaching it, from this perspective, requires considering the principles of complexity (Morin 2000), by including it as a management strategy, considering teaching, research, outreach, and institutional administration.

For Vallaeys (2006), the social responsibility of universities can be understood as a policy of ethical quality that encompasses all activities of the university community, through responsible management of educational, cognitive, social, and environmental services. This implies establishing a participatory dialogue between the university and society, with a view to promoting sustainability. The Brazilian Environmental Ministry (*Ministério do Meio Ambiente*—MMA 2017) points out that the public administration, for the development of its middle and final activities, acts as a major consumer of natural resources, goods, and services. Thus, it is urgent to review its production and consumption patterns based on the adoption of new socio-environmental sustainability benchmarks, in order to induce new criteria and practices. Therefore, the Brazilian Environmental Agenda in Public Administration can be an important tool to be adopted by public HEIs to pursue sustainability aims.

4 Brazilian Environmental Agenda in Public Administration (A3P)

The Brazilian public management of environmental issues is based on the Brazilian Law N. 6,938/81, which established the National Environmental Policy and the National Environment System, which was expanded by the Brazilian Federal Constitution of 1988. The State must play an active role, as it is the main agent of society in the participation of public institutions in the corporate social responsibility process.

To assist the National Environmental Policy, the Brazilian Environmental Ministry created the Brazilian Environmental Agenda in Public Administration (A3P), an environmental management system subdivided into different topics (Almeida et al. 2020; Nascimento et al. 2015). Although it is not mandatory, the A3P is recommended by the Ministry, considering that public institutions should be a reference in the adoption of practices aimed at reducing negative socio-environmental impacts and the rationalisation of public money (MMA 2017). At the same time, public HEIs, as members of the public administration, can also adhere to the A3P guidelines (Freitas et al. 2011). However, according to MMA data (2023), a few HEIs are part of it.

The A3P was already recognised for its relevance and positive results by receiving, in 2002, the award 'The best of examples', in the Environment category, from the UN for Education, Science and Culture (UNESCO). Given its importance, the A3P was included in the 2004/2007 Multiannual Plan as an action within the scope of the environmental education program. This measure guaranteed resources so that A3P could be effectively implemented and become a new benchmark for the socio-environmental sustainability of public activities.

In 2007, the A3P became part of the Department of Citizenship and Social and Environmental Responsibility, of the Secretariat for Institutional Articulation and Environmental Citizenship. In this new institutional arrangement, the A3P was strengthened as the government's Environmental and Social Responsibility Agenda and became one of the main actions for proposing and establishing a new standard of responsibility in public management (Monteiro and Camelo 2015). The MMA (2009, 2011) highlights the 5 R's (reduce, rethink, reuse, recycle and refuse) that support the A3P, which were created to be used in environmental education projects, with the objective of changing society's attitude towards consumption and waste.

Based on these principles and seeking to stimulate the perception and behavioural change of public employees towards sustainability, the MMA defined the A3P's objectives as: (1) making public managers aware of environmental issues; (2) encouraging the adoption of new sustainability criteria within the scope of public administration; (3) increase management efficiency, promote savings in natural resources and institutional expenses; and (4) contribute to the review of production and consumption patterns.

To achieve the proposed objectives, the A3P was divided into six thematic topics that guide the application of sustainable actions: (1) rational use of natural resources and public goods; (2) proper management of generated waste; (3) quality of life in the work environment; (4) awareness and training of employees; (5) sustainable public procurement; and (6) implementation of criteria for sustainable construction.

According to the MMA (2017), the first topic encompasses the rational use of energy, water, wood, consumption of paper, plastic cups and other office materials, and implies using them economically and rationally, avoiding their waste. The second topic refers to the prevention and reduction of waste generation, seeking to practice sustainable consumption habits and a set of instruments to promote increased recycling, reuse of solid waste and the environmentally appropriate disposal of waste. The third topic aims to facilitate the health and performance of the employees by improving aspects of ergonomics, comfort, and well-being, among others, in the workplace. The fourth seeks to create and consolidate citizen awareness of socio-environmental responsibility in employees through the development of institutional and individual skills. The fifth topic refers to bids that consider the acquisition of sustainable products and services, which not only contribute to the conservation of the environment but also present a better cost/benefit ratio in the medium or long term when compared to those that use the lowest price criterion. Finally, the last topic is related to a set of measures adopted at all stages of the work aimed at the sustainability of the building.

In 2012, a Brazilian policy (Brazil 2012b) established the obligation for the direct federal public administration, autarchic and foundational and the dependent state companies, to prepare and implement Sustainable Logistics Management Plans (PLS), which must provide, at least: (1) updated inventory of goods and materials and identification of similar products with a lower environmental impact for replacement; (2) sustainability practices and rationalisation of the use of materials and services; (3) responsibilities, implementation methodology, and plan evaluation; and (4) dissemination, awareness and training actions.

Subsequently, another Brazilian policy (Brazil 2012a) was published by the Brazilian Ministry of Planning, Budget and Management, which established some rules for the preparation of the PLS. Thus, sustainability practices and rationalisation of the use of materials and services should cover, at least, the following topics: (1) consumables, including, at least, printing paper, disposable cups, and printing cartridges; (2) electrical energy; (3) water and sewage; (4) selective collection; (5) quality of life in the work environment; (6) sustainable procurement and contracting, including at least works, equipment, surveillance, cleaning, telephony, data processing, administrative support and building maintenance services; and (7) displacement of personnel, considering all means of transport, with a focus on reducing expenses and emissions of polluting substances.

In 2017, another Brazilian policy was launched, complementing the PLS guides. It is observed that the PLS agree with the A3P Agenda and reinforces it, instituting more sustainable actions by public institutions and the Public Administration.

It is observed that A3P, throughout its history, has been incorporating several updates, due to the need for public institutions to collaborate for sustainable development, in an increasingly complex environment. The partner institutions receive technical support for the implementation and operation of the Agenda, training courses, participation in events, educational content, monitoring and supervision by the MMA. This relationship with the Ministry and other partner institutions allows for the exchange of experiences and good practices, as well as the promotion of more ambitious results towards a sustainable management model. Therefore, public HEIs, especially because they work with knowledge, have a key role in accelerating these demands, and should themselves be models of sustainability for society.

5 Conclusions

Faced with a scenario of climate change, war, tragedies, pandemics, and socio-environmental catastrophes, among other issues, the discussion on the topic of sustainability is central. In this regard, public institutions play a leading role in achieving goals and objectives in favour of sustainable development.

In Brazil, the Brazilian Environmental Agenda in Public Administration (A3P) was instituted towards an active role of the State in the environmental and social responsibility process—promoting its application in the different public institutions. The public HEIs, as part of the public administration, can also adhere to the A3P guidelines. Even without consequences if public universities do not commit to the A3P Agenda, joining the Agenda can become a competitive advantage for public universities by generating a positive marketing, which puts them in positions of notoriety when students choose a university for their education. The cause becomes even more urgent due to the relevance of these institutions for the education of environmentally responsible citizens, the development of academic research on the area, and the management and execution of activities on the campuses aligned with these demands, serving as a model of sustainable management for society.

Thus, this study sought to analyse the sustainability topic in HEIs through the lens of the Environmental and Social Responsibility and University Social Responsibility, seeking to reflect on the A3P as a possible tool for the implementation of practices and management models which collaborate with sustainable development. However, the changes brought by a greater incorporation of sustainability in HEIs challenge the traditional models of management of these institutions and thus, require a commitment from the entire university community.

References

Andrade ROBD, Tachizawa T, Carvalho ABD (2002) Gestão ambiental: enfoque estratégico aplicado ao desenvolvimento sustentável [Environmental management: strategic approach applied to sustainable development]. In Gestão ambiental: enfoque estratégico aplicado ao desenvolvimento sustentável (232–232)

Arruda L, Rodrigues RN (2018). Agenda ambiental na administração pública (A3P): uma análise da aderência das diretrizes da A3P em uma companhia hidroelétrica [Environmental agenda in public administration (A3P): an analysis of the adherence to the A3P guidelines in a hydroelectric company]. REFAS: Revista FATEC Zona Sul, 5(2):10

Baker-Shelley A, van Zeijl-Rozema A, Martens P (2017) A conceptual synthesis of organisational transformation: how to diagnose, and navigate, pathways for sustainability at universities?. J Cleaner Production 145:262–276

Baron DP (2001) Private politics, corporate social responsibility, and integrated strategy. J Econ Manag Strategy 10(1):7–45

Becker WS, Carbo JA, Langella IM (2010) Beyond self-interest: integrating social responsibility and supply chain management with human resource development. Human Resour Dev Rev 9(2):144–168

Beynaghi A, Trencher G, Moztarzadeh F, Mozafari M, Maknoon R, Leal Filho W (2016) Future sustainability scenarios for universities: moving beyond the United Nations decade of education for sustainable development. J Cleaner Prod 112:3464–3478

Brazil (2012a) Ministério do Planejamento, Orçamento e Gestão. Instrução Normativa nº 10, de 12 de novembro de 2012. Estabelece regras para elaboração dos Planos de Gestão de Logística Sustentável de que trata o art. 16, do Decreto nº 7.746, de 5 de junho de 2012, e dá outras providências [Normative Instruction No. 10, of November 12, 2012]. Diário Oficial da União, Poder Executivo, Brasília, DF

Brazil (2012b) Decreto nº 7.746, de 5 de junho de 2012. Regulamenta o art. 3º da Lei nº 8.666, de 21 de junho de 1993, para estabelecer critérios e práticas para a promoção do desenvolvimento nacional sustentável nas contratações realizadas pela administração pública federal direta, autárquica e fundacional e pelas empresas estatais dependentes, e institui a Comissão Interministerial de Sustentabilidade na Administração Pública—CISAP [Decree No. 7,746, of June 5, 2012]. Diário Oficial da União, Poder Executivo, Brasília, DF

Camelo GLP, de Oliveira Monteiro M (2015) Agenda Ambiental Na Administração Pública-A3P nos campi do IFRN: um olhar a partir dos gestores [Environmental Agenda in Public Administration-A3P in IFRN's campuses: a look from the managers]. EmpíricaBR-Revista Brasileira de Gestão, Negócio e Tecnologia da Informação 1(1):26–42

Cavalcante MLSA (2012) Administração Pública e Agenda Ambiental–A3P-Considerações sobre a implementação nos órgãos públicos [Public Administration and the Environmental Agenda-A3P-Considerations on the implementation in public institutions]. Revista Controle: Doutrinas e Artigos 10(1):193–216

Cohen, M, Cavazotte, FDSCN, Costa, TMD, Ferreira, KCS (2017). Corporate social-environmental responsibility as an attraction and retention factor for young professionals. BBR. Brazilian Bus Rev 14:21–41

Daly HE (2006) Sustainable development—definitions, principles, policies. In: The future of sustainability (39–53). Springer, Dordrecht

De Almeida VF, Neto CY, Vidal TCS, Damasceno LA, de Souza ALC, Simão MODAR (2020) Agenda ambiental da administração pública (A3P) e sua aderência: o caso do Instituto Federal do Amazonas [Environmental Agenda for Public Administration (A3P) and its adherence: the case of the Federal Institute of Amazonas]. Revista Ibero-Americana de Ciências Ambientais 11(5):677–693

De Maio Nascimento M, de Oliveira Virgínio MV, Lopes LR (2015) Educação Ambiental na administração pública: a implantação da A3P na Universidade Federal do Vale do São Francisco/ Univasf-PE [Environmental Education in public administration: the implementation of A3P at the Federal University of São Francisco Valley/Univasf-PE]. Revista Eletrônica em Gestão, Educação e Tecnologia Ambiental, 493–501

Detomasi DA (2008) The political roots of corporate social responsibility. J Bus Ethics 82(4):807–819

Elkington J, Rowlands IH (1999) Cannibals with forks: the triple bottom line of 21st century business. Alternatives J 25(4):42

Figueiredo G, Silva G, Soares F (2019). Sistema de gestão ambiental na administração pública: estudo em uma universidade federal do interior da Amazônia [Environmental management system in public administration: a study in a federal university in the interior of the Amazon]. Enciclopédia Biosfera 16(29)

Fouto ARF (2002) O papel das universidades rumo ao desenvolvimento sustentável: das relações internacionais às práticas locais [The role of universities towards sustainable development: from international relations to local practices]. 2002. Thesis (Mestrado em Gestão e Políticas Ambientais, Relações Internacionais do Ambiente)—Universidade Nova de Lisboa, Lisboa

Freitas CLD, Borgert A, Pfitscher ED (2011) Agenda Ambiental na Administração Pública: uma análise da aderência de uma IFES as diretrizes propostas pela A3P [Environmental Agenda in Public Administration: an analysis of the adherence of one IFES to the guidelines proposed by the A3P]. In: XI Colóquio Internacional sobre Gestão Universitária na América Latina & II Congresso Internacional Iglu, 2011, Florianópolis. Florianópolis: Instituto de Pesquisa em Administração Universitária—UFSC, 1–16

Gazzoni F, Scherer FL, Hahn IS, de Moura Carpes A, dos Santos MB (2018) O papel das IES no desenvolvimento sustentável: estudo de caso da Universidade Federal de Santa Maria [The role of HEIs in sustainable development: a case study of the Federal University of Santa Maria]. Revista Gestão Universitária na América Latina-GUAL 11(1):48–70

Hrdlicka HA, Kruglianskas I (2010) Influência das práticas de gestão socioambiental na internacionalização de empresas brasileiras [Influence of social and environmental management practices on the internationalization of Brazilian companies]. Gestão empresarial para a internacionalização das empresas brasileiras 1:253–289

Kania J, Kramer MR (2011) Collective impact. In: Stanford social innovation review, [s.l.], pp 35–41

Kruger SD, Pfitscher ED, de Freitas CL, Petri SM (2011) Gestão ambiental em Instituição de Ensino Superior-Uma análise da aderência de uma instituição de ensino superior comunitária aos objetivos da agenda ambiental na administração pública (A3P) [Environmental Management in Higher Education Institutions-An analysis of the adherence of a community-based higher education institution to the objectives of the environmental agenda in public administration (A3P)]. Revista Gestão Universitária na América Latina-GUAL 4(3):44–62

Lanzarin J, de Camargo TF, Mazzioni S, Zanin A (2018) Agenda ambiental da administração pública em instituições federais de ensino superior [Environmental agenda of the public administration in federal institutions of higher education]. Brazilian J Dev 4(3):1020–1044

Lozano R, Ceulemans K, Alonso-Almeida M, Huisingh D, Lozano FJ, Waas T, Lambrechts W, Lukman R, Hugé, J (2015) A review of commitment and implementation of sustainable development in higher education: results from a worldwide survey. J Cleaner Prod 108:1–18

MMA—Ministério do Meio Ambiente. Agenda Ambiental na Administração Pública—A3P [Environmental Agenda in Public Administration—A3P]. Disponível em: http://a3p.mma.gov.br/

MMA—Ministério do Meio Ambiente. Cartilha A3P: Agenda ambiental na administração pública [A3P Primer: Environmental Agenda in Public Administration]. 5. ed. Brasília: Ministério do Meio Ambiente, 2009

MMA—Ministério do Meio Ambiente. Como implantar a A3P [How to implement A3P]. Brasília: Ministério do Meio Ambiente, 2011

MMA—Ministério do Meio Ambiente. Curso de capacitação sustentabilidade na administração pública [Training course on sustainability in public administration]. Brasília: Ministério do Meio Ambiente, 2017

Morin E (2000) A cabeça bem-feita [The well-made head]. Rio de Janeiro: Bertrand Brasil, 99

Morin E, Almeida MDC, Carvalho EDA (2002) Educação e complexidade: os sete saberes e outros ensaios [Education and complexity: the seven knowledges and other essays]. 4 ed. São Paulo: Cortez

Nunes EBLDLP, Pereira ICA, Pinho MJD (2017). A responsabilidade social universitária e a avaliação institucional: reflexões iniciais [University social responsibility and institutional assessment: initial reflections]. Avaliação: Revista da Avaliação da Educação Superior (Campinas) 22:165–177

Rockström J, Steffen W, Noone K, Persson Å, Chapin FS, Lambin EF, Lenton TM, Scheffer M, Folke C, ..., Foley JA (2009) A safe operating space for humanity. Nature, [s.l.] 461:472–475

Rosa LABD (2015) Agenda ambiental na administração pública A3P: uma análise na Universidade Federal de Santa Maria/RS [Environmental Agenda in Public Administration A3P: an analysis at the Federal University of Santa Maria/RS]

Santos FL, de Freitas Carneiro A, de Souza JA, de Souza RMS (2017). Análise da adesão à agenda ambiental na administração pública (A3P) pela cidade portal da Amazônia [Analysis of the adherence to the environmental agenda in public administration (A3P) by the portal city of Amazonia]. Gestão e Sociedade 11(28):1583–1610

Scherer AG, Palazzo G, Matten D (2009) Introduction to the special issue: Globalization as a challenge for business responsibilities. Bus Ethics Quart 19(3):327–347

Soderbaum P (2008) Understanding sustainability economics: towards pluralism in economics. London: Earthscan

Sustainable Development Solutions Network (SDSN)—Australia/Pacific (2020) Accelerating education for the SDGS in universities: a guide for universities, colleges, and tertiary and higher education institutions. [s.l.]: SDSN

Tauchen J, Brandli LL (2006) A gestão ambiental em instituições de ensino superior: modelo para implantação em campus universitário [Environmental management in higher education institutions: a model for campus implementation]. Gestão & Produção 13:503–515

Trencher G, Yarime M, McCormick KB, Doll CN, Kraines SB (2014) Beyond the third mission: exploring the emerging university function of co-creation for sustainability. Sci Public Policy 41(2):151–179

Trevino LK, Nelson KA (2021) Managing business ethics: Straight talk about how to do it right. Wiley

Vallaeys F (2006) Que significa responsabilidade social universitária? [What does university social responsibility mean?]. Estudos, Brasília 24(36):35–55

Velazquez L, Munguia N, Platt A, Taddei J (2006) Sustainable university: what can be the matter? J Cleaner Prod [s.l.] 14(9):810–819

Viegas SFSS, Cabral ER, Gomes SC, Carvalho AC (2015) Agenda ambiental na administração pública A3P: estudo da adesão em uma universidade pública [Environmental agenda in public administration A3P: study of adherence at a public university]. AOS—Amazônia, Organizações e Sustentabilidade, [s.l.] 4(2):7–28

World Commission On Environment And Development—WCED (1987) Our common future. Oxford: Oxford University Press

Yarime M, Tanaka Y (2012) The issues and methodologies in sustainability assessment tools for higher education institutions: a review of recent trends and future challenges. J Educ Sustain Dev [s.l.] 6(1):63–77

Zeijl-Rozema AV, Cörvers R, Kemp R, Martens P (2008) Governance for sustainable development: a framework. Sustain Dev [s.l.] 16:410–421

Zeitoune B, Trigo JA, Trigo A, Maruyama U (2019) Práticas sustentáveis: adoção de cultura institucional em IES [Sustainable practices: adoption of institutional culture in HEIs]. RPCA—Revista Pensamento Contemporâneo em Administração, Rio de Janeiro 13(1):150–168

Zilahy G, Huisingh D (2009). The roles of academia in regional sustainability initiatives. J Cleaner Prod [s.l.] 17:1057–1066

A Guide on Environmental Education: Current Pedagogical Problems and Recommendations for Its Implementation in the University English Classroom

Estefanía Sánchez Auñón

Abstract Education is a valuable tool for achieving a more sustainable world, especially when it comes to environmental protection. Still, teachers hardly address ecological aspects in their lessons, and the few times they do it, the environmental programmes applied often turn out to be ineffective. This is the case for University English courses, which rarely delve into environmentalism. In this paper, the author highlights the major reasons for such failure of environmental initiatives, which are related to (i) the nature of classroom pedagogy, (ii) the didactic methods used, (iii) the contents targeted, (iv) and the perspective from which those contents are tackled. Furthermore, the researcher compiles multiple recommendations based on the available academic production on this field which need to be considered when implementing this type of programmes in University education in general, and in particular, in the University English class, providing an informed guide on the application of environmental education.

Keyword Sustainability · Environmental education · Challenges · Guidelines · English teaching

1 Introduction

Notable thinkers such as John Muir, Aldo Leopold, or Henry David Thoreau foretold and denounced humanity's detrimental impact on nature as far back as the nineteenth century. What seemed like dystopian fiction which might or might not come true is now part of our reality. Desertification, wildlife extinction, climate change, air and

E. S. Auñón (✉)
Faculty of Education of the University of Murcia, Campus Universitario Street 12, 30100 Murcia, Spain
e-mail: estefania.sanchez8@um.es

W. Leal Filho et al. (eds.), *The Contribution of Universities Towards Education for Sustainable Development*, World Sustainability Series,
https://doi.org/10.1007/978-3-031-49853-4_7

water pollution, and resource shortage are just a small sample of the serious environmental issues which are already noticeable across the globe and which are starting to compromise our daily life (Rands 2009; Robbins et al. 2022; Valera-Losada et al. 2016). As noted by James (2015), the consequences of environmental degradation have an even more adverse effect in disadvantaged countries, widening social gaps. In line with this, seven out of the 17 Sustainable Development Goals included in the *2030 Agenda for Sustainable Development* are related to ecological matters—6. Clean water and sanitation, 7. Affordable and clean energy, 11. Sustainable cities and communities, 12. Responsible consumption and production, 13. Climate action, 14. Life below water, and 15. Life on land (United Nations 2022); and it is urged that today's economic, social, and environmental problems are addressed with priority, in an integrated manner, so that a sustainable future is possible (Purvis et al. 2019). Nevertheless, according to plentiful scholars, the actions taken so far to tackle the environmental dimension of sustainability are not enough. Nature's and society's well-being are often dealt with separately despite being interdependent (Caillon et al. 2017; Richardson et al. 2021); and environmental care has been mainly limited to political reforms and eco-friendly technological advances, which are rather superficial measures. Most worryingly, citizens tend to feel that politicians and field experts are the only capable agents of making the difference, whereas, since every person's behaviour affects the biosphere, deep changes both at a community and at an individual level are needed (Steinhorst and Beyerl 2021). As summarized by Timothy and Obiekezie (2019), the whole prevailing value system must be transformed to solve this ecological crisis.

Education is a powerful weapon for such purpose as it is one of the most influential aspects of people's lives and contributes to shape personal values. In early and middle education, students build their personalities and start interacting with their ecological surroundings. It has been proved that positive environmental experiences at those ages is critical to children's lifelong relationship with the natural world (Gill 2014; Le Hebel et al. 2014). Then, University instruction forms our future leaders and intellectuals of the social, economic, and academic spheres. At this stage, learners receive specific training which allows them to broaden their knowledge and develop the required skills to take an active part in society. Added to this, Universities may establish sustainable campus policies and stimulate environmental research, playing a particularly determining role in the fight for a healthier, greener planet (Leal 2022; Miranda et al. 2021). Education should, therefore, encompass the human being's environmental duties (Wulf 2022). This has been continuously stressed in international conferences and official reports like the *Belgrade Charter* (1975) and *Agenda 21* (1992), respectively, with the result that ecological matters have become an interdisciplinary curricular content and that a new teaching paradigm has been created: environmental education (EE).

Environmental education emerged in the mid 1960s, and, moulded by the conservation movement of that time, it involves a diverse mixture of domains, namely, nature study, fieldwork, urban studies, action research, post-modernism, philosophy, and sociology (Pedretti and Nazir 2014). Consequently, it has been a complex term to define. As asserted by Fang et al. (2022), the most widely accepted description of EE goes as follows:

> The process of recognising values and clarifying concepts in order to develop skills and attitudes necessary to understand and appreciate the interrelatedness among man, his culture, and his biophysical surroundings. Environmental education also entails practice in decision-making and self-formulation of a code of behaviour about issues concerning environmental quality. (UNESCO 1970, p 11)

With regard to its major goals, in essence, environmental programmes must increase learners' awareness and consideration of the links between socioeconomic, political, and ecological factors; allow the acquisition of the necessary knowledge, values, and competences to take care of nature; and promote eco-friendly behavioural patterns both individually and as a community (UNESCO 1978). Although the definition and objectives of environmental education have been clearly established, it is still an ambiguous concept as scholars and educators hold distinct views on its core features (Eilam and Trop 2010; Stables 2001).

Furthermore, research has revealed that few educators address ecological matters in their lessons (Sukma et al. 2020; Zaradez et al. 2020). They have highlighted various challenges they usually face when designing environmental programmes, and which disincline them to do it: the lack of resources, the vast curriculum, the difficulty to use the outdoors, and their unreadiness to teach these specialist contents (Barrett 2007; Pedretti and Nazir 2014; Tan and Pedretti 2010). Cotton et al. (2007) and Zaradez et al. (2020) also remark instructors' perceived irrelevance of EE to some disciplines despite being a cross-curricular aspect. Many of the teachers participating in these studies regarded EE as being subjugated to science subjects. Hence, a minority of them targeted environmentalism, and they did it as a peripheral, additional content rather than as an approach per se from which any discipline may be taught. In tune with these beliefs, it has been shown that EE is tackled in science and geography courses more often than in the other fields, something which students themselves have reported (Harness and Drossman 2011; Marpa 2020; Timothy and Obiekezie 2019). This applies to English formal instruction, which rarely covers ecological issues. Besides the reasons previously detailed, English teachers are not aware of the potential of this discipline to achieve environmental aims (Reid 2019; Stavreva and Kirova 2013). In point of fact, the scant implementation of environmental education is often unsuccessful because, as Hossain (2018) states, most programmes are neither suitably planned nor properly put into practice.

Given the significant contribution of higher education to attaining ecological sustainability and the general failure of the few isolated environmental initiatives carried out in class, it is necessary, firstly, to study the implementation of EE in order to determine any shortcomings and solve them; and, secondly, to encourage the incorporation of this approach into all fields. Thus, this paper explores the common pedagogical problems of environmental programmes and provides a theory and research grounded set of guidelines accordingly.

In the next section, an analysis of the most common pedagogical practices followed when addressing environmentalism is presented, highlighting the reasons why they are ineffective as well as field experts' recommendations to improve them. Afterwards, the significant implications drawn from the analysis, the limitations of this work, and several avenues for future research are detailed in the conclusions' section. This paper especially focuses on University English courses, however, most of the information presented hereafter is applicable to all University disciplines.

2 Pedagogical Analysis

The different shortcomings detected and the suggestions proposed to solve each of them have been classified according to the following typology: the nature of classroom pedagogy, the didactic methods used, the contents targeted, and the perspective from which those contents are tackled.

2.1 The Nature of Classroom Pedagogy

One of the most hindering factors for the efficiency of pro-environmental didactic actions is the general characteristics of the current educational system. To begin with, today's classroom pedagogy relies on competitiveness and standardized testing. Instructors adjust their teaching practices to the subsequent assessment of students so that they are able to pass exams, tasks that merely demand memorization and cognitive understanding (Bhattacharyya et al. 2013; Kostova and Atasoy 2008). This emphasizes individualism and spreads the erroneous belief that studying synthetic information results in learning. It seems as if the only goal of education was to prepare students to enter the working world (Garrard 2010). Additionally, the fact that syllabuses are strictly discipline-based makes it harder to implement EE in a cross-curricular manner. The available research reviewed has indicated that, when addressing environmentalism, teachers normally work independently on any ecological information included in the course curriculum in question (Alyaz et al. 2016). Contrary to this, environmental education must be cooperative in order to foster a stronger sense of community and it must be based on real-world practical situations, requiring constant interdisciplinarity (Tan and Pedretti 2010).

It needs to be noted that most instructional processes are designed to take place indoor, even the courses which revolve around ecological matters, in which youths usually learn about the environment by means of technology-mediated tasks. Hence, students are forced to be secluded in a building for the duration of their whole training period and they experience the natural world mostly virtually (Hung 2017). Numerous environmentalists have condemned this, arguing that the educational system as it operates these days isolates students physically, spatially, and emotionally (Gruenewald 2003). Against such backdrop, they suggest replacing the classroom by outdoor spaces and natural settings so that students can feel closer to the environment (Edwards and Larson 2020). Green (2012) even proposes that learners themselves design and build some of the infrastructures in an eco-friendly way, and that they are allowed to move freely as they wish during the instruction process. In her opinion, educational programmes, including EE, will be more successful if students feel comfortable. Overall, learners have to be in direct and constant contact with their natural surroundings for environmental education to have an actual impact on their behavioural patterns (Marpa 2020).

2.2 The Didactic Methods Used

The outcome of environmental education initiatives depends to a large extent on the type of methodology applied. The fundamental problem here is that teachers tend to stick to passive methods, which, on their own, fail to address the three goals of this approach. Broadly speaking, instructors explain scientific facts on nature with the help of a textbook, and, afterwards, implement controlled environment-related activities, turning students into mere receivers of theoretical information (Fernández-Manzanal et al. 2015; Hart 2010). As described in the introduction of this paper, the objectives of environmental education go beyond the transmission of factual knowledge since they also involve the development of a values ethic and of certain skills to take ecological action. Therefore, it is necessary to implement more active pedagogical strategies that enable students to think critically and analytically. In environmental education, the student must be an autonomous subject and play a central role in their learning process whereas the teacher acts as a facilitator and mentor (Green 2012; Palmer 2002).

The most recommended active methods to incorporate ecological aspects into education are problem-based learning, project-based learning, collaborative learning, and field work (Özalemdar 2021; Selçuk and Yilmaz 2020). In the concrete case of English courses, instructors should opt for English teaching methodologies which complement the above mentioned ones such as content and language integrated learning (CLIL), which facilitates the introduction of environmental content; communicative language teaching (CLT), which fosters deep reflection through oral expression; task-based language teaching (TBLT); and cooperative language learning (CLL).This methodological framework allows teachers to achieve all the goals of environmental education and to design inspiring lessons, and it contributes to overcome the shortcomings discussed in the previous sub-section. Scholars have provided diverse examples of activities that may be implemented within the proposed methods—role-plays, experiments, case-studies, expert learning, and discussions in response to a multimedia stimulus (Cotton et al. 2007; Eilam and Trop 2010). Considering the common isolating nature of instructional processes, teachers should also plan numerous tasks in which students come into profound contact with the environment like field trips or gardening (Marpa 2020). Moreover, they are advised to make use of interactive didactic resources such as audio-visual material, games, posters, or photographs (Nasibulina 2015). Regardless of the set of methods applied, EE initiatives must comprise David Ausubel's "meaningful learning" (Harianto et al.

2019; Le Hebel et al. 2014). Thus, educators have to examine students' previous ecological knowledge and awareness and adjust the subsequent teaching steps to that information.

2.3 The Contents Targeted

Another shortcoming identified is the limited range of contents teachers usually cover throughout environmental programmes. It is already known that there is a tendency to over-emphasize theoretical knowledge. Not surprisingly, then, educators concentrate on background scientific information and they leave many practical environmental aspects out (Hart 2010), which are crucial to foster the adoption of eco-friendly attitudes. EE learners have acknowledged that such lack of practical training makes them feel unprepared to participate in greening action (Fernández-Manzanal et al. 2015). In fact, the theoretical content taught is too simple and general because, for one thing, it just deals with the global effects of environmental harm, and, besides, it is based on the anthropocentric view of nature, reinforcing it (Jensen 2002; Kopnina 2013).

Several orientations have been offered in this regard. Given that environmental education entails a balance between theory and practice, teachers have to target both factual content and a series of key competences (investigative skills, information technology skills, naturalist skills, personal and social skills, creative and critical thinking, and decision-making capacities) (Hart 2010). English instructors have been further urged to tackle all specific competences in the discipline as they seem to neglect some of the four language skills when implementing EE (Neville et al. 2021; Setyowati and Widiati 2014). Moreover, field experts underline that these programmes cannot be limited to the worldwide effects of environmental degradation (what). They need to cover all the factors involved in this situation for students to understand its complexity. Hence, teachers also have to address the causes of the current ecological crisis (why), the change strategies at our disposal (how), as well as the local, regional, national, and international levels of ecological matters and the connections between them (where) (Finger and Verlaan 1995; Jensen 2002). As Freire (2011) elucidates, teachers must personalize these environmental contents. Students should be firstly taught about local ecological features like the main natural resources or the animal and plant species they can find in their area, and, only once they are fully conscious of this, should the teacher move on to national and global issues. Likewise, it is crucial that the content targeted transcends the human-centred conception of nature as a resource provider so that learners can generate an emotional attachment with the environment and start caring about it. To this end, teachers should include philosophical concepts such as the intrinsic worth of nature or all living beings' interconnectedness, which is part of the objectives of this approach (Pereira and Forster 2015; Tan and Pedretti 2010).

2.4 *The Perspective from Which Those Contents Are Tackled*

As exposed in the foregoing Sect. 2.3, most knowledge students acquire in EE programmes is about the effects of ecological harm, and, according to some researchers, the information is transmitted from a very pessimistic perspective. Educators often stress the gravity of these issues and humanity's responsibility for them in the hopes that students feel obliged to do something. Nonetheless, far from triggering action, this "apocalyptic doomsday narrative," in Strife's words (2010, p. 181), has been proved to disempower students. When receiving so much negative input on environmental matters, learners start feeling afraid, worried, guilty, and frustrated. They get the idea that it is too late for them to try to save our home planet (Jones and Davison 2021; Tsevreni 2011). This teaching perspective turns out to be problematic for environmental education because, as learners feel unable to act, they are more resistant to change. It has been verified in a number of empirical studies like, for instance, Lowe et al. (2006). Their findings revealed that, after watching the catastrophist film *The Day After Tomorrow* (2004), the youths participating in the research felt overwhelmed, and that this affective response reduced their perceived self-efficacy to take individual action.

Thus, instructors must convey a positive environmental message that stimulates learners to react. They have to show that a change for the better is still possible and highlight students' power by teaching them the numerous manners in which they may contribute to mitigate ecological concerns (Freire 2011). Strife (2010) and Zylstra et al. (2014) suggest teachers to centre their discourse on the physical, emotional, and cognitive benefits learners can obtain when protecting nature to boost action.

Furthermore, since the teacher becomes a role model for learners, instructors should practice what they preach and go green themselves, which has been shown to influence students' ecological attitudes positively (Cotton et al. 2007). As exemplified by environmentalists, they may ask students to submit all assignments by email to minimize the use of paper, exploit electronic didactic materials in place of handouts, recycle in class, or choose low-emission means of transport to go to the educational centre (Setyowati and Widiati 2014; Timothy and Obiekezie 2019).

3 Conclusions

Ecological sustainability is an ever-increasing global concern in today's day and age, and, despite having been integrated into formal instruction through the curriculum and the development of environmental education, teaching initiatives in this direction are still rare and hardly effective, especially in University English courses. This paper was mainly aimed at detecting the common shortcomings of EE programmes and at providing a research-based set of orientations to solve them. After scrutinizing the currently available academic production on this matter, the author identified diverse pedagogical problems: (i) the conservative and isolating nature of classroom pedagogy, (ii) the passive didactic methods used, (iii) the narrow range of contents targeted, and (iv) the discouraging perspective from which the contents are tackled. Such unsuitable teaching practices when addressing environmentalism might be due to the lack of a standardized guide to implement the approach, as affirmed by Eilam and Trop (2010).

This paper compiles the most relevant recommendations scholars have proposed so far. In brief, it has been shown that environmental education must be participatory, interdisciplinary, based on real-world aspects and ecological principles, awareness-raising, values-driven, multi-method, learner-centred, theoretical and practical, and holistic; include all factors and space dimensions; and approach learners to nature, both physically and emotionally. Besides, in light of citizens' common feeling of inability to act pro-environmentally, it is necessary to empower students by giving them all the tools necessary and also by spreading an optimistic environmental message, for which University instruction has been shown to be particularly advantageous. Hence, it has become clear that introducing EE into every University course would be extremely helpful. Considering all the information reviewed, the main lesson learnt and on which teachers and researchers should continue working is that, apart from adjusting English teaching as well as the other University disciplines to the methodological requirements of environmental education, a radical change of the educational system is urgently needed for these initiatives to come to fruition.

This set of orientations on environmental education is intended to serve instructors as a reference point, guiding their teaching practice in this respect so that they may overcome the obstacles they have underlined and avoid the common problems detected; and encouraging them, therefore, to address ecological concerns more frequently and comprehensively in their lessons. Despite being relevant and useful, this work has some limitations that need to be addressed. Since the study has been limited to a theoretical approach, the findings cannot be generalisable. Accordingly, the major avenue for future studies addressing the implementation of environmental education is putting this synthesis of recommendations into practice and testing their efficiency to confirm whether or not they contribute to EE programmes' success. Indeed, scholars could concentrate on one of the typologies established in this paper to conduct more focused studies and reach more comprehensive conclusions. Once the effectiveness of these recommendations has been verified, they may be standardized, which is necessary too for a suitable homogeneous pedagogical practice. Comparing the efficiency of enhanced EE initiatives in each discipline and exploring instructors' and learners' views on the proposed modifications would be revealing too.

References

Alyaz Y, Isigicok E, Gursoy E (2016) The impact of the environmental documentary movies on pre-service German teachers' environmental attitudes. J Educ Training Studies 5(1):159–170. https://doi.org/10.11114/jets.v5i1.1976

Barrett M (2007) Homework and fieldwork: investigations into the rhetoric-reality gap in environmental education research and pedagogy. Environ Educ Res 13(2):209–223. https://doi.org/10.1080/13504620701284928

Bhattacharyya S, Junot M, Clark H (2013) Can you hear us? Voices raised against standardized testing by novice teachers. Creat Educ 4(10):633–639

Caillon S, Cullman G, Verschuuren B, Sterling E (2017) Moving beyond the human-nature dichotomy through biocultural approaches: including ecological well-being in resilience indicators. Ecol Soc 22(4):1–10

Cotton D, Warren M, Maiboroda O, Bailey I (2007) Sustainable development, higher education and pedagogy: a study of lecturers' beliefs and attitudes. Environ Educ Res 13(5):579–597. https://doi.org/10.1080/13504620701659061

Edwards R, Larson B (2020) When screens replace backyards: strategies to connect digital-media-oriented young people to nature. Environmental Education Research.https://doi.org/10.1080/13504622.2020.1776844

Eilam E, Trop T (2010) ESD Pedagogy: a guide for the perplexed. J Environ Educ 42(1):43–64. https://doi.org/10.1080/00958961003674665

Fang W, Hassan A, LePage B (2022) Research methods for environmental education. In: Series SDG (ed) Pp. Springer, Singapore, pp 49–90

Fernández-Manzanal R, Serra L, Morales M, Carrasquer J, Rodríguez-Barreiro L, del Valle J, Murillo M (2015) Environmental behaviours in initial professional development and their relationship with university education. J Cleaner Production 108(A):830–840. https://doi.org/10.1016/j.jclepro.2015.07.153

Finger M, Velaan P (1995) Learning our way out: a conceptual framework for social-environmental learning. World Dev 23(1):503–513. https://doi.org/10.1016/0305-750X(94)00128-L

Freire H (2011) Educar en verde: ideas para acercar a niños y niñas a la naturaleza. Editorial Graó, Barcelona

Garrard G (2010) Problems and prospects in ecocritical pedagogy. Environ Educ Res 16(2):233–245

Gill T (2014) The benefits of children's engagement with nature: a systematic literature review. Child Youth Environ 24(2):10–34. https://doi.org/10.7721/chilyoutenvi.24.2.0010

Green M (2012) Place, sustainability and literacy in environmental education: frameworks for teaching and learning. Rev Int Geogr Educ Online 2(3):326–346

Gruenewald D (2003) Foundations of place: a multidisciplinary framework for place-conscious education. Am Educ Res J 40(3):619–654

Harianto E, Nursalam L, Ikhsan F, Zakaria Z, Damhuri D, Sejati A (2019) The compatibility of outdoor study application of environmental subject using psychological theories of intelligence and meaningful learning in senior high school. Geosfera Indonesia 4(2):201–216

Harness H, Drossman H (2011) The environmental education through filmmaking project. Environ Educ Res 17(6):829–849. https://doi.org/10.1080/13504622.2011.618626

Hart P (2010) No longer a "little added frill": the transformative potential of environmental education for educational change. Teacher Educat Quart 37(4):155–177

Hossain M (2018) Enacting green pedagogy in the EFL classrooms in Bangladesh: prospects, challenges and pragmatic breakthroughs. Language Educat Forum 1(1):1–5

Hung R (2017) Towards ecopedagogy: an education embracing *ecophilia*. Educat Stud Japan Int Yearbook 11:43–56

James S (2015) Environmental philosophy: an introduction. Polity Press, Cambridge

Jensen B (2002) Knowledge, action and pro-environmental behaviour. Environ Educ Res 8(3):325–334. https://doi.org/10.1080/13504620220145474

Jones C, Davison A (2021) Disempowering emotions: the role of educational experiences in social responses to climate change. Geoforum 118:190–200. https://doi.org/10.1016/j.geoforum.2020.11.006

Kopnina H (2013) Schooling the world: exploring the critical course on sustainable development through an anthropological lens. Int J Educ Res 62:220–228. https://doi.org/10.1016/j.ijer.2013.10.002

Kostova Z, Atasoy E (2008) Methods of successful learning in environmental education. J Theory Pract Educat 4(1):49–78

Le Hebel F, Montpied P, Fontanieu V (2014) What can influence students' environmental attitudes? Results from a study of 15-year-old students in France. Int J Environ Sci Educat 9(3):329–345

Leal W (2022) Pursuing sustainable development across disciplines. In: World Sustainability Series, pp 495–504. Switzerland: Springer. https://doi.org/10.1007/978-3-031-04764-0_27

Lowe T, Brown K, Dessai S, de França DM, Haynes K, Vincent K (2006) Does tomorrow ever come? Disaster narrative and public perceptions of climate change. Public Underst Sci 15(4):435–457. https://doi.org/10.1177/0963662506063796

Marpa E (2020) Navigating environmental education practices to promote environmental awareness and education. Int J Studies Educat 2(1):45–57

Miranda L, Sánchez J, Viloria J (2021) Environmental sustainability in higher education: mapping the field. Revista Electrónica de Investigación Educativa (Redie) 23(9):1–16. https://doi.org/10.24320/redie.2021.23.e09.4053

Nasibulina A (2015) Education for sustainable development and environmental ethics. Procedia—Social and Behavioral Sciences 214:1077–1082. https://doi.org/10.1016/j.sbspro.2015.11.708

Neville I, Petrass L, Ben F (2021) The impact of an outdoor learning experience on the development of English creative writing skills: an action research case study of year 8 and 8 secondary school students in Australia. J Adventure Educat Outdoor Learning.https://doi.org/10.1080/14729679.2021.1983445

Özalemdar L (2021) The effect on environmental attitude of the active learning method applied in teaching the biology topic "current environmental issues and human" for 10th grade students. J Turkish Sci Educat 18(2):276–289

Palmer J (2002) Environmental education in the 21st century: theory, practice, progress and promise. Routledge, New York

Pedretti E, Nazir J (2014) Tensions and opportunities: a baseline study of teachers' views of environmental education. Int J Environ Sci Educat 9:265–283

Pereira M, Forster P (2015) The relationship between connectedness to nature, environmental values, and pro-environmental behaviours. Reinvention Int J Undergr Res 8(2)

Purvis B, Mao Y, Robinson D (2019) Three pillars of sustainability: in search of conceptual origins. Sustain Sci 14:681–695. https://doi.org/10.1007/s11625-018-0627-5

Rands G (2009) A principle-attribute matrix for environmentally sustainable management education and its application: the case for change-oriented service-learning projects. J Manag Educ 33(3):296–322. https://doi.org/10.1177/1052562908323191

Reid (2019) Curriculum and environmental education: perspectives, priorities and challenges. Oxfordshire: Routledge

Richardson M, Passmore H, Lumber R, Thomas R, Hunt A (2021) Moments, not minutes: the nature-wellbeing relationship. Int J Wellbeing 11(1):8–33. https://doi.org/10.5502/ijw.v11i1.1267

Robbins P, Hintz J, Moore S (2022) Environment and society: a critical introduction. Wiley, Oxford

Selçuk A, Yilmaz M (2020) The effect of constructivist learning approach and active learning on environmental education: a meta-analysis study. Electron J Environ Educat 10(1):44–84

Setyowati L, Widiati U (2014) Integrating environmental education into a genre-based EFL writing class. English Teaching Forum 52(4):20–27

Stables A (2001) Who drew the sky? Conflicting assumptions in environmental education. Educ Philos Theory 33(2):245–256. https://doi.org/10.1111/j.1469-5812.2001.tb00266.x

Stavreva S, Kirova S (2013) Blending the teaching of environmental matters and English as a second or foreign language in Macedonia. Natura Montenegrina 12(3–4):1065–1071

Steinhorst J, Beyerl K (2021) First reduce and reuse, then recycle! Enabling consumers to tackle the plastic crisis—qualitative expert interviews in Germany. J Clean Prod 313:1–14. https://doi.org/10.1016/j.jclepro.2021.127782

Strife S (2010) Reflecting on environmental education: where is our place in the green movement? J Environ Educ 41(3):179–191

Sukma E, Ramadhan S, Indriyani V (2020) Integration of environmental education in elementary schools. In: IOP Conference series: journal of physics. https://doi.org/10.1088/1742-6596/1481/1/012136

Tan M, Pedretti E (2010) Negotiating the complexities of environmental education: a study of Ontario teachers. Can J Sci Math Technol Educ 10(1):61–78. https://doi.org/10.1080/14926150903574320

Timothy A, Obiekezie E (2019) Green English: environmentally responsive pedagogy for Nigerian secondary schools. Prestige J Counselling Psychol 2(1):35–48

Tsevreni I (2011) Towards an environmental education without scientific knowledge: an attempt to create an action model based on children's experiences, emotions and perceptions about their environment. Environ Educ Res 18(1):53–67. https://doi.org/10.1080/13504621003637029

UNESCO (1970) International working meeting on environmental education in the school curriculum. Final Report, at Foresta Institute, Carson City, Nevada, US. https://files.eric.ed.gov/fulltext/ED045490.pdf

UNESCO (1978) Intergovernmental conference on environmental education. Final report, at Tbilisi, USSR, Georgia. https://unesdoc.unesco.org/ark:/48223/pf0000032763

United Nations (2022) The sustainable development agenda. Sustainable development goals. https://www.un.org/sustainabledevelopment/development-agenda/

Valera-Losada M, Vega-Marcote P, Pérez-Rodríguez U, Álvarez-Lires M (2016) Going to action? A literature review on educational proposals in formal environmental education. Environ Educ Res 22(3):390–421. https://doi.org/10.1080/13504622.2015.1101751

Wulf C (2022) Education as human knowledge in the Anthropocene—an anthropological perspective. Routledge, Oxfordshire

Zaradez N, Sela-Sheffy R, Tal T (2020) The identity work of environmental education teachers in Israel. Environ Educat Res, 1–18.https://doi.org/10.1080/13504622.2020.1751084

Zylstra M, Knight A, Esler K, Grange L (2014) Connectedness as a core conservation concern: an interdisciplinary review of theory and a call for practice. Springer Science Reviews 2:119–143

The Integration of ESD in Geography Syllabi at Higher Education Level in Malta

Rebecca Cassar and Mark Mifsud

Abstract Education for sustainable development (ESD) plays a crucial role in guaranteeing a sustainable future for both present and future generations. This paper explores how ESD is integrated into the present geography syllabi at post-secondary level in Malta, a small Mediterranean island state. It seeks to make a primary contribution to research in ESD by understanding how ESD is connected to geography education at post-secondary level, a form of higher education. The Geography Advanced and Intermediate syllabi are analysed through the UN Economic Commission for Europe (UNECE) Competences in ESD. The method includes a desk study analysis of the syllabi and semi-structured interviews with geography educators for further data triangulation. The results show that several ESD competences feature in the syllabi confirming the link between ESD and geography education. Furthermore, educators at this level are able to integrate ESD competences and skills using transformative practices. In response to these findings, a number of recommendations that are internationally relevant, are drawn up to further enhance the effectiveness of geography syllabi in infusing Education for Sustainable Development (ESD) principles in teaching and learning.

Keywords ESD · Education for sustainable development · ESD competences · Geography education · Higher education

R. Cassar (✉)
Geography Lecturer at St. Martin's College Sixth Form, Msida, Malta
e-mail: rebeccacassar@smc.edu.mt

M. Mifsud
Centre for Environmental Education and Research, University of Malta, Msida, Malta
e-mail: mark.c.mifsud@um.edu.mt

W. Leal Filho et al. (eds.), *The Contribution of Universities Towards Education for Sustainable Development*, World Sustainability Series,
https://doi.org/10.1007/978-3-031-49853-4_8

1 Introduction

The objectives of Target 7 within the fourth Sustainable Development Goal, the obtainment of Education for Sustainable Development (ESD), can be reached through the development of competences within this form of education. Competences in ESD are developed and honed to transform the way both educators and students think and act. Both partners in the educational journey require the knowledge, skills, values and attitudes to empower them to contribute to sustainable development.

Environmental Education (EE) was first integrated within the Maltese formal education with its inclusion in formal education in 1999 through the National Minimum Curriculum. Several ESD themes like tackling 'the threats posed by contemporary societies to the … environment', understanding 'the social and environmental effects of different patterns of consumption', appreciating 'the values of cooperation and solidarity', 'safeguarding the natural environment in a sensitive manner', and engaging students through 'curiosity' and 'critical and creative thinking' were incorporated into the curriculum (Ministry of Education 1999, p. 15). EE was then formally taught through subjects at primary and secondary level whilst also being integrated into schools by means of educational programmes such as EkoSkola and Dinja Waħda (translated into One World) initiated by NGOs. Mifsud (2012) argues that EE in Malta truly began through the work carried out by environmental NGOs, as a response to major environmental issues on the island. He also states that the effect of overt environmental issues are factors that have a direct influence on public awareness on such issues. He further concludes that schools must be the main stakeholders in both educational, but also environmental policies, both at the regional and national level.

ESD was eventually included in the National Curriculum Framework launched in 2012 which led to its official incorporation into the educational system by involving knowledge, skills and attitudes of students to understand their surroundings, respect and value diversity (MEE 2012), as well as empower them through creative and critical thinking skills (Cassar Cordina 2015).

At Higher Education level in Malta's educational institutions, EE is included in subjects such as Geography, Environmental Science and Systems of Knowledge at post-secondary level. Geography creates the opportunity for ESD knowledge and skills to be integrated into the teaching. Although the knowledge and skills are present within these syllabi, it is still up to each individual teacher or student whether these will be translated into the ESD values. Students are being taught values in sustainability and environmental issues that are of concern within the Maltese Islands (MATSEC 2019d; Cassar and Mifsud 2023). Environmental Science allows students to study environmental issues objectively from a scientific perspective and provides them with the basic environmental literacy to be able to pursue careers related to the environment. Its syllabus promotes environmental issues holistically and approaches them with an interdisciplinary lens. It is also crucial to note that sustainable development is a 'cross-cutting theme' within the syllabus (MATSEC 2019a). The compulsory subject Systems of Knowledge also promotes attitudes and values in sustainable development. Here, students develop knowledge in the environmental issues and sustainable development whilst fostering attitudes and values pertaining to environmental responsibility and participation in the environment's protection (MATSEC 2019b).

These values and attitudes pertained to ESD can be achieved through a holistic and transformational educational system that promotes and integrates ESD. Educational institutions that boast ESD within their curricula allow learners to become engrossed in matters that are relevant within their everyday challenges. This also ensures that the material learnt, pedagogical techniques and learning environment promote ESD through interactive, learner-centred approaches. The educator and learner competences within the the United Nations Economic Commission for Europe (UNECE) competences for ESD support education in this such described transformative, independent, yet participative and collaborative manner.

The aim of this research study was to analyse the inclusion of ESD within the geography syllabi at post-secondary level in Malta using the UNECE ESD Competences. The research will explore the opportunities and strengths, whilst also taking note of the challenges and weaknesses in implementing ESD across the geography syllabi and within the teaching of geography through these syllabi.

Beyond this introduction, the paper is divided into another four sections. This introduction is followed by a brief section describing the UNECE competences in order to give the reader an understanding of how they are divided and why they are used. This is followed by the methodology which lays down the ways in which the syllabi were analysed and the post-secondary educators were interviewed in order to gather information to triangulate the data. The methodology is followed by the results section where a comprehensive report of the analytical findings within both syllabi is given. A thematic analysis of the content within the interviews allows the views of the educators to be compared and triangulated with the syllabi analysis. This concludes the results section. A final section concludes the chapter by summarising the paper's findings and explains its contributions.

2 The UNECE ESD Competences

The Global Action Programme highlighted these learning outcomes or competences to stimulate learning and promote 'critical and systemic thinking, collaborative decision-making, and taking responsibility for present and future generations' (UNESCO 2014, p. 12). The abilities established through such a competence approach allow the learner to act with sustainable development in mind in the real world and partake in socio-political processes to work towards a more sustainable society (Rieckmann 2012; Wiek et al. 2011). Ultimately, dividing the goals of ESD into competences allows the educator to explore key areas vital for the education's success (Rieckmann 2018).

The goal of the UNECE Competences is to help to transform unsustainable societies into ones with sustainable futures (UNECE 2012). The framework of competences in this report act as a guideline for educators for them to play a role in transforming their society towards greater sustainability. The competences framework is divided into three main groups which are then subdivided into a further four learning outcomes. It enlists outcomes on what the educator should understand, be able to do, ways in which they should work with others and overall all characteristics of what they should be as an educator. These outcomes are then divided further into ones that take a holistic approach, that envision change and achieve transformation (Wals 2014).

3 Methodology

The research study takes on a mixed method research strategy. The qualitative data was collected through both a desk review study of the syllabi and interviews with post-secondary geography educators, which can be characterised as qualitative content. The quantitative data also features in the syllabi analysis through the desk review study. The quantitative and qualitative data collection methods occurred in a sequential explanatory strategy. The quantitative data was first collected and analysed followed by the qualitative data phase. The conclusions from the desk review study of the ESD competences found in the syllabi gave the necessary information to create the interview questions to triangulate the data obtained from the syllabi analysis.

3.1 The Review of the Syllabi

The reading and analysis of the syllabi can be characterised as qualitative content analysis because it involved searching for patterns and categories and summarising trends (Jóhannesson et al. 2011). This is identical to the way the featured ESD competences were systematically recorded in the syllabi. A novel approach was developed to conduct the desk study review of the syllabi and therefore can be considered as a unique method to reviewing and analysing a syllabus document.

The documents Geography AM syllabus and Geography IM syllabus (MATSEC, University of Malta 2019c) were used to study the degree to which ESD features within the syllabi at these levels. These syllabi were updated for assessment use in 2019 across Maltese and Gozitan post-secondary colleges teaching students between the ages of sixteen and eighteen. Both syllabi aim to educate learners on the disciplines of physical and human geography, whilst emphasising the inter-relationships and inter-actions between human activities and the environment. The global and local geographical content is intertwined with the learning and practice of skills required in the field.

The ESD Competences listed and discussed in the UNECE's Strategy for Education for Sustainable Development report titled 'Learning for the Future—Competences in Education for Sustainable Development' published in 2012 were used to review the syllabi. The following quote from the document supports why they were used.

> The Competences should be a basis for the review of curriculum documents. While the Competences are those of educators, they also suggest approaches to educational practice. In order for educators to practice the Competences they should be supported by a curriculum which reflects such educational approaches (p. 7).

Upon carefully studying the competences, they were individually categorized into five themes that would be used to analyse the syllabi. The themes and the competences that were placed into each theme are shown in Table 1.

The results from the syllabi review were subsequently used to determine what to include in the interview questions for further triangulation. A summary of the syllabi desk review analysis has been created in Fig. 1.

3.2 The Interviews

The research's qualitative method continues with semi-structured interviews to triangulate the data. Semi-structured interviews were chosen as the interview questions were written through the data obtained from the syllabi review. The initial stage involved planning and forming the interview questions. The research areas for the interview surfaced through the literature on the UNECE ESD Competences and the analytical review of the syllabi. As the featured ESD competences were listed and counted, the elements of ESD that needed to be investigated in the interview became evident.

4 Results

4.1 The Syllabi Analysis

4.1.1 Findings in the Intermediate Level Syllabus

The task of coding the syllabus text placed individual parts from the syllabus into the self-created themes listed in Table 1. Overall, more of the Intermediate syllabus fell under Theme 1: Understanding Sustainable Development with 13 occurrences, however Theme 3: Problem setting, critical reflection, visioning and creative thinking and Theme 2: Understanding different worldviews and perspectives came in close with 10 and 9 occurrences, respectively. Figure 2 shows the extent to which these ESD competences feature within this syllabus document. The colours in the bar charts

Table 1 The Themes and competences used to analyse the syllabi

Theme	Competences
Understanding Sustainable Development	How the interrelated social and economic systems function
	How generations, rich and poor and humans and nature are interdependent
	The connection between sustainable futures and the way we think, live and work
	That sustainable development is evolving
	The urgent need for change from unsustainable practices towards advancing quality of life, equity, solidarity and environmental sustainability
	The importance of scientific evidence in supporting sustainable development
	Critically assess processes of change in society and envision sustainable futures Are students encouraged to negotiate alternative futures?
	The causes of unsustainable development Facilitate the evaluation of potential consequences of different decisions and actions
Understanding different worldviews and perspectives	Know their personal world view and cultural assumptions and understanding those of others. Create opportunities for sharing ideas and experiences from different disciplines/places/cultures/generations without prejudice and preconceptions
	Work with the different perspectives on dilemmas, issues, tensions and conflicts
	Connect the learner to their local and global spheres of influence
	Are students facilitated to understand the emergence of new worldviews that address sustainable development?
	Inclusive of different disciplines, cultures and perspectives, including indigenous knowledge and worldviews
Problem setting, critical reflection, visioning and creative thinking	The importance of problem setting, critical reflection, visioning and creative thinking in planning the future and affecting change
	Why it is important to prepare learners to meet new challenges

(continued)

Table 1 (continued)

Theme	Competences
	Are students actively engaged across generations, cultures, places and disciplines? Engage in real-world issues as this enhances learning outcomes and helps learners to make a difference in practice
	The importance of building on the experience of learners as a basis for transformation
	Use the natural, social and built environment, including their own institution, as a context and source of learning
Transformative Education	Why there is a need to transform the education systems that support learning
	Why there is a need to transform the way we educate/learn
	Why it is important to prepare learners to meet new challenges
	The importance of building on the experience of learners as a basis for transformation
	Facilitate the evaluation of potential consequences of different decisions and actions
	Are students motivated to make a positive contribution to other people and their social and natural environment, locally and globally?
	Are students challenged on their assumptions underlying unsustainable practices?
	Are students actively engaged across generations, cultures, places and disciplines?
Advocating for change and inspiring hope	The urgent need for change from unsustainable practices towards advancing quality of life, equity, solidarity and environmental sustainability
	The importance of preparedness for the unforeseen and a precautionary approach
	Communicate a sense of urgency for change and inspire hope

represent the colours used when coding the syllabus with the following key. Theme 1—Yellow, Theme 2—Red, Theme 3—Blue, Theme 4—Green.

The introduction to the syllabus features several ESD competences. The description of the syllabus features the competences from Theme 1: Understanding Sustainable Development as it includes the competences *Know how humans and nature are interdependent* and *Critically assess processes of change in society and envision sustainable futures* as it specifically states 'Emphasis on the inter-relationships and

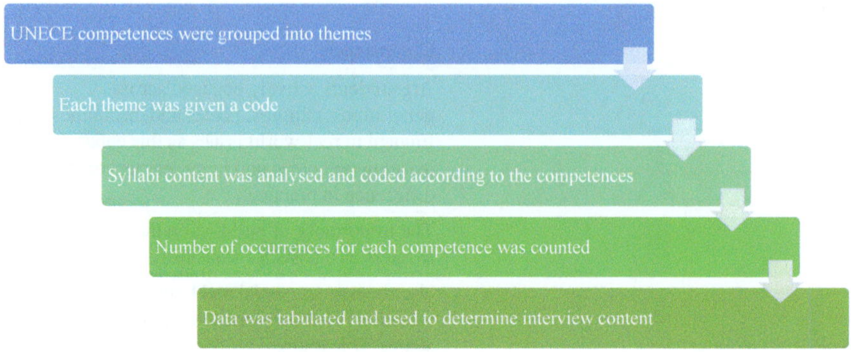

Fig. 1 Summary of the syllabi desk review analysis method

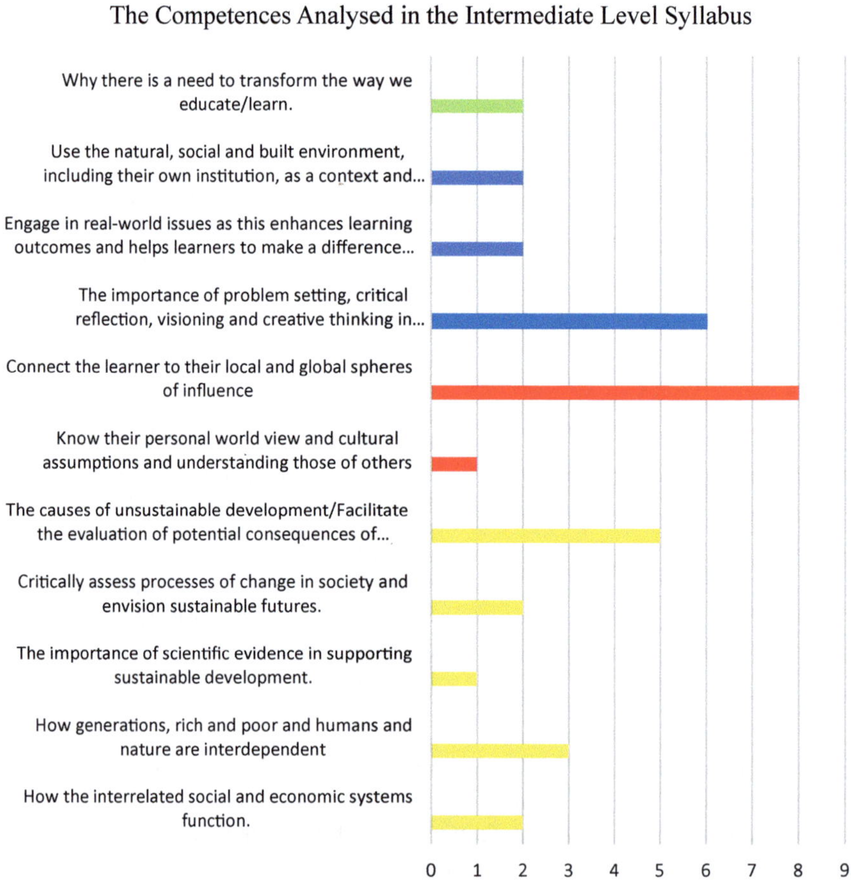

Fig. 2 The number of occurrences for Competences in the Intermediate Level Syllabus

inter-actions between human activities and the environment should form an important aspect of the teaching and learning of Geography' and 'acquire a range of skills to be able to gather evidence for policy making and environmental management.' (p. 2).

The aims of the syllabus go on to include the competences *Connect the learner to their local and global spheres of influence* and *Actively engage students across generations, cultures, places and disciplines* as it states, 'the inclusion of application in the syllabus is meant to highlight the links between the key ideas and a particular part of the real world'. This is repeated as it states that the course should enable 'a considerable, spatially referenced knowledge of issues regarding the human race and the environment'. These competences are featured in the second theme, *Understanding different worldviews and perspectives.*

Another major competence which also features in this introductory section is *The need to transform the way we educate/learn.* This section states that the course should enable 'a different structure of geographical knowledge which should act as a stimulus as a challenge to the investigation of principles' and 'an opportunity to accommodate a range of teaching methods'. This competence falls under the fourth theme, Transformative Education. The introductory section also includes the competence *Use the natural, social and built environment, including their own institution, as a context and source of learning* as it makes 'a strong emphasis on the learning and practice of skills through an interaction with the field' (p. 2). This was categorised under the third theme, Problem setting, critical reflection, visioning and creative thinking.

The assessment objectives largely emphasise the theme Problem setting, critical reflection, visioning and creative thinking as they test the application of geographical knowledge, to 'describe, explain and critique' and 'analyse decisions, suggest solutions to problems and assess their likely consequences', which once again also reflects the competence *Critically assess processes of change in society and envision sustainable futures.* They also test 'human–environment problems at a global and/or local scale' reflecting the competences *How humans and nature are interdependent* and *Connect the learner to their local and global spheres of influence* (p. 3).

The learning content in the syllabus largely feature competences from the theme Understanding Sustainable Development. When it comes to the themes Understanding different worldviews and perspectives and Problem setting, critical reflection, visioning and creative thinking, the competences feature as the content requires students to refer to local and global issues, whilst having a 'critical understanding' of the issue learnt (p. 5). Fieldwork skills make a sufficient part of the discipline's skills allowing the students to *Use the natural, social and built environment, including their own institution, as a context and source of learning.*

It is important to note that although the syllabus does feature several competences, it centralises around a mere 11 from the 40 UNECE Competences. Particularly those competences placed in Theme 5: Advocating for change and inspiring hope, do not feature at all. Competences encouraging students to take on a participative role are amongst those that are not included. These include competences such as *The urgent need for change from unsustainable practices towards advancing quality of life, equity, solidarity and environmental sustainability; Prepare learners to meet new challenges, Build on the experience of learners as a basis for transformation.* These competences would be vital for constructive and transformative teaching methods. In this way, the syllabus creates a very prescriptive content on which to hone geographical skills and knowledge.

4.1.2 Findings in the Advanced Level Syllabus

The Advanced level syllabus featured far more occurrences of the competences. The syllabus is longer and has the possibility to include far more content and skills. Once again it emphasises the evaluation of consequences of unsustainable practices (37 occurrences) and connects the learner to local and global issues (27 occurrences). It also attempts to instil problem setting, critical reflection, visioning and creative thinking in students through analytical and interpretative skills (15 occurrences). The division of the syllabus text into the various competences is presented in Figs. 3 and 4.

The fundamental knowledge and skills within this geography course overlap with those from the Intermediate level syllabus, leading to similar competences to be mentioned in the Introduction description and the Assessment Objectives. There is a repeated emphasis on *how humans and nature are interdependent; the evaluation of potential consequences of different decisions and actions; connecting the learner to their local and global spheres of influence; the importance of problem setting, critical reflection, visioning and creative thinking* and, *preparing learners to meet new challenges.* The syllabus states how geography at this level aims for students to apply their geographical knowledge to their everyday situations and hence encourage life-long learning, a principle strongly encouraged as part of ESD. This will be done through the education of various people, places and environments; the skills translated from the interpretation of maps and other geographic tools and technologies such as charts, graphs and photographs, and systematic problem-solving techniques in using geographic methods.

The skill competences *problem setting, critical reflection, visioning and creative thinking* (15 occurrences) continue to be featured throughout the Assessment Objectives as students are asking to apply theories; 'analyse, organise, and present information in a coherent essay or structured response', 'collate, present and interpret data … or evaluate evidence in problem solving situation(s)' (p. 3). These skills are also gained through the content dealing with 'personal and group fieldwork' where a variety of suitable geographical techniques are used. A few of the fieldwork techniques listed in the syllabus include sampling techniques, the use of quadrats and transects, measuring slope form and angle with the use of pantometers and clinometers, soil data collection techniques, various surveys and the use of questionnaires and interviews. Students are also trained in the use of cartographic and statistical tools to represent data. Moreover, the students are assessed through two group and one individual fieldwork report where students apply all the geographical research skills to obtain and analyse data in a critical and holistic manner. The assessment

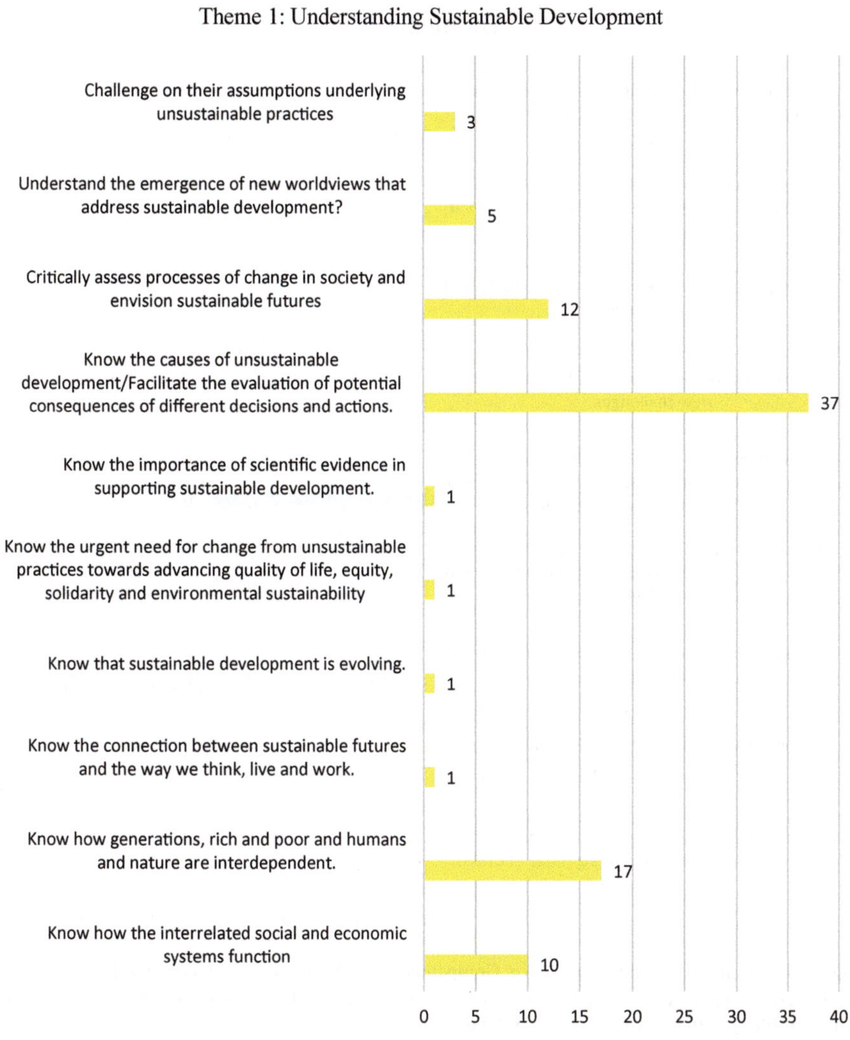

Fig. 3 The number of occurrences for Competences in the theme Understanding Sustainable Development in the Advanced level syllabus

criteria for these graded reports specifically states 'Effective, coherent and independent analysis and interpretation' (p. 19). This would be crucial to obtain the ESD competence mentioned above.

These skills are put to the test when students are asked to apply and analyse the use of models. These include fundamental models in the learning of Geography to teach relationships between social and economic factors, such as the Demographic Transition Model, the Central Place Theory, the Gravity Model, Von Thunen or even

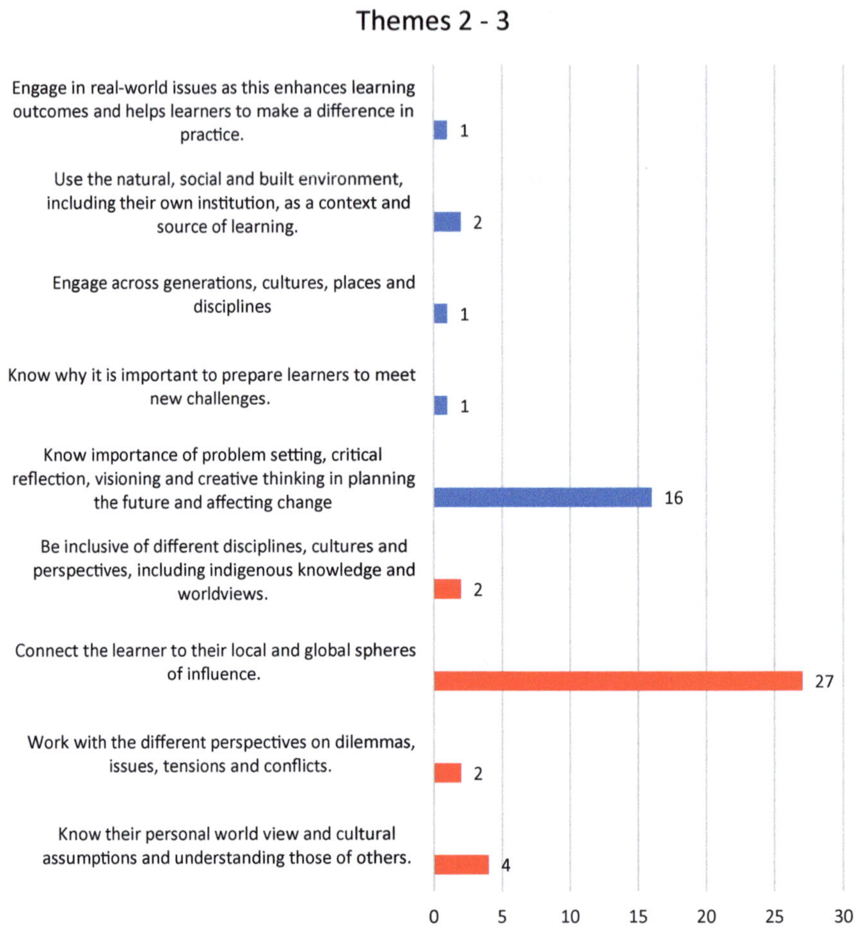

Fig. 4 The number of occurrences for Competences in the themes 2: Understanding different worldviews and perspectives (red) and 3: Problem setting, critical reflection, visioning and creative thinking (blue) in the Advanced level syllabus

Weber's Model. Here they not only must understand their use, but also critique them, enabling students to reflect and envision different scenarios.

As geography deals with issues on a global scale the Assessment Objectives highlight *being inclusive of different disciplines, cultures and perspectives* (6 occurrences) and refer to this as having 'an awareness of the significance of the attitudes and values of others and of the impact of cultural differences on geographical issues.' This features throughout the learning content of the syllabus in topics such as global populations, world and indigenous agriculture types and international tourism. It must be said that this competence features far more in topics in the human geography section than it does in that of physical geography.

This competence ties in with that where students *connect to their global spheres of influence* (27 local and global occurrences together). Geographical content is optimum to instil and teach this feature of ESD, say through the topic of global population distribution. Notably, stress is also placed on the education of Maltese geography. This allows students to *connect to their local spheres of influence.* Within the physical geography context, students appreciate the climate, geology and geomorphology, hydrology, coastal processes, ecosystems, and soils of the Maltese Islands. With regards to human geography, students learn about the demography, settlements, agriculture and fisheries, energy production and demand, and tourism of Malta.

Simultaneously, students can appreciate the *interdependence between humans and nature* (17 local and global occurrences) as they observe the connections between the two branches of geography and the environmental issues of human activity of the environment in Malta. Here it is important for educators to *facilitate the evaluation of potential consequences of different decisions and actions,* whilst students understand *the causes of unsustainable development* (placed together, 37 local and global occurrences). This allows one of the most crucial principles of ESD to be transmitted, where learners are exposed to the concept of sustainability and understand what and how human activities influence the environment. A few of the issues discussed in the syllabus are global warming and climate change, environmental hazards, coastal instability, flood hazards, exploitation of resources including energy and water resources, degradation of habitats and ecosystems, soil erosion and desertification, impacts of agriculture, fisheries and manufacturing and also waste management.

At the same time, *students are encouraged to negotiate alternative futures* (12 local and global occurrences), allowing the knowledge they are learning to enable them to make a difference in their lives and society. This is imperative for the successful integration of ESD. At this level, students are made aware of 'local environmental measures and legislation' (p. 10) to protect the environment, planning and development control of the local built environment and various renewable sources of energy. Students also learn about 'the Club of Rome; the Bruntland report; Lovelock's 'Gaia' theory, the concept of Sustainable Development; the United Nations Sustainable Development Goals (SDGs); Climate Change Conventions and agreements' (p. 13) that all show that *students are facilitated to understand the emergence of new worldviews that address sustainable development* (5 occurrences).

Overall, it can be concluded that the Advanced level syllabus creates the opportunity for ESD knowledge and skills to be integrated into the teaching. It must also be said that it always depends on the depth at which the teacher or student decides to enter into each issue discussed. Ultimately, the knowledge and skills are present, however it is up to each individual teacher or student whether these will be translated into the ESD values. Students are being taught values in sustainability and environmental issues that are of concern, however whether students act on these values in their daily life is a personal choice.

As the UNECE ESD competences are directed towards educators, there is an emphasis on transformative education which is not included in the syllabus. Again, this is a matter left in the hands of the educator to realise *the importance of building on the experience of learners as a basis for transformation.* Similarly, the syllabus makes no mention of *communicating a sense of urgency for change and inspire hope* or, *motivate students to make a positive contribution to other people and their social and natural environment, locally and globally.*

4.2 The Thematic Analysis—Triangulation of Data

In order to triangulate the data obtained through the syllabi analysis, educators teaching these syllabi were interviewed. The interview participants were asked whether, and in what way, they thought a number of ESD competences featured in the syllabi they taught.

The competence *Knows the causes of unsustainable development* educates students to critically analyse past human developments and their root causes and understand their successes and failures from a cultural, social, economic and environmental standpoint (UNECE 2012). Three, or 60%, of the participants believed that this competence features in several topics in the syllabi whereas two doubted its proper integration. Therefore, there is a similarity between the findings on this competence between the syllabi analysis and the interviewee's perceptions of its integration.

The three participants explained that since many environmental issues are covered in the syllabi, their underlying causes also feature greatly. They listed population growth, the use of non-renewable energy sources, overfishing, and local issues like waste management and transport as examples of topics that educate students on this competence that are featured in the syllabi.

> *For example, if we are talking about population growth then you would automatically mention how population growth has its impacts on the environment and resource demand. If we are talking about energy and the dependency on fossil fuels, then that is another underlying cause, so that would feature a lot.* (Participant 2).

On the other hand, two, or 40%, of the other participants stated that the syllabi were not clear on the implementation of this competence and thought that the syllabi ought to focus on it more. Both participants felt that many parts of the syllabi covered issues of the past and not the present or future, even stating, "*there is a large historical element in the syllabus, without necessarily looking forward*" (Participant 1). Therefore, by stating this, there is the belief that the competence is not fulfilled according to its aim to envision change (UNECE 2012).

The competence *Connects the learner to their local and global spheres of influence* encourages integrative thinking in response to the challenges of sustainable development that are both global and local and leads students to be cognitive on how issues in one country or region of the world can influence other regions (UNECE 2012). Four, or 80%, of the participants firmly believed that this competence features in topics in the syllabi whereas one participant argued against its occurrence. Again, there is a similarity between the findings on this competence between the syllabi analysis and the interviewee's perceptions of its integration.

The four participants expressed how the integration of this competence in the subject was only natural. Many stated that it was included because topics feature issues from a global, regional (the Mediterranean) and local sphere. One participant stated how it was important that the syllabi focus on the Maltese Islands as it places the geography in context that students can understand. Similarly, another participant stated that including Maltese geography was imperative for students to connect the realities they experience and for both the teacher and the students to draw examples from. Simultaneously, it was also stated that case studies from "*for example … Brazil, China*" (participant 2) increased the students' awareness from a global perspective.

On the contrary to these opinions, one participant felt that the syllabi were not clear on the integration of this competence and felt that it was more dependent on how the teacher addressed it in the classroom. According to this lecturer, since this competence was not "*spelt out*" (Participant 1) in the syllabi, it may not actually be taught to students and once again argued that this was because the content included is often dealing with matters of the past, instead of the future.

The competence *Know how generations, rich and poor and humans and nature are interdependent* inspires students to think in integrative ways on these interrelationships and the dangers of inharmonious relationships (UNECE 2012). Four, or 80%, of the participants strongly believed that this competence features in topics in the syllabi whereas one was of an opposite viewpoint. Again, there is a similarity between the findings on this competence between the syllabi analysis and the interviewee's perceptions of its integration.

The four participants shared their views on how each relationship was integrated in different topics. The topics population, migration, multiculturalism, tourism, manufacturing, sustainable development, the dependence of more developed nations on the less developed nations for the trade of goods, and the overall exploitation of resources were particularly mentioned. However, the fifth participant once again argued that the inclusion of this competence depended very much on the individual teacher. The educator expressed the wish for it to be included further, though also stated that the structure of the examination limited the ability for the connections between these issues to be discussed. This participant's views are in line with Whitby's who states that the implementation of such ESD competences depends on the teacher's flexibility to move away from the regular curriculum (2019).

The competence *Know how the interrelated social and economic systems function* further provides the learner with connections related to the three pillars of sustainable development. Only two, or 40%, of the participants believed that this competence features in topics in the syllabi. The overall perceptions resembled the findings on this competence in the syllabi analysis.

The two participants quoted topics like the use of fossil fuels, forestry, and fish resources as examples where the competence featured in the syllabi. It is possible that the interviewees did not understand the meaning behind the competence and the knowledge it should educate students with. The remaining participants argued that the social and economic systems were not related in the syllabi and could feature more in a greater interrelated manner.

The competence *Allows for sharing ideas and experiences from different disciplines/places/cultures/generations without prejudice and preconceptions* provides valuable insights on different worldviews to connect the learner to other people, locally and globally, and to their natural environment (UNECE 2012). Only two, or 40%, of the participants believed that this competence features in topics in the syllabi. The overall perceptions were comparable to the findings on this competence in the syllabi analysis.

Those participants who felt that this competence featured in syllabi topics stated that it helps to discuss the relationship between different people, and hence, understand different people coming from different countries minimising prejudice. This is exemplified as Participant 2 said,

> *Unfortunately, at times, the students would not understand why people are actually fleeing their country, but once they understand this notion, then they start to empathise. Or if we talk about the apartheid in South Africa is helps us to understand the experience that people went through due to different racial backgrounds.* (Participant 2).

Most participants then gave two varying opinions on why they thought this competence is not included in the syllabi. Firstly, one participant once again mentioned that its integration depends on the values and personal decisions taken by each teacher and concluded that the extent to which students are exposed to different cultures is not specified in the syllabi. A second participant stated that the syllabi do not refer to specific case studies, which, in their opinion, would help in incorporating this competence into the syllabi.

In conclusion, most of the participants tended to agree with the extent to which certain ESD competences featured in the syllabi through the research's syllabi analysis. This can be seen both in the instances where the competences were established to be featured greatly and when they were not. Figure 5 provides the main findings from the analysis of all the data gathered and analysed in this study.

5 Conclusion

The study and its method presented a few limitations to acquiring full validity. IPA researchers argue how it may be difficult to achieve reliability and validity in qualitative research analysis studies. This is because this form of research cannot be 'subject to the traditional ways of assessing validity and reliability' as purely quantitative research would (Flick 2009, p. 384).

The mixed method approach produced a large amount of data that was lengthy to analyse. This could be said for both the syllabi analysis and the analysis of the interviews. The syllabi analysis through the desk study was a novel method, and therefore lacks external validity. As the study researched the integration of ESD within the teaching content of the subject geography at post-secondary level in the

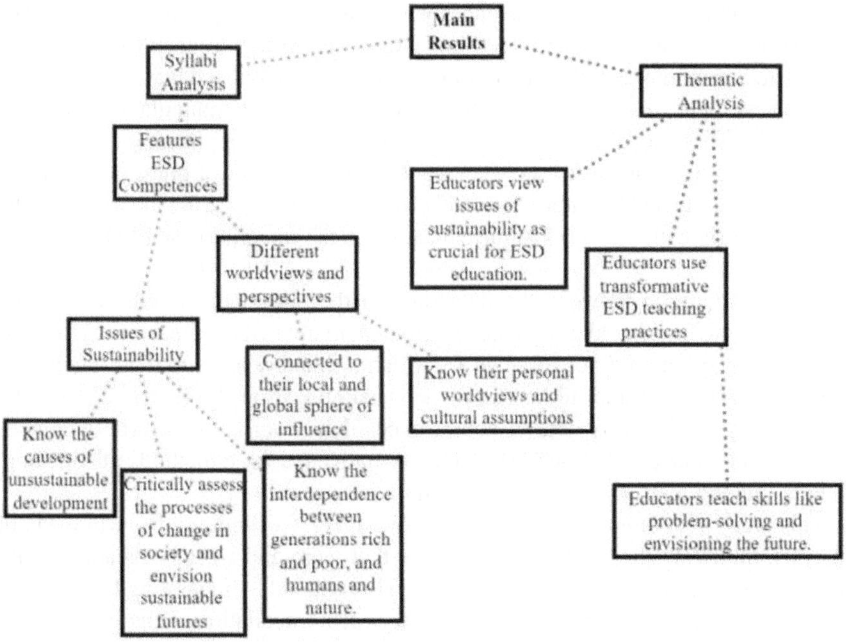

Fig. 5 Main results from the study

country of Malta, it targets a specific research niche. The main reason for this is that the study was conducted in the small island state of Malta. Nevertheless, the study can still be replicated to study the content analysis of other syllabi documents within or beyond the Maltese educational system. This was a prime example of how ESD can be studied within subject material, and can therefore be replicated towards any subject, in any educational level in any country.

When specifically discussing the interview method, validity is also challenged through its subjective nature as it is subjected to bias from both the interviewer and interviewee. The data was obtained from a small sample size as the number of post-secondary geography teachers in the country is a small one. Secondly, the interviews had to take place via video calls due to social distancing procedures due to the Covid-19 pandemic. This was both advantageous and limiting. On the one hand, the participants preferred to conduct the meeting online and participated at any time wherever they were available, whether it be at work or home. On the other hand, a different rapport could have been created face to face, where the participant's expressions could have been read better.

Notably, validity was achieved through consistency and triangulation. Firstly, attention was paid to consistency both while conducting the syllabi analysis to ensure that the method was replicated in the same way in both syllabi, and while conducting the interviews. With regards to the quantitative data obtained through the syllabi analysis, it was imperative to remain consistent with the type of syllabus content that was counted as part of occurrences where the ESD competence featured. The semi-structured element of the interview method permitted to keep to the specific formulated questions as well as to minimise interviewer input and bias. Secondly, the research attempted to achieve further trustworthiness through triangulation of aspects of the data. This was achieved via iterative questioning by rephrasing questions to compare participant answers and drawing the subject matter of questions from the syllabi desk study and literature. It was also achieved by comparing the results of the syllabi analysis with the views of the educators on the integration of ESD within the syllabi.

The triangulation between the data showed that although the educators often highlighted the fact that although they implement many ESD competences into their teaching, the syllabi did not stipulate so, which ultimately leads to their integration into the education system to be one of personal choice on the teacher's part. The syllabi analysis also took note of competences that never featured. These competences could make a difference to the transformative teaching methods of educators and the lifelong learning process of students.

This study recommends that the MATSEC takes the findings from the syllabi analysis, including those ESD competences that were and were not featured, into consideration during the next syllabi review. This will aid in integrating ESD further into the next update of the syllabi. A further recommendation is the compilation and publication of a complimentary document with the MATSEC syllabi, aimed at educators to instil ESD values directly connected to the content laid out in the syllabi. This would encourage transformative education in the teaching of the values and issues related to sustainable development and the gaining of further skills to make a difference in student lives and the lives of others around them.

The area studied by this research still has a number of research gaps that future research could attempt to fill. The following are some recommendations. Further research could explore:

- The views of post-secondary geography students on the integration of ESD in their learning.
- ESD integration in a cross-curricular manner at post-secondary institutions.

The results obtained show that the Advanced and Intermediate level syllabi both teach several universal competences. The analysis highlighted understanding issues of sustainability, particularly UNECE ESD competences such as, *Know the causes of unsustainable development, Know how generations, rich and poor and humans and nature are interdependent* and *Critically assess processes of change in society and envision sustainable futures.* The syllabi also addressed different worldviews and perspectives, as well as allowing students to be *Connected to their local and global sphere of influence.* UNECE ESD competence skills were also analysed to be integrated into the syllabi where students *Know the importance of problem setting, critical reflection, visioning, and creative thinking in planning the future and affecting change.*

References

Cassar R, Mifsud M (2023) Moving towards sustainability through geographical education in Malta. In: Filho WL, Salvia AL, Coathe B, Pallant E, Pearce K (Ed), Educating the sustainability leaders of the future. Springer

Cassar Cordina M (2015) Maltese primary school teachers: their perceptions towards education for sustainable development. [Master's thesis, University of Malta]. https://www.um.edu.mt/library/oar/handle/123456789/52046

Flick U (2009) An introduction to qualitative research. Sage Publications Ltd., London

Jóhannesson IÁ, Norðdahl K, Óskarsdóttir G, Pálsdóttir A, Pétursdóttir B (2011) Curriculum analysis and education for sustainable development in Iceland. Environ Educ Res 17(3):375–391. https://doi.org/10.1080/13504622.2010.545872

MATSEC Examinations Board, University of Malta (2019a) Syllabus (2019): Geography AM. https://www.um.edu.mt/__data/assets/pdf_file/0008/396260/AM_13.pdf

MATSEC Examinations Board, University of Malta (2019b). Syllabus (2019) Geography IM. https://www.um.edu.mt/_data/assets/pdf_file/0009/396234/IM_13.pdf

MATSEC Examinations Board, University of Malta (2019c) Syllabus (2019): Environmental Science IM. https://www.um.edu.mt/_data/assets/pdf_file/0005/396401/IM11.pdf

MATSEC Examinations Board, University of Malta (2019d) Syllabus (2019): Systems of Knowledge IM. https://www.um.edu.mt/_data/assets/pdf_file/0008/396242/IM_32.pdf

Mifsud M (2012) Environmental Education Development in Malta: a contextual study of the events that have shaped the development of environmental education in Malta. J Teacher Educ Sustain 14(1):52–66. https://doi.org/10.2478/v10099-012-0004-x

Ministry of Education: Malta (1999) Creating the future together: national minimum curriculum. http://www.curriculum.gov.mt/docs/nmc_english.pdf

Ministry of Education and Employment: Malta (2012) A national curriculum framework for all 2012. https://education.gov.mt/en/Documents/A%20National%20Curriculum%20Framework%20for%20All%20-%202012.pdf

Rieckmann M (2012) Future-oriented higher education: which key competences should be fostered through university teaching and learning? Futures 44(2):127–135

Rieckmann M (2018) Chapter 2—learning to transform the world: key competences in ESD. Academia.edu. https://www.academia.edu/35941134/Chapter_2_Learning_to_transform_the_world_key_competences_in_ESD

UNECE (2012) Learning for the Future: Competences in Education for Sustainable Development. Geneva: UNECE. www.unece.org/fileadmin/DAM/env/esd/ESD_Publications/Competences_Publication.pdf

UNESCO (2014) UNESCO roadmap for implementing the global action programme on education for sustainable development. UNESCO, Paris

Wals AEJ (2014) Sustainability in higher education in the context of the UN DESD: a review of learning and institutionalization processes. J Clean Prod 62:8–15. https://doi.org/10.1016/j.jclepro.2013.06.007

Whitby A (2019) Advancing education for sustainable development. Key success factors for policy and practice. https://www.worldfuturecouncil.org/wp-content/uploads/2019/01/Handbook-ADVANCING-EDUCATION-FOR-SUSTAINABLE-DEVOLOPMENT-by-Alistair-Whitby-WFC_2019.pdf

Wiek A, Withycombe L, Redman CL (2011) Key competences in sustainability: a reference framework for academic program development. Sustain Sci 6(2):203–218

References

Ecological Design as Framework of the Sustainable Systems (MS3) Curriculum: A Historical Review of Achievements, Challenges and Community Outreach in Architecture for Sustainability

Bruno Borsari

Abstract This work aims at recognizing contributions of ecological design projects when these were applied to the construction, or remodeling of buildings, including the campus landscape, at the Robert A. Macoskey Center for Sustainable Systems Education and Research (RAMC), of Slippery Rock University of Pennsylvania. This is a 33.6 hectares facility on the eastern edge of the university campus, that since 1990 has been offering a plethora of academic courses, community events and similar programs to spur education in sustainability. Keystone features of this living laboratory are its two buildings: Harmony House and adjacent timber-frame barn. The construction of the barn and four, main renovation projects at the house were driven by students in the MS3 program under their professors' guidance, between 1991 and 2004. These endeavors remain tangible historical records of this unique master's curriculum, its passionate community of learners and the university support for an ecological design approach that fostered experiential learning for all students (especially, those enrolled in the built-environment track), of the MS3 program. The buildings demonstrate architectural versatility and applications of the regenerative principles of sustainable design. These included using natural materials and alternative construction methods, whereas participatory team design approach was adopted as a distinctive curriculum emphasis.

Keywords Architecture · Built environment · Climate change · Ecological design · MS3 curriculum · Outreach · Sustainable systems

---

B. Borsari (✉)
Department of Biology, Winona State University, Winona, MN 55987, USA
e-mail: bborsari@winona.edu

W. Leal Filho et al. (eds.), *The Contribution of Universities Towards Education for Sustainable Development*, World Sustainability Series,
https://doi.org/10.1007/978-3-031-49853-4_9

1 Introduction

Ecological design is a well-known paradigm that for forty years in the U.S. has been influencing architecture and closely related disciplines (e.g.: from interior to industrial design, environmental, civil and more specialties within engineering), to consider studying living beings and their communities, for possible applications in the design process of built environments and associated infrastructures (Todd and Josephson 1996). Also, biologist E.O. Wilson's concept of biophilia has been valued for decades in conservation and restoration efforts, while inspiring designers to study nature because aspects of organismal autecology can become effective components of architectural and engineering constructions (Van Der Ryn and Cowan 1996). In addition to this, biomimicry is another concept that together with biophilia evolved into extraordinary notions, that when adopted in sustainable architecture, boost the resiliency and adaptability of built structures, even when these are constructed in uncongenial spaces (Pawlyn 2016). However, Bennett (2017) pointed out that biophilia applies more directly to urban, interior design and architecture, whereas biomimicry is a more widely used concept in technology and product-making. In essence, both concepts derive from studying biology where biomimicry recognizes the innovation potential of technologies that have already been tested by nature, whereas biophilia acknowledges the benefits for people who maintain strong ties and exposure to nature.

The relevance of ecological design can be enormous for sustainable development and most relevant for climate change education (CCE), to the point that failure of achieving sustainability is often due to poor design. For example, the oil crisis of the early 1970s was a wake-up call for architects and engineers because only at this time they had to begin reckoning with the environmental costs of construction projects and building materials, in addition to the use of resources, waste production and energy needs, to bring these to completion (Van Der Ryn and Cowan 1996). Concurrently, a steady growth of the human population, which by the middle of this century could expand to ten billion, has determined drastic changes to landscapes and expansions of urban environments swelling worldwide, the footprint of megacities (Rockström et al. 2021). Orsini et al. (2013) estimated that the 55% of the human population lived in urban environments in 2020 and that by midcentury this percentage may grow to 70%. Urban development requires massive amounts of resources necessary to build and maintain the needed infrastructure (power plants, roads, gas, water and sewage lines, buildings), making the reductionist design approach that humanity inherited from the industrial revolution, unsustainable and oblivious of its ill effects on global climate change. Buckminster Fuller's "dymaxion" concept (the idea that rational action in a rational world demands the most efficient performance per unit of input), materialized into geodesic structures that demonstrated already in the 1960s a more sustainable design, according to the knowledge of construction technology in those years (Buckminster Fuller and Marks 1973). More extreme changes have been occurring to landscapes surrounding the urban environments that continue to be converted into industrial farms, to produce the food needed by growing multitudes of city dwellers (Foley et al. 2011; Hooke et al. 2012). These fast conversions of ecosystems are the outcome of human activities, that continue acting as massive forces, comparable to those causing continental drifts and similar geologic events (e.g.: volcanic eruptions, earthquakes, tsunamis), increasing the rate of pollution, soil erosion and biodiversity loss (Wilting et al. 2017), that threaten more than ever to jeopardize human life as we know it. Therefore, there is a compelling need of constraining human activities within the physical boundaries of planet Earth as Rockström and his collaborators (2009) recommended, so that the uncertainties brought about by global climate change in the Anthropocene may become more predictable and controllable.

In the early 1970s and before further expansions of present planetary crises, at Slippery Rock University of Pennsylvania (SRU), the need for developing more sustainable living styles arose from a popular environmental ethics course (later called: Philosophy and Alternative Futures), that was developed and taught by professor Robert A. Macoskey (Anderson 2001). From this experience Macoskey envisioned the Alternative Living Technology and Energy Research (ALTER) project as an inclusive community of individuals in western Pennsylvania, interested in studying and practicing "natural alternatives to food and energy production, while tailoring those technologies and ideas to the bioregion" (Macoskey 1988, p. 2). This idea materialized into a proposal, which approved by the university became established on land available on the SRU campus (33.6 ha.), in the late 1980s. This property included a residential center (Harmony House), that later became the Robert A. Macoskey Center for Sustainable Systems Education and Research (RAMC). Also, a remarkable undertaking of the ALTER project evolved into developing a master's curriculum in sustainable systems (MS3 program). This work was led by the philosophy professor in collaboration with, ALTER members and more SRU faculty, so that in the fall of 1990 the graduate program began enrolling students, despite the unfortunate and premature demise of professor Macoskey, who passed away on May 7 of the same year (Anderson 2001). This chapter contribution is a tribute to the pioneering vision and work of professor Macoskey, ALTER members, students, faculty, and graduates of the MS3 curriculum. The inclusiveness of this program as envisioned by its leader, brought together a community of brilliant minds for 15 years, in the first, multidisciplinary plan of studies in sustainability education. The latter was framed within a theory of ecological design and permaculture, which infused with this education philosophy all the three academic emphases (agroecology, built environment, natural resource management), offered by the curriculum. This work presents also selected projects from the built environment track to demonstrate an alignment of this academic program to sustainable architecture and to the 17 sustainable development goals (SDGs) of agenda 2030.

2 The MS3 Curriculum and Its Built-Environment Track

The Master of Science in Sustainable Systems (MS3) curriculum at Slippery Rock University of Pennsylvania was a >30 credit program charged with preparing its graduates to better face pressing environmental and social challenges while, using principles of sustainability as an underlying framework for action. This program offered a core set of mandatory courses and three emphasis tracks (Agroecology, Built Environment, Sustainable Resource Management), which allowed students to acquire extra knowledge and skills in either one of these professional fields (Table 1).

The built environment track was a keystone emphasis of the curriculum in sustainable systems, as demonstrated by the two design courses that were part of the core courses of this program. These were strongly inspired by the tenets of ecological design, inclusive of permaculture principles and their applications to sustainable living (Mollison 1999). Knowledge about design for the built environment was employed also in the landscape surrounding the RAMC, either to grow food, or to manage open space and its resources (e.g.: soil, water, timber, wildlife). Ultimately, whether a student was interested in pursuing this curriculum track or not, an emphasis on ecological design remained ubiquitous and overarching in most study subjects that were offered by this program. This education philosophy inspired the learning, the thinking and the actions of students interested in developing sustainable communities and living styles, as envisioned by Dr. Macoskey and his early collaborators.

Ecological Design, Permaculture and their Educational Role for Sustainable Development

Permaculture was considered the harbinger of ecological design in the MS3 curriculum. This concept was conceived as an environmentally, functional paradigm

Table 1 List of courses in the sustainable systems (MS3) curriculum

Professional Core Courses	Track
• Design for Sustainable Landscapes • Foundations of Sustainability • Applied Ecology • Ecosystem Management • Design Graphics and Problem Solving • Soils as a Resource • Research Methods • Thesis, or Internship option	**Agroecology** • Sustainable Agriculture Techniques • Plant and Animal Husbandry • Soil Fertility Considerations
	Built Environment • Design and Resource Development for Energy Conservation • Healthy Building Systems and Materials • Energy and Engineering for Sustainable Systems • Site and Building Feasibility Studies
	Sustainable Resource Management • Open Space Planning • Sustainable Forest Management • Restoration Ecology

of developing permanent, agriculturally productive systems, where the built environment serves as human shelter (Mollison 1999). Yet, its addition to the landscape is harmonious in the design and contributive to sustainability (Hemeneway 2015). The construction materials derived always from resources already available on-site, or from locally available sources (Kobet 1993; Welton 2002). Also, the design process was inclusive of vernacular knowledge in architectural techniques and technologies of western Pennsylvania (Leininger et al. 2004). Through ecological design the curriculum was solidly embedded in a philosophy of community building and ecocentrism that are pivotal principles for an education in sustainable development (Borsari 2012).

Ecological design and permaculture foster equal participation in the process in which the technical aspects and constraints of design ideas are shared, in a communal understanding of problems that demand for solutions through a new, innovative design. Henfrey (2017) added that permaculture education is learner-focused, inclusive and environmentally conscious. It fosters individuals to find and activate their skills, in ways that enables people's realization of their potential in fulfillment of societal and ecological needs. Therefore, within a framework of ecological design and permaculture, sustainable development becomes a cultural process where everyone can participate because we humans make decisions every day and these, shape the future of others, including our own future (van Der Ryn and Cowan 1996). The MS3 curriculum complied with this approach to education and learning since its beginning, enmeshing in its holistic thinking all its academic courses, research projects and community activities. This strong ethical 'know-how' is the *anthropoharmonic phronesis* embodied by the wisdom of permaculture (Hathaway 2015), that pervaded the whole curriculum. Through this pedagogy the sustainable systems program was ahead of its time, urging its host university to be more permissive of higher levels of sustainability inclusions across all other curricula. This strong interdisciplinary focus envisioned an expansion of the graduate program into an undergraduate curriculum in sustainable development. This effort would have assisted also with students' retention and recruitment for the MS3 program. Obviously, those times were premature for engaging more deeply the whole institution into a transformative education reform in support of sustainability, as described by Borsari and Mora (2020). Nonetheless, the emphasis towards sustainable development of the built environment track of the MS3 curriculum remained the leading and steady force of this graduate program, throughout its 15-year lifespan.

3 A Brief History of Robert A. Macoskey Center for Sustainable Systems Education and Research (RAMC)/Harmony House

The Robert A. Macoskey Center for Sustainable Systems Education and Research (RAMC) is a 33.6 ha. homestead adjacent to the east of the SRU campus that had been donated to the university in the 1960s by the Patterson family (Vivas Macoskey, personal communication, August 29, 2000). The initial plan was to establish a golf course on this land yet, the property remained unused for more than 20 years when professor Macoskey envisioned this as the site for the realization of his ALTER project (Fig. 1).

The diversity of this landscape and the farmhouse (including their vicinity to the university campus), were ideal conditions to accommodate the needs that had been deemed necessary for the MS3 curriculum, community extension programs and research (Anderson 2001). The restoration design that was envisioned for the Patterson homestead laid the foundation for developing the site as a model of permaculture design for western Pennsylvania (Kobet 1993). The space near the house would accommodate gardens, a diversified, multistoried orchard, a storage building, and more elements, demonstrating the process of integrated permaculture/ecological design in self-sustaining food, fiber, recycling, and energy producing systems (Fig. 2). The space farther away from the house would demonstrate more sustainable land uses like crops cultivation (including silviculture, agroforestry), large scale composting and more garden plots available to community members.

The landscaping was accomplished between 1986 and 1988 however, more substantial funding was needed to undertake the very expensive renovations of the old farmhouse. After several, unsuccessful attempts, a $ 55,000 grant was received by the Pennsylvania Energy Office in the spring of 1989. This award was matched by the university with an additional $ 25,000 and a donation of $ 5,000 by a Pennsylvania senator the same year. These funds sufficed for the planned renovations at the house, which included the construction of the passive solar greenhouse, the masonry stove, and the photovoltaic array (Dale Stewart, personal communication, October 4, 2000).

4 Harmony House

The iconic building at the Macoskey Center (RAMC) is the Harmony House, the site's original Patterson's family farmhouse that was renovated in 1989–1990 and transformed into a residential laboratory/classroom with living quarters for two MS3 students. This building beautifully integrated the academic knowledge from the MS3 curriculum with the practical experience gained during the restoration of the Harmony House. The landscape surrounding this building was restored through a permaculture approach, which included elements of ecological design and engineering

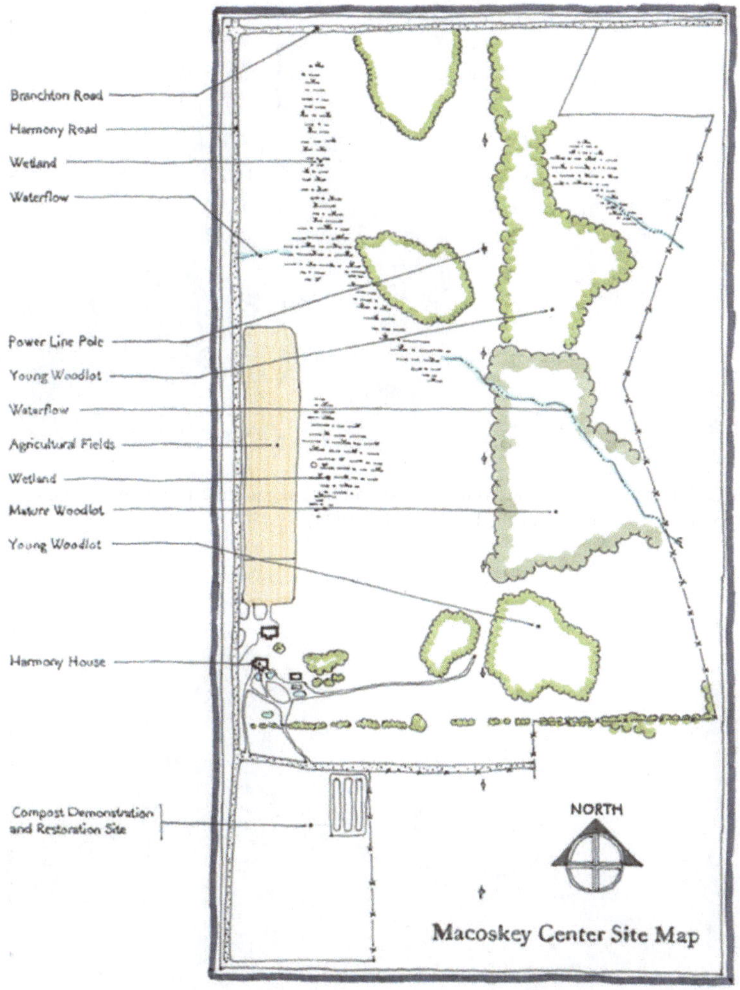

Fig. 1 Permaculture design plan for the MC/Harmony House, which includes space for the barn, gardens and orchard (Kobet 1993)

as recommended by Odum (1989). Within the boundaries of this design framework every component served multiple functions and in reciprocity, every function supported multiple components (Mollison 1999). Therefore, it is impossible to sever the components of the Harmony House from the whole system (house-landscape-people). Instead, it is essential to describe each component, as each one relates to the functioning of the whole system in tune with the space allocated for the planned design and nature (Fig. 3).

The participatory execution of this project together with academic learning in sustainability skills that were required to complete successfully the MS3 program,

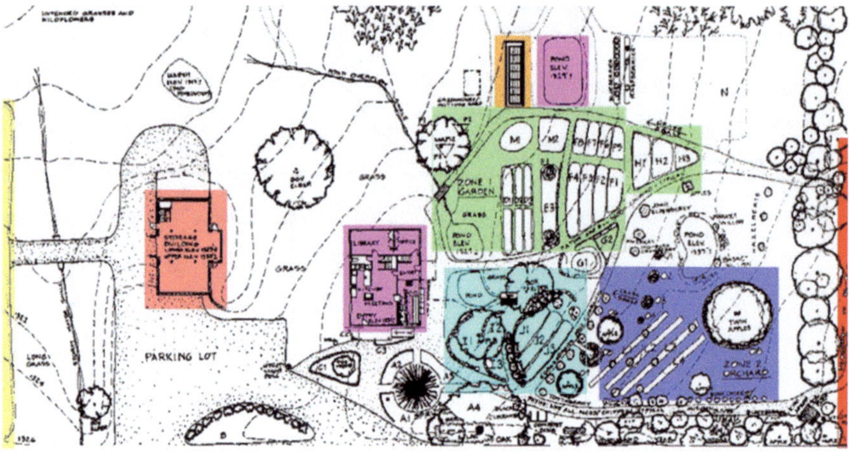

Fig. 2 MC/Harmony House view from the food production gardens. The photovoltaic system (on the right) produces electricity for the house (Kobet 1993)

Fig. 3 Harmony House view from the food production gardens with photovoltaic system (on the right), that produces electricity for the building

demonstrated the didactic excellence of this pioneer, graduate curriculum in sustainability (Kobet 1993). Thus, the Harmony House inspired a desire to connect the human shelter with the cycles of nature by harnessing energy from renewable sources like sun, earth, wind, water, so that the building became part of the cycling occurring within nature, as in a biogeochemical cycle (Odum 1971).

5 The Barn

The barn is located north of the Harmony House and is the product of an MS3 student's thesis. Its construction needs were legitimized by the idea of having a storage shelter for agricultural and gardening tools, in addition to space to host community events, workshops and similar programs. A timber framing workshop organized in 1992 at the Harmony House sparked the interest of a student to design and construct this building, while being in pursuit of the built environment track of the MS3 program. The barn aimed at showcasing a variety of sustainable building concepts according to vernacular architecture approaches (Tawayha et al. 2019). These were achieved through an employment of locally produced lumber like red oak (*Quercus rubra*) and hemlock (*Tsuga canadensis*), recycled slate for its roof, and straw-bale walls with recycled windows (Anderson 2001).

Also, the idea for this project was intended to employ the selected study topic as a vehicle to direct research into a feasible demonstration of alternative building materials and methods, as part of an inclusive education in climate change and sustainability for everyone, at Slippery Rock University. Workshops, presentations and hands-on activities were offered on topics such as: surveying, excavation, soil identification, timber framing, reinforced concrete foundation construction, clay straw and straw-bale wall construction, stucco applications and slate roofing, while the student completed the site feasibility study and designed the barn (Leininger et al. 2004). These activities attracted and engaged hundreds of interested individuals from other universities and groups, yielding articles and press releases that gave visibility to the MS3 curriculum and SRU. These complementary programs to the barn planning phase also brought to campus experts in the field of ecological design and construction. Many of these professionals shared their knowledge and experience with more seminars and workshops that contributed effectively to participatory learning in the design process, while setting the pace for the construction of the building (Fig. 4).

As Leininger (2004) reported, the participatory and team-oriented approach to the barn design was a unique learning experience, despite the challenge of complying with many demanding construction standards as demanded by ecological design principles and approaches. This process became an integral, growth component that was undertaken concurrently by the MS3 curriculum, the Harmony House and the ALTER project. The supervision and management of structures and policies at the Harmony House posed additional challenges to the process of designing and constructing the storage space however, these and similar disputes never hijacked the project goals, from their originally intended trajectory.

Fig. 4 Straw-bale, timber-frame barn with recycled slate roof at RAMC

6 Clivus Multrum Composting Toilet

When guests at the Harmony House use the bathroom facilities and prop open the toilet seat they are looking into a darkened bin, finding themselves in a quandary when they realize that there is no water tank, nor a handle with it to flush the toilet. That is because the house toilet called Clivus Multrum is not a waste disposal device, but rather a resource (humanure), making unit. In fact, the Clivus stores nutrients like nitrogen, phosphorous, potassium in human waste as they are undergoing composting aerobically, by a plethora of microorganisms added with the feces and sawdust (carbon source), that feeds them (Jenkins 1999). In addition to converting human waste into humanure, the Clivus saves on average 22,712 L of water every year, from becoming polluted thus, demonstrating a viable pedagogy to overcome fecophobia, while fostering awareness in water conservation practices (Borsari and Borsari 2021). This system operates with minimal electric power to run a modest fan in the collecting receptacle that maintains aeration, which together with adequate temperature are the two important conditions that trigger and maintain the composting process. No other inputs are needed to manage the Clivus. Two toilet chutes and a kitchen chute supply the waste to the Clivus chamber, located in the basement of the house (Fig. 5). Water vapor (H_2O) and carbon dioxide (CO_2) are the main outputs of composting, which are vented outside, through the Clivus ventilation system. Microbial activity reduces the volume of the biomass in the composting chamber and the mature compost accumulates to the front of it, in the collection bin, where it will be removed (every six months) and used as soil amendment for forestry, or ornamental plants grown at the RAMC.

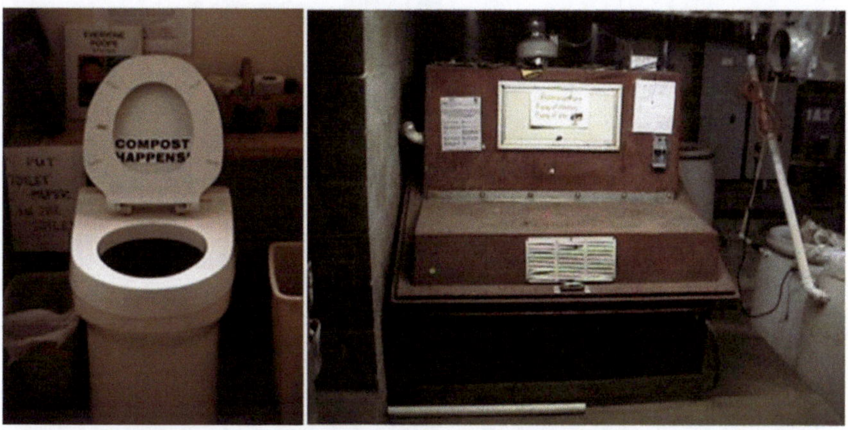

Fig. 5 Composting toilet (Clivus Multrum). The photo on the right is the collecting tank for the human feces, urine and sawdust where composting occurs. This is located in the Harmony House basement, under the waterless toilet (Borsari and Borsari 2021)

7 Firewood Stove

Winters in western Pennsylvania can be cold and linger into late spring thus, making the stove a necessary device to amplify the heating granted by the Harmony House, passive solar design. Wood and stones are abundant natural resources in this region making, the masonry stove become a sustainable choice for heating this building. The design of the thermal mass fire stove, known as "masonry heater" in the U.S., was adopted for this project (Fig. 6). The benefits of this device consist in the heat being stored in the stones (beside the one generated by combustion) of the small, iron fire chamber. Stoves like this one are very effective (90% combustion efficiency), and limited is their pollution caused by smoke, creosote, or solid particulates because combustion temperature may reach 1,093 °C thus burning thoroughly, most particulate matter and gases (Wilson 2012).

The efficient function of the stove depends on giving careful attention to the thermal mass concept. A concrete foundation was necessary to sustain this structure that together with chimney and pier pads weighs 7.7 tons. Besides its substantial weight the masonry stove needs to be lighted 2–3 times a day as compared to a typical woodstove where a continuous fire may be maintained for the entire winter season and revived from its embers, every day (Vivian 1992). Also, the reduced size of the fire box requires small wood pieces for this kind of stove. A small oven for baking was included in this design to optimize the energy produced during combustion and radiated by the stones. Two cords (1,814 kg.) of hardwood fuel may be burned

Fig. 6 Masonry wood stove at the RAMC, Harmony House

annually to heat the Harmony House. Tree species like willow (*Salix spp.*), poplar (*Populus spp.*), black locust (*Robinia pseudoacacia*), are grown at the RAMC for fuelwood, habitat restoration, crafts and building materials.

8 Greywater Recycling System

The southside of the Harmony House has a built-in greenhouse space that accommodates the greywater recycling system. The water theme has always been a pivotal study topic in the MS3 program because life depends on it! When learning that wastewater from a typical family home in the U.S. yields both black water from flushing toilets and greywater, from appliances, sinks, or showers, something had to be done to find a solution for these problems. Morris (2007) conceded that government regulators consider wastewater a major threat to public health due to its likelihood for spreading waterborne diseases and parasites, making it mandatory for waste waters to be disposed through underground sewage, or septic systems. However, these 'modern' systems are not free from limitations because they isolate wastewater from ecological processes, thus obstructing a decomposition of organic residues (Jenkins 1999). In fact, these continue to be toxic because they are separated from water at the wastewater treatment facility to be disposed in landfills, instead of returning to the soil where they could nourish bacteria and similar microbiota like algae, protozoa, micro invertebrates to sustains all plants communities and ecosystems (Borsari and Borsari 2021).

For these reasons Welton (2002) studied the existing greywater system of three water purification beds, or artificial, micro wetlands, that one of his student predecessors had designed in the early 1990s. He replaced this system with a more versatile one, which consisted of vertical fiberglass cylinders, connected by plastic (pvc) pipes (Fig. 7).

The new design was inspired by the living machine prototype, which included a diverse assortment of organismal populations inhabiting the cylinders where the greywater flowed through, from prokaryotes to aquatic plants and eventually, fish species (Todd and Josephson 1996). This greywater recycling system was built with the assistance of a group of MS3 students and faculty at the RAMC of Slippery Rock University. This new system considered many other design ideas and their applications to restore water quality at various levels of sanitation for re-use (Welton 2002). Its design versatility is inclusive of expanding research in wastewater management for small-scale applications and was achieved by facilitating different, possible reconfigurations of the system. A preliminary assessment of its ecological efficiency demonstrated that the functionality and effectiveness of this design were encouraging to spur sustainability in greywater management and recycling at the Robert A. Macoskey Center (Welton et al. 2004). The design and building processes of this greywater system served also as a valuable hands-on experience for interested students who chose to assist with this project, thus fulfilling the curriculum objectives

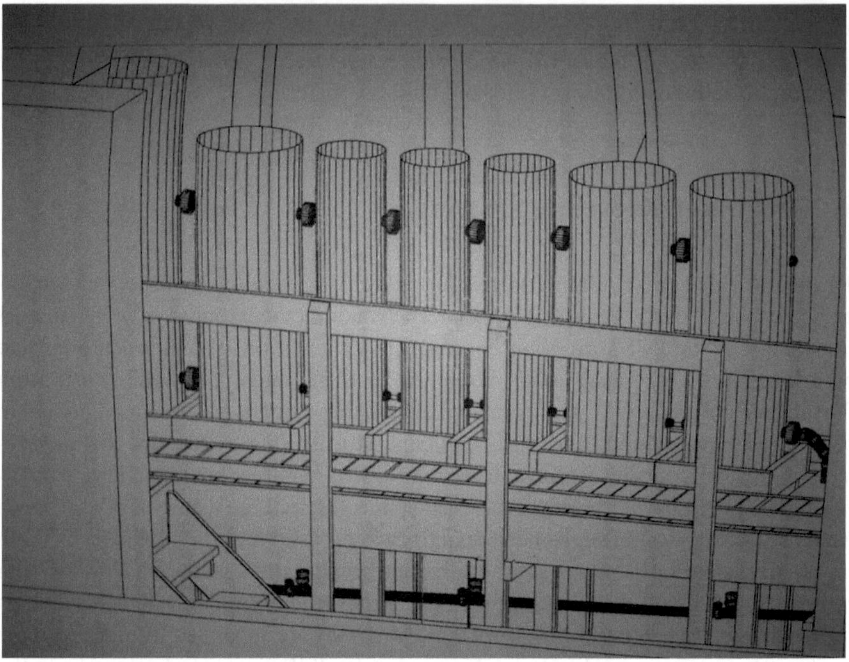

Fig. 7 Graywater system diagram with fiberglass tanks aligned in series to facilitate maintenance and system reconfiguration. This unit was built and maintained in the greenhouse space at the Harmony House (Welton 2002)

of the MS3 program and the envisioned service provided by the Harmony House, for research and education purposes.

9 Alternative Energy: Passive Solar Greenhouse & Photovoltaic System

Effective passive solar design involves the adoption of two important concepts: the collection of solar energy while passing through glazing materials and the storage of the same using thermal mass (Mollison 1999). When beginning renovation works Kobet (1993) observed that the house orientation was already well located to receive an adequate south-facing, sun exposure. Thus, to avoid a quick loss of heat energy, the greenhouse was designed to accommodate large windows whose function consisted in maximizing in-coming, solar radiation (Fig. 7).

In addition to providing a valuable source of heat, energy from the sun can also be harvested and converted into electricity through photovoltaic systems (Ivanko & Kivirist 2018). To offset the need for electricity that conventionally would have been generated through a combustion of non-renewable, fossil fuels, a 28-panel

Fig. 8 Photovoltaic array (1.5 kWh capacity) and solar reflection pond that maximizes exposure to solar radiation of the system through the seasons

photovoltaic (PV) array was installed in the vicinity of the Harmony House in 1991 (Fig. 8).

Although PV systems produce renewable energy, both the cost of such systems and the energy used to build their components are legitimate concerns that should be weighed carefully, prior to making similar decisions (Voas 1991). However, on-going research in this sector of renewable energy technology is demonstrating consistent improvements in efficiency in energy conversion, durability, and versatility of solar energy systems, in various scales of applications (Saghaleini et al. 2011; Mughal et al. 2018). A notable example among these is the solar powered cultivator that was another successful project accomplished by MS3 students and faculty for weed control at the RAMC gardens (Borsari et al. 2004).

10 Conclusion and Final Reflections

Sustainable development remains a multifaceted endeavor, nonetheless its pursuit through curricula and education in sustainability and climate change, remain compelling and most necessary (Borsari 2016), also to achieve excellence in education, as outlined by SDG#4 (Ferguson and Roofe 2020; Borsari and Mora 2020). Despite these efforts however, the architectural culture is not yet free of barriers neither quandaries for adapting to a multidisciplinary vision that supports and contributes to a sustainable development (Bonenberg and Kapliński 2018).

The ALTER project at Slippery Rock University was first in the U.S. to instill a passion for education in sustainability, that was accomplished through its MS3 curriculum. This pioneer master program (MSc.), was intended to prepare the next generation of leaders and professionals to resolve the many uncertainties brought about by extractive economies, excessive affluence, and the resistive inertia that to present days remains a constraint to sustainable development. This can be tackled through improvements in human activities that aim at lessening the sustainability deficit index, although the laws of thermodynamics impede an achievement of 100% sustainability, as Onwueme and Borsari (2007) demonstrated. Since 2015 the SDGs of agenda 2030 have been inspirational to set standards and a timeline for achieving sustainability. More guidelines for helping universities direct their efforts in pursuit of the SDGs more thoroughly, are available at present, also to delineate the significance about an attainment of these goals in teaching, research and management, at institutions of higher learning around the world (Leal Filho et al. 2021).

Modern architecture has been devoted for several decades to fulfill the needs of designing built environments that are more energy efficient, resilient, and capable of contributing in part to ecological services, aimed at improving quality of life and wellbeing among city dwellers. In this context, the experience of the University of Applied Sciences in Vienna, Austria is notable for having developed a curriculum with focus on green buildings, that since 2016 has been demonstrating an outstanding adaptability to modern environmental challenges, like climate change, in a timely fashion (Aichholzer et al. 2018). Also, vernacular approaches in architecture can be effective in achieving sustainability when these are employed in the design of new buildings and landscapes. Tawayha and his collaborators (2019) suggested that these strategies should be adopted in conjunction with contemporary design methods and knowledge to fulfill people's needs and expectations. What is lacking instead are research data (quantitative) in sustainable architecture, directed at measuring the quality of vernacular design, in addition to qualitative data that should focus on collecting responses from buildings users.

These data are most relevant to better understand how traditional design can contribute to satisfy human needs in sustainable built environments. Furthermore, the opinions of students of architecture are necessary as well to this pursuit because they will be the soon-to-become designers and decision makers of future built environment developments. A recent study about this topic indicated that most students of architecture regard sustainable development a keystone feature of their education however, the variance in students' views is wide-ranging with respect to sustainability outcomes in design courses, because these depend on the objectives set by each curriculum (Boarin et al. 2020). This knowledge suggests that a better understanding about the effects produced by outcomes in every course across curricula in architecture should contribute to the evaluation of teaching and learning approaches, from universities offering architecture and design programs against global trends. Therefore, generous administrative support should be granted for formative and summative curriculum assessments and adjustments, for curricula in sustainable architecture.

Finally, a dearth of works bridging ecological concepts with practical examples of ecosystem services design for a built environment emerged as a compelling need from a recent study, that explored key limitations for a more widespread consideration of biomimicry in architecture (Pedersen Zari and Hecht 2020). Hopefully, this chapter will suffice to counteract this claim, through the evidence provided by what has been accomplished by the MS3 curriculum, even though new paradigms always struggle to emerge as new ideas and worldviews, bring with them inevitable, cultural crises (Kuhn 1996). In this regard, one more example worth mentioning is the bat conservancy that was installed on the Slippery Rock University campus in 2004.

This project epitomizes an excellent endeavor in ecological design that blended an appreciation for the biology and ecological services provided by native chiropterans in western Pennsylvania, where these are hosted in a hibernaculum on campus, whose construction met also all the criteria for LEED certification (Doherty et al. 2005).

On a larger scale, recent trends in ecology have become supportive of "rewilding" ecosystems, which refers to conservation efforts aimed at restoring and protecting natural processes and wilderness areas through restoration approaches, requiring minimal management. Successful long- term rewilding projects can need little ongoing human attention, as successful reintroduction of keystone species creates self-regulatory and self-sustaining ecosystems (Perino et al., 2019). As rewilding may include provisions for connectivity among such areas, while protecting, or reintroducing top predators and keystone species, in a similar manner architecture education could expand its cultural boundaries to allow a full re-establishment of vernacular design principles and building materials, framed within a philosophy of ecological design and permaculture. This recommendation is supported by the success of the projects carried out by MS3 students and faculty, presented with this work. From these experiences it was learned that although the principles of architecture education may be universally taught yet, their applicability is limited by the ecosystem attributes where the design is going to be realized. In a similar fashion an architecture for sustainability should value the knowledge and wisdom of vernacular design to create built environments that are resilient and adaptive to a changing climate, as rewilded ecosystems, to the inevitable shifts and uncertainties that time will continue to bring.

From an expanded, pedagogical perspective, Leal Filho and team argued (2018) that, the sustainable development education construct has not yet become fully enmeshed with the idea of transformation at institutions of higher learning. This gap affected the MS3 curriculum that although well connected to the ALTER group and many members of local communities, it continued operating in isolation of other curricula available at Slippery Rock University. This limitation is suggestive that leadership and long-term, administrative support remain key elements to boost an appreciation for the inclusive, multicultural, and multidisciplinary vision that sustainability and an education for climate change embody. These are all intended as study topics, but also as inquiry paths for education and development of ecologically sound technologies (Borsari 2016), that are conducive to initiate the needed transformation at institutions of higher education (Wamsler 2019). These personal reflections converge also toward a re-examination need of ethics in education, as values remain essential for forging transformative potentials in students who, while preparing to become agents of change, can amplify support and mobilization for socio-environmental improvements, towards more sustainable futures. Through students' actions the path for sustainable development will become clearer of residual barriers when the transformation needs advocated in this work will have occurred within universities. At this point, institutions of higher learning will become fully capable to serve society. Their main contribution will consist in modeling social justice and championing stewardship, while continuing to expand an indefatigable passion for education in sustainable development, way beyond current knowledge limits and campus boundaries.

Acknowledgements The author wishes to dedicate this work to Dr. Steven Doherty (MS3 professor from 2001 to 2005), Mrs. Vivas Macoskey, members of the ALTER project, in addition to all students and graduates of the MS3 program for their enthusiasm and commitment to sustainability. Although the curriculum is gone, their projects and products remain available to learn about sustainable development at Slippery Rock University of Pennsylvania.

References

Aichholzer M, Fischer H, Holzl C, Osterreicher D, Pfleger MP, Spitzenberger E, Vill M, Ploch A (2018) Teaching and research on sustainable architecture at the university of applied sciences Vienna–FH campus Wien. J Green Build 13(3):158–178. https://doi.org/10.3992/1943-4618.13.3.158

Anderson CM (2001) Harmony homestead and the Macoskey center for sustainable systems education and research. In: Daniels MR (ed) Creating sustainable community programs. Examples of collaborative public administration. Praeger, Westport, CT

Bennett A (2017) Biomimicry versus Biophilia: what's the difference? Terrapin Bright Green. https://www.terrapinbrightgreen.com/blog/2017/02/biomimicry-versus-biophilia/#:~:text=Biomimicry%20is%20more%20heavily%20used%20in%20technology%20and,recognizes%20the%20innovation%20potential%20of%20life%E2%80%99s%20tested-and-true%20%E2%80%9Ctechnologies.%E2%80%9D

Boarin P, Martinez-Molina A, Juan-Ferruses I (2020) Understanding students' perception of sustainability in architecture education: a comparison among universities in three different continents. J Clean Prod 248:119237. https://doi.org/10.1016/j.jclepro.2019.119237

Bonenberg W, Kapliński O (2018) The architect and the paradigms of sustainable development: a review of dilemmas. Sustainability 10(1):100. https://doi.org/10.3390/su10010100

Borsari B, Elder T, Reynolds T (2004) Assessing the educational opportunities from a solar powered cultivator at Slippery Rock University of Pennsylvania. Int J Sustain High Educ 5(2):190–198

Borsari B, Borsari G (2021) Water use, fecophobia, and implications for public health. In: Leal Filho W, Azul AM, Brandli L, Lange Salvia A, Wall T (eds) Clean water and sanitation. Encyclopedia of the UN Sustainable Development Goals. Springer, Cham. https://doi.org/10.1007/978-3-319-70061-8_183-1

Borsari B (2016) Sustainability education as a curriculum emphasis. Acad Exch Q 20(4):31–37

Borsari B (2012) Curriculum framework for sustainability education. Acad Exch Q 16(1):74–78

Borsari B, Mora C (2020) Education reform through a systems approach for sustainable development. In: Leal Filho W, Azul A, Brandli L, Özuyar P, Wall T (eds) Quality education. Encyclopedia of the UN Sustainable Development Goals. Springer, Cham

Buckminster Fuller R, Marks R (1973) The dymaxion world of Buckminster Fuller. Anchor Books Press, South Shore, MA

Doherty S, Davis J, Borsari B, Butchkoski C (2005) An ecological engineering approach to pest management: Slippery Rock University's Bat Conservancy. In: Engineering sustainability conference. Next generation technology for green construction and sustainable water use. University of Pittsburgh, Pennsylvania, Pittsburgh, PA, 10–12 Apr 2005 (Extended Abstract)

Ferguson T, Roofe CG (2020) SDG4 in higher education: challenges and opportunities. Int J Sustain High Educ 21(5):959–975. https://doi.org/10.1108/IJSHE-12-2019-0353

Foley JA, Ramankutty N, Brauman KA, Cassidy ES, Gerber JS, Johnston M, Mueller ND, O'Connell C, Ray DK, West PC, Balzer C, Bennett EM, Carpenter SR, Hill J, Monfreda C, Polasky S, Rockström J, Sheehan J, Siebert S, Tilman D, Zaks DP (2011) Solutions for a cultivated planet. Nature 478:337–342

Hathaway M (2015) The practical wisdom of permaculture: an anthropoharmonic phronesis for moving towards an ecological epoch. Environ Ethics 37(4):445–463. https://doi.org/10.5840/enviroethics201537442

Hemeneway T (2015) The Permaculture City. In: Regenerative design for urban, suburban, and town resilience. Chelsea Green Publishing, White River Junction, VT

Henfrey T (2017) Permaculture education as ecology of mind: the head, hands and heart of transformation. In: Hall R, Winn J (eds) Mass intellectuality and democratic leadership in higher education, pp 171–183. Bloomsbury Academic, London, UK

Hooke R, Le B, Martín-Duque J, Pedraza J (2012) Land transformation by humans: a review. GSA Today 22(12):4–10

Ivanko JD, Kivirist L (2018) Living with solar and wind. Mother Earth News:14–19

Jenkins J (1999) The humanure handbook: a guide to composting human manure, 2nd edn. Grove City, PA, Jenkins Publishing

Kobet RJ (1993) The design and construction of Harmony House for use in the Master of Science in sustainable systems program at Slippery Rock University. Unpublished Masters Thesis, Slippery Rock University of Pennsylvania, Slippery Rock, PA

Kuhn TS (1996) The structure of scientific revolutions, 3rd edn. The University of Chicago Press, Chicago, IL

Leal Filho W, Raath S, Lazzarini B, Vargas VR, de Souza L, Anhlon R, Quelhas OLG, Haddad R, Klavins M, Orlovic LV (2018) The role of transformation in learning and education for sustainability. J Clean Prod. https://doi.org/10.1016/j.jclepro.2018.07.017

Leal Filho W, Frankenberger F, Lange Salvia A, Azeiteiro U, Alves F, Castro P, Will M, Platje J, Orlovic LV, Brandli L, Price E, Doni F, Mifsud M, Veiga Avila L (2021) A framework for the implementation of the sustainable development goals in university programmes. J Clean Prod 299:1–12. https://doi.org/10.1016/j.jclepro.2021.126915

Leininger CP, Borsari B, Doherty S, Stewart D (2004) Ecological design process for the construction of a storage building at slippery rock University of Pennsylvania, pp 60–68. In: Onwueme I, Fisher G (eds) Sustainability: diverse perspectives for the path ahead. Proceedings of the Richard Alsina Fulton conference on sustainability and the environment, Wilson College. Chambersburg, PA, 26–27 Mar 2004

Macoskey RA (1988) The why of it. The Alternator 2(5):1–3

Mollison W (1999) Permaculture: a designer's manual. Tagari Publications, Tyalgum, NSW, Australia

Morris RD (2007) The blue death: disease, disaster and the water we drink. Harper Collins Publishers, New York

Mughal S, Sood YR, Jarial RK (2018) A review on solar photovoltaic technology and future trends. Int J Sci Res Comput Sci Eng Inf Technol 4(1):227–235. ISSN: 2456-330

Odum HT (1989) Ecological engineering and self-organization. In: Mitsch WJ, Jorgensen SE (eds) Ecological engineering: an introduction to ecotechnology. John Wiley & Sons, New York, NY, pp 79–102

Odum EP (1971) Fundamentals of ecology, 3rd edn. W. B. Saunders Co., Philadelphia, PA

Onwueme I, Borsari B (2007) The sustainability asymptogram: a new philosophical framework for policy, outreach and education in sustainability. Int J Sustain High Educ 8(1):44–52. https://doi.org/10.1108/14676370710717580

Orsini F, Kahane R, Nono-Womdim R, Gianquinto G (2013) Urban agriculture in the developing world: a review. Agron Sustain Develop. https://doi.org/10.1007/s13593-013-0143-z

Pawlyn M (2016) Biomimicry in architecture, 2nd edn. RIBA Publishing, Newcastle upon Tyne, NE1 1RH United Kingdom. ISBN: 978-1-85946-628-5

Pedersen Zari M, Hecht K (2020) Biomimicry for regenerative built environments: mapping design strategies for producing ecosystem services. Biomimetics 5(2):18. https://doi.org/10.3390/biomimetics5020018

Rockström J, Gupta J, Lenton TM, Qin D, Lade SJ, Abrams JF et al (2021) Identifying a safe and just corridor for people and the planet. Earth's Future 9:e2020EF001866. https://doi.org/10.1029/2020EF001866

Rockström J, Steffen W, Noone K, Persson Å, Chapin III FS, Lambin E et al (2009) Planetary boundaries: exploring the safe operating space for humanity. Ecol Soc 14(2):32. http://www.ecologyandsociety.org/vol14/iss2/art32/

Saghaleini M, Kaviani AK, Hadley B, Mirafzal B (2011) New trends in photovoltaic energy systems. In: 10th International conference on environment and electrical engineering, pp 1–4. https://doi.org/10.1109/EEEIC.2011.5874728

Tawayha FA, Braganca L, Mateus R (2019) Contribution of the vernacular architecture to the sustainability: a comparative study between the contemporary areas and the old quarter of a Mediterranean City. Sustainability 11(3):896. https://doi.org/10.3390/su11030896

Todd J, Josephson B (1996) The design of living technologies for waste treatment. Ecol Eng 6:109–136

Van Der Ryn S, Cowan S (1996) Ecological design. Island Press, Washington, D. C.

Vivian J (1992) Guide to wood and coal stoves. Mother Earth News. https://www.motherearthnews.com/homesteading-and-livestock/wood-burning-stoves-zmaz93djztak/

Voas S (1991) Harmony House offers living without harm. Pittsburgh Post-Gazette

Wamsler C (2019) Education for sustainability. Fostering a more conscious society and transformation towards sustainability. Int J Sustain High Educ 21(1):112–130. https://doi.org/10.1108/IJSHE-04-2019-0152

Welton S (2002) A renovated system design for greywater treatment at the Robert A. Macoskey Center: approach and assessment. Master's Thesis, Slippery Rock University of Pennsylvania, Slippery Rock, PA

Welton S, Borsari B, Doherty S, Kefeli V, Reynolds T (2004) Graywater treatment at the Robert A. MacoskeyCenter: assessment and approaches to a renovated system design, pp 35–46. In: Onwueme I, Fisher G (eds) Sustainability: diverse perspectives for the path ahead. Proceedings of the Richard Alsina fulton conference on sustainability and the environment, Wilson College, Chambersburg, PA, 26–27 Mar 2004

Wilson A (2012) Masonry heaters burn hot and clean. Green Building Advisor. https://www.greenbuildingadvisor.com/article/masonry-heaters-burn-hot-and-clean

Wilting HC, Schipper AM, Bakkenes M, Meijer JR (2017) Quantifying biodiversity losses due to human consumption: a global-scale footprint analysis. Environ Sci Technol 51:3298–3306

Environmental Sustainability in Traditional University Libraries: Contributions from BIM/BES Integration to the Thermal Performance and Books Conservation

Carolina Fernandes Vaz, Ana Carolina Fernandes Maciel, Iago Piccello, and André Luís de Araujo

Abstract As a spreader of knowledge, universities must be practical examples of sustainable applications, including building or intervening in their buildings. However, some authors point out that universities focus more on the social dimension of sustainability than the environmental one, emphasizing the need for research aimed at environmental sustainability in these places. In this context, university campuses present emblematic buildings with historical and architectural relevance with more than 20 years old and still need analyses in terms of thermal performance. Among these traditional buildings on university campuses are the libraries, usually proposed in open floor plans, with spaces for collaborative study and an integrated collection of books. However, both spaces must meet different criteria of thermal performance and conservation of books, which is associated with the internal microclimate and consequently with the adequacy of climatic conditions. The main objective of this research was to analyze the environmental sustainability in iconic libraries through transdisciplinary procedures, where the thermal performance and conservation of books from the Library of the Federal University of Uberlândia (UFU Library), Brazil, were jointly evaluated. The analyzed spaces have open floor plans with natural ventilation, a communal study area, and a collection of related books for consultation. The method adopted was the hygrothermal simulation

C. F. Vaz (✉) · I. Piccello · A. L. de Araujo
Faculty of Architecture, Urban Planning and Design of the Federal University of Uberlândia, Uberlândia, Brazil
e-mail: carolina.vaz@ufu.br

I. Piccello
e-mail: iagopiccello@ufu.br

A. L. de Araujo
e-mail: andre.araujo@ufu.br

A. C. F. Maciel
Faculty of Civil Engineering of the Federal University of Uberlândia, Uberlândia, Brazil
e-mail: anamaciel@ufu.br

W. Leal Filho et al. (eds.), *The Contribution of Universities Towards Education for Sustainable Development*, World Sustainability Series,
https://doi.org/10.1007/978-3-031-49853-4_10

based on the BIM/BES data transfer, having as parameters the operating temperature ranges according to (ASHRAE 55 (2020) Thermal environmental conditions for human occupancy. ASHRAE, Atlanta) and the isoperm method of (Sebera in Isoperms an environmental management tool. Comission on Preservation and Access, Washington (1997)). We concluded that the hygrothermal simulation contributes to environmental sustainability since it allows the analysis of the current scenario of the library, as well as the creation of new scenarios. Regarding the thermal performance, we observe that the appropriate scenario is the current one, as it presented service between 82.15% and 86.88% of the occupied periods throughout the year. The proposed scenario with double glazing did not present significant temperature differences concerning the current scenario. Regarding the book conservation, we concluded that the current scenario is unsatisfactory, with isosperms between 0.28 and 1. That corresponds to the longevity of the paper between 12.6 and 45 years. Unfortunately, the data transfer between BIM/BES could have been finer since there were errors in the transfer of geometry, thermal properties, and information such as usage, occupation, and scheduling.

Keywords Simulation-based architectural design · Campus architecture · Library · Building information modeling · Building energy simulation

1 Introduction

After World War II, with the spending on energy and the oil crisis in the 1970s, it became necessary to invest in innovations for energy efficiency in several areas, including buildings and constructions (Mitchell 2010; Papadopoulos 2016). In the same period, the first conferences about energy and environment at a global level emerged, such as the Stockholm Conference in 1972 and the Brunktland Commission in 1980, which corroborated the concept of sustainability (Keeler and Burke 2009). In the 1990s, the first BES (Building Energy Simulation) software for energy simulations rose to assist the decision-making in the design process, which helped in sustainable design alternatives (Crawley et al. 2000).

In 2015 the 2030 Agenda was created with 17 sustainable development goals to be achieved globally by 2030. In this sense, universities as a spreader of knowledge must be practical examples of sustainable applications, including building or intervening in their buildings. However, Leal Filho et al. (2021) point out that universities focus more on the social dimension of sustainability than the environme'ntal one, emphasizing the need for research aimed at environmental sustainability in these places. In this context, several University Campuses have iconic buildings with historical and architectural relevance, which, mainly because they are more than 20 years old, have not been analyzed in terms of thermal performance. In addition, these buildings must be suited to functional realities and sustainable goals to be enjoyed today and preserved for future generations. Among the sustainable objectives, which include intervention in existing assets, is target 11.4, to protect the world's cultural and natural heritage, and 7.3, which values global energy efficiency. It should be noted that in addition to aesthetic and historical values, these buildings must meet various regulations and their peculiarities, and according to (Keeler and Burke 2009), sustainable projects must be, above all, integrated, that is, they must have a holistic view and multidisciplinary design solutions.

Among these traditional buildings on university campuses are the libraries, usually proposed in open floor plans, with spaces for collaborative study and an integrated collection of books. However, both spaces must meet different criteria of thermal performance and conservation of books, which is associated with the internal microclimate and consequently with the adequacy of climatic conditions. For the analysis of existing libraries, some authors used energy simulation, as is the case of the research by Tronchin and Fabbri (2017), who analyzed the conservation of books in a historical library in Italy using the IES VE software, and the research by Waddicor et al. (2016) who analyzed the thermal and energy performance of a library in Italy in different scenarios through simulation with the IDA ICE software. In addition, BIM (Building Information Modeling) platform software, pointed out by authors such as Jalaei et al. (2020) and Haidar et al. (2018), is efficient for sustainable analysis and decisions based on parametric modeling.

In this sense, the main objective of this research was to analyze the environmental sustainability in iconic libraries through transdisciplinary procedures, where the thermal performance and conservation of books from the Library of the Federal University of Uberlândia (UFU Library), Brazil, were jointly evaluated. The analyzed spaces have open floor plans with natural ventilation, a communal study area, and a collection of related books for consultation. The method adopted was the hygrothermal simulation based on the BIM/BES data transfer, having as parameters the operating temperature ranges according to ASHRAE 55 (2020) and the isoperm method of Sebera (1997).

2 Recent Developments

Recent scientific research has sought to analyze library buildings, or even specific sectors of these buildings, in order to compare their thermal performance measured in loco with the limits proposed both by the standards: ASHRAE (2007) and EN 15757 (2010), and by other comparison índices (Silva and Henriques 2014). Recent measurements produced in more severe conditions of temperate climates, such as in Greece (Drougka et al. 2020), demonstrate how temperature and humidity, albeit restricted to summer, have proved to be the most critical threats to book conservation. Recent research in Italian libraries has established comparisons with different standards, demonstrating that 80% of the acceptable values concerning humidity (Andretta et al. 2016), but that, thus, the climatic conditions monitored typically in the summer could lead to the loss of the collections in 300 years (Verticchio et al. 2020). In the same region, recent estimates at the Biblioteca Classense in Ravenna showed that the predicted durability of the most recent books is about half that of

19th-century collections (Coppola et al. 2020). Diulio et al. (2015) and Diulio et al. (2019) observed similar results in libraries in Argentina. They presented better conditions in the autumn and spring seasons than UNI 10,829, concluding that the most influential parameters on thermal performance were the adjacency with other spaces, followed by the global transmittance and the transparency of the openings.

Also, having on-site measurement and comparisons with indexes as a fundamental scientific method, studies have verified inadequate conditions for the conservation of books and, therefore, have proposed architectural solutions to improve these conditions. In the European context, Sahin et al. (2017) and Boeri et al. (2022) have considered passive design alternatives without introducing heating or air conditioning systems. They found situations where archival assets were stored without any air conditioning system, producing microclimatic conditions to ensure their preventive conservation. In simultaneous searches in libraries in Portugal, Schito et al. (2019a) demonstrated the effectiveness of thicker walls in smoothing and delaying changes in the external climate but that the need to reduce relative humidity at specific and identified parts represents a limit for the production of excellent conditions Schito et al. (2019b). In the Asian context, Liu and Lu (2009) argued that although HVAC systems can be expensive, monitoring systems for temperature and humidity data are simple, reliable, and inexpensive solutions, which can decisively contribute to the selection of sustainable design strategies.

Compromising human comfort levels with thermal performance to conserve books is not a simple task. As evaluated by Krüger and Diniz (2011) in full-scale models reproducing three Brazilian library buildings in cities with different climates, there is a significant difference between the conditions of book storage, although there are similarities in the conditions of internal thermal comfort. Recently, Turhan et al. (2021) shared a similar inference that about 90% of library users in Turkey are satisfied with thermal sensation, while at the exact location, thermal performance results point to a high risk of degradation. In certain regions with some design alternatives, there may be a convergence between thermal comfort and book conservation, which raises the need for further studies on this subject.

Computer simulations aiming to delineate the condition of indoor climates have allowed relevant conclusions regarding the reduction of energy expenditure with cooling and heating in libraries. Simulations performed directly in BIM modeling applications have demonstrated potential tools for reducing building heating and cooling loads (Sadeghifam et al. 2021; Liu and Wang 2022). In applications aimed explicitly at thermo-energetic simulations, research with the IDA ICE (Equa, Inc.) has sought to reduce the computational effort through simplifications without significant distortions about more complex models (Verticchio et al. 2020) or with metamodels (Bracht et al. 2021; Amaral et al. 2022) leading to objective conclusions on reduced energy consumption (Waddicor et al. 2016). Likewise, the design driven by simulation models in IES VE (Integrated Environmental Solutions Limited, Inc.) clearly outlines the energy consumed by the environment (Hussin et al. 2019) and proposes allocating books according to conditioning criteria attested by the application (Tronchin and Fabbri 2017). In addition, analyses of the impact of specific systems on building energy consumption, such as lighting power density, could be more accurately outlined in eQUEST, Inc. (Song et al. 2015).

Considering that humidity and temperature are the variables that most interfere with the conservation of books, recent studies have revealed crucial impacts of air exchanges on indoor environments' humidification and dehumidification processes (Kupczak et al. 2017). In this sense, passive strategies such as controlling natural ventilation by opening doors and windows in the summer can represent a strategy that determines the collection's durability (Stoakes et al. 2011). Among the most critical factors to be considered as design strategies is the determination of the risks of books degradation by chemical, mechanical or biological means (Coşkun et al. (2017) and the definition of air renewal rates in line with thermal comfort and indoor air quality (Conceição et al. 2019).

3 Qualitative Sample: UFU Library, Brazil

The library of the Federal University of Uberlândia at Santa Mônica Campus (UFU Library) is located in Uberlândia, a countryside city in Brazil (18.92S, 48.26W), a tropical wet and dry climate in Köppen–Geiger climate classification system. Finished in 1991, designed by the architects Paulo Zimbres and Luís Antônio Almeida Reis, it is an iconic example of the campus and the city, with architectural and social relevance (Fig. 1).

According to Luís Antônio Almeida Reis, the designers propose natural ventilation, and subsequently, new environments were added and artificially conditioned. The spaces analyzed in the research (Fig. 2) are located in open floor plans with book collections and communal study areas interconnected and naturally ventilated. Among the passive strategies of the building are: shading provided by the ceramic cobogós that permeate the entire glazed facade, trees on the facades that receive sunlight, cross ventilation through large openings and atriums, as well as the chimney effect provided by several dormer vents. The collection of the analyzed

(a) (b)

Fig. 1 **a** UFU Library overview, **b** second-floor view with spaces for collaborative study and books storage. (Authors 2022)

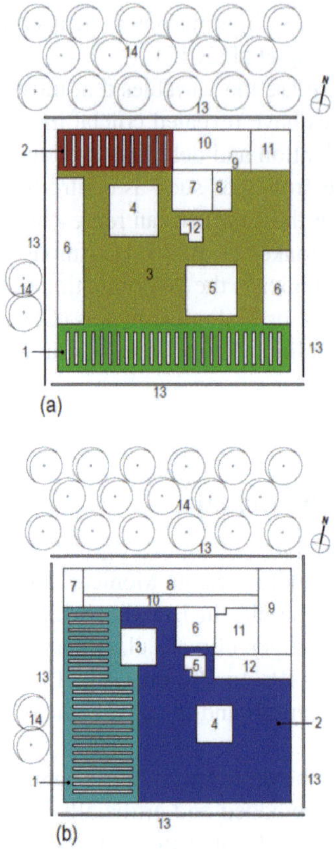

1.	Book collection 1
2.	Book colletcion 2
3.	Collective study area
4.	Mezzanine
5.	Atrium with windows
6.	Individual study rooms
7.	Bathrooms
8.	Audiovisual room
9.	Circulation
10.	Offices
11.	Records and book colletcion
12.	Stairs
13.	Brise soleil
14.	Trees

1.	Book collection
2.	Collective study area
3.	Mezzanine
4.	Atrium with windows
5.	Stairs
6.	Bathroom
7.	Meeting room
8.	Offices
9.	Cataloguing room
10.	Circulation
11.	Deposit
12.	Rare collection
13.	Brise soleil
14.	Trees

Fig. 2 Spaces analyzed in the hatch, **a** 1st-floor plan, **b** 2nd-floor plan. (Authors 2022)

spaces consists of books for loan, with rare books being found indoors with artificial control of temperature and humidity.

4 Method

This research sought to verify the joint service of thermal performance and conservation of books in libraries with passive strategies in the qualitative sample UFU Library. Based on hygrothermal simulation methods with BIM/BES workflow, three steps were defined: the definition of thermal performance requirements, the definition of book conservation requirements, and hygrothermal simulation using Revit 2021 (Autodesk, Inc.) and DesignBuilder software version 7.0.1.006 (Software Ltd.), the last one being validated by ASHRAE 140 (2017). Three scenarios were analyzed:

the current scenario with open windows, the current scenario with closed windows, and the scenario with double glazing and open windows.

4.1 Thermal Performance Requirements

Chapter 5.4.2.2 of ASHRAE 55 (2020) presents thermal performance requirements for naturally ventilated buildings by determining an acceptable operating temperature range. We define the operating temperatures with the acceptability limit of 80% of users from Eq. 01.

To define the acceptable operating temperature ranges for the library, the average monthly temperatures outside Uberlândia were collected based on the simulation in DesignBuilder from the meteorological file provided by (ABNT TR 15,575–1-1 (2021), and applied in the sequence with Eq. 01. As a result, we define the acceptable operating temperature ranges for each month, according to Table 1.

$$\begin{aligned}\text{Upper acceptability limit of } 80\% &= 0,31\,\text{tpma(out)} + 21,3\\ \text{Lower acceptability limit of } 80\% &= 0,31\,\text{tpma(out)} + 14,3\end{aligned} \tag{1}$$

Table 1 Acceptable operative temperature ranges for naturally ventilated spaces in Uberlândia. (Authors 2002)

Month	Acceptability limit OF 80% (°C)	
	Upper limit	Lower limit
Jan	0,31.22,97 + 21,3 = 28,42 °C	0,31.22,97 + 14,3 = 21,42 °C
Feb	0,31.23,39 + 21,3 = 28,55 °C	0,31.23,39 + 14,3 = 21,55 °C
Mar	0,31.24,65 + 21,3 = 28,94 °C	0,31.24,65 + 14,3 = 21,94 °C
Apr	0,31.23,77 + 21,3 = 28,66 °C	0,31.23,77 + 14,3 = 21,66 °C
May	0,31.21,05 + 21,3 = 27,82 °C	0,31.21,05 + 14,3 = 20,82 °C
Jun	0,31.20,65 + 21,3 = 27,70 °C	0,31.20,65 + 14,3 = 20,70 °C
Jul	0,31.20,39 + 21,3 = 27,62 °C	0,31.20,39 + 14,3 = 27,62 °C
Aug	0,31.21,97 + 21,3 = 28,11 °C	0,31.21,97 + 14,3 = 21,11 °C
Sep	0,31.24,69 + 21,3 = 28,95 °C	0,31.24,69 + 14,3 = 21,95 °C
Oct	0,31.24,24 + 21,3 = 28,81 °C	0,31.24 + 14,3 = 21,81 °C
Nov	0,31.23,42 + 21,3 = 28,56 °C	0,31.23,42 + 14,3 = 21,56 °C
Dez	0,31.24,22 + 21,3 = 28,80 °C	0,31.24,22 + 14,3 = 21,80 °C

4.2 Requirements of Books Conservation

The combinations between temperature and relative humidity are internal variables that directly influence book conservation (Sebera 1997; Michalski 2016). Sebera (1997) presents methods of isoperms elaborated from pairs of temperature and relative humidity that are associated with the longevity of the paper, that is, the time required for its deterioration. In this research, the Sebera isoperm method was used, which uses enthalpy ranges of 20–35 kcal to create various combinations of temperature and relative humidity, associating these with the permanence index (PI), an index mentioned by Krüger and Diniz (2011) obtained from Eq. 2. Sebera (1997) defines permanence as the time required for the paper to deteriorate.

$$\frac{P_2}{P_1} = \left(\frac{UR_1}{UR_2}\right)\left(\frac{T_1 + 460}{T_2 + 460}\right)10^{+394\Delta H^{\neq}\left(\frac{1}{T_2+460} - \frac{1}{T_1+460}\right)} \tag{2}$$

For instance, PI = 1.00 corresponds to temperature and relative humidity combinations of 20 °C and 50% and paper longevity of 45 years, as well as the combination of 22.2 °C and 30%, presents the same result as shown in Fig. 3. On the other hand, combinations that result in a PI of 5 indicate a book life expectancy of 225 years and PI of 0.33, an expectation of 15 year.

The enthalpy of 25 kcal/mole was adopted for the study, referencing the research by Assis and Bastos (2007). Thus, the Sebera chart shows an enthalpy of 25 kcal/mol, adapted for this research, as shown in Fig. 4. The Sebera diagram also

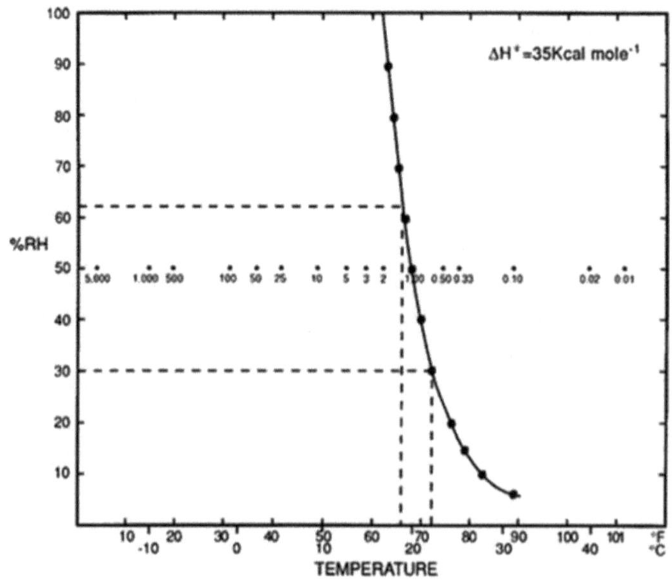

Fig. 3 Combinations of temperature and relative humidiy with PI = 1,00. (Sebera 1997)

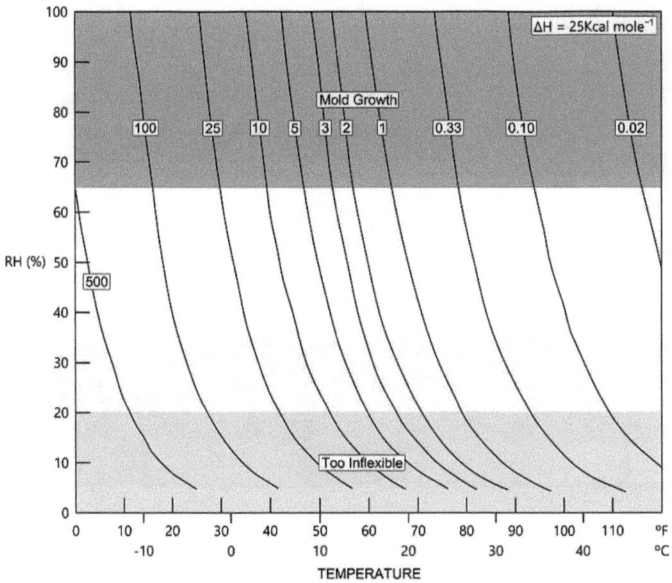

Fig. 4 Sebera diagram used on this research. (Adapted from Sebera 1997)

shows the zones where indoor relative humidity influences mold development or stiffness on paper.

From the hygrothermal simulation, pairs of internal temperature and relative humidity could be plotted on the graph to verify the longevity of the papers.

4.3 Hygrothermal Simulation Based on BIM/BES Workflow

Hygrothermal simulations simulate indoor temperatures and relative humidity. In this research, these variables were calculated according to the BIM/BES workflow from modeling and data transfer from a BIM software, Revit, and exported to the selected BES software, DesignBuilder, for hygrothermal simulation. Thus, the simulation was approached in 3 steps: modeling in BIM, input data in BIM, and exporting to BES software. Figure 5 shows the building and models in 4 moments: (a) photograph of the building, (b) detailed modeling in Revit, (c) simplification of the model in Revit, and (d) model imported into DesingBuilder.

4.3.1 Building Information Modeling

To produce the hygrothermal simulation of the spaces analyzed in the library, we model the detailed geometry in Revit from consultations with the original project

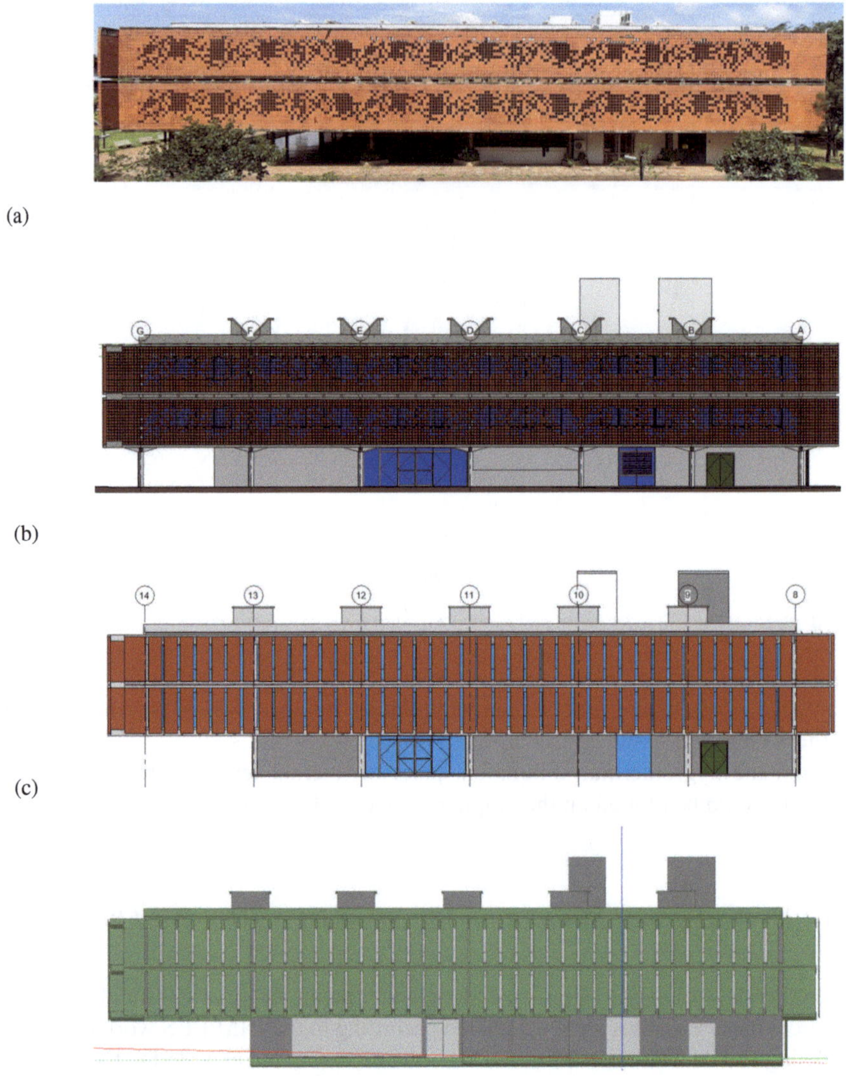

Fig. 5 **a** Photo of the building, **b** detailed modeling in Revit, **c** simplification of the model in Revit, **d** model imported into DesignBuilder. (Authors 2022)

and on-site. This step required considerable time since it was necessary to investigate the materials used and their thermal properties since the original projects provided by the University Campus City Hall were incomplete. In order to model the brise soleil in ceramic, it was necessary to map the hollow elements from on-site visits.

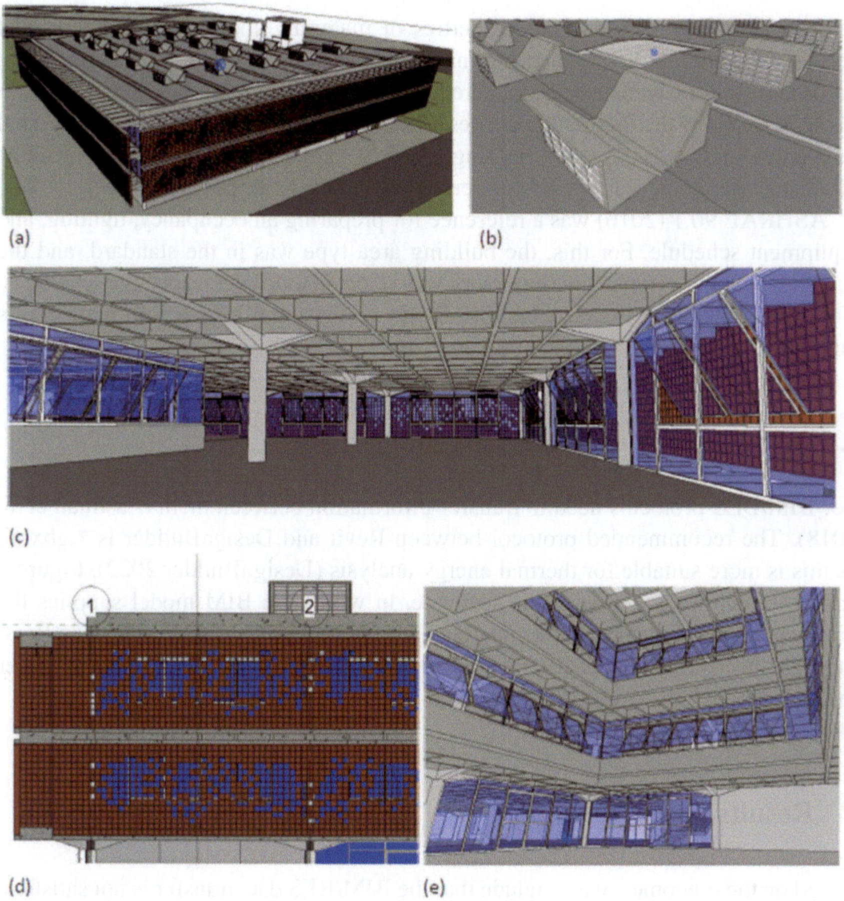

Fig. 6 Detailed geometry modeled in Revit software. **a** extern view, **b** roof with dormer vents and skylight, **c** internal view of the open plan of the first floor, **d** detail of the brise soleil with cobogós, and **e** external atrium. (Authors 2022)

Figure 6 shows the detailed model in BIM/BES data transfer using an existing BIM model.

Figure 6 shows the details of the external geometry, such as hollow brick walls and a roof with dormer vents and skylight, as well as the open floor plan of the interior with frames with fixed and tilting leaves, pillars with capitals, and an atrium with frames.

4.3.2 Input in Revit (Autodesk, Inc.)

Input data is information entered into the BIM model for further hygrothermal simulation in the BES software. In Revit, spaces with their respective uses and information such as geometry with thermal properties of materials, location, orientation, and spaces with internal heat gains. International standards, as well as Brazilian standards and authors, were used as a reference to better adapt to the Brazilian reality. In Table 2, we show the thermal properties of the materials, and in Table 3, the internal heat gains such as occupation, lighting power densities, and equipment, as well as air change per hour, outdoor air per person, and outdoor air per area.

ASHRAE 90.1 (2016) was a reference for preparing an occupancy, lighting, and equipment schedule. For this, the building area type was in the standard, and the schedule index C referring to the library was selected to fill the respective schedules. The frame operation is unavailable in Revit, so the input of this information was entered into the DesignBuilder software.

4.3.3 Protocols

For BIM/BES protocols need to transfer information between them (Eastman et al. 2018). The recommended protocol between Revit and DesignBuilder is *.gbxml, as this is more suitable for thermal energy analysis (DesignBuilder 2022). Figure 7 shows the workflow between the software, in which the BIM model supplies the data necessary for the thermal simulation. The energy analysis was modeled in Revit through the spaces and environments mode and then exported to DesignBuilder through the *.gbxml protocol.

5 Results and Discussion

Based on the outcomes, we conclude that the BIM/BES data transfer is not satisfied. There were several failures on the first attempt when exporting the detailed model from Revit, including:

- geometry transfer errors (as shown in Fig. 8)
- failure to transfer thermal properties data
- no transfer of occupation and gains schedule of internal heat.

Thus, we performed several tests to simplify the model in Revit. Figure 9 compares the detailed model, simplification, and imported into DesignBuilder. The main simplifications were: roofing, louvers, external frames, and dormer vents. Different alternatives to model the louvers in Revit were exported to DesignBuilder to verify if the geometry simplification would be compatible with the BES software. Thus, we modeled the louver with the same permeability as the detailed model. Despite simplifying geometry, the Venetian blind of the dormer vents requires their modeling in DesignBuilder with a central wall to simulate the sloping roof of the dormer vents since this provides a ventilation block between the two shutters. As for the coverage, it had to be elaborated in a single slab to be modeled in Revit and calculated manually, taking into account all the layers and their thermal properties, applying the

Table 2 Library´s material thermal properties. (Authors 2022)

System	Material	Thichness (mm)	λ (w/m.k)	c (kJ/kg.k)	ρ (kg/m³)	FS	Tvis	U (w/m².k)	α
8-hole brick wall plastered on both sides (a)	Mortar	25	1,15	1	2000	–	–	–	0,2
	Ceramic	13	0,9	0,92	1600	–	–	–	
	Air space	64	0,356	10,035	–	–	–	–	
	Ceramic	13	0,9	0,92	1600	–	–	–	
	Mortar	25	1,15	1	2000	–	–	–	
Ceramic brises	Ceramic	–	1200	0,92	0,7	–	–	–	
Dormer vents with policarvonate	Alvelolar policarbonate [b] [d]	6	0,2	1,26	1200	0,565	0,42	5	0,2
Glass windows	Colourles glass [b]	6	–	–	–	0,87	0,87	5,7	0,3
Concrete colunm	Concrete [b]	150	1,75	1	2300	–	–	–	0,7
MDF partitions	MDF [b]	40	0,2	2,3	850	–	–	–	–
Drywall	Plasterboard [b]	12,5	0,35	0,84	750	–	–	–	–
	Air space	75	0,468	1	1,2	–	–	–	
	Plasterboard [b]	12,5	0,35	0,84	750	–	–	–	
Interior slabs	Vinyl floor [c]	2	0,27	1,5	2300	–	–	–	–
	Subfloor [b]	20,8	1,15	1	2000	–	–	–	
	Concrete [b]	10	1,15	1	2000	–	–	–	
Simplified roof = thermoacoustic tile + waterproofed slab	Steel [b]	0,43	55	0,46	7800	–	–	–	0,3962
	Expanded polystyrene [b]	30	0,04	1,42	35	–	–	–	
	Steel [b]	0,43	55	0,46	7800	–	–	–	
	Air space	4	0,356	10,035	–	–	–	–	
	Layer of protection [b]	30	1,15	1	2000	–	–	–	
	Waterproofing [b]	0,4	0,23	1,46	1000	–	–	–	
	Regularization [b]	20	1,15	1	2000	–	–	–	
	Concrete [b]	100	1,75	1	2300	–	–	–	

(continued)

Table 2 (continued)

System	Material	Thichness (mm)	λ (w/m.k)	c (kJ/kg.k)	ρ (kg/m³)	FS	Tvis	U (w/m².k)	α
	White alvelolar Policarbonate [b] [d]	6	0,2	1,26	1200	0,565	0,42	5	–
Simplified roof = skylight	Air space	75	0,356	10,035	–	–	–	–	
	Simple glass [b]	6	1,1	0,84	2500	–	–	–	

[a] equivalent to an eight-hole ceramic block, as Weber et al. (2017), p. 45
[b] ABNT NBR 15,220–2 (2005)
[c] Weber et al. (2017), p. 44
[d] Balsamo et al. (2019)

Table 3 Occupancy, power density, air changes per hour, outdoor air per person, and outdoor air per área. (Authors 2022)

Input	Value	Unity	Source
Occupancy	9,29	m^2/person	ASHRAE 90.1 (2016)
*Lighting power density	8,1	W/m^2	André et al. (2022)
*Equipment power density	9,7	W/m^2	
Air changes per hour	1	ac/hour	ISO 17772–1 (2017)
Outdoor air per person	14	l/s	
Outdoor air per area	1,4	l/(sm^2)	

*It was used by André et al. (2022) in offices in Brazil and adopted as a reference for the research, as they are better suited to the Brazilian reality

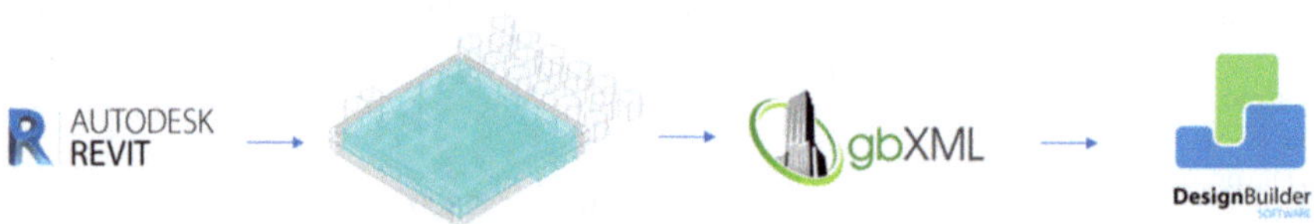

Fig. 7 The BIM/BES workflow. (Authors 2022)

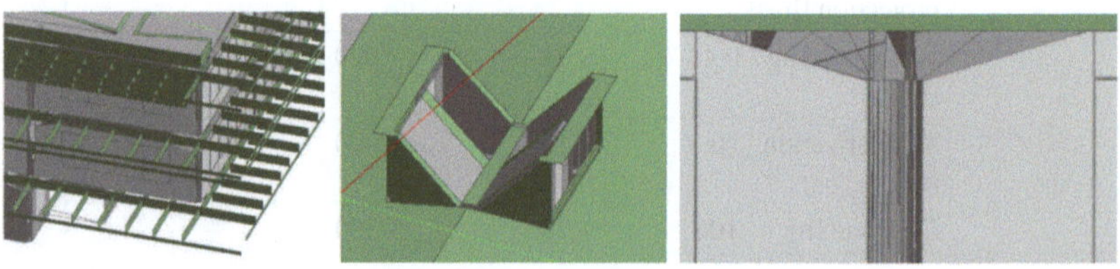

(a) (b) (c)

Fig. 8 Geometry transfer errors between Revit and DesignBuilder. **a** Brises were not imported, **b** Dormer vents with errors, **c** Errors in the colunms. (Authors 2022)

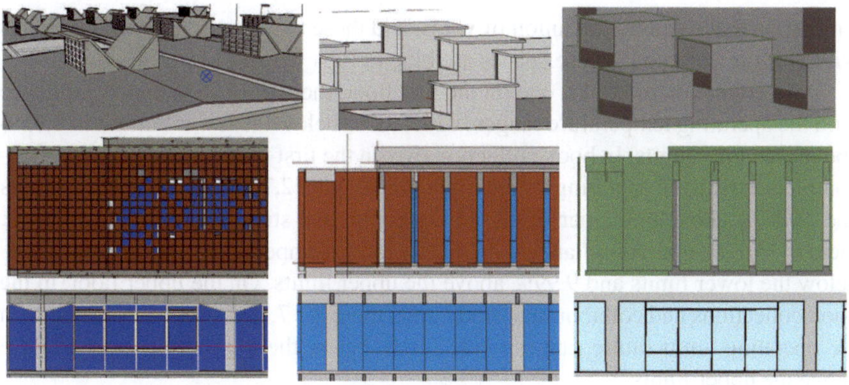

(a) (b) (c)

Fig. 9 **a** Detailed model in Revit, **b** Simplifying geometry in Revit, **c** Simplified BIM model imported into DesingBuilder. (Authors 2022)

weighted average of the roofs with waterproofed slabs, and the waterproofed slab with thermoacoustic tile. Another error was in the import of voids from the mezzanine, atriums, and dormer vents necessary to open the holes in the energy software itself. As for the import of thermal properties, there was a failure to transfer only the air chambers and absorbances, requiring manual correction of these items in the BES software.

The tipper windows were set for opening according to the library's occupancy schedule with an opening percentage of 30%, temperature setpoint of 21,45 °C (temperature indicated for natural ventilation for Uberlândia by the Climate Consultant 6.0 software, and operation of the windows when the internal temperature is higher than the external one. Another limitation of the BIM software was the impossibility of inserting the weather file. Therefore, this was inserted directly into the DesignBuilder software, as well as it was necessary to model. Regarding compliance with Thermal Performance, the spaces analyzed did not fully meet the ranges recommended by ASHRAE 55 (2020), but their performance was considered satisfactory since percentage of operative temperature in the occupied periods reached 82,15% to 86,88%. There are high-temperature peaks in the afternoon, mainly in March and September.

Figure 10 shows the spaces that presented the best percentage of service to the operating temperature range in ocupation periods are located on the first floor and are, respectively: reading area with 86.88%, book collection 2 with 86.73%, and book collection 1 with 85.75%. The fact that the study area showed better thermal performance might be due to the lower glazing on the facades. The spaces that presented lower thermal performance were located on the upper floor: book collection and collaborative study area, with 82,15%. Lower performance may be due to proximity to coverage. These two spaces presented the same result due to a limitation in the simulation that required the union of voids, and these two spaces have dormer vents with voids in common. These spaces are blue in Figs. 10 and 11.

In the collection of books, one on the first floor, there is 14,25% non-compliance with the operating temperature ranges, with 5,60% below the lower limits and 8,65% above the upper limits. In book collection two on the first floor, there is 13,27% non-compliance with the operating temperature ranges, with 2,70% below the lower limits and 10,57% above the upper limits. In the communal study area on the first floor, there is 13,12% non-compliance with the operating temperature ranges, with 3,12% below the lower limits and 9,99% above the upper limits. On the upper floor, in the book collections and collaborative study area, there is 17,85% non-compliance with the operating temperature ranges, with 2,01% below the lower limits and 15,84% above the upper limits.

Due to the lack of full compliance with the operating temperature ranges, we simulate two more scenarios for comparison with the current scenario, shown in Fig. 11, namely: single glass/open windows (current scenario); single glass/closed windows; and double glazing/open windows (with change from standard clear glazing ($U = 5.7$ w/m^2.k) to double glazing with solar control ($U = 1.9$ w/m^2.k, FS = 0.37, and Light Transmission = 0.47), according to Inmetro 50 (2013). The proposed scenario

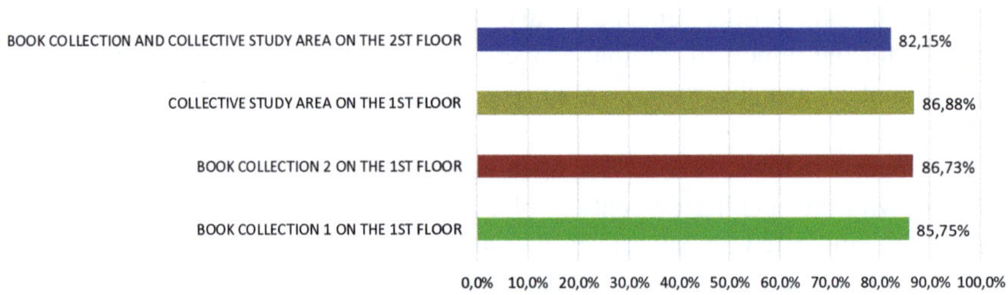

Fig. 10 Accordance of operative temperature range in the occupied periods of each space in the current scenario. (Authors 2022)

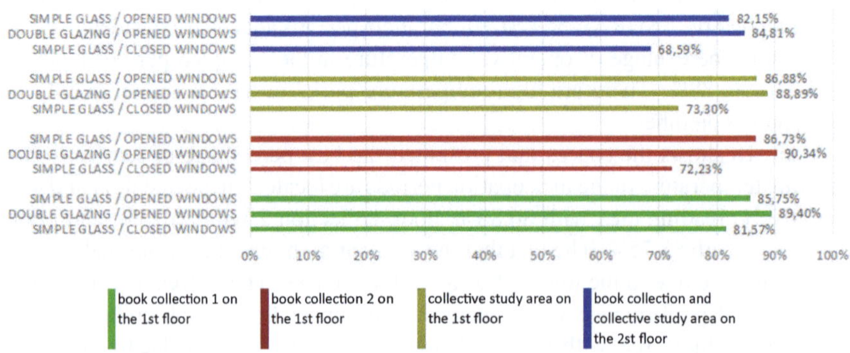

Fig. 11 Accordance of operative temperature range in the occupied periods in the three scenarios of the analyzed spaces. (Authors 2022)

took into account the lack of characterization of the building. We proposed a scenario preserving the translucent elements. Thus, in this scenario, the one-layer glass was replaced by the glass with lower thermal transmittance.

It can be seen in Fig. 11 that the scenario with closed windows had the worst thermal performance, as well as showing the importance of natural ventilation in the building. Although the scenario with double glazing improves the thermal performance of the analyzed spaces, it still did not fully meet the operating temperature range since this scenario did not significantly attenuate the maximum and minimum operating temperatures. The hottest periods are in March and September, so, as a way of evaluating this scenario, the operating temperatures in the periods occupied in September in the collection of books two on the first floor in the three scenarios shown in Fig. 12. The double-glazed/open window scenario was excluded from the analysis as the results were very close to the current scenario. Note the importance of natural ventilation in the scenario with closed windows. It can be seen in Fig. 12 that there was a difference in the operating temperature between the scenario with open and closed windows, reaching a difference of 1.56 °C, causing the scenario with open windows not to reach the upper limit at times. Even with high temperatures in September, this space attended 72.73% of the time in the current scenario. There were no hours below the lower operating temperature limit and 27,27% above the upper operating temperature limit.

Regarding book conservation, the results are the internal operating temperatures of 8760 h of the three collection spaces in the current scenario (single glass/open windows) plotted on the Sebera graph (Figs. 13, 14, and 15). In this, the year was divided into seasons through four colors: winter (blue), spring (green), autumn

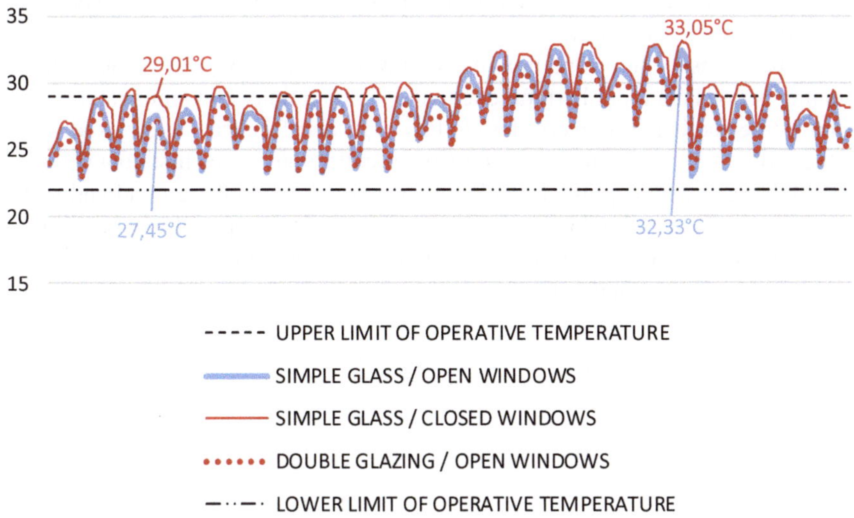

Fig. 12 Operative temperature in occupied periods in the book collection 2 on the 1st-floor in the three scenarios. (Authors 2022)

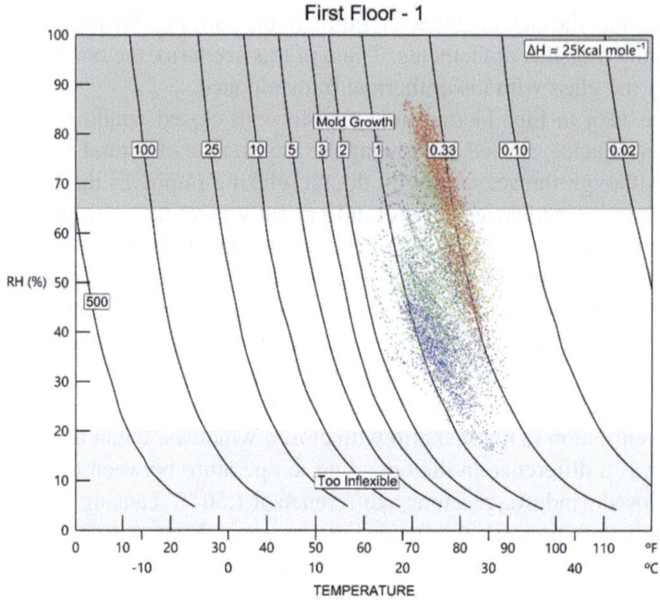

Fig. 13 Sebera diagram for book collection 1 on the 1st floor. (Authors 2022)

(yellow), and summer (red). It is observed that the internal microclimate is not suitable for the conservation of books since all spaces have surpassed the mold zone due to the high humidity inside, which may be due to the shading strategies using brise-soleil and trees, as well as ventilation blocking due to the louvers, bookshelves, and partitions allocated later.

The best conservation conditions occur in winter and spring, and the autumn and summer periods are more harmful due to the high humidity and temperatures. The temperature and humidity pairs close to PI = 0.28 and 1 indicate a low expectation of the books, around 12,6 to 45 years. Also, the collections on the first floor (Figs. 13 and 14) are more susceptible to mold, as they are closer to the ground and suffer from solar and ventilation blocks. It happens in periods of paper of low humidity in winter. Besides, we observe lower humidity in the second-floor collection (Fig. 15), mainly in summer and autumn.

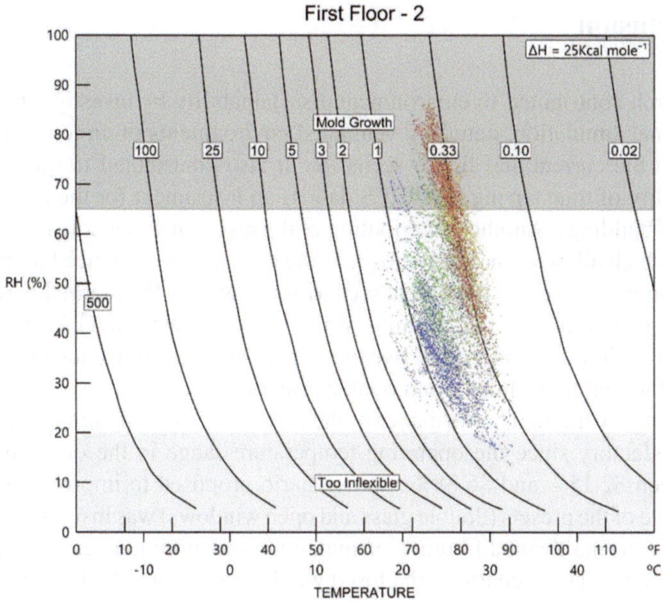

Fig. 14 Sebera diagram for book collection 2 on the 1st-floor. (Authors 2022)

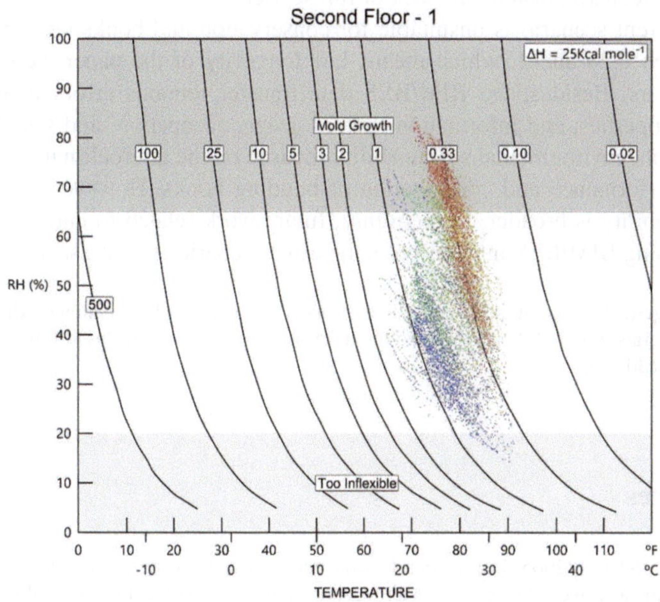

Fig. 15 Sebera diagram for book collection 1 on the 2st floor. (Authors 2022)

6 Conclusion

This research contributed to environmental sustainability by investigating, through hygrothermal simulation, naturally ventilated environments of an iconic university building in the current and future scenarios. It also contributed to the analysis of the feasibility of transferring BIM/BES data as an instrument for the microclimatic analysis of buildings. Another contribution of the research was a holistic view of the property, which allowed analyzing not only the thermal performance but the conservation of books, as well as proposing scenarios to improve the thermal performance without de-characterizing the building, which is in line with SDG 11.4 and 7.3. A hygrothermal simulation is an effective tool for environmental sustainability analysis regarding thermal performance and book conservation.

Regarding the thermal performance of the current scenario of the spaces analyzed, it was satisfactory since the operating temperature range in the occupied periods was between 82.15% and 86.88%. The scenario proposed to improve the thermal performance of the presets (double glass and open windows) was insignificant since it did not present significant differences in maximum and minimum temperatures in the current scenario. The scenario with closed windows presented higher temperatures, demonstrating the importance of natural ventilation for thermal performance. Thus, the appropriate scenario for the thermal performance of the analyzed building was the current scenario, despite the lack of full service.

The current scenario is unsuitable for conservation and books since isosperms are between 0.28 and 1, which means low longevity of the paper, between 12.6 and 45 years. Besides, the BIM/BES data transfer remains errors in geometry, thermal properties, and information such as usage, occupation, and scheduling. In this research, environmental sustainability focused on the microclimatic analysis of thermal performance and conservation of building books. However, the meaning of sustainability is broader. In this sense, future work related to energy efficiency analysis using BIM/BES integration in the exact scenarios is necessary.

Acknowledgements The authors would like to thank the architect Luís Antônio Almeida Reis, the Minas Gerais Research Foudation and the infrastructure of the Form-Based Information Lab: www.informa3d.xyz.

References

ABNT NBR 15220-2 (2005) Desempenho térmico de edificações-Parte 2: Métodos de cálculo de transmitância térmica, da capacidade térmica, do atraso térmico e do fator solar de elementos e componentes de edificações [Thermal performance of buildings-part 2: methods for calculating thermal transmittance, thermal capacity, thermal lag and solar factor of building elements and components]. ABNT, Rio de Janeiro

ABNT TR 15575-1-1 (2021) Residencial buildings-performance part 1-1: standard database of weather files for the evaluation of thermal performance using the computer simulation procedure. ABNT, Rio de Janeiro

Amaral JVS, Montevechi JAB, Miranda RDC, Sousa Junior WT (2022) Metamodel-based simulation optimization: a systematic literature review. Simul Model Pract Theory 114:1–21. https://doi.org/10.1016/j.simpat.2021.102403

André M, Kamimura A, Bavaresco M et al (2022) Achieving mid-rise NZEB offices in Brazilian urban centres: a control strategy with desk fans and extension of set point temperature. Energy Build 259:1–22. https://doi.org/10.1016/j.enbuild.2022.111911

Andretta M, Coppola F, Seccia L (2016) Investigation on the interaction between the outdoor environment and the indoor microclimate of a historical library. J Cult Herit 17:75–86. https://doi.org/10.1016/j.culher.2015.07.002

ASHRAE (2007) Museums, galleries, archives, and libraries. In: Handbook-heating, ventilating, and air-conditioning applications (I-P Edition), SI. W. Stephen Comstock, Atlanta, pp 21.1–21.23

ASHRAE 55 (2020) Thermal environmental conditions for human occupancy. ASHRAE, Atlanta

ASHRAE 90.1 (2016) Energy standard for buildings except low-rise residential buildings. ASHRAE, Atlanta

ASHRAE 140 (2017) Standard method of test for the evaluation of building energy analysis computer programs. ASHRAE, Atlanta

Assis ES de, Bastos RV (2007) Análise das condições de conservação do acervo da Escola de Arquitetura da UFMG [Analysis of the conservation conditions of the UFMG School of Architecture collection]. Paper presented at the IX Encontro Nacional e V Latino Americano de Conforto no Ambiente Construído. ENCAC, Ouro Preto, 08–10 Aug 2007

Balsamo RAC, Cabral BV, Sanfelice RC (2019) Estudo comparativo entre materiais a base de vidro e policarbonato em projetos de superfícies transparentes [Comparative study between glass and polycarbonate materials in transparent surface projects]. Revista Brasileira De Ciência, Tecnologia e Inovação 4:327–345

Boeri A, Longo D, Fabbri K et al (2022) Library Indoor microclimate monitoring with and without heating system: a bologna university library case study. J Cult Herit 53:143–153. https://doi.org/10.1016/j.culher.2021.11.012

Bracht MK, Melo AP, Lamberts R (2021) A metamodel for building information modeling-building energy modeling integration in early design stage. Autom Constr 121:2–19. https://doi.org/10.1016/j.autcon.2020.103422

Conceição E, Gomes J, Awbi H (2019) Influence of the airflow in a solar passive building on the indoor air quality and thermal comfort level. Atmosphere (basel) 10:1–18. https://doi.org/10.3390/atmos10120766

Coppola F, Brown N, Amicucci F et al (2020) Non-destructive collection survey of the historical classense library part II: conservation scenarios. Herit Sci 8:2–10. https://doi.org/10.1186/s40494-020-00430-y

Coşkun T, Gülhan Ö, Sahin DC et al (2017) The effect of spatial interventions on historic buildings' indoor climate (Case study: Tire Necip Paşa Library, Izmir-Turkey). Energy Procedia 133:358–356. https://doi.org/10.1016/j.egypro.2017.09.362

Crawley DB, Lawrie LK, Pedersen CO, Winkelmann FC (2000) EnergyPlus: energy simulation program. ASHRAE J:49–56

DesignBuilder (2022) DesignBuilder Software Ltda. https://designbuilder.co.uk/. Accessed 20 Oct 2022

de la Paz diulio M, Mercader-Moyano P, Gómez AF (2019) The influence of the envelope in the preventive conservation of books and paper records. Case study: libraries and archives in La Plata, Argentina. Energy Build 189:727–738. https://doi.org/10.1016/j.enbuild.2018.11.048

de la Paz Diulio M, Santa Cruz MG, Gómez A (2015) Preventive conservation plan for library buildings in La Plata, Argentina. Energy Proc 78:1293–1298. https://doi.org/10.1016/j.egypro.205.11.143

Drougka F, Liakakon E, Sakka A et al (2020) Indoor air quality assessment at the library of the national observatory of Athens, Greece. Aerosol Air Qual Res 889:889–903. https://doi.org/10.4209/aaqr.2019.07.0360

Eastman C, Sacks R, Lee G, Teicholz P (2018) BIM handbook: a guide to building information modeling for owners, designers, engineers, contractors, and facility managers, 3rd edn. Wiley, New Jersey
EN 15757 (2010) Conservation of cultural property-specifications for temperature and relative humidity to limit climate-induced mechanical damage in organic hygroscopic materials. European Committee for Standardization, Brussels
Haidar A, Underwood J, Coates P (2018) Smart proceses for smart buildings:'sustainable processes', "recyclable processes" and "building seeds" in parametric design. Architect Eng Des Manage:1–28. https://doi.org/10.1080/17452007.2018.1564645
Hussin N, Razak AA, Baharum F, Yaakob Y (2019) Simulation of energy consumption in the library building. Paper presented at the 5th international conference on green design and manufacture (IConGDM 2019). AIP Publishing, Bandung, 29–30 Apr 2019
INMETRO (2013) Portaria n.º 50, de 01 de fevereiro de 2013. Anexo Geral V-Catálogo de propriedades térmicas de paredes, coberturas e vidros [Ordinance no. 50, February 1, 2013. General Annex V - Catalog of thermal properties of walls, roofs and glass]. INMETRO, Rio de Janeiro
ISO 17772-1 (2017) Energy performance of buildings-indoor environmental quality-part 1: indoor environmental input parameters for the design and assessment of energy performance of buildings. ISO, Geneva
Jalaei F, Jalaei F, Mohammadi S (2020) An integrated BIM-LEED application to automate sustainable design assessment framework at the conceptual stage of building projects. Sustain Cities Soc 53:1–18. https://doi.org/10.1016/j.scs.2019.101979
Keeler M, Burke B (2009) Fundamentals of integrated design for sustainable building, 1st edn. US Green Building Council, Washington
Krüger EL, Diniz W (2011) Relationship between indoor thermal comfort conditions and the time weighted preservation index (TWPI) in three Brazilian archives. Appl Energy 88:712–723. https://doi.org/10.1016/j.apenergy.2010.03.011
Kupczak A, Sadlowska-Salega A, Krzemien L et al (2017) Impact of paper and wooden collections on humidity stability and energy consumption in museums and libraries. Energy Build 158:77–85. https://doi.org/10.1016/j.enbuild.2017.10.005
Leal Filho W, Frankenberger F, Salvia AL et al (2021) A framework for the implementation of the sustainable development goals in university programmes. J Clean Prod 299:1–12. https://doi.org/10.1016/j.jclepro.2021.126915
Liu Q, Wang Z (2022) Green BIM-based study on the green performance of university buildings in Northern China. Energy Sustain Soc 12:1–17. https://doi.org/10.1186/s13705-022-00341-9
Liu YM, Lu WY (2009) A library's environmental monitoring system based on Wireless Sensor Networks. Paper presented at the 2009 international conference on networking and digital society, ICNDS 2009. IEEE Computer Society Press, Guiyang, 30–31 May 2009
Michalski S (2016) Climate guidelines for heritage collections: where we are in 2014 and how we got here. Canadian Conservation Institute, Ottawa
Mitchell T (2010) The resources of economics. J Cult Econ 3:189–204. https://doi.org/10.1080/17530350.2010.494123
Papadopoulos AM (2016) Forty years of regulations on the thermal performance of the building envelope in Europe: achievements, perspectives and challenges. Energy Build 127:942–952. https://doi.org/10.1016/j.enbuild.2016.06.051
Sadeghifam AN, Joo MS, Pauzi NNM, Namini SSB (2021) Energy retrofitting and thermal comfort improvement of the public buildings in Malaysia. Paper presented at the 2021 international conference on green energy, computing and sustainable technology, GECOST 2021. Institute of Electrical and Electronics Engineers Inc., Miri, 07–09 July 2021
Sahin CD, Coşkun T, Arsan ZD, Gökçen Akkurt G (2017) Investigation of indoor microclimate of historic libraries for preventive conservation of manuscripts. case study: tire Necip Paşa Library, İzmir-Turkey. Sustain Cities Soc 30:66–78. https://doi.org/10.1016/j.scs.2016.11.002

Schito E, Pereira LD, Testi D, Silva MG (2019a) A procedure for identifying chemical and biological risks for books in historic libraries based on microclimate analysis. J Cult Herit 37:155–165. https://doi.org/10.1016/j.culher.2018.10.005

Schito E, Pereira LD, Testi D, Silva MG (2019b) Data of temperature and relative humidity in a historic library in Portugal. Data Brief 24:1–14. https://doi.org/10.1016/j.dib.2019.103788

Sebera DK (1997) Isoperms an environmental management tool. Commission on Preservation and Access, Washington

Silva HE, Henriques FMA (2014) Microclimatic analysis of historic buildings: a new methodology for temperate climates. Build Environ 82:381–387. https://doi.org/10.1016/j.buildenv.2014.09.005

Song J, Zhang X, Meng X (2015) Simulation and analysis of a university library energy consumption based on EQUEST. Procedia Eng 121:1382–1388. https://doi.org/10.1016/j.proeng.2015.09.028

Stoakes P, Passe U, Battaglia F (2011) Predicting natural ventilation flows in whole buildings. Part 1: the Viipuri library. Build Simul 4:263–276. https://doi.org/10.1007/s12273-011-0045-4

Tronchin L, Fabbri K (2017) Energy and microclimate simulation in a heritage building: further studies on the Malatestiana library. Energies (basel) 10:1–14. https://doi.org/10.3390/en10101621

Turhan C, Topan C, Arsan ZD, et al (2021) Comfort based investigation on historic libraries for user satisfaction and preservation of paper-based collections. Paper presented at the IOP conference series: earth and environmental science. IOP Publishing Ltd, Bolzano-Bozen, 14–16 Apr 2021

Verticchio E, Frasca F, Cornaro C, Siani AM (2020) Investigation on the use of hygrothermal modelling for paper collections. Paper presented at the IOP conference series: materials science and engineering. IOP Publishing Ltd, Florence, 14–16 Oct 2020

Waddicor DA, Fuentes E, Sisó L et al (2016) Climate change and building ageing impact on building energy performance and mitigation measures application: a case study in Turin, northern Italy. Build Environ 102:13–25. https://doi.org/10.1016/j.buildenv.2016.03.003

Weber FS, Melo AP, Marinoski DL et al (2017) Desenvolvimento de um modelo equivalente de avaliação de propriedades térmicas para elaboração de uma biblioteca de componentes construtivos brasileiros para o uso no programa EnergPlus [Development of an equivalent model for evaluating thermal properties to create a library of Brazilian building components for use in the EnergPlus program]. LabEEE, Florianópolis

Implementing Sustainability Programs in Universities –An Overview of Evolution, Challenges from a Global, Brazilian Perspective

Gustavo José Rodrigues Lopes and José Maria Franco de Carvalho

Abstract Based on the principle that the academic environment should excel by setting the example of commitment, knowledge transmission and opinion and thought formation, making the University Campus more sustainable is mandatory but is also a challenge that many researchers have been reporting. This chapter aims to compile information (state of the art) on sustainable activities carried out in Universities in the different aspects of the theme, whether social, environmental or economic, through a systematic bibliographic review of articles on "Sustainable University (SU)" published in recent years. Among the 228 articles found, 33 published from 2017 to 2022 were selected, and 15 other papers considered relevant to the research were included. The different experiences and challenges encountered in national and international institutions are reported and discussed, showing how the knowledge acquired can assist in the understanding and application of actions for the implementation of practices that lead us to a more sustainable campus and in the path of the implementation process of the Sustainable Development Goals (SDGs). Recognizing the importance of implementing methods that bring Higher Education Institutions (HEIs) closer to the SDGs is already a big step on the right path toward achieving a SU.

Keywords Sustainability · Sustainable university · Sustainable development goals · Higher education

G. J. R. Lopes (✉) · J. M. F. de Carvalho
Universidade Federal de Viçosa, Av. Peter Henry Rolfs, s/n-Campus Universitário, Viçosa, MG 36570-900, Brazil
e-mail: gustavo.lopes@ufv.br

J. M. F. de Carvalho
e-mail: josemaria.carvalho@ufv.br

W. Leal Filho et al. (eds.), *The Contribution of Universities Towards Education for Sustainable Development*, World Sustainability Series,
https://doi.org/10.1007/978-3-031-49853-4_11

1 Introduction

The word "sustainability" is defined as "planning based on the use of resources and the implementation of industrial activities so as not to deplete or degrade natural resources" (Houaiss and Villar 2015). It can also be defined as human actions and activities that aim to meet the current humans needs without compromising the future of the next generations (Keeble 1988). In 1979, during the United Nations General Assembly, the term "sustainable development" was introduced to the world, but it became known and widespread with the publication of the report "Our Common Future" in 1987 (Torgal and Jalali 2010).

The "Sustainable University" (SU) has been debated and studied for some time now. Amaral et al. (2015) brought in their research a comparison between the evaluation systems of sustainable programs in universities. The authors highlighted the ISO 14001 systems and the Eco-Management and Audit Scheme (EMAS), emphasizing that these are considered more generic, and stressed the programs created specifically for Higher Education Institutions (HEIs), particularly the Sustainability Tracking, Assessment and Rating System (STARS) and the Graphical Assessment of Sustainability in Universities (GASU).

To the researchers Washington-Ottombre et al. (2018), STARS has multiple advantages for HEIs, because it has a more simplified reporting process; thus, it assists managers in planning and guides the implemented actions with an evaluation of the progress achieved. Lauder et al. (2015) compare STARS with other tools, for example, the Holcim Awards for Sustainable Construction or the GREENSHIP programs and highlight that the former includes a series of indicators related to education and, according to Lidstone et al. (2015), the results related to the environmental area are consistent, but the answers regarding the economic and social optics are questionable. Ceulemans et al. (2015) report that HEI engagement is most strongly evidenced by STARS.

According to Sallit (2020), Brazil has 28 universities ranked among the most sustainable in the world. The Federal University of Viçosa (UFV) was ranked 162nd in the world (17th in Latin America and 6th in national rankings). In the current ranking, UFV is the 285th in the world (GREENMETRIC 2021). The challenge of implementing sustainable actions at the Pontifical Catholic University of the Rio Grande do Sul State was described by Wachholz (2014), presenting the main environmental actions as part of its environmentalization process and its responsibility as an institution facing environmental issues. Maio (2017) evaluated how sustainable are the actions and the activities in the Federal University of Rondônia, highlighting the existence of four dimensions of sustainability: economic, environmental, social, and cultural. He framed the institution in the range considered "unsustainable", especially in the environmental and social dimensions. Academic performances and sustainability were described by Atici et al. (2021) after measuring universities efforts in sustainable actions, also mentioned by Heravi et al. (2019) with statistical assessments of SU performance. Fissi et al. (2021) evaluated the performance of the University of Florence, emphasizing the ways to build a more sustainable society (promoting the reduction of negative impacts on the economy, social and environmental aspects and promoting the adoption of sustainable practices in research programs and curricula), corroborating the opinions of Dagiliūtė et al. (2018).

In the engineering field, sustainability can be measured through: (i) environmental indicators, which take into account energy consumption, raw material consumption, and waste emissions and destinations (solid and atmospheric emissions); (ii) social indicators, which encompass working conditions, partnership policies and community relations (communication and participation); (iii) economic indicators, where the fixed costs of operations and the estimated cost of investment are analyzed.

Thus, this area shows itself as a promising field for research linked to the use of indexes for the application of sustainable methods in enterprises and sectors of public management, aiming at a better use of natural, financial, and social resources.

In order to have a more specific view of the theme addressed, a bibliometric review was conducted in order to select works that addressed the theme of sustainability in universities. The searches were conducted in April 2022 through the academic platform Google Scholar, searching for editorials published from 2017 to 2022. In addition to these works, the review included articles published in previous years but considered necessary for the research. The keywords "sustainable university campus" and "sustainable university" were used with the occurrence of 228 related articles, of which 33 were selected. Other 15 references complemented the chapter.

2 Implementation of Sustainability Programs in Universities and Cities Around the World

Amaral et al. (2015) demonstrated the importance of including sustainable concepts in higher education courses and in our everyday life of HEIs, when they evidenced the large consumption of resources and energy, in addition to the enormous generation of waste.

The minimization of impacts must be one of the main concerns of institutions, whether economic, environmental or social, in order to serve as a showcase to society and, consequently, of excellence in their teaching, research and extension. The concept of "Sustainable University" is not a novelty since several studies are related to the subject in the world. Velazquez et al. (2006) report that the process of implementing sustainable programs in HEI is incremental and continuous.

The quest to reduce the negative impacts caused by its social, environmental and economic activities and the implementation of sustainable curricula, has made the University of Florence in Italy an example for societies (Fissi et al. 2021). HEIs have and cause significant impacts on society and the materialization of sustainability programs in society (Dagiliūtė et al. 2018). Zheleznyak and Korelina (2022) addressed the sustainability problems encountered at Irkutsk University in Russia and concluded that university campuses should serve as models regarding environmentally sustainable development in the current stage of civilizational development.

This will allow the experience of campus design and operation to be used to create sustainable cities. Based on this perception, the notion of SU emerges in the Cuban context, with its science, technology and innovation centered on the so-called Cuban Local Sustainable Development (CLSD). The proposal starts in the State with environmental concern and no commercial bias, without the concern of obtaining profit or any form of individual enrichment. The CLSD seeks to improve the quality of life and the population participation by implementing Municipal University Councils that works as concentrators and disseminators of human and innovative patrimony of the place, acting as agents of knowledge and improvement. Thus, they participated in elaborating development of strategies and training human resources (Santos 2019a). The same strategic profile was verified by Bizerril et al. (2018) from the case of a Portuguese university that suggested strategic actions for HEIs seeking the transition to SU, with the application of questionnaires and document analysis, including the adoption of a holistic perspective through discussion and official documents, the practice of democratic and participatory management and the creation of formal channels of dialogue with society.

The United Nations (UN) launched the Sustainable Development Goals (SDGs), it became a new universal set of goals and target indicators that were adopted on September 25th, 2015, to "eradicate poverty in all its forms by 2030" and "balance the three dimensions of sustainable development, the economic, the social, and the environmental" (Guterres 2020; Leal Filho et al. 2022b). Following such goals, Leal Filho et al. (2019) and Leal Filho et al. (2022a) report that HEIs must align their curricula and research with the SDGs and the many commitments they entail because, in them, unique opportunities are offered to combine course content with the goals to be achieved, enriching the learning experience. However, as reported by Fissi et al. (2021), research is mostly carried out with specific purposes in relation to sustainable perspectives in HEIs, rarely taking into account the human activities found on campus.

An increasing number of universities around the world have adopted the SU concept. The UI GreenMetric World University Ranking (UIGMWUR) (GREENMETRIC 2021) was created in 2010 on the initiative of the University of Indonesia (Universitas Indonesia-UI), located in Depok, West Java province. The main purpose of ranking is to provide the current status of an institution in relation to its sustainable policies in a global scale. The UIGMWUR criteria include configuration and infrastructure, energy and climate change, waste, water, transport and education. Each of these criteria was broken down into indicators compiled by mutual agreement between the ranking's coordinators and initiators.

Atici et al. (2021) revealed a relationship between universities' academic performances using sustainability scores published by UIGMWUR, which seeks to measure universities' efforts in sustainable actions by adopting the concept of sustainability that encompasses environmental, economic, and social dimensions and four main academic ranking systems. The research results confirmed that university administrations that implemented environmental policies aimed in a more sustainable environment directly impacted the ranking of universities, which is currently emerging as a competitive differential for universities around the world. Hakiki (2019) reported the steps listed as priorities in the President University in Cikarang, Indonesia, for implementing sustainable methods. Highlighted the fact that the UIGMWUR is not only a ranking for a green campus, but also an encouragement in understanding the importance and reflections that sustainable management can bring to the academia and its surroundings. This fact can be observed by Abtahi (2021) in the Isfahan University of Technology, one of the most respected and reputable universities in Iran, after adopting measures necessary to implement "Green Management", achieving recognition in this regard, including a significant ranking by UIGMWUR. Among the measures taken, there were no novelties, but only the implementation of extensive educational and environmental research activities, as well as executive measures for expansion of clean energy consumption, energy efficiency, improvements in water and wastewater consumption and treatment, and solid waste management.

In another case study in Iran in the University of Tehran, Heravi et al. (2019) developed a performance evaluation framework using statistical techniques. The researchers identified areas of sustainable practices using questionnaires and structural equation modeling techniques. The results identified four critical areas (water consumption, energy expenditure, sustainable behavior and purchasing procedures) for implementing sustainable practices and identified that "sustainable behavior" and "purchasing procedures" were the critical and key areas for implementing "green" practices.

The vision of universities as major responsible for the development of society was addressed by Alonso-Almeida et al. (2015), reporting that it is crucial that HEIs get involved in projects linked to sustainable development, mainly with the purpose of breaking old paradigms and they also reinforce that the improvement of the means of interlocution between the academic and social spheres occurs so as to help in the amplification and dissemination of the models. They also expose that little research related to this theme has been conducted.

In Sweden, Sammalisto et al. (2015) researched the importance of faculty and staff training in preparing students for the difficulties of being a sustainable society. The researchers emphasized the existence of little research related to the topic. They showed different understandings and ignorance regarding some of the pillars that guide sustainability, similar to the results found by Nousheen et al. (2020) in Pakistan, who advocate the implementation of sustainability-oriented academic programs. Another important factor reported by Dagiliūtė et al. (2018) is about the low incidence of research comparing the performances of universities that have sustainable programs with others that do not have it. This perception points to the need for specific activities on the topic to be conducted along with information campaigns on campuses.

In China, Yuan et al. (2013) researched the implementation of sustainability at Shandong University, one of the largest academic institutions in the country. Over short- and long-term horizons, there was increasing commitment from the academic community, from directors, faculty, staff and students. The research was based on the factors that most impacted the achievement of an SU's goals and were grouped into seven categories: enhancement of internal management systems; environmental sustainability; alignment of subject curricula; investigations and implementations of practices aimed at sustainable development; staff incentives in capacity building; offering opportunities for students to engage in sustainable issues; social responsibility.

Washington-Ottombre et al. (2018) present a chronological study of the implementation of sustainability models in HEIs in the United States of America (USA), divided into distinct periods and approaches. At first, the environmental issue begins to be debated at international levels and arouses interest in the topic. University campuses begin to debate the topic of "sustainability," even though the term was not used at the time. In a second stage, the environmental issue became the focus of US HEIs with international dissemination. Similar to the first stage, the SU concept is being promoted internationally, but now focusing on different and increasingly comprehensive pillars.

With a specific focus on transportation within a campus, researchers Fadhil and Waheeb (2021) report the experience of creating sustainable routes (greenways) in Diyala University in Baqubah, Iraq. The goal was to identify routes or pathways that would contribute to transportation sustainability on the university campus. This being one of the aspects that allow us to achieve the three pillars of sustainability. Sundram et al. (2021) published a similar study in Malaysia and Indonesia, following the same research line. Growing awareness of the impact of carbon dioxide (CO_2) emissions and fossil fuel consumption on the environment has led researchers to focus on sustainable transportation solutions for their campuses, where high dependence on private vehicles was reported to cause traffic jams in their areas, especially at peak times, leading to parking concerns. Alternative transportation solutions are cited as alternatives to minimize environmental consequences. Mustafa et al. (2022) researched ways to neutralize carbon emissions at the NED—Karachi University of Engineering and Technology in Pakistan. The authors cited as necessary interventions the use of non-polluting vehicles (bicycles or electric motors) but also focused on alternatives to achieve carbon balance, such as using energy-efficient appliances, planting trees on and off-campus, and changing habits with the use of renewable energy. At Chulalongkorn University (CU), Thailand, eco-friendly modes of transport have been available on campus since 2014 through a bike-sharing system, but there is still a lack of an institutional program to educate students about the use of this more sustainable mode of transport (Yan 2020).

3 Cities and Universities with Sustainability Projects and Programs in Brazil

Several institutions have been researching sustainability on their campuses. As an example, researchers Marques, Bachega & Tavares (2019) from the Federal University of Goiás, Goiás State, developed a study on the concept of Sustainable Information Technology (IT), which refers to ecologically correct IT practices, with the proposition of creating a database to assess environmental impacts and direct HEI decisions towards a sustainable bias. The research was conducted by measuring the consumption and cost of electricity and CO_2 emissions in selected computer labs. With the measurement of environmental metrics, it was possible to compile the information and constantly remedy the impacts and compare the results obtained with the indices prior to the implementation of the system. For the University of Brasília–Planaltina Faculty (UnB–Federal District), Ribeiro et al. (2019) indicate that the campus has been increasing its energy and water consumption since 2006 due to the increase in staff and campus infrastructure expansion. However, 67% of the waste generated is classified as recyclable. The study indicates improvements in adopting programs for efficient water and electric energy use and implementing selective waste collection with a destination of the recyclable fraction to recycling cooperatives. Ottoni and Amaral (2021), analyzing data on solid waste generated in the State University of Rio de Janeiro (UERJ), describe the participation of public universities as still timid when it comes to sustainable solid waste management, but highlight many benefits, mainly focusing on the reduction of negative environmental impacts and a considerable socioeconomic return for the institution.

By the Federal University of Rondônia (UNIR), Maio (2017) accents the importance of Educational Institutions in the performance of the social role and influence in the society that they are inserted, impacting economic, political and social growth, having its importance in achieving sustainability in organizations. The evaluation through the perception of university managers was based on the Method of Assessment of Sustainability Indicators, analyzing four dimensions of sustainability: social, environmental, economic and cultural. The survey answered by UNIR managers placed the University in the range considered "unsustainable" and the indicators that contributed most to this result were those of the environmental and social dimensions. The author also pointed out that analyzing these dimensions and their indicators makes it possible to list corrective or preventive actions in the search for improvements related to sustainable development. A similar study was conducted by Gazzoni et al. (2018) in the Federal University of Santa Maria (UFSM) and sought to investigate the degree of knowledge the institution's staff had about sustainable themes of Public Administration and how the SD practices proposed for HEIs are adopted on campus. It was analyzed through HEI's documents, archives and a questionnaire applied to 649 UFSM employees. The results showed that the participants were unaware of the sustainable themes focused on Public Administration. It was also found that it was not common practice for most of the interviewees to be concerned with rationalizing the resources they use. There are individual efforts, but they do not reflect the group's behavior. The authors concluded that the search for SD at UFSM moves slowly, a fact confirmed in a new study conducted by Falcão and Silveira (2021), showing that the institution did not reach a satisfactory performance by the UIGreenMetric measurement method. However, the authors highlighted that within the context of the ranking and the reality of the participating Brazilian universities, one could not consider that UFSM obtained a bad performance at all.

In the Federal University of Mato Grosso do Sul at Aquidauana (UFMS/CPAQ), Santos et al. (2021) evaluated how implementing sustainable policies is influenced by organizational culture. The authors concluded that when developing a culture focused on sustainability, campus management must discuss the topic and, thus, take the lead in institutional actions to be carried out with the definition of objectives to be achieved from the presentation of a plan of action guidelines.

Lemke and Barroso (2021) studied how the concept of Sustainability and Sustainable Development is understood by students of technical courses at the Federal Institute of Education, Science and Technology of the State of Rio Grande do Sul (IFSul) in Pelotas. The authors found that the vast majority of students understand the importance of the concept of Sustainable Development for their course and their career. However, they perceive that the institution has little focus on environmental issues. The professors have autonomy in executing the menus, so these issues are diffused during the contents given to the students, losing the expected potentiality. The research also indicates that more emphasis on the concept of Sustainability may favor the school community so that its students continue to enjoy the institution's potential.

In Minas Gerais State, Fernandes et al. (2019) present a case study on the implementation phases of the sustainability program (Licensing, Management, Green Campus and Education for Sustainability) in Pontifical Catholic University - PUC Minas, in Betim. They concluded that the institution still can not be classified as sustainable, but they indicate progress in daily activities on campus and advances compared to periods before implementation. In Juiz de Fora (MG), Rosas (2020) conducted a documentary survey and highlighted that the Federal University of Juiz de Fora is still in the early stages of implementing SD practices. Not all dimensions are mentioned in its Institutional Development Plan and suggested improvements. In a survey conducted in the State University of the Paraná Central-West region (UNICENTRO) in Guarapuava, Paraná State, actions aimed at sustainability through extension projects were verified. However, no internal policy is implemented and the topic is not treated with the importance it needs. Many activities already proposed to the institution never left the design stage and only a few actions were taken in a very timid way, which has not been contributing effectively to the academic community's awareness of the subject (Silveira 2017). Campos and Almeida (2019) descriptively analyzed the institutionalization of sustainable management in the Federal University of Tocantins, going through three methodological steps: bibliographic; documentary research and questionnaire application to the university staff. They lay emphasis on the little knowledge about socio-environmental practices.

Borja et al. (2019) concluded that the SDGs must be interconnected with the population's aspirations for a sustainable city and that the objectives of public policies to provide for the total development of urban property and its social functions must be defined. In this sense, Moura & Barreto Filho (2018) studied the case of Rafael Fernandes, a municipality in the Rio Grande do Norte State, where they analyzed the public policies applied and their relationship with Sustainable Development.

The authors point as an example the city of Curitiba, Paraná State, considered one of the leading sustainable cities in Latin America, evidencing that the successful trajectory of urban planning has positive effects in promoting a more qualified urban environment (Bichueti et al. 2017). In addition to HEIs, other Brazilian cities have sustainability programs in progress or in the research phase. In the city of Uberlândia, Minas Gerais, Santos (2019a, b) studied the impacts and paths that must be taken with the management of solid waste, having as a concern the inadequate disposal in the municipal sanitary landfill. In this sense, reintroducing in the production cycle is the most effective destination to ensure efficient control and management.

In research conducted by Souza et al. (2021), a study on the use and reinsertion of construction and demolition waste (CDW) was conducted in the city of Rio Branco, Acre State. According to the authors, besides the reduction of impacts (environmental, economic and social) caused, the city lacks aggregate deposits, which causes a direct effect on the construction costs. Thus, the production and use of CDW have become an interesting alternative. With substitutions ranging from 50 to 100% of aggregates, the mortars and concretes produced with CDW showed to be satisfactory for use. Despite a slight decrease in the mechanical strength, other characteristics, such as lower density and higher resistance to chloride penetration, were improved compared to the conventional mixture. In a previous study, Souza et al. (2020) pointed out the inefficient management of CDW by the municipal administration, which, despite being concerned with the collection, transportation and final disposal of the debris, lacks adequate sites, which serve mainly the small volume generators.

4 The Federal University of Viçosa in the Context of Sustainability

As part of the great universities of the world, the Viçosa campus of the Federal University of Viçosa (UFV), located in the Zona da Mata of the State of Minas Gerais. It has a large infrastructures such as classroom buildings with auditoriums, restaurants, sports practice pavilions, student housing, library, several departments equipped with modern research laboratories, as well as workshop structures and maintenance sectors to ensure the proper functioning of the institution. Its total physical area is 2,353.94 ha and the total built area is 431,001.94 m^2, according to the Pro-Rector of Planning and Budget (Silva 2017). It can be considered analogous to a small town, but with problems of large urban centers.

Currently, UFV is ranked 285th in the UIGreenMetric (Overall ranking 2021) (GREENMETRIC 2021). Since 2016 (the first year of participation in the ranking), UFV has kept falling, as seen in Table 1. An increase is observed in the total score over the years, but a drop in the overall ranking. In part, such a drop is attributed to the increase in the number of institutions participating in the ranking since (it is a voluntary action of the institutions). In addition, it is reasonable to believe that a HEI is willing to present its data after implementing programs that enhance its scores in the evaluated aspects. However, it is worth mentioning that the continuous decreases in the score related to "Configuration and infrastructure" from 2019 to 2021 occurred in the point related to "Transportation". These are not the most relevant topics, but they account for 33% of the institution's evaluation.

According to Brianezi et al. (2014), the UFV Viçosa campus has a large vegetation cover but not enough to offset the emissions and removals of Greenhouse Gases (GHG) emitted by the activities of the energy sector (electricity, biomass and Liquefied Petroleum Gas - LPG), transportation, waste (solid and effluent) and livestock, which were considered in their research. According to the researchers, the total GHG emissions of the UFV Viçosa campus were estimated at 2,962.54 $tCO_2eq.\ year^{-1}$, if livestock is not included. The afforestation of the campus contributes to the mitigation of only 4.24% of the total amount. The forest, grassland and reforestation areas on campus presented an annual carbon fixation rate of 5,181.47 $tCO_2eq.$, a value higher than the GHG emissions from livestock (3,071.64 $tCO_2eq.$). To neutralize only the emissions from livestock, an area of 311.31 ha would be necessary, about 2/3 of the total area of sinks. However, even having a huge green area, the offset of the emission of 6,034.18 $tCO_2eq.$ is not achieved. There is a deficit of 727.02 $tCO_2eq.$ Carbon sinks show their importance and must be increased to at least achieve neutrality of the GHG emissions generated.

The institution's structure linked to the Administration Pro-Rectory (PAD) includes several directorates, including are the Projects and Works Directorate (DPO) and Building Maintenance Directorate (DIM), which are responsible for the renovation works and maintenance of building facilities. The Environment Directorate (DAM) is in charge of green area maintenance, services and disposal of waste generated on campus.

The activities developed on the campus generate different types of waste, from common waste to hazardous waste that has its destination controlled by environmental agencies (toxic waste from laboratory activities).

To the common garbage, there is a collection and disposal routine performed by employees using vehicles from the fleet itself. The organic and non-recyclable waste is collected in a compactor truck and sent to the sanitary landfill of the city of Viçosa, located at a distance of approximately twelve kilometers from the campus. The recyclable waste is destined to the Associations of Collectors of Recyclable Materials (ACAMARE and ACAT), located at the Screening and Recycling Plant of Viçosa, which is closer to the campus (six kilometers). In both cases, these services generate costs to be destined. In this context, the residues generated by building maintenance activities are also included. The construction and demolition residues (concrete, mortar, plaster, granite, masonry, ceramics and residues from water and sewage treatment) have the same destination as the common garbage, that is, the sanitary landfill. The transportation is carried out in buckets, at polycrane trucks (also called Buck trucks) by an outsourced company.

Table 1 Overall ranking of the Federal University of Viçosa, Minas Gerais, Brazil, according to the UIGreenMetric ranking from 2016 to 2021 (GREENMETRIC 2021)

Universal targets of the Sustainable Development Goals (SDGs)									
Year	Classification	Total institutions analyzed	Overall score	Configuration and infrastructure	Energy and climate change	Wastage	Water	Transportation	Education and research
2016	139	515	5352	953	856	849	486	555	1653
2017	187	619	5074	905	1054	1077	505	513	1020
2018	219	718	5500	675	1125	900	625	600	1575
2019	162	780	6325	1025	1450	825	675	1200	1150
2020	206	911	6725	800	1525	1050	725	1200	1425
2021	285	956	6775	725	1425	1275	750	1050	1550

5 Concluding Remarks

The sustainability programs implemented in HEIs bring with them the fundamental essence of universities, to train citizens capable of transmitting and multiplying practices and thoughts in an attempt to minimize the damage that disorderly growth and industrialization cause to the planet. Small modifications in the organizational structure can reflect measurable gains and significant improvements in the UIGreenMetric ranking.

Each institution has demands and limitations related to sustainable issues. The peculiarities and local cultures are limiting factors for standardization in implementing techniques and programs aimed at sustainability in HEIs. However, the growing number of publications on the subject leads to the creation of systematic databases on specific topics, which, in a long run, can become a means of consultation for future implementation actions.

This research has shown that:

- In the world scenario, some HEIs emerge with implemented sustainable programs, in the most diverse areas, with positive reflections, which should serve as an example for other universities.
- The concern of Brazilian institutions with the improvement of their practices and facilities is evident and has been gaining strength every year, which can be noticed by the growing number of research and publications reporting efforts and advancements in implementing sustainable practices.
- Like other institutions in Brazil, the UFV has sought to introduce green conduct in recent years, even though it is not yet reflected as expected in the international rankings.

The concern with the institutional management should focus on the three pillars of sustainability (social, economic and environmental). Promoting the training of university staff (professors and administrative) is essential for raising awareness and training students and is one of the main roles in preparing qualified professionals for the future. The option for purchasing and consuming sustainable items assists in appropriate logistics and better use of invested resources. Last but not least, the concern with managing water and energy consumption, incentives to use more ecological means of transportation, correct waste disposal, and implementing internal recycling programs are fundamental processes for classifying the academy as a SU.

References

Abtahi SM (2021) Isfahan university of technology (IUT): towards a green campus energy, climate and sustainable development initiatives at IUT. J Sustain Perspect 1:424–430. https://doi.org/10.14710/jsp.2021.12035

Alonso-Almeida MDM, Marimon F, Casani F, Rodriguez-Pomeda J (2015) Diffusion of sustainability reporting in universities: current situation and future perspectives. J Clean Prod 106:144–154. https://doi.org/10.1016/j.jclepro.2014.02.008

Amaral LP, Martins N, Gouveia JB (2015) Quest for a sustainable university: a review. Int J Sustain High Educ 16:155–172. https://doi.org/10.1108/IJSHE-02-2013-0017

Atici KB, Yasayacak G, Yildiz Y, Ulucan A (2021) Green University and academic performance: an empirical study on UI greenmetric and world university rankings. J Clean Prod 291:125289. https://doi.org/10.1016/j.jclepro.2020.125289

Bichueti RS, Gomes CM, Kneipp JM et al (2017) Cidades sustentáveis no contexto brasileiro: a importância do planejamento para o desenvolvimento urbano sustentável [Sustainable cities in the Brazilian context: the importance of planning for sustainable urban development]. XIX Encontro Int sobre Gestão Empres e Meio Ambient, p 17

Bizerril M, Rosa MJ, Carvalho T (2018) Construindo uma universidade sustentável : uma discussão baseada no caso de uma universidade portuguesa [Building a sustainable university : a discussion based on the case of a Portuguese university]. Avaliação Rev Da Avaliação Da Educ Super 23:424–447

Borja PC, dos Santos MEP, Moraes LRS (2019) Direito à cidade sustentável e à qualidade urbanoambiental: um estudo em Salvador, Bahia, Brasil [Right to sustainable city and urban-environmental quality: a study in Salvador, Bahia, Brazil], pp 174–193

Brianezi D, Gonçalves Jacovine LA, Boechat Soares CP, et al (2014) Balanço de emissões e remoções de gases de efeito estufa no campus da universidade federal de viçosa [Balance of emissions and removals of greenhouse gases on the campus of the federal university of viçosa]. Floresta e Ambient 21:182–191. https://doi.org/10.4322/floram.2014.017

Campos MC, Almeida L de A (2019) Institucionalização da gestão sustentável em instituições de ensino superior: um estudo com servidores da Universidade Federal do Tocantins [Institutionalization of sustainable management in higher education institutions: a study with employees of the Federal University of Tocantins]. Rev Bras Gestão Ambient e Sustentabilidade 6:819–840. https://doi.org/10.21438/rbgas.061414

Ceulemans K, Molderez I, Van Liedekerke L (2015) Sustainability reporting in higher education: a comprehensive review of the recent literature and paths for further research. J Clean Prod 106:127–143. https://doi.org/10.1016/j.jclepro.2014.09.052

Dagiliūtė R, Liobikienė G, Minelgaitė A (2018) Sustainability at universities: students' perceptions from green and non-green universities. J Clean Prod 181:473–482. https://doi.org/10.1016/j.jclepro.2018.01.213

Fadhil A, Waheeb R (2021) A greenway for sustainable transportation of the university campus: Diyala University as a case study. In: IOP conference series: earth and environmental science, p 754. https://doi.org/10.1088/1755-1315/754/1/012013

Falcão MFS, Silveira ADO (2021) A UFSM é uma universidade sustentável [Is UFSM a sustainable university]? Rev Gestão e Avaliação Educ 9:1–16. https://doi.org/10.5902/2318133840372

Fernandes IS, Laureano FV, Leite EB et al (2019) Histórico de ações por uma universidade sustentável na PUC Minas Betim [History of actions for a sustainable university at PUC Minas Betim]. Sinapse Múltipla 8:8–21

Fissi S, Romolini A, Gori E, Contri M (2021) The path toward a sustainable green university: the case of the University of Florence. J Clean Prod 279:123655. https://doi.org/10.1016/j.jclepro.2020.123655

Gazzoni F, Luciane Scherer F, Schneider Hahn I, Brachak dos Santos M (2018) O papel das IES no desenvolvimento sustentável: Estudo de caso da Universidade Federal de Santa Maria [The role of HEIs in sustainable development: A case study from the Federal University of Santa Maria], pp 48–70

GREENMETRIC U (2021) Overall rankings 2021. https://greenmetric.ui.ac.id/rankings/overall-rankings-2021. Accessed 19 May 2022

Guterres A (2020) The sustainable development goals report 2020. United Nations Publ issued by Dep Econ Soc Aff:1–64

Hakiki R (2019) Identification of green and sustainable campus indicators in its implementation at president university. In: ICSECC 2019 International conference on sustainable engineering and creative computing new idea new innovation processing, pp 287–292. https://doi.org/10.1109/ICSECC.2019.8907154

Heravi G, Aryanpour D, Rostami M (2019) J Pre Proof Sci Total Environ:135907. https://doi.org/10.1016/j.jobe.2021.102798

Houaiss A, Villar M de S (2015) Pequeno Dicionário Houaiss da língua portuguesa, 1st edn [Small Houaiss dictionary of the Portuguese language, 1st edn]. Moderna, São Paulo

Keeble BR (1988) The brundtland report: "our common future." Med War 4:17–25. https://doi.org/10.1080/07488008808408783

Lauder A, Sari RF, Suwartha N, Tjahjono G (2015) Critical review of a global campus sustainability ranking: greenmetric. J Clean Prod 108:852–863. https://doi.org/10.1016/j.jclepro.2015.02.080

Leal Filho W, Shiel C, Paço A et al (2019) Sustainable development goals and sustainability teaching at universities: falling behind or getting ahead of the pack? J Clean Prod 232:285–294. https://doi.org/10.1016/j.jclepro.2019.05.309

Leal Filho W, Vidal DG, Chen C et al (2022a) An assessment of requirements in investments, new technologies, and infrastructures to achieve the sdgs. Environ Sci Eur:1–17. https://doi.org/10.1186/s12302-022-00629-9

Leal Filho W, Wall T, Barbir J et al (2022b) Relevance of international partnerships in the implementation of the UN sustainable development goals. Nat Commun 13:20–23. https://doi.org/10.1038/s41467-022-28230-x

Lemke CS, Barroso PF (2021) Percepções sobre sustentabilidade dos estudantes do IFSul Campus Pelotas-Visconde da Graça (CaVG) [Students' perceptions about sustainability at the IFSul Campus Pelotas-Visconde da Graça (CaVG)]. Rev Insignare Sci (RIS) 4:189–205. https://doi.org/10.36661/2595-4520.2021v4i1.12100

Lidstone L, Wright T, Sherren K (2015) Canadian STARS-rated campus sustainability plans: priorities, plan creation and design. Sustain 7:725–746. https://doi.org/10.3390/su7010725

Maio GF (2017) Práticas de Gestão Sustentável na Universidade Federal de Rondônia. Universidade Federal de Rondônia [Sustainable management practices at the federal university of Rondônia. Federal University of Rondônia], p 93

Moura RSC, Barreto Filho BDF (2018) Políticas Públicas Para a Promoção Do Desenvolvimento Sustentável Na Cidade De Rafael Fernandes-Rn [Public policies for the promotion of sustainable development in the city of Rafael Fernandes-Rn]. Rev Geotemas 7:17–44. https://doi.org/10.33237/geotemas.v7i2.2079

Mustafa A, Kazmi M, Khan HR et al (2022) Towards a carbon neutral and sustainable campus: case study of NED university of engineering and technology. Sustain 14:1–17. https://doi.org/10.3390/su14020794

Nousheen A, Yousuf Zai SA, Waseem M, Khan SA (2020) Education for sustainable development (ESD): effects of sustainability education on pre-service teachers' attitude towards sustainable development (SD). J Clean Prod 250:119537. https://doi.org/10.1016/j.jclepro.2019.119537

Ottoni AB, Amaral M (2021) Gestão Sustentável de Resíduos Sólidos em Universidade Públicas Brasileiras [Sustainable Solid Waste Management in Brazilian Public Universities]. Alta Paulista. XVII Fórum Ambiental, pp 491–504

Ribeiro EN, Carneiro RL, Da Silva Galdino OP et al (2019) Environmental diagnosis of a university campus as a strategy for a sustainable practices proposal. Urbe 11:1–17. https://doi.org/10.1590/2175-3369.011.e20190029

Rosas FR (2020) Análise das ações de Desenvolvimento Sustentável implementadas na Universidade Federal de Juiz de Fora-Um estudo de caso. Universidade Federal de Juiz de Fora [Analysis of sustainable development actions implemented at the Federal University of Juiz de Fora-a case study. Federal University of Juiz de Fora], p 122

Sallit M (2020) universidades-brasileiras-estao-entre-as-mais-sustentaveis-do-mundo @ querobolsa.com.br. https://querobolsa.com.br/revista/universidades-brasileiras-estao-entre-as-mais-sustentaveis-do-mundo. Accessed 3 Jun 2021

Sammalisto K, Sundstrom A, Holm T (2015) Implementation of sustainability in universities as perceived by faculty and staff e a model from a Swedish university. J Clean Prod 106:45–54. https://doi.org/10.1016/j.jclepro.2014.10.015

Santos LIV dos, Freitas WR de S, Mello MMM, et al (2021) O papel da cultura organizacional para o desenvolvimento da sustentabilidade: o caso de um campus universitário brasileiro [The role of organizational culture in the development of sustainability: the case of a Brazilian university campus]. Desafio Online 9:212–239

Santos NAG (2019a) A universidade com sua ciência, tecnologia e inovação e as cooperativas no desenvolvimento local sustentável cubano. In: IX Seminário Internacional sobre Desenvolvimento Regional [The university with its science, technology and innovation and the cooperatives in the Cuban sustainable local development. In: IX International seminar on regional development]. Santa Cruz do Sul, p 23

Santos TBF (2019b) Coleta seletiva de resíduos sólidos urbanos em Uberlândia (MG): desafios e possibilidades de boas práticas para uma cidade sustentável [Selective collection of urban solid waste in Uberlândia (MG): challenges and possibilities of good practices for a sustainable city], p 127

Silva LMP (2017) UFV em números [UFV in numbers]. http://www.ppo.ufv.br/wp-content/uploads/2018/11/UFV-EM-NÚMEROS-2018-Gráfica.pdf. Accessed 7 Jun 2021

Silveira JHP (2017) Sustentabilidade e Responsabilidade Social [Sustainability and social responsibility], vol 3

Souza FS, Carvalho JMF, Silveira GG et al (2021) Application of construction and demolition waste in civil construction in the brazilian amazon—case study of the city of Rio Branco. Materials (Basel) 14. https://doi.org/10.3390/ma14092247

Souza F, Oeste B, Lima E, Peixoto R (2020) Gestão dos resíduos de construção e demoliçãoem rio branco, acre [construction and demolition waste management in rio branco], acre. South Am J 7:741–764

Sundram VPK, Hashim N, Shariff SH, Pujiati A, Ardiansari A (2021) Sustainable Transportation on University Campus: A Case at UiTM Selangor, Puncak Alam Campus, Malaysia and Universitas Negeri Semarang, Indonesia. Asian J. Univ. Educ. 17. https://doi.org/10.24191/ajue.v17i2.13407

Torgal FP, Jalali S (2010) Considerações sobre a Sustentabilidade dos Materiais de Construção [Considerations on the sustainability of building materials]. C Mater

Velazquez L, Munguia N, Platt A, Taddei J (2006) Sustainable university: what can be the matter? J Clean Prod 14:810–819. https://doi.org/10.1016/j.jclepro.2005.12.008

Wachholz CB (2014) A sustentabilidade na universidade : o desafio da ambientalização na pontifícia universidade católica [Sustainability at the university : the challenge of environmentalization at the pontifical catholic university]. 1–17

Washington-Ottombre C, Washington GL, Newman J (2018) Campus sustainability in the US: environmental management and social change since 1970. J Clean Prod 196:564–575. https://doi.org/10.1016/j.jclepro.2018.06.012

Yan J (2020) Encouraging sustainable environment using bicycle sharing system in university campus, Bangkok. Built Environ J 17:85. https://doi.org/10.24191/bej.v17i2.10118

Yuan X, Zuo J, Huisingh D (2013) Green Universities in China e what matters? J Clean Prod 61:36–45. https://doi.org/10.1016/j.jclepro.2012.12.030

Zheleznyak OY, Korelina MV (2022) University campus as a model of sustainable city environment. In: Proceedings of the 5th international conference on construction, architecture and technosphere safety, pp 375–385. https://doi.org/10.1007/978-3-030-91145-4_36

Gustavo José Rodrigues Lopes Civil Engineer from Federal University of Viçosa (2006), Master in Civil Engineering from Federal University of Viçosa (2008). Currently a civil engineer at the Federal University of Viçosa and a Ph.D. student at the Civil Engineering Department of the Federal University of Viçosa. He has experience in Civil Engineering, with emphasis on environmental sanitation and sustainable materials.

José Maria Franco de Carvalho is a Civil Engineer from the Federal University of Viçosa (2004), a Master in Civil Engineering from the Federal University of Ouro Preto (2007) and a Doctor in Civil Engineering from the Federal University of Ouro Preto with a sandwich at the Federal Institute for Materials Research and Testing Berlin (2019). Currently, he is an Adjunct Professor at the Federal University of Viçosa and his research interests include cementitious and composite materials, eco-efficient concrete and the reuse of industrial and mineral waste.

Energy Consumption Evaluation for an Experimental Supermarket Located on a University Campus

Thiago Toledo Viana Rodrigues and Joyce Correna Carlo

Abstract University campuses can be testing environments that emulate other realities. This promotes ideas that can be replicated in society, especially concerning sustainable development practices. A campus typology characterized by scattered buildings provides a landscape with unshaded buildings. Given the current concerns about climate change, research applied to these spaces enables studies to be further used on a large scale, such as in buildings and their energy use. This study aims to identify the potential to generate photovoltaic energy by characterizing the energy end-use in a supermarket located inside the Federal University of Viçosa's campus. Besides its commercial purpose, the supermarket is an experimental laboratory for internships and practical lectures for courses in the areas of nutrition, economy, management, and civil engineering. The method consists in analysing the energy bills and measuring the energy use on site, enabling a calibrated energy model for building performance simulation on EnergyPlus, upon which a roof-mounted photovoltaic system simulation had been proposed. The results showed that the main end-use is 74% for refrigeration, followed by 18% for equipment, 6% for lighting, and 2% for HVAC (Heating, Ventilating and Air Conditioning) consumption. The photovoltaic system generated 31% of the supermarket's annual energy demand, which indicated an improvement in the building's energy performance provided by the unshaded area on the roof. It is important to emphasize solar energy is one of the most sustainable forms of renewable energy generation, avoiding carbon dioxide emissions, noise, and power grid losses since it can be generated at the site of consumption.

Keywords Energy auditing · Energy efficiency · Sustainability · Photovoltaic energy

---

T. T. V. Rodrigues (✉) · J. C. Carlo
Federal University of Viçosa, Viçosa, Brazil
e-mail: thiago.t.rodrigues@ufv.br

Departamento de Arquitetura e Urbanismo, Av. P. H. Rolfs, S/no., Centro, Viçosa, Minas Gerais, Brazil

J. C. Carlo
e-mail: joycecarlo@ufv.br

W. Leal Filho et al. (eds.), *The Contribution of Universities Towards Education for Sustainable Development*, World Sustainability Series,
https://doi.org/10.1007/978-3-031-49853-4_12

1 Introduction

University campuses are frequently employed to develop critical thinking and spur social engagement. This promotes ideas that can be replicated in society, especially concerning sustainable development practices.

The object of this study is a supermarket located on the main campus of the Federal University of Viçosa. In addition to functioning as a normal supermarket, this building serves as a space for internships and practical classes related to the food sector and is considered an experimental laboratory for research in marketing, economics, food sector and civil engineering.

Energy efficiency research in the 2000s and 2010s in Brazil focused on addressing residential and office buildings. As a result, some specific typologies such as supermarkets still lack representation in scientific literature, despite their ubiquity and the high energy budget they demand (Barbosa and Labaki 2013).

According to the Brazilian Association of Supermarkets (*Associação Brasileira de Supermercados—ABRAS*) in 2018 the country had about 89,700 retail stores in operation, 38,300 of which had two or more check-outs (cash registers) covering supermarkets and hypermarkets (ABRAS 2019). The remaining 51,400 units, which had only one check-out, were characterized as grocery and convenience stores.

Supermarkets are among the types of buildings with the highest energy consumption density, demanding more than twice the energy consumed by a hotel or office of the same size (Iyer et al. 2015). In Europe, supermarkets are responsible for 9% of energy consumption by buildings (Building Performance Institute Europe 2011), between 500 and 1,000 kWh/m^2/year, which corresponds to three times the consumption density of residential buildings and five times that of office buildings (Martos et al. 2013).

Supermarkets consume about 4% of electricity in industrialized countries (Eurelectric 2011). If supermarkets could save 50% of their total energy use, through more efficient solutions, the savings would amount to 64 TWh in electricity consumption in the European Union (Lindberg et al. 2011).Supermarkets cover several systems and services, which makes the analysis and forecasting of energy demand a difficult task. The various exchanges of heat between the building, the HVAC system and the refrigeration systems, along with the many different storage requirements for products, hours of operation and transitory occupancy patterns demand in loco studies and surveys to establish parameters that feed dynamic simulation models (Mylona et al. 2017). As an example, maintaining the quality of the food on display requires refrigeration equipment, which makes this typology quite different from other commercial buildings (Barbosa and Labaki 2013); this system has a high energy consumption and remains in constant operation. International research indicates that the largest share of energy consumption in a supermarket occurs by refrigeration and lighting systems (kolokotroni et al. 2019; Mylona et al. 2018; Pather-Elias et al. 2012; Timma et al. 2016). Furthermore, about 60% of the total energy demand in a supermarket is used to store and display refrigerated and frozen foods (Suzuki et al. 2011).

Nevertheless, there is a lack of studies showing the typical consumption of this typology, the contribution of each system to the total energy consumption, and that propose measures to increase its energy efficiency, mainly for hot countries such as Brazil. Most of the studies found on supermarkets are for predominantly cold climates (Braun et al. 2014; Czajkowski et al. 2004; Kolokotroni et al. 2019).

The National Electric Energy Conservation Program (*Programa Nacional de Conservação de Energia Elétrica—PROCEL*) prepared a report on equipment ownership and usage habits for supermarkets and hypermarkets in Brazil based on data collected in 2005 (PROCEL 2008). Regarding end uses, the report points out that refrigeration, air conditioning and lighting systems are the main energy consumers in supermarkets, accounting for 89% of electricity budgets. End uses were: refrigeration (38%); air conditioning (32%); lighting (19%); equipment (3%); heating (2%); and others (6%). It is important to notice the research refers to 2005, and no other surveys have been published since then, thus, the authors understand the data may be noncurrent.

In Brazil, the PBE Edifica program through the National Energy Conservation Etiquette (*Etiqueta Nacional de Conservação de Energia—ENCE*) defines performance levels for building design, however, there is still a need to better understand the energy consumption of buildings in operation to allow consumption management and a more efficient operation.

Moreover, this typology shows great potential for solar energy generation. Since 2012, with the regularization of micro and mini-distributed generation, the number of buildings with photovoltaic installations has grown, following regulations RN 482 (ANEEL 2012) and RN 687 (ANEEL 2015), on micro and mini-renewable energy distribution. Direct solar radiation appears among all Renewable Energy Sources (RES), as the most abundant, inexhaustible and clean (Jelle et al. 2012). Additionally, photovoltaic modules do not emit pollutants or noise when generating energy (Kang et al. 2012).

Although renewable energy contributes to 43.5% of the Brazilian energy matrix, 82% of this energy comes from centralised hydroelectric plants (BEN 2017). These, despite being a renewable energy source, are not sustainable because they have limited growth capacity, as their growth causes a series of economic, environmental and social impacts. In addition, the irregularity of rainfall generates periods of prolonged drought that combined with the increasing restrictions on the construction of new hydroelectric plants and the rapid growth in demand, leads to the activation and construction of new plants powered by coal, natural gas, and other pollutant fuels (BEN 2017).

In this scenario, and given their vastness, supermarket roofs offer some prime potential for distributed energy generation, mainly photovoltaic (PV) energy. Generally, medium-sized supermarkets have a ground floor for sales and a mezzanine for administration, topped with a roof with similar dimensions to the ground floor. When surrounded by low-rise buildings, this typology also presents a broad sky view factor, which contributes to the effectiveness of PV modules.

Some works have evaluated the potential of installing PV systems in supermarkets. Barbosa et al. (2015) presented a technical and economic study on the use of photovoltaic energy for lighting in two supermarkets in the city of Bogotá (Colombia), while Renno et al. (2018) analyzed the insertion of PV systems in a supermarket located in Naples (southern Italy).

Likewise, this work also investigates the potential for photovoltaic energy generation as well as proposes an on-site measurement and calibration process that exposes the consumption profile of the building in operation.

2 Methodology

This research is based on an analysis of consumption data from electricity bills, electricity consumption data measured in loco and electricity consumption data generated by computer simulation. In addition, it seeks to observe, analyze and collect specific information about the building uses and consequent energy consumption of supermarkets. The authors collected the building usage modes from an assessment of electricity consumption. These were processed to create a computational model which was calibrated using on-site measurements of energy consumption so that the two surveys made it possible to model and understand the profiles of electricity consumption and end uses of the building.

2.1 Description of the Case Study

This study analyzes the in-campus supermarket of the Federal University of Viçosa, in the Minas Gerais state of Brazil. The city of Viçosa (Table 1) belongs to bioclimatic zone 3, according to NBR 15,220 (ABNT 2005a).

Despite being located on campus, the company is open to all customers and supplies Viçosa and surrounding towns. The construction of the supermarket dates from 2003, however, the building underwent renovations in 2008, 2011 and 2016. Opening hours are from 8 am to 8 pm from Monday to Saturday and from 9 am

Table 1 Weather data for Viçosa, MG, Brazil. Based on Guimarães and Carlo (2015) e CRESESB (2021)

Location	Latitude	Longitude	Altitude (m)	Mean annual temperature (2005 to 2010) (°C)	Mean annual RH (2005 to 2010) (%)	Mean annual daily solar irradiation totals (1999 to 2015) (kWh/m^2)
Viçosa	20.77° S	42.87° O	712.2	19.7	78.5	4.72

to 1 pm on holidays, remaining closed on Sundays. Most employees work from 07:00 to 20:30 from Monday to Saturday, except for the administrative sector which works from 08:00 to 18:00 on weekdays and from 08:00 to 13:00 on holidays. There are 91 employees in total, which characterizes it as a medium-sized supermarket, according to the Brazilian Institute of Geography and Statistics (*Instituto Brasileiro de Geografia e Estatistica–IBGE*) classification for commercial/service companies.

The establishment's immediate surroundings are sparsely built, with the building being close to green areas and one of the reservoirs that supply the campus (Fig. 1). The building is in full sun throughout the year, with the main entrance facing 29° N and the biggest facades facing Southeast and Northwest. These solar orientations promote thermal gains and can increase heat discomfort in hot climates. This is especially true on the Northwest façade, which receives direct solar radiation during the afternoon, even if primarily on the equinoxes and in winter.

The authors see some potential for the installation of photovoltaic systems on the roof, which in addition to generating energy, could contribute to mitigating the thermal gains of the building.

The building has a sawtooth type roof, with clerestories that allow natural lighting inside (zenith lighting). A side effect of this is direct sunlight in the afternoon, which causes heat discomfort (Fig. 2). This discomfort is felt mainly in the administrative sector, which is located on the mezzanine and receives direct solar radiation, making an air conditioning system necessary.

The supermarket has a total built area of 2,996 m^2, divided into two floors, the ground floor with 1,863 m^2 and the mezzanine with 1,133 m^2 (Fig. 3). The sales area has approximately 1,100 m^2. Only the administrative area has artificial air conditioning. There are 11 rooms equipped with autonomous split-type air

Google Earth Image ©2022 Maxar Technologies

Fig. 1 Aerial photograph of Supermercado Escola (Google Earth 2022)

Fig. 2 Nortwest perspective of Supermercado Escola

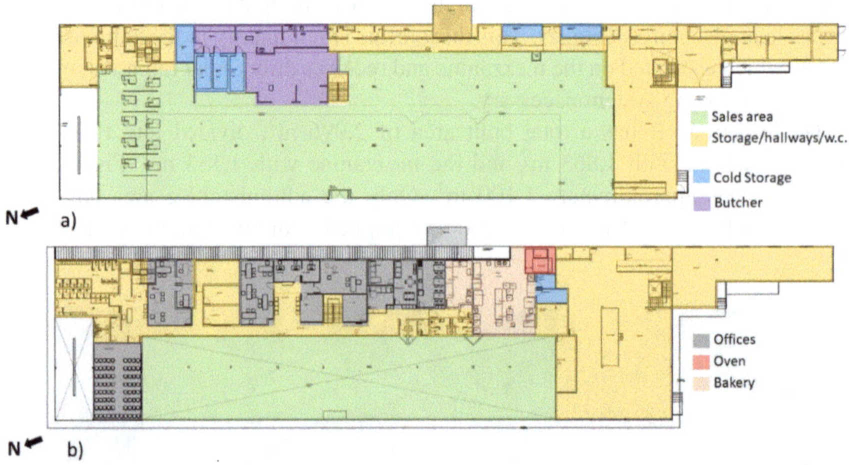

Fig. 3 Floor plans and sectors of Supermercado Escola, in Viçosa-MG, Brazil (2021): **a** ground floor e **b** 1st Floor

conditioning and typical office equipment. The lighting system consists of LED (Light-Emitting Diode) and fluorescent lamps in luminaires for tubular lamps. The only systems in operation at night are the cold rooms and refrigerators. There is a diesel generator for power outages, whose Scania engine has a maximum power of 250 kVA/h.

The authors collected the building's monthly electricity consumption for 2018 and 2019 (Fig. 4). Its annual consumption (Sep. 2018—Aug. 2019) was 706,880

Fig. 4 Annual energy consumption (set. 2018—ago. 2019)

Table 2 Average supermarket characteristics based on *ABRAS* (2019)

Ranking ABRAS	Stores	Check-outs/store	Sales area(m^2) /store	Employees/store
924 replying companies	8,345	9	1,243	79.7 (medium-sized)
Supermercado Escola	1	11	1,100	(medium sized)

kWh, which represented about 4.5% of the total consumption of the University for that year, or 236 kWh/m^2/year and corresponded to an annual cost of R$ 384,929.00.

The consumption of Supermercado Escola differs from the data in the available international literature (Rodrigues et al. 2021), mainly due to climatic variations and differences in the use of artificial air conditioning systems.

Supermercado Escola is a relevant case study for Brazil due to its similarity to the norm. According to the *ABRAS* ranking data (2019), its main characteristics, such as the number of check-outs; sales area; and number of employees (Table 2), fall close to the average of the 8,345 voluntarily responding establishments.

2.2 *Simulation of the Case Study*

The building consumption was simulated in EnergyPlus version 8.7, with a TMY (Typical Meteorological Year) file for the city of Viçosa-MG (Guimarães and Carlo 2015). The simulation generated electricity consumption data for a full year period. The virtual model has two floors (Ground floor + Floor 1), a sawtooth roof (Fig. 5), and the sales area has a double-height ceiling.

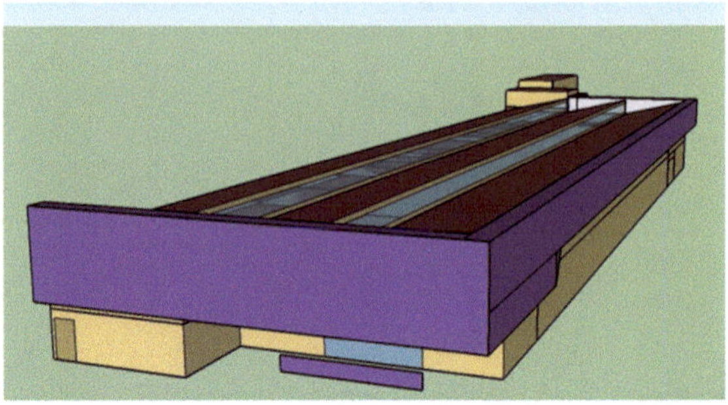

Fig. 5 Northeast perspective of the virtual model

The model has dimensions of 100.90 m × 21.63 m × 7.47 m, with a ceiling height of 3.15 m on the Ground floor and 2.75 m on Floor 1, plus 1.57 m sawtooth height. The main entrance facing 29° N. The model was configured with 49 thermal zones (Figs. 6 and 7). On the roof, each zone represents one sawtooth (Z45, Z46 and Z47).

The building exhibits different construction systems, mainly due to its various renovations. The on-site survey aimed to identify these components (Table 3).

The usage and occupation patterns followed working hours, considering the monthly average of check-outs. On-site observations helped determine the distribution of customers throughout the day. Similarly, work positions determined the distribution of employees by sector. ASHRAE 55 (ASHRAE 2013) was used to specify the metabolic rate (Table 4).

The manufacturer websites for the lamps and equipment provided the average consumption for its products. The whole lighting system was considered to be on during the hours of use of the building, while equipment used one of two schedules: full-workday and occasional use (Table 5). The refrigeration system was excluded

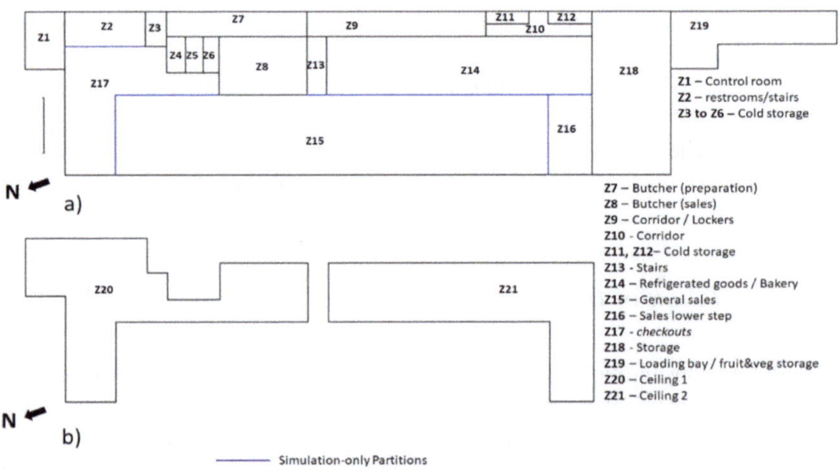

Fig. 6 Thermal zones: **a** Ground floor and **b** Ceilings

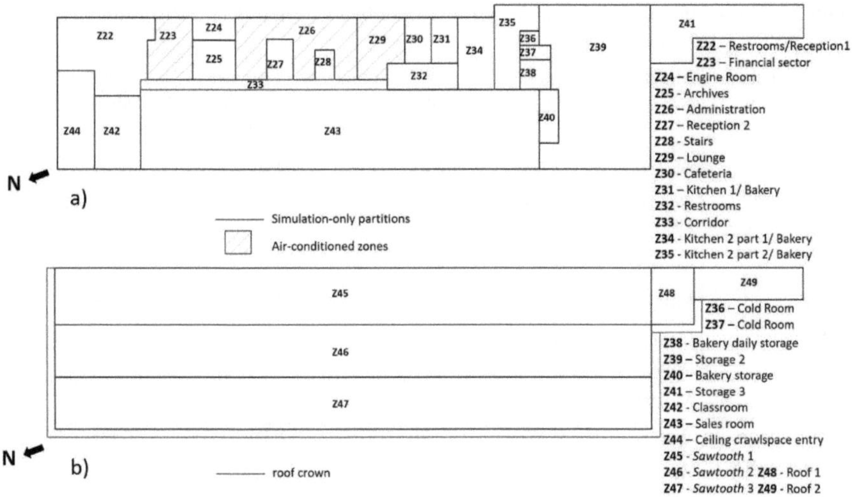

Fig. 7 Thermal zones: **a** Ground floor and **b** Attic

from these schedules since it operates 24 hours a day, and EnergyPlus offers specific models for simulating refrigerators and walk-in freezers.

Ventilation was configured with an air conditioning system in the offices and natural ventilation in the rest of the building. The windows were considered to be closed throughout the year for the offices and open during operation hours in the rest of the building. Wind pressure coefficients (Cp) were obtained using CpSimulator (2020), which employs Computational Fluid Dynamics (CFD) to derive Cp data for different construction scenarios and geometries. This method was described and validated in a wind tunnel (Bre et al. 2018; Gimenez and Bre 2019; Gimenez et al. 2018). Pressure coefficients were calculated for all external openings and entered into EnergyPlus, with wind incidence angles at 15° intervals.

Supermercado Escola's split air conditioner is located only in the administrative area, consisting of 11 rooms of different sizes, which total 200.19 m^2 (about 6.7% of the total area). The administrative part is represented by three zones: Z23 (47.39 m^2); Z26 (105.83 m^2); and Z29 (46.97 m^2). In these areas, a heat pump-type air conditioning system was configured using self-sizing with an 18°C heating setpoint in winter and a 24 °C cooling setpoint in summer.

2.3 *Energy Consumption Calibration*

Electricity bills from 2018 and 2019 were used to calibrate the model since the present research was carried out during the atypical years of 2020 and 2021. The main steps for calibration included discriminating electrical energy consumption by end-use and measuring on the electrical panel to identify the building's daily electric

Table 3 Construction materials and properties, according to Anexo Geral V (Instituto…, 2013) and to NBR 15.220 (ABNT 2005b)

Components	Materials	Thickness (cm)	Transmittance (W/m^2K)	Thermal capacity (KJ/m^2K)	Solar absorptance
Nortwest exterior wall	Mortar	2.50	2.46	150	0.15
	Ceramic brick	14.00			
	Mortar	2.50			
Exterior and Interior walls	Mortar	2.50	1.85	161	0.15
	Ceramic brick	9.00			
	Mortar	2.50			
Interior doors and partitions	Wood	0.50	2.70	8.04	0.60
	Air	3.00			
	Wood	0.50			
Cold storage	Steel cladding	0.08	0.25	11.06	0.25
	EPS	15.00			
	Steel cladding	0.08			
Ground floor slab	Ceramic	2.00	4.00	309.44	0.15
	Subfloor	3.00			
	Concrete	10.00			
Upper floor slab	Ceramic	2.00	3.57	423.74	0.15
	Subfloor	3.00			
	Concrete	15.00			
	Steel deck	0.12			
Forro PVC	PVC	0.80	1.12	22.22	0.15
	Rock wool	3.00			
	PVC	0.80			
Wooden ceiling	Wood	0.50	1.09	10.29	0.60
	Rock wool	3.00			
	Wood	0.50			
Roof	Metallic roofing	0.08	0.68	7.51	0.25
	EPS	5.00			
	Metallic roofing	0.08			

Table 4 Metabolic rates (Watts/person) according to ASHRAE 55 (ASHRAE 2013)

Clients	Employees		
	Office	Low-effort (lounge)	Medium effort
180	130	108	206

Table 5 Lighting power density (DPIL) and installed internal load density of equipment

	Lighting system	Equipment	
Sectors	DPIL (W/m^2)	Full-workday (W/m^2)	Occasional use (W/m^2)
Sales area	5.24	2.53	7.98
Offices	4.58	18.71	2.84
Production area	5.31	138.60	97.84
Storage and others	3.32	1.17	47.58

load curve. The premise for the calibration is that the sensitivity of the simulated model tends to follow the end uses that consume the most energy, as shown by Lam and Hui (1996).

The model was calibrated by making adjustments until two simulated parameters matched their measured counterparts. First, the hourly consumption measured during a period of time served as a goal for the simulation's hourly consumption for that same period. Similarly, monthly energy bills served as a goal for the simulation's monthly consumptions for each month. At the end of the calibration process, statistical indices were used to validate the calibrated model.

(a) Monitoring of actual consumption

A Fluke 435 power and energy quality analyzer was employed. It features a measuring range from 1 to 600 V with an accuracy of ±0.1% of the rated voltage. For electric energy measurement, the accuracy is ± 1% ± 10 points, with a measuring range depending on clamp scale and rated voltage.

Two measurements of one week each were performed. This procedure provided a typical working day, a typical Saturday and a Sunday as a reference for the calibration. In the first week of measurement, between January 28 and February 4, 2021, the analyzer was installed in the supermarket's main circuit board and the monitoring accounted for all the electricity consumption of the building over the measured days.

The second measurement was carried out between February 4th and 11th, 2021, on a section of the circuit board controlling a part of the refrigeration system, and produced the consumption of cold storage, butcher's refrigerators and larger vertical refrigerators (dairy products and vegetables).

After measurements, the model was calibrated by comparing simulated hourly consumption with consumption recorded in the measurement period (daily load curve), and also by comparing simulated monthly consumption with actual monthly electricity bills for a full year. The time scale of analysis varied for each version, starting in the Base Case (BC) with hourly and annual consumptions, and ending in the last version with monthly consumptions.

The weather file was altered to compare the measured data to the simulated hourly consumptions for the monitored period. That period's hourly meteorological variables were taken from the National Institute of Meteorology (*Instiuto Nacional de Metereologia—INMET*) database and replaced in the climate file, which was measured by the automatic station in Viçosa (A510). The variables are atmospheric pressure; dry bulb air temperature (TBS); dew point temperature; relative humidity; wind direction; and wind speed.

Maximum variation intervals between measured and simulated consumptions, as indicated by Westphal (2007), were used as a reference during the calibration process. These values are $\pm$ 5% for annual consumption and $\pm$ 20% for monthly consumption. Thus, modifications were made to the simulated model until it reached the ranges indicated by the literature. When the differences between the simulated and measured data met these intervals, the model was considered calibrated. Additionally, statistical indices were used to validate this model, with the intent of achieving greater scientific rigour in the analyses after the calibration process.

(b) Validation through statistical indices (MBE and CVRMSE)

The calibration was validated statistically, using the indices Mean Bias Error (*MBE*) and Coefficient of Variation of the Root Mean Square Error (*CVRMSE*). The *MBE* represents the mean of the errors in a certain sample, indicating the general behaviour of the simulated values regarding its regression line. When the value is positive, the model is underestimating real consumption, when it is negative, it is overestimating. The *CVRMSE* measures the variability of the errors between the simulated and measured values, indicating the capability of the model to predict the profile shape of the building´s total energy consumption.

By combining both methods, it was possible to avoid a validation error from the error compensation (ASHRAE 2014). For validation, hourly and monthly limits were adopted based on ASHRAE Guideline 14–2014 (2014) (Table 6).

2.4 Photovoltaic System Simulation

After calibrating the simulated model, a DAH-Solar DHM72X photovoltaic system with 19% efficiency according to Standard Test Conditions (STC) was added to its roof. The roof fields face southeast, which is not ideal for solar energy generation

Table 6 Limits for *MBE* and *CVRMSE* established by ASHRAE Guideline 14–2014 (2014)

Statistical indices	Hourly calibration limits	Monthly calibration limits
MBE (%)	$\pm$10	$\pm$5
CVRMSE (%)	30	15

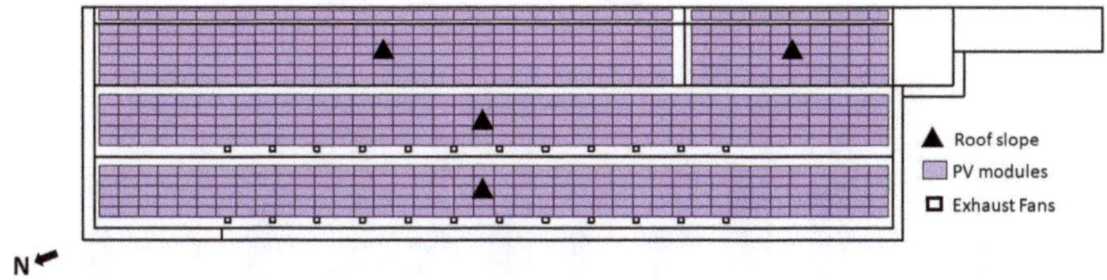

Fig. 8 Positions of PV modules on the roof

since facing north would be the most efficient for the southern hemisphere. Similarly, the 12° roof pitch is also subpar for that latitude.

The simulation was run with 593 modules on the roof of the supermarket. It is important to emphasize that the weight of the modules was ignored and that in a real situation a structural study must be carried out. Conversely, lightweight thin film alternatives are available on the market, such amorphous silicon. On roof Z45, it was possible to cover the entire roof face with modules since the building's surroundings cast no shadows on the roof during the energy generation period (Fig. 8).

However, module positioning on the Z46 and Z47 roofs adopted a distance of 2.22 m from the bottom of the slope to avoid the shadow of the next slope during the energy generation period. That distance was calculated using the solar altitude at its lowest point in the year (June 21st) at 09:00 in the morning since the energy generation before that hour is non-significant. The maximum number of modules possible was set up on roofs Z46 and Z47 in the simulation, avoiding the shadowed area and the roof exhaust fans.

3 Results

3.1 Results of the Base Case Energy Simulation

After the first simulation of the model, henceforth Base Case (BC), the simulated annual energy consumption was compared to the energy consumption from utility bills (Fig. 9). The simulation displayed 31.7% higher annual energy consumption than the 2018/2019 energy bills, consuming 930,888 kWh, while the energy bills amounted to 706,880 kWh. This demonstrates the necessity of calibration since the maximum annual difference should be 5% according to (Westphal 2007).

Dashed lines have been added to the energy consumption graph (Fig. 9) to indicate the maximum monthly variation in consumption according to the literature, which is 20% (Westphal 2007). Differently from the energy bills, monthly energy consumption did not fluctuate in the BC because no month-specific usage patterns had been specified.

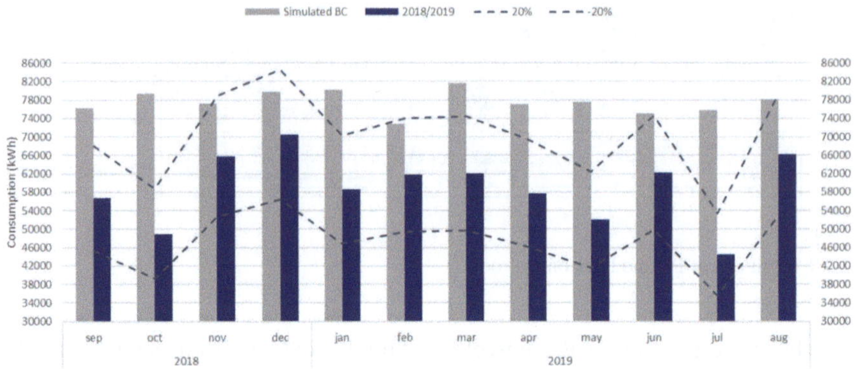

Fig. 9 Monthly simulated energy consumption of uncalibrated simulation (Base Case) versus consumption from energy bills

3.2 *Base Case End-Uses*

The simulated energy consumption was discriminated according to end-use. The results showed that the main end-use is 74% for refrigeration, followed by 19% for equipment, 6% for lighting, and 2% for HVAC consumption.

The lighting system was assessed on-site, along with the specifications of all the lamps and light fixtures. This implies the lighting system is the least uncertainly simulated. Conversely, the refrigeration system bears the most uncertainty, due to its large percentage in the total energy consumption and the lack of manufacturer specifications. Consequently, the authors resorted to generic models from the EnergyPlus database, adjusting only the operative temperatures to those indicated on the equipment displays, and dimensions (Fig. 10).

Fig. 10 Energy consumption by end use on the Base Case (uncalibrated)

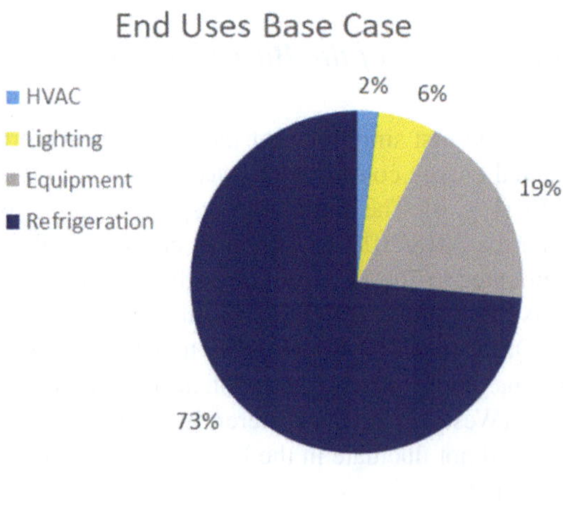

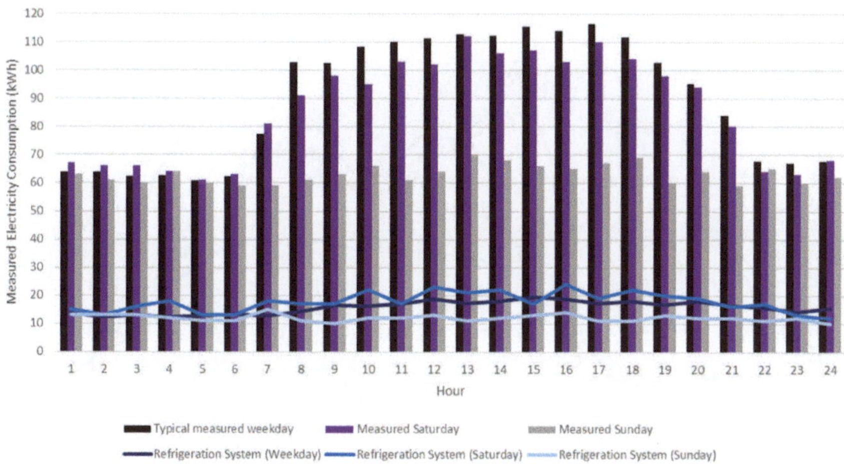

Fig. 11 Measured electricity consumption

3.3 Monitoring of Actual Electricity Consumption

The hourly consumption for 24 h of a typical weekday, a Saturday and Sunday is shown in Fig. 11. The consumption curve measured at the energy board matched the expectations for hourly variations in equipment usage and space occupancy by employees and shoppers.

The refrigeration system did not exhibit significant consumption variations throughout the day, but a rise in consumption was observed during the opening hours (on weekdays and Saturdays). This rise reflects the opening of the refrigerator doors by consumers and employees since on Sundays and at night the consumption remains low and stable.

3.4 Calibration Process

Calibration was carried out in three steps, starting with the analysis of the Base Case so the authors could adjust it.

Base Case (BC) of the simulated model: The BC exhibited a 31.7% higher annual consumption than the one year comprised by the energy bills from September 2018 to August 2019. After comparing the daily load curves (Fig. 12) the authors realized the model was exaggerating the energy consumption for the supermarket's opening hours since the simulated and measured values were similar at nighttime. Furthermore, the simulation displayed two consumption peaks, at 09:00 h and at 19:00 h, which differed from the measured charge behavior. They were due to the programmed defrost of all the refrigeration equipment in the BC, set to be simulated twice daily.

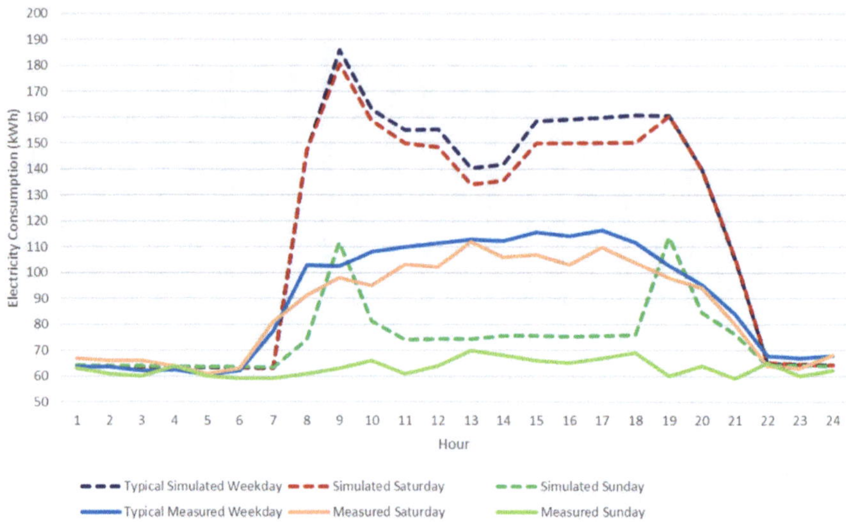

Fig. 12 Daily measured consumption versus simulated consumption (Base Case)

Version 1 (V1) of the simulated model: Adjustments in the refrigeration system configuration: With the removal of the defrost programming, the annual consumption difference between simulation and measured data fell to 12.7% (Fig. 13).

In Version 1 the daily simulated consumption on Sundays was close to the measured data, if slightly lower and more stable. Nevertheless, the consumption on Saturdays and typical weekdays remained above the measured values (Fig. 13).

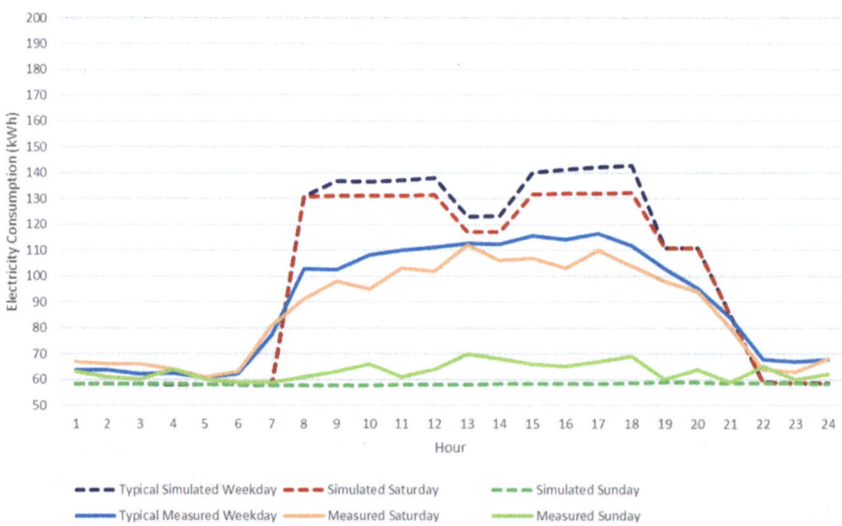

Fig. 13 Daily measured consumption versus simulated consumption (Version 1)

Some more commonly used pieces of equipment, along with the lighting system, were configured to function at 100% capacity during work hours, but the on-site survey showed that not every lamp, room and piece of equipment obeyed this usage pattern.

It was also evident that the simulated consumption values for Saturdays were closely similar to the weekdays', while Saturday's measured values were lower. This is explained by the administration's work schedule, which only encompasses weekdays and had been misconfigured. Additionally, occupancy hours needed to be adjusted, firstly because the measured energy consumption began growing at 06:00 am and the simulated one at 07:00 am, and secondly, because a reduction in consumption at lunchtime (noon h to 14:00 h) had been simulated but was not measured (Fig. 13).

Version 2 (V2) of the simulated model: Adjusting usage and occupancy patterns: After adjusting the general and administrative sector use and occupancy patterns, and removing lunchtime reduction in consumption, Version 2 displayed a 7.4% higher annual consumption than the energy bills.

In Version 2, the simulated consumption during work hours was still above the measured load curve. An analysis of the monthly variation in the energy bills evidenced the need for usage schedules that considered monthly variations. It was noticed that further lowering the use schedule uniformly throughout the year would underestimate consumption in some months, like November and December, while other months like October, May and June still needed simulated consumption values reduced.

In Version 2, usage equipment and lighting usage schedules were respectively 80% and 90% for every month. Since the actual building does not seasonally alter its lighting usage, only the equipment schedule incorporated these new monthly variations. Beyond that, an error was found in the simulation of the exhaust fans, which were being simulated as electrical devices when the actual equipment does not consume energy.

Thus, for Version 3 of the simulated model, a monthly calibration was carried out, adding variance to the equipment use schedule. The usage pattern increased for November and December (90% capacity) due to higher bakery output and decreased for May, July and October (20% capacity) due to vacations and an increased number of holidays. In the other months of the year, intermediate values were adopted (from 40 to 80%) according to the monthly variations in the energy bill (Fig. 14).

Version 3 (V3) of the simulated model: Monthly consumption and exhaust fan simulation adjustments: The simulated annual consumption was 4.0% higher than energy bills, which is within the 5% acceptable error threshold proposed by Westphal (2007). Therefore, we have a calibrated model.

With the monthly adjustments, simulated consumption variations greatly approached energy bill values (Fig. 15), and the simulated load curves better matched the measured week (Fig. 16).

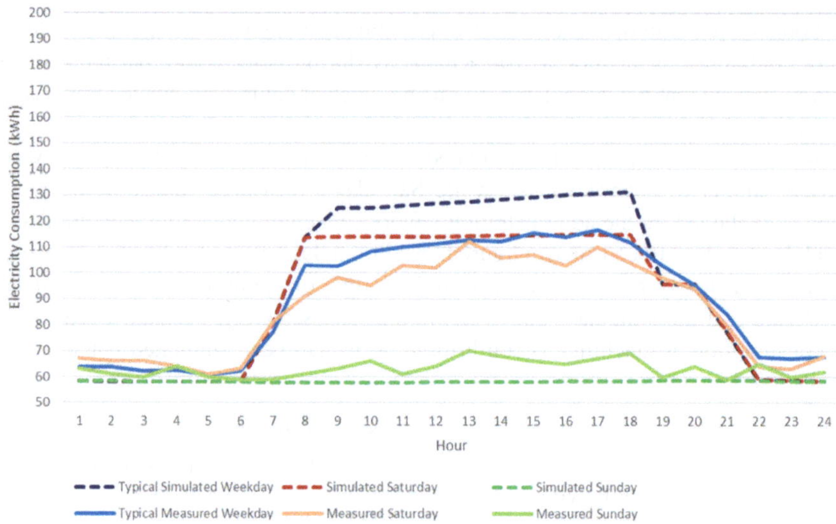

Fig. 14 Daily measured consumption versus simulated consumption (Version 2)

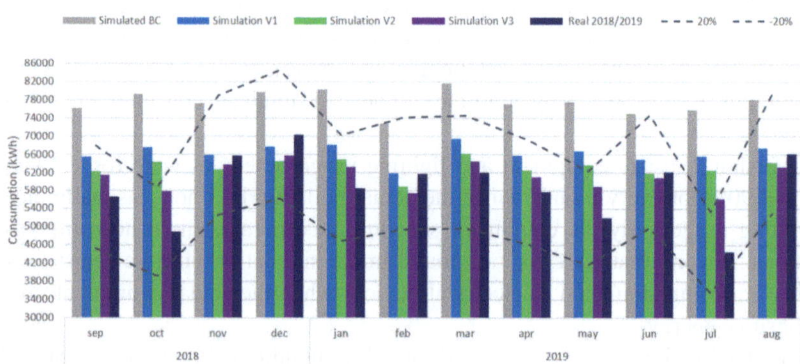

Fig. 15 Monthly Variation in simulations Base Case through Version 3 versus monthly variation in the actual energy bills

3.5 Validation

The *MBE* and *CVRMSE* found between the calibrated model and the measured hourly consumption and monthly energy bills are within the limits established by ASHRAE Guideline 14– (2014) (Table 7), validating the calibration process.

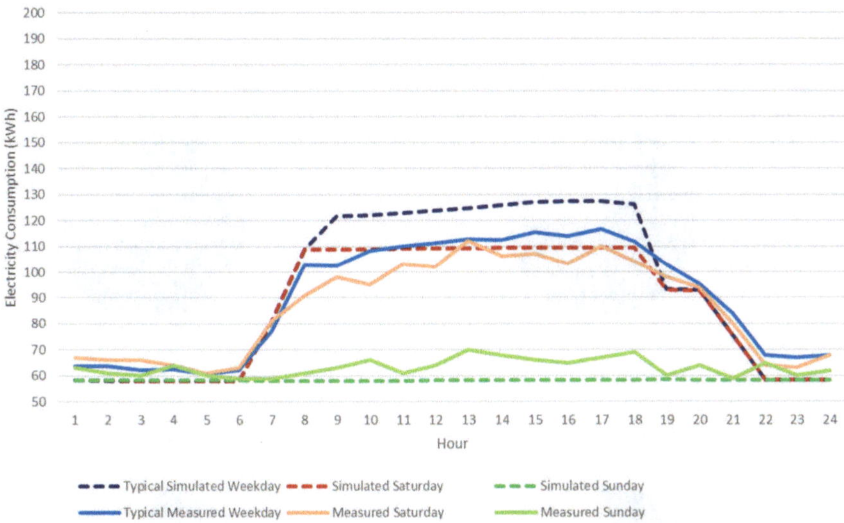

Fig. 16 Daily measured consumption versus simulated consumption (Version 3)

Table 7 Differences between reference consumption rates and the calibrated model's consumption (Version 3)

Statistical indices	ASHRAE hourly calibration limits	Simulation Version 3			ASHRAE Monthly Calibration Limits	Simulation Version 3
		Weekday	Saturday	Sunday		Month
MBE (%)	±10	3.22	0.03	−7.74	± 5	3.99
CVRMSE (%)	30	11.21	8.44	9.23	15	9.60

3.6 *End Uses—Calibrated Model*

There were no significant changes in the end uses of the model throughout the calibration process (Fig. 17). Changes were made to the building's main consumption components, except for the HVAC system, as it only contributed to 2% of consumption.

Knowing the end uses is essential in an energy efficiency assessment to identify potential improvements and thus achieve greater efficiency in future retrofits. In addition, knowing the percentage of manageable and unmanageable energy can help propose energy efficiency measures. Unmanageable loads correspond to the periods when the supermarket is closed, and are mostly generated by the refrigeration system. These represent 33% of annual consumption, so the remaining 67% are manageable loads, which offer an energy-saving opportunity in the form of measures

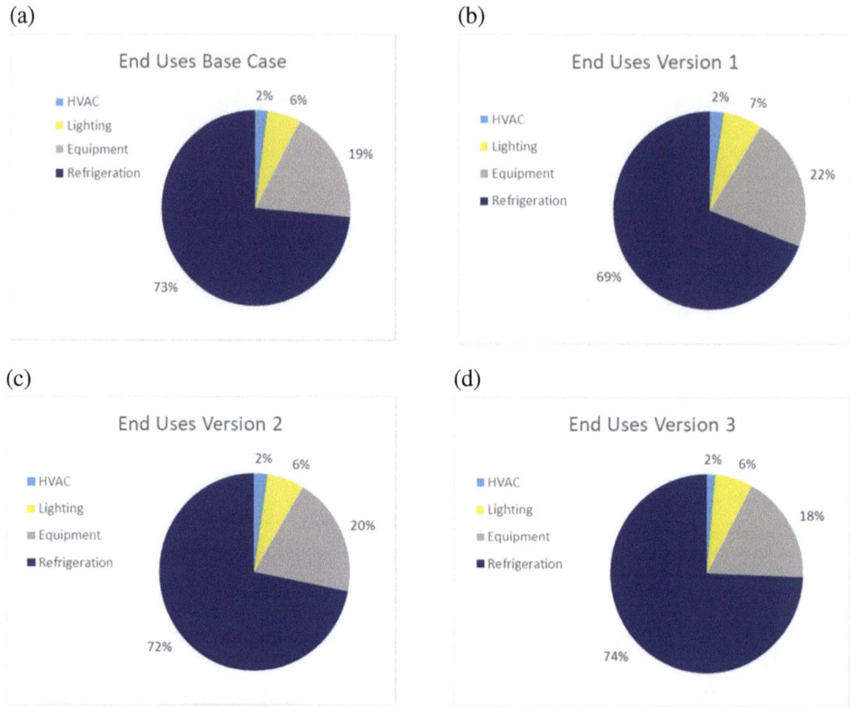

Fig. 17 Energy end-use comparison: **a** Base Case; **b** Version 1; **c** Version 2; **d** Version 3

such as discouraging unnecessary door openings, acquiring refrigerators with doors that cause no thermal loss when opened, and optimizing lights and equipment use.

The refrigeration system was responsible for the largest share of energy consumption (74% in V3), which is higher than the 60% indicated in the literature for artificially conditioned supermarkets (Suzuki et al. 2011). Since Supermercado Escola has air conditioning only in offices, this increase was expected. The refrigeration system has significantly more closed refrigerators (48 units) than open refrigerated display shelves (9 units), which are pointed out in studies as major energy consumers. Thus, in addition to refrigerator opening types, it is possible that adopting passive strategies could reduce energy consumption by the refrigeration system, by reducing heat gain through the walls and roof.

Lighting consumption represented only 6% of total consumption, below what is indicated by the literature for commercial buildings, which is about 22% (Mangiapelo et al. 2011), however, that paper was published before the increased use of LEDs in the building sector. Supermercado Escola's energy management has already replaced fluorescent lamps with more economical LED models and makes use of natural lighting through the clerestories on the roof. Motion detectors in the administrative sector and dimmers in the sales area can increase electric energy savings by taking advantage of natural light.

The HVAC system showed negligible consumption (2%) since only the administrative area of the supermarket is artificially conditioned, about 200 m^2 or 6.7% of the total area. Similarly to the refrigeration system, energy efficiency measures for the envelope could possibly further reduce this consumption by reducing cooling loads.

The supermarket's management team seems generally interested in energy efficiency and, even without an energy consumption benchmark for the typology, it is believed that the building in question is energy efficient, with the use of natural light and an efficient lighting system. There was a great emphasis on the refrigeration system, which is the main energy consumer within the typology, with new equipment and updated technology. Identifying manageable and unmanageable loads can bring new options for rationalizing the use of energy for this final use.

As most of the systems identified already operate at high levels of efficiency, we investigated photovoltaic generation on the roof, which provides new options for saving electricity and significant operational cost reduction for the supermarket, besides contributing to the sustainable use of energy.

3.7 Solar Energy Generation Potential

Despite the facing and roof pitch being subpar, the photovoltaic system delivered good performance due to the lack of shading by the surroundings. The simulated PV system generated 225,319 kWh in one year, which represented 30.65% of the supermarket's annual consumption (Table 8).

The consumption intensity of the simulated supermarket model was 245 kWh/m^2/year (Fig. 18), considering the total area of the building. By subtracting the electrical energy produced by the PV system inserted in the roof, this consumption intensity is reduced to 170 kWh/m^2/year. We underline that the energy generated by the photovoltaic system on the roof would be able to supply the demand of all the supermarket's final uses except for the refrigeration system. This would increase the building's sustainability, emphasizing a cleaner energy source and serving as an example of the implementation of this technology. This could inspire its dissemination and stimulate the adoption of more sustainable energy sources in the Brazilian energy matrix.

4 Conclusion

This work's object of analysis is located on a university campus and serves as an environment for experimentation and proofing, where the research developed can be used to sensitize both the academic community and the local population. In addition, there is scant literature on energy efficiency in supermarkets, despite the typology presenting a significant energy consumption density worldwide. Therefore,

Table 8 PV generation on the roof (c-Si/EOD)

Generator	Module area	Dimensions (cm)	Strings	Modules/*string*	Modules/generator	Total modules	Max. Simulated eff. (%)	PV generation/total consumption (%)
1	1.98	100.2 × 197.9	4	40	160	593	21.38	30.65
2			4	40	160			
3			6	29	174			
4			1	29	29			
5			6	10	60			
6			1	10	10			

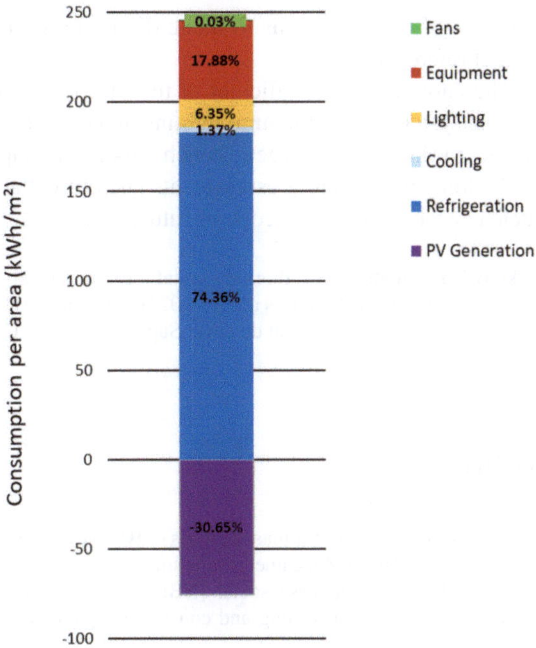

Fig. 18 Consumption intensity (total area) by end-use

the importance of knowing its typical consumption to help managers decide about energy-efficient strategies is emphasized.

The method consisted in analyzing the energy bills and measuring the energy use in loco enabled a calibrated energy model for building performance simulation, whereupon the simulation of PV systems on the model's roof using EnergyPlus was proposed. The energy diagnosis of the building showed results in partial agreement with the consumption indicated in the literature, with the refrigeration system being the largest energy consumer for the typology, taking climate differences into account. Differences were observed in the main end uses: refrigeration system (74%), equipment (18%), lighting system (6%) and HVAC (2%).

Regarding the generation of solar energy, the potential of the studied building for the insertion of photovoltaic systems on the roof was confirmed. The implementation of distributed generation generated about 30.65% of the building's annual electricity demand. This could contribute to cleaner, more sustainable energy generation, avoiding greenhouse gas emissions and power line losses. This study can serve as an example to promote and disseminate the use of photovoltaic technology in supermarkets in the local community, both for establishments already in operation and for new projects, contributing to sustainable development practices.

As for limitations, even though the studied building has relevant characteristics compared to the supermarkets covered by the *ABRAS* survey, it is understood that a study that takes into account other aspects, mainly architectural, is still needed. The objective of this paper was to incite discussions on the subject and not to create a market reference.

The case study was efficient in the face of the systems used, such as updated cooling systems, use of natural lighting and the use of efficient lamps (LED). Nevertheless, the lack of an energetic benchmark for the supermarket typology complicates the building performance assessment. Therefore, the authors suggest the creation of such a benchmark as the scope of future work.

Acknowledgements The authors are grateful for the support of FAPEMIG (Convênio para PD&I de Concessão de Cota de Bolsas nº 5.12/2022). This study was financed in part by the Coordenação de Aperfeiçoamento de Pessoal de Nível Superior—Brasil (CAPES)—Finance Code 001.

References

Associação Brasileira de Normas Técnicas (ABNT) (2005a) NBR 15220: Desempenho térmico de edificações–Parte 3: Zoneamento bioclimático brasileiro e diretrizes construtivas para habitações unifamiliares de interesse social [NBR 15220: Thermal performance of buildings-Part 3: Brazilian bioclimatic zoning and constructive guidelines for single-family housing of social interest]

Associação Brasileira de Normas Técnicas (ABNT) (2005b) NBR 15220-2: Desempenho térmico de edificações–Parte 2: Métodos de cálculo da transmitância térmica, da capacidade térmica, do atraso térmico e do fator solar de elementos e componentes de edificações [NBR 15220-2: Thermal performance of buildings-part 2: methods for calculating thermal transmittance, thermal capacity, thermal delay and solar factor of building elements and components]

Associação Brasileira de Supermercados (ABRAS) (2019) Ranking Abras 2019-ano base 2018. Superhiper, v.45, n. 514. https://www.abras.com.br/economia-e-pesquisa/ranking-abras/. Accessed May 2020

Agência Nacional de Energia Elétrica (ANEEL) (2012) Resolução Normativa nº. 482 de 2012 [Normative Resolution n° 482 of 2012]. Brasília. http://www2.aneel.gov.br/cedoc/ren2012482.pdf. Accessed 22 Aug. 2020

Agência Nacional de Energia Elétrica (ANEEL) (2015) Resolução Normativa nº. 687 de 2015 [Normative Resolution n° 687 of 2015]. Brasília. https://www2.aneel.gov.br/cedoc/ren2015687.pdf. Accessed 22 Aug 2020

American Society of Heating, Refrigerating and Air Conditioning Engineers (ASHRAE) (2013) ANSI/ASHRAE 55: thermal environmental conditions for human occupancy. Atlanta, Georgia

American Society of Heating, Refrigerating and Air Conditioning Engineers (ASHRAE) (2014) ASHRAE guideline 14-2014: measurement of energy, demand and water savings. Atlanta, Georgia

Barbosa UJ et al (2015) Iluminación con Energía Solar Fotovoltaica para Autoservicios en Bogotá [Lighting with photovoltaic solar energy for self-services in Bogotá]. Ingenio Magno 5(1):81–94. ISSN: 2422-2399

Barbosa EFT, Labaki LC (2013) Conforto térmico em ambientes de um supermercado de médio porte [Thermal comfort in environments of a medium-sized supermarket]. XII ENCAC-Encontro Nacional de Conforto no Ambiente Construído. Anais...Brasília, DF

Braun MR, Altan H, Beck SBM (2014) Using regression analysis to predict the future energy consumption of a supermarket in the UK. Appl Energy 130(June):305–313. https://doi.org/10.1016/j.apenergy.2014.05.062

Balanço Energético Nacional (BEM) (2017) Balanço Energético Nacional: ano base 2016 [National Energy Balance: base year 2016]. Ministério de Minas e Energia. https://ben.epe.gov.br/BENRelatorioFinal.aspx?anoColeta=2017&anoFimColeta=2016. Accessed 20 Jan 2018
Bre F, Gimenez JM, Fachinotti VD (2018) Prediction of wind pressure coefficients on building surfaces using artificial neural networks. Energy Build 158:1429–1441
Building Performance Institute Europe (2011) European buildings under the microscope
Cpsimulator. https://cpsimulator.cimec.org.ar/. Accessed 21 Oct 2020
Centro de Referência para Energia Solar e Eólica Sérgio Brito (CRESESB) (2021) http://www.cresesb.cepel.br/. Accessed 15 Jul 2021
Czajkowski J et al (2004) Evaluación de las condiciones ambientales en un supermercado [Evaluation of environmental conditions in a supermarket]. Avancos en Energias Renovables y Medio Ambiente, v. 8. ISSN: 0329-5184
Union of the Electricity Industry (Eurelectric) (2011) Power production EU-27. www.eurelectric.org/content/Default.asp?PageID=948
Gimenez JM, Bre F (2019) Optimization of RANS turbulence models using genetic algorithms to improve the prediction of wind pressure coefficients on low-rise buildings. J Wind Eng Ind Aerodyn 193
Gimenez JM, Bre F, Nigro NM, Fachinotti VD (2018) Computational modeling of natural ventilation in low rise non rectangular floor plan buildings. Build Simul 11:1255–1271
Google Earth (2022) Vista aérea do Supermercado Escola, Viçosa/MG [Aerial photograph of Supermercado Escola]. Location 20°45'56.87" S, 42°52'08.60" W, elev 2164 ft, eye alt 2866 ft
Guimarães IBB, Carlo JC (2015) Comparação estatística entre arquivos climáticos desenvolvidos com métodos diferentes [Statistical comparison between climate files developed with different methods]. Euro ELECS 2015-latin American and European conference on sustainable buildings and communities. Anais
Instituto Nacional de Metrologia, Normalização e Qualidade Industrial (INMETRO) (2013) Portaria nº 50, de 01 de fevereiro de 2013. Aprova o aperfeiçoamento dos Requisitos de Avaliação da Conformidade para a Eficiência Energética de Edificações [Ordinance n° 50, of February 1, 2013. Approves the improvement of the Conformity Assessment Requirements for the Energy Efficiency of Buildings]. Diário Oficial da República Federativa do Brasil, Poder Executivo, Brasília, DF, 06 feb. 2013. Seção 1, p. 87. http://www.inmetro.gov.br/legislacao/detalhe.asp?seq_classe=1&seq_ato=1961. Accessed 16 Oct 2020
Iyer SR et al (2015) Energy disaggregation analysis of a supermarket chain using a facility-model. Energy Build 97:65–76. https://doi.org/10.1016/j.enbuild.2015.03.053
Jelle BP, Breivikb C, Rokenesb HD (2012) Building integrated photovoltaic products: a state-of-the-art review and future research opportunities. Solar Energy Mater Solar Cells, Elsevier 100:69–96
Kang S, Yoo S, Lee J, Boo B, Ryu H (2012) Experimental investigations for recycling of silicon and glass from waste photovoltaic modules. Renew Energy 47:152–159
Kolokotroni M et al (2019) Supermarket energy use in the UK. Energy Proc 161(2018):325–332. https://doi.org/10.1016/j.egypro.2019.02.108
Lam JC, Hui SCM (1996) Sensitivity analysis of energy perfomance of office buildings. Build Environ 31(1):27–39
Lindberg U, Axell M, Rolfsman L (2011) Energy efficiency in supermarkets—implication of Eu efficiency directives. Energy
Mangiapelo LBS, Francato AL, Andreasi WA (2011) Avaliação da eficiência energética nos sistemas de iluminação natural e artificial em dois hipermercados [Evaluation of energy efficiency in natural and artificial lighting systems in two hypermarkets]. IV Congresso Brasileiro de Eficiência Energética. Anais...Juiz de Fora, MG
Martos JLG, Styles D, Schoenberger H (2013) Identified best environmental management practices to improve the energy performance of the retail trade sector in Europe. Energy Policy 63:982–994. https://doi.org/10.1016/j.enpol.2013.08.061

Mylona Z, Kolokotroni M, Tassou SA (2017) Frozen food retail: Measuring and modelling energy use and space environmental systems in an operational supermarket. Energy Build 144:129–143. https://doi.org/10.1016/j.enbuild.2017.03.049

Mylona Z, Kolokotroni M, Tassou SA (2018) A study of improving energy efficiency of small supermarkets by modelling interactions between building, Hvac, Refrigeration and display product. In: 5th IIR conference on sustainability and the cold chain. Anais...Beijing, China. https://doi.org/10.18462/iir.iccc.2018.0004

Pather-Elias S, Davis S, Cohen B (2012) A techno-economic study of energy efficiency technologies for supermarkets in South Africa. J Energy South Afr 23(3):2–8. https://doi.org/10.17159/2413-3051/2012/v23i3a3166

Programa Nacional de Conservação de Energia Elétrica (PROCEL) (2008) Relatório da Pesquisa de Posse de Equipamentos e Hábitos de Uso: Classe Comercial–Alta Tensão–Supermercados/Hipermercados [Survey report on possession of equipment and usage habits: commercial class-high voltage-supermarkets/hypermarkets]. Ano base 2005. http://www.procelinfo.com.br/main.asp?ViewID=%7B5A08CAF0-06D1-4FFE-B335-95D83F8DFB98%7D&params=itemID=%7BCCF153AF-6BD3-4F2B-B373-3E514A6C2AAF%7D;&UIPartUID=%7B05734935-6950-4E3F-A182-629352E9EB18%7D. Accessed 28 Apr 2020

Renno C et al (2018) Performance analysis of a CPV/T-DC integrated system adopted for the energy requirements of a supermarket. Appl Therm Eng 149(November):231–248

Rodrigues TTV, Oliveira DF, Carlo JC, Guimarães HS (2021) Uma discussão sobre indicadores energéticos para supermercados [A discussion on energy indicators for supermarkets]. XVI ENCAC-Encontro Nacional de Conforto no Ambiente Construído. Anais...Palmas, To

Suzuki Y et al (2011) Analysis and modeling of energy demand of retail stores. Proceedings of building simulation 2011: 12th conference of international building performance simulation association. Anais

Timma L, Skudritis R, Blumberga D (2016) Benchmarking analysis of energy consumption in supermarkets. Energy Proc 95:435–438. https://doi.org/10.1016/j.egypro.2016.09.056

Westphal FS (2007) Análise de incertezas e de sensibilidade aplicadas à simulação de desempenho energético de edificações comerciais [Uncertainty and sensitivity analysis applied to the simulation of energy performance of commercial buildings]. Tese (Doutorado)–Programa de Pós-Graduação em Engenharia Civil, Universidade Federal de Santa Catarina, Florianópolis

Thiago Toledo Viana Rodrigues Architect and Urbanist (2016), Master's degree in Architecture and Urbanism (2018) and Ph.D. in Architecture and Urbanism (2023) at the Federal University of Viçosa. Has experience in Energy Efficiency, with scientific initiation by the Laboratory of Technologies in Environmental Comfort and Energy Efficiency-Latecae.

Joyce Correna Carlo Architect and Urbanist at the Federal University of Minas Gerais (1997), Specialist in Urban Analysis at the Federal University of Minas Gerais (1999), master's degree (2002) and Ph.D. (2008) in Civil Engineering at the Federal University of Santa Catarina. Post-doctorate at the Federal University of Santa Catarina and professor at the Department of Architecture and Urbanism at the Federal University of Viçosa.

Mapping the Environmental Quality of University Campus Areas for the Needs of Blue-Green Infrastructure Planning

Jan Kopp, Jiří Preis, Pavel Žák, and David Vogt

Abstract The aim of this chapter is to analyze the methodologies for assessing the environmental quality of areas and to show their application on the territory of the University of West Bohemia campus in Pilsen. The environmental assessment methods of areas for planning the overall development of university campuses have been applied. At the level of cities, coefficients, which enable the evaluation of green ecosystem functions, are most often used, e.g., Biotope Area Factor or Green Space Factor. The evaluation of the campus area makes it possible to compare the environmental quality of the university campus within the territory of the entire city. The runoff coefficient and Helsinki Green Factor were the main environmental parameters for evaluating blue-green infrastructure. Detailed mapping of areas is the basis for responsible assessment of the environmental status and planning of campus development with regard to greenery ecosystem services and rainwater management. Specifically, there are plans to establish green roofs or to change a drainage system. The advantages and limitations of mapping the environmental quality of campus areas for planning the development of blue-green infrastructure on the university campus are discussed. The basic methodological questions of the use of environmental parameters involve the selection of a suitable index for evaluation, a suitable scale for mapping some environmental properties and the application of the results for planning practice.

Keywords Blue-green infrastructure · Urban planning · Rainwater management · Higher education · Sustainable cities · Environmental indicators

---

J. Kopp · J. Preis (✉) · P. Žák · D. Vogt
Faculty of Economics, University of West Bohemia, Univerzitni 8, 301 00 Pilsen, Czech Republic
e-mail: jpreis@fek.zcu.cz

W. Leal Filho et al. (eds.), *The Contribution of Universities Towards Education for Sustainable Development*, World Sustainability Series,
https://doi.org/10.1007/978-3-031-49853-4_13

1 Introduction

University campuses have a specific position within the urban landscape. By the structure of the areas and the typology of the building, authors can classify university campuses among sites similar to, for example, hospitals or administrative centers. But the unique importance of university campuses results from their role in science, research and education (Kestin et al. 2017). By using the example of the involvement of university campuses in the blue-green infrastructure of cities, the authors of the chapter wants to show the innovative potential of university campuses. It can be done not only in terms of the development of new technologies and trends in the field of rainwater management (Gunasekara 2006) but also in the practical application of new approaches to the sustainable development of the territory (Alshuwaikhat et al. 2017; Göçer et al. 2018; Herzner and Hommerová 2022). University campuses should be developed as suitable model areas where the testing of new approaches in urban planning and landscape design management is combined, with strengthening the importance of education for sustainable development, with environmental education and green marketing of universities (Gulwadi and Scholl 2017; Melo et al. 2020; Wojciechowska et al. 2021).

The development of blue-green infrastructure (BGI) is supported worldwide by innovations in rainwater management technologies and new trends in growing urban greenery. The need for the development and enforcement of BGI is part of the adaptation of cities to climate change (Morison and Brown 2011; Voskamp and van de Ven 2015). Various terminology has been used, most often blue-green infrastructure (BGI) (Thorne 2016) and green–blue infrastructure (Bacchin et al. 2016). Blue-green infrastructure is defined as a new system of rainwater management supporting water retention and its quality in the urban landscape, together with benefits for public space, for adaptation to climate change and for biodiversity (Kopp et al. 2021a). In contrast to traditional approaches oriented towards technical solutions that drain rainwater as quickly as possible through the sewage system (Stránský and Kabelková 2015), BGI is conceived as a system close to the natural circulation of water, connecting green infrastructure (an organized system of urban greenery) and rainwater management in the city (Hoang and Fenner 2016). The distinction between close to nature and an artificial or technical element can be problematic. On a practical level, for example, different degrees of ecological quality of surface retention reservoirs can be distinguished from water bodies close to nature to concrete reservoirs (Kopp and Preis 2019). Furthermore, it must be recognized that elements of rainwater management imitate or support natural processes such as infiltration or evapotranspiration, but they are essentially established as artificial elements according to technical standards. Therefore, compromise terms blue-green-grey (Alves et al. 2020) or hybrid or mixed infrastructure (Depietri and McPhearson 2017) already appear in the literature.

The involvement of university campuses in the development of the blue-green infrastructure of cities fulfills several Sustainable Development Goals (SDGs) (Kestin et al. 2017; Leal Filho 2011), such as the Goal 9 (*Build resilient infrastructure, promote inclusive and sustainable industrialization and foster innovation)* and the

Goal 11 (*Make cities and human settlements inclusive, safe, resilient and sustainable*) (Gulwadi and Scholl 2017; Ferguson et al. 2013). Blue-green infrastructure should be resistant to extreme situations, so it recommended to take into account the whole spectrum of rainfall-runoff situations, from long-term drought to heavy torrential rainfall. In terms of the links between the urban landscape and natural systems, the development of BGI also contributes to the fulfillment of the Goal 13 (*Take urgent action to combat climate change and its impacts*) (Depietri and McPhearson 2017; Lehnert et al. 2021b) and the Goal 6 (*Ensure availability and sustainable management of water and sanitation for all)* (Kopp et al. 2021a). In the environment of university campuses, not only the efficient processing of rainwater, but also the involvement of water in the quality of public spaces must be addressed (Wojciechowska et al. 2021). Within the urban landscape, BGI planning of university campuses depends on their location (Simperler et al. 2018; Kopp et al. 2021b). There are limits to the creation of BGI by universities within the dense development of urban centers, while campuses in suburban areas have significant potential for strengthening recreational and natural functions (Iojă et al. 2014; Alshuwaikhat et al. 2017). University campuses should be at the forefront of building sustainable cities that economically efficiently develop environmental quality for the benefit of their residents or visitors (De Jong et al. 2015; Baynes and Wiedmann 2012). In this regard, it should be remembered that university campuses are both work and recreation areas with a specific mode of daily and annual commuting rhythms not only for the people of the city itself, but also for students and academics from a wide commuting area (Melo et al. 2020; Wojciechowska et al. 2021).

This chapter deals with the possibilities of BGI planning using environmental quality indicators of university campuses. In general, the goal is to analyze the methodologies for assessing the environmental quality of areas. On a practical level, the goal is to present their application on the campus of the University of West Bohemia in Pilsen (Czech Republic). In the conclusion, the advantages and limits of the evaluation of the environmental quality of campus areas, usable for planning the development of BGI, are subsequently discussed.

2 Blue-Green Infrastructure Planning

BGI planning at the city level is based on strategic documents with a practical impact on territorial and regulatory plans. In many countries, as well as in the Czech Republic, the concept of BGI remains insufficiently anchored in legislation (Kopp et al. 2021a; Depietri and McPhearson 2017). Planning processes therefore use a combination of tools to support urban greenery and to build decentralized water management infrastructure to support BGI (Hoang and Fenner 2016; WEF 2014; Woods-Ballard et al. 2015). The basic principles and rules of rainwater management in a given category of territory are often enshrined in the spatial plan, as well as, for example, defined areas for emergency water retention and emergency drainage routes (UPDICP 2020). Possible environmental parameters include runoff coefficients or specific runoff limits (amount of water flowing out per second from a unit area) (Smets et al. 2019; Vítek et al. 2015). If the definition of blue-green infrastructure is not used in planning practice, it is partially replaced by green infrastructure planning (Hansen et al. 2017). More detailed conditions for the development of the territory with regard to BGI can be established for individual localities by the regulatory plan. Appropriately defined environmental parameters related to the required amount of greenery etc. can be applied in it (Mattanovich et al. 2018).

In the area of strategic planning, in addition to city strategies for adapting to climate change, plans for implementing a rainwater management system can also be an example, which creates framework conditions for the coordination of construction activities and the development of the territory in accordance with the priorities of the development of BGI (UPDICP 2020). An example is the City of Copenhagen's Cloudburst Management Plan (Ziersen et al. 2017). The aim of the transformation of the stormwater management system in Copenhagen is to ensure that during rainfall with the probability of occurring on average once every 100 years, no part of the city of Copenhagen is flooded with water higher than 10 cm. In Copenhagen, the retention and emergency drainage of all problem areas is gradually being solved according to the plan: capacity systems for underground rainwater drainage are being built, and water retention is being supported in lowered areas of public spaces, in rain gardens or in small dry polders integrated into the urban area (Lerer et al. 2017).

3 Environmental Indicators of Areas

BGI development planning is based on the current state and potential of the territory, for which the design is made with regard to the target requirements of environmental quality (Voskamp and van de Ven 2015; Bacchin et al. 2016; Derkzen et al. 2015). Therefore, the environmental indicators of the areas are a key element in this process (Pauleit and Duhme 2000; Antoszewski et al. 2020). The environmental indicators of the areas make it possible to quantify both the condition and the target limit requirements of the development area. Of course, it is necessary to distinguish whether it is planning a new construction "on the green field" or supplementing and improving the current state in favor of BGI innovations (Szulczewska et al. 2014; Mattanovich et al. 2018).

A basic example of an environmental indicator in the field of BGI is the maximum specific runoff, which limits the load on downstream drainage systems. For example, the Czech technical standard recommends 3 $l.s^{-1}.ha^{-1}$ (Vítek et al. 2015). The requirements of individual cities are, however, usually more moderate in practice. A tabular comparison of the functional types of territory enables the runoff coefficient used in practice, indicating the share of direct surface runoff in the value of the total amount of precipitation (Baiamonte 2020). Determining the runoff coefficient also makes it possible to calculate the volume of runoff rainwater and, based on this, to design the capacity of blue-green infrastructure measures, for example retention tanks, objects enabling infiltration or safety elements of emergency drains (Woods-Ballard et al. 2015).

Other indicators of the quality of areas are coefficients linked to greenery, or green infrastructure. Typical examples are Biotope Area Factor (Becker 1990), Green Space Factor (McCulloch and Robertson 2015; Kruuse 2011) or other similar indicators (Szulczewska et al. 2014; Keeley 2011; Mattanovich et al. 2018). The Biotope Area Factor (BAF) is used as a reference index for evaluating the green infrastructure of an urbanized area. The tool was developed already in 1997 in Berlin, and several other modifications were later created (Peroni et al. 2020). The categorization of areas is not only related to the vegetation cover on the surface but also related to the natural soil cover or the artificial bed. Therefore, BAF can only be used as a guideline for evaluating areas by remote sensing methods (Kopp et al. 2017). BAF parameters describe the level of greenery´s ecosystem services (Keeley 2011).

The coefficient of greenery is sometimes used in Czech planning practice, for example it was already introduced in the methodology for the needs of the capital city of Prague (Prague Institute of Planning and Development 2002). The coefficient of greenery determines the minimum share of countable green areas in the territory, in practice the value also depends on the number of floors of buildings. The calculation of areas of greenery in the territory results from a detailed calculation of greenery on vegetated terrain and other greenery outside vegetated terrain, on buildings, etc. Therefore, a detailed survey of the situation or a detailed development project is needed for evaluation.

In recent years, green infrastructure coefficients have been expanded according to the increase in the importance of connecting blue and green infrastructure with parameters that take into account water management. An example is the Helsinki Green Factor (City of Helsinki 2016). The Helsinki Green Factor is a comprehensive indicator that consists of an individual score and an expert score. In addition to the categories of ecology, functionality, urban landscape and maintenance, the individual score also includes the category of rainwater (capture and treatment of rainwater). In Czech practice, there is an effort to define and use a similarly comprehensive Index of blue-green infrastructure in settlements. However, in our case study we use the Helsinki Green Factor, which is already published internationally (Juhola 2018).

4 Methodology

4.1 Model Area—the Campus of the University of West Bohemia

Mapping and evaluation of the environmental quality of areas was processed for the campus of the University of West Bohemia in Pilsen (UWB) with the aim of identifying the possibilities of developing blue-green infrastructure as part of the planned revitalization of the campus. The University of West Bohemia was founded in 1991 by the merger of seveal colleges located in Pilsen. The gradual relocation of some faculties to the university campus, which began as early as 1992, contributed significantly to the development of the university.

The campus of the University of West Bohemia is still located on the periphery of the city's compact development, in a suburban area called Borská pole. At the time of the laying of the foundation stone of the university campus, in 1985, there was almost no development on the given territory. The land use corresponded to the suburban location, with extensive fields and a local airport. Currently, the campus is used by five faculties, there is an administration building of the university management, a modern library building, capacity sports facilities and catering facilities. The campus includes three new research centers built in 2012–2015. Since 2019, the campus has been accessible from the city center by a new tram line.

The area of interest for mapping with a total area of 265,951 m^2 includes not only the university's land, but in the north the demarcation was expanded to include development areas bounded by the tram line (Žák 2021). The center of the campus is the university park. The space in front of the gym offers sports activities, connected to the football and workout fields. The southern part of the area of interest is occupied by large parking lots. On the edges of the area of interest, there are hitherto unused areas overgrown with extensive vegetation, the further use of which is part of the current campus development plans (Fig. 1).

Fig. 1 Area of interest—the campus of the University of West Bohemia. *Source* own elaboration; *Geodata* Czech Office for Surveying, Mapping and Cadastre

4.2 *Mapping and Evaluation of the Environmental Quality of Areas*

The first step was an accurate inventory of all surfaces on campus in 2020. The background source for this part of the work was the General Development of the UWB in the format for Autodesk AutoCAD with drawings of buildings, parking lots, sidewalks, trees, etc. For analytical work, the general development was converted to the ArcGIS Desktop program. Other sources were Ortofoto of the Czech Republic in the form of a WMS service and aerial photographs from the geoportals Mapy.cz, Google Earth and Czech Office for Surveying, Mapping and Cadastre. The boundaries of the generated polygons were modified according to the current state based on terrain mapping. At the same time, exact categories of surfaces were determined by specifying their typology. Thus, a type of greenery, roof material, pavement surface permeability, etc. have been assigned. Campus inventory point features were targeted in the field using GPS. On the basis of field research, an accurate inventory of greenery was created, including individual solitary trees (Žák 2021).

The main environmental parameters used were (a) runoff coefficient—describing hydrological conditions (Baiamonte 2020) and (b) Helsinki Green Factor—evaluating the level of blue-green infrastructure (City of Helsinki 2016). It was, of course, possible to derive other parameters on the basis of a detailed inventory, such as the share of green areas.

Surface parameters according to the runoff coefficient are based on the Czech technical standard (No. 75 6760 Drainage systems inside buildings). Since some types of surfaces are not listed in this standard, it was necessary to supplement these elements from studies carried out for the city of Pilsen (Kopp et al. 2017). The runoff coefficient values depend not only on the surface, but also on its slope. In the Czech standard, surface slopes are divided into three categories: slopes up to 1%, in the range 1–5% and above 5% (Kopp et al. 2021b). Detailed data from the Digital Terrain Model of the Czech Republic of the 5th generation was used to calculate the slope on a certain surface. It was created in 2009–2013 based on remote scanning the Earth's surface. The output is the categorization of the campus surfaces according to the runoff coefficient, as a basic indicator of the potential formation of the surface runoff of rainwater in a given location. Neither the influence of trees, nor their ability to intercept precipitation, is taken into account in the calculation of the total surface runoff potential.

The Helsinki Green Factor (HGF) indicator was chosen especially for the comprehensiveness of the BGI evaluation (Juhola 2018). Helsinki Green Factor was created in 2012–2014 as part of the Climate-Proof City—Tools for Planning project. The goal was to create tools that will help urban planners develop a climate-resilient city. The so-called "individual score" and "expert score" are involved in determining the HGF factor of individual elements.

Individual score includes categories (a) ecology (such as stormwater purification and reduction, biodiversity); (b) functionality (such as pleasant microclimate, recreation, and environmental education); (d) landscape (such as amenity value and prominent landscape features) and (d) maintenance (City of Helsinki 2014). Points are determined in the methodology on the basis of studies on the ecological and social impact of individual elements. The expert score is based on a survey of the opinions of specialists in the fields of spatial planning, landscape architecture, regional development and environmental protection. The evaluation parameter of the type of surface or element of the BGI is determined as the weighted average of the individual score multiplied by the expert score. Parameter values (for map elements) are listed in the results table (Table 1).

An Excel calculator was used to calculate the total score, which is freely downloadable from the Iwater Project website (City of Helsinki 2016). The tool requires information about the surfaces and the number of individual elements that have already been identified as part of the inventory of all surfaces on the campus. The application allows you to calculate the total value of the Helsinki Green Factor for the entire territory and compare it with the recommended minimum level of the given functional type of the city territory. It is also possible to calculate the total runoff coefficient and derive the potential volume of runoff from precipitation. However, the tabular values of the runoff coefficient in the application partially differ from the values of the Czech standard, and therefore were not included in the calculations and results.

The overall assessment of the results of the evaluation of the campus territory took place on two levels: (1) from the point of view of the internal territorial structure of the campus and (2) from the point of view of the overall level of the campus area in the context of the BGI of the city of Pilsen. The evaluation of the territory within the city was processed on the basis of a comparison of the summary indicators of the runoff coefficient, the share of built-up areas and the share of greenery. The Helsinki Green Factor could not be used within the city, as it requires the entry of detailed information on the elements of the blue-green infrastructure. The runoff coefficient and the proportion of built-up areas and greenery of the university campus were compared with selected types of functional areas in the territory of the city of Pilsen (Kopp et al. 2017). To compare the level of the campus area, average values were used for (1) residential areas in apartment buildings, (2) residential areas in single-family houses, (3) industrial and store areas, (4) shops and services areas, and (5) public green areas. Geodata from the project of ecohydrological assessment of the urban landscape of the city of Pilsen was used for the analytical processing of statistical data on the structure of areas of individual types of development. Geodata includes, in addition to the functional typology of the territory, also the necessary resolution of the surface structure for the entire territory of the city cadastre, which enabled the processing of a representative statistical sample of data. The identification and determination of the properties of elementary areas with a resolution of 1 to 0.5 m was carried out for the original needs of creating ecohydrological maps on the basis of underlying geodata (Kopp et al. 2021b). For the analysis of the structure of surface types, grid squares with an area of 50 × 50 m were chosen as the basic units of the statistical file (Fig. 2). The selection of grid areas of the relevant type for statistical analysis was aimed at the territory where construction was built in Pilsen in

Table 1 Calculation of the Helsinki green factor for the UWB campus

Element group	Element description	Unit	Area or quantity	Weighting	Weighted area (m^2)
Preserved vegetation and soil	Preserved large (fully grown > 10 m) tree in good condition, at least 3 m (25 m^2 each)	pcs	898	3.5	77 842.3
	Preserved natural meadow or natural ground vegetation	m^2	58 104	2.2	128 232.3
Planted/new vegetation	Large tree species, fully grown > 10 m (25 m^2 each)	pcs	83	2.8	5 832.3
	Small tree species, fully grown $\leq$ 10 m (15 m^2 each)	pcs	469	2.3	16 119.4
	Other shrubs	m^2	11 321	1.4	16 053.8
	Meadow or dry meadow	m^2	15 296	1.8	27 092.0
	Lawn	m^2	68 963	1.1	75 979.0
Pavements	Semipermeable pavements	m^2	1 108	1.0	1 130.7
	Permeable pavements	m^2	5 438	1.4	7 754.8
	Impermeable surface	m^2	104 637	–	–
Stormwater management solutions	Intensive green roof/roof garden, depth of substrate 20–100 cm	m^2	160	2.0	325.3
	Pond. wetland or water meadow with natural vegetation	m^2	924	2.8	2 625.4
Bonus elements	Permeable surface designated for play or sports	m^2	1 676	0.7	1 192.4
Total weighted area					360 179.7
Total area					265 951.0
Green factor					1.35

Source The authors

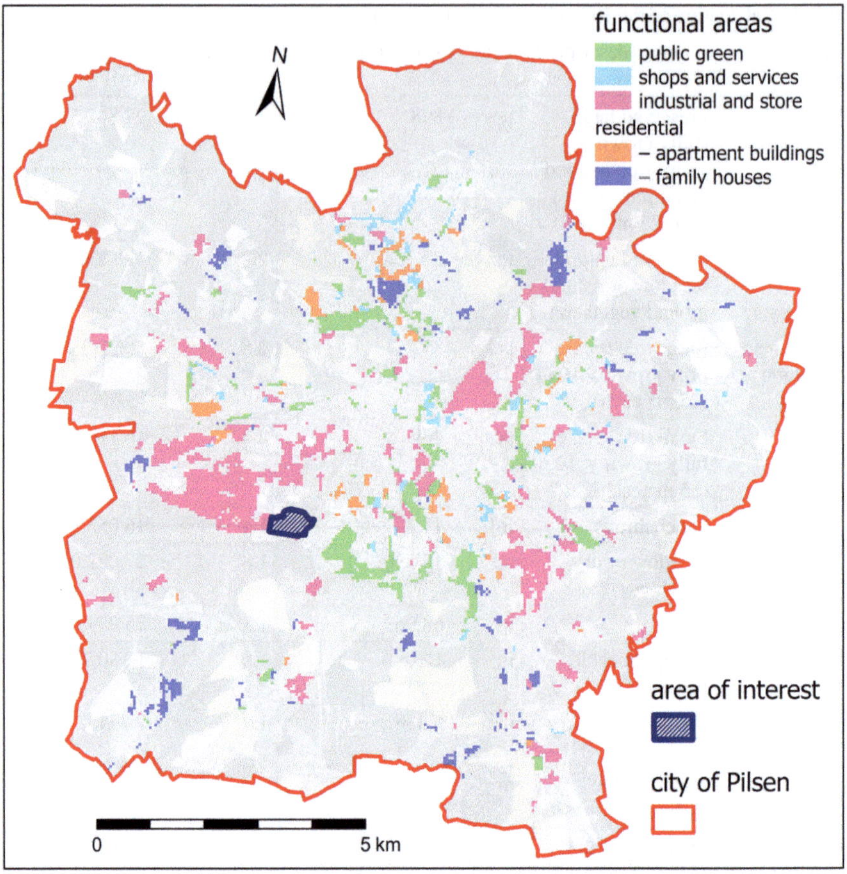

Fig. 2 Selected types of functional areas of the city of Pilsen for land cover comparing with the area of interest—the campus of the University of West Bohemia. *Source* own elaboration; *Geodata* Kopp et al. (2017, 2021b)

a more recent period, e.g., for the typology of mass housing areas, areas of modernist housing estates were not used, with exceptions.

5 Results

5.1 Evaluation of the Areas of the University Campus

The distribution of stormwater runoff conditions on the UWB campus is quite different between individual areas, which differ not only in the degree of permeability, but also in slope (Fig. 3). Altimetry analysis yielded information on where rainwater naturally accumulates. These are the lowest places in the eastern part of the area, where water can soak in naturally within areas close to nature. The overall aggregate coefficient of campus area runoff reaches a value of 0.364 (36.4%). The highest runoff coefficients are tied to gray infrastructure, specifically to impervious surfaces of roofs, roofs with a layer of gravel and asphalt or concrete surfaces. Low values are tied to permeable surfaces, especially grassy areas and areas covered with trees and bushes. These are in the campus both in the central location, where the university park is located, and mainly in the development areas on the outskirts The development areas significantly reduce the overall runoff coefficient for the entire campus.

Water from impermeable areas of roads, parking lots and roofs of buildings with a high proportion of runoff is drained into the municipal sewer system. The area's drainage is aided by drainage channels. In order to slow down the runoff, grass blocks or occasionally a green roof are partially applied in the area of interest. However, the share of these measures is very small. In the case of newly built buildings, the load of the sewage system during intense rainfall is limited by the temporary accumulation of water in underground reservoirs. Unfortunately, no equipment has yet been installed to use the temporarily accumulated water. Evaluating campus areas using the runoff coefficient makes it possible to calculate the volumes of runoff water, and thus to design systems that efficiently manage rainwater.

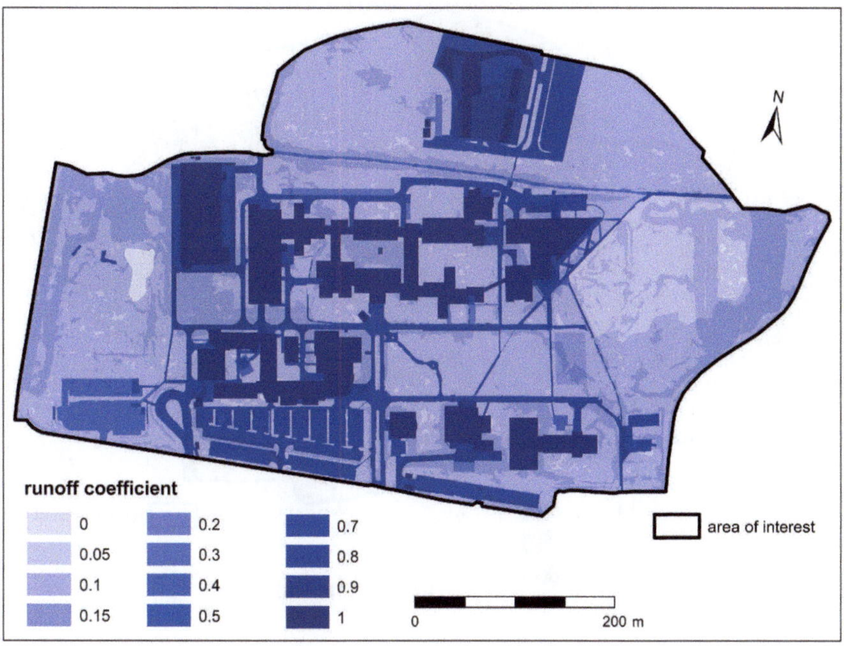

Fig. 3 Classification of UWB campus areas according to runoff coefficient. *Source* own elaboration; *Geodata* Czech Office for Surveying, Mapping and Cadastre; Žák (2021)

In detail, the areas of the campus were evaluated according to the Helsinki Green Factor calculation. The calculation includes not only individual surface types, but also individual BGI elements (Table 1). The score of the area of interest reaches a value of 1.35, significantly exceeding the target score of 0.8 set by the methodology for areas of services and offices (City of Helsinki 2014). If the HGF was calculated only on university-owned land, the score would be reduced to 1.10. The high HGF score is achieved due to the significant proportion of green space in the campus. The trees, which are also located in the central part of the campus, contribute to the high quality. The development areas of the university contribute significantly to the overall grade (Fig. 4). It is precisely in development locations that there are stands of trees, shrubs, water bodies or areas of extensive greenery. Therefore, the issue of further development of the university campus is very important, where the construction and accessibility of paved roads will reduce the overall quality of the BGI in the campus. When designing new development plans, it is therefore necessary to make maximum use of the principles of blue-green infrastructure. Further potential improvements can be brought by the conversion of existing impervious surfaces to semi-permeable surfaces (e.g., parking lots), the implementation of green roofs and new elements of rainwater management (e.g., natural retention tanks and wetlands). The hydrogeological conditions of the territory do not allow the application of rainwater infiltration elements.

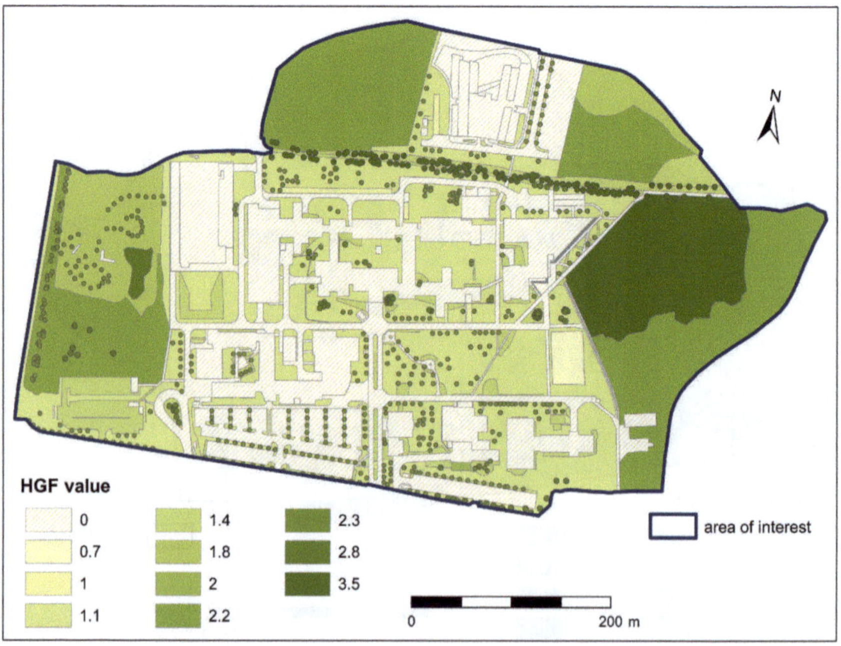

Fig. 4 Evaluation of UWB campus areas according to the Helsinki Green Factor. *Source* own elaboration; *Geodata* Žák (2021)

5.2 *Evaluation of the Campus Within the Territory of the City of Pilsen*

The overall environmental parameters for the campus area with the average values for selected functional types of the territory of Pilsen were compared (Fig. 5; Table 2). The comparison in all three benchmarks shows that the overall level of the campus is better than that of production and storage areas or areas of commercial civic amenities with a small proportion of public spaces. The value of the runoff coefficient and the structure of the campus areas corresponds to the level of the housing area, i.e. it ranges between the average of the area of individual and collective type of housing.

When evaluating the level, one must again perceive the great internal differentiation of conditions inside the campus. The overall assessment is influenced by development areas that are covered by extensive vegetation. With the use of the Helsinki Green Factor assessment methodology, development plans can be further modeled for the use of these areas and variants can be compared from the point of view of BGI quality. At the same time, the evaluation of areas does not take into account the spatial structure of the BGI from the point of view of functionality and connectivity. It is therefore necessary to use the detailed results of the mapping also for the effective design how to include the campus into the BGI suburban landscape. In this context, the connection of the area in the southern direction to the suburban forest complexes is important. The BGI in the campus area forms an important hotspot of biodiversity even within the wider locality, where otherwise production and storage areas, or civil amenities of a commercial type predominate.

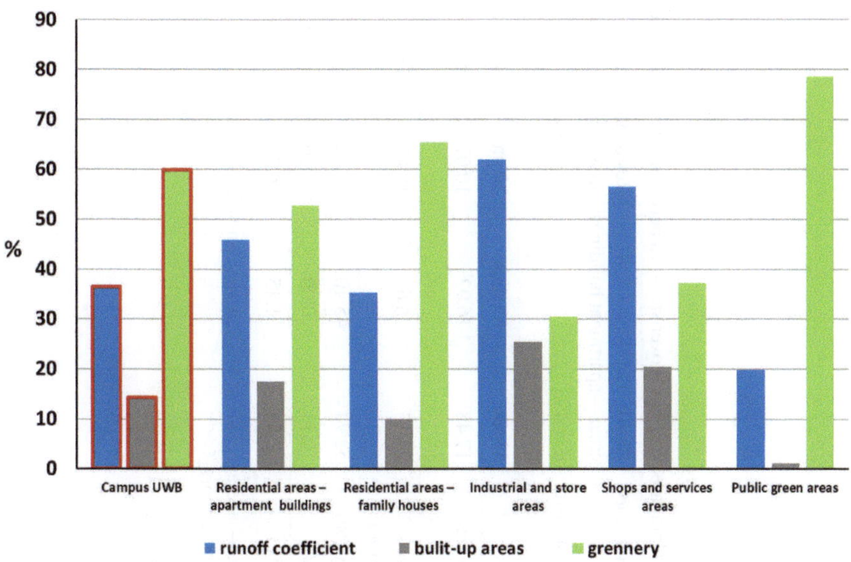

Fig. 5 Comparison of environmental parameters between campus UWB area and the average values of selected functional types of the territory of the city of Pilsen. *Source* own elaboration

Table 2 Comparison of land cover statistics between campus UWB area and the average values of selected functional types of the territory of the city of Pilsen

Type of territory	Residential areas—apartment buildings		Residential areas—family houses		Industrial and store areas		Shops and services areas		Public green areas		Campus UWB
Number of raster samples	$n = 478$		$n = 798$		$n = 2664$		$n = 378$		$n = 1251$		
Statistics	Mean	Median	Mean	Median	Mean	Median	Mean	Median	Mean	Median	
Built-up areas (%)	17.48	16.92	10.07	9.73	25.40	18.48	20.40	15.71	1.15	0.00	14.28
Impervious surfaces (%)	27.13	26.68	18.07	16.64	39.18	38.33	38.62	37.50	11.29	3.81	17.31
Greenery (%)	52.74	45.40	65.45	61.42	30.47	19.13	37.24	27.27	78.62	58.25	59.85
Water (%)	0.11	0.00	0.20	0.00	0.31	0.00	0.72	0.00	6.95	0.00	0.35
Others (%)	2.54	1.45	6.22	5.14	4.64	2.23	3.03	1.13	1.99	0.36	8.22

Note Mean—average values; Median–median values
Source The authors

6 Discussion

6.1 Using the Results for Campus Development

The strategic plan of UWB for the period 2021–2025 declares sustainable development and the use of smart technologies in the development plans of the university campus (UWB 2022; Herzner and Hommerová 2022). The carried out mapping and assessment of areas became the basis for processing the campus development study and the design of a new rainwater management system. Another analytical basis is the results of a survey of the opinions of campus users (CpKP 2018). Information was gathered through workshops as well as a emotional mapping where campus users could directly record their likes and dislikes and what development students and staff envisioned.

Considering the results of our mapping, attention should be paid to strengthening green elements, permeable surfaces and retention areas in the parts of the campus that are the worst rated. Specifically, it is planned to establish green roofs on selected buildings and to change the system from centralized drainage to the city sewer system to a system that collects and uses rainwater in a decentralized manner (Fránek et al. 2022). The development study proposes the use of an area of extensive tree cover in the eastern part of the campus as a park-like recreation area. An important element of the park should be a retention reservoir, created as part of a system for managing rainwater from the roofs of university buildings.

But the discussion is ongoing about the intensity of the transformation of the current vegetation with the character of "urban wilderness". Considering that the importance of the vegetation area for the system of blue-green infrastructure of the territory was confirmed by mapping, authors recommend giving preference to a solution with the maximum possible use of the existing vegetation (Čibik et al. 2020). Urban greenery planned using succession processes is discussed in the current literature because it delivers ecosystem services but has low public support (Qiu et al. 2013; Carrus et al. 2015; Palliwoda and Priess 2021; Aznarez et al. 2022). A survey of opinions among students and university staff also showed that the development of areas for sports activities or modified park areas without extensive vegetation close to nature is more supported (Fig. 6). Unmaintained areas of dense bushes and trees are more likely to cause concerns about safety at night or the feeling of neglected space (CpKP 2018). A suitable solution is the use of a park with a water area to demonstrate ecosystem services and its involvement in the teaching of the fields of architecture, environmental design, spatial planning or eco-pedagogy (Melo et al. 2020; Wojciechowska et al. 2021).

An important part of the development study of the campus is the green axis of the area in the west–east direction, developed as a pedestrian promenade, but also as a biocorridor of wider significance within the suburban landscape. The connectivity of the urban green system is a necessary condition for its functionality (Hansen et al. 2017; Kohout and Kopp 2020). However, the approaches used for the environmental assessment of the quality of areas do not take connectivity into account. As an

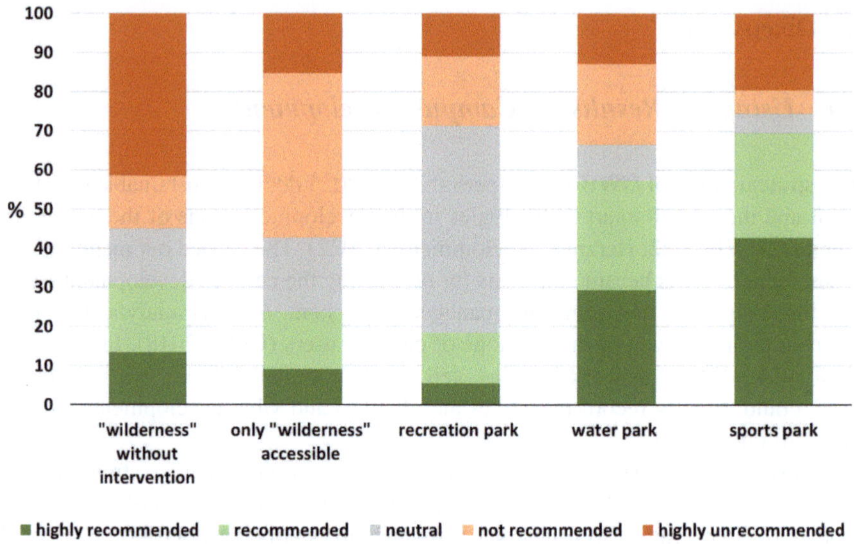

Fig. 6 Recommendations of UWB students and staff for planning the development of an area with extensive vegatation in the eastern part of the campus. *Source* own elaboration based on project of geography students, 2022, $n = 164$

extension of methodological approaches to assessment, the use of other landscape-ecological parameters can therefore be recommended (Smets et al. 2019; Liu et al. 2022).

6.2 *Benefits and Limits of Using Environmental Indicators of Areas*

The mapping of the campus area using the environmental parameters of the areas showed their advantages and disadvantages. The authors consider it beneficial to describe the differences between individual areas in detail and thus more effectively propose their further management or transformation within the framework of campus development. The use of standardized indicators (runoff coefficient, Hesinki Green Factor) is also important, which enables comparison either within the city territory or with recommended limit values (City of Helsinki 2014). In relation to a specific development area, they can be part of the regulatory conditions (Prague Institute of Planning and Development 2002). Environmental indices can also fulfill the function of a condition for economic support, be part of the eco-labelling of buildings or premises or serve to define the objectives of the territory's development strategy (Mattanovich et al. 2018). When presenting the total value for a larger area, such as a campus, the result should always be supplemented with characteristics of internal differences in the quality of the area (Göçer et al. 2018; Alshuwaikhat et al. 2017).

The total value is significantly dependent on the demarcation of the territory, as shown in the case of the UWB campus by the inclusion of development areas on the edges of the campus (Žák 2021).

The use of environmental indicators of blue-green infrastructure is limited to the use for a larger area of the entire city (Szulczewska et al. 2014; Peroni et al. 2020; Juhola 2018). The choice of indicator details is a necessary compromise between the professional perfection of complex index constructions and the possibility of obtaining sufficiently detailed information for a larger area (Mattanovich et al. 2018). In particular, the inclusion of elements of rainwater management does not allow the use of remote sensing geodata alone (Derkzen et al. 2015; Sulzer 2016). The environmental indicators of areas are an important indicator of the quality of the territory, but the methodology used does not take into account spatial links, e.g., the connectivity of the BGI system (Liu et al. 2022). In addition to mapping the physical state of the urban landscape, attention must also be paid to the perception of the needs of the users of the territory (Bacchin et al. 2016; Macháč and Louda 2019). The planning requirements that result from the physical condition of the territory may not coincide with the requirements of the city's inhabitants (Lehnert et al. 2021a).

7 Conclusion

University campuses have an important potential for the development of BGI, which is part of the planning of sustainable cities. The use of environmental quality indicators of areas based on detailed mapping of university campuses provides important data for BGI planning. The selection, implemented application and discussed limits of environmental quality indicators of areas on the example of the University of West Bohemia campus offer inspiration for other university campuses. The transferability of the case study is partly limited by the context of local environmental conditions. Although the indicators used (runoff coefficient, Helsinki Green Factor) are widespread in the international scientific and planning community, local conditions must be taken into account when interpreting them for the planning of university campuses. Based on accurate mapping and assessment of the territory of the University of West Bohemia campus using selected environmental parameters, the basis for targeted development planning was created.

The results serve to locate the elements of effective rainwater management using measures close to nature in the proposed development plan. Concrete aspects of implementation must meet technical standards, correspond to genuine relationships in the environment, and be accepted by the academic community. Participatory planning is therefore recommended as an essential step, with the involvement of students and university staff in the planning and implementing of the new vision of campus development according to the principles of the blue-green infrastructure.

Connecting university campuses to BGI is important not only for the development of cities, but also from the point of view of the development of the universities themselves. Innovative approaches to campus development planning are an integral part of student professional education, development and research. Efficient management of rainwater and high-quality public spaces contribute to improving the operating conditions of the university. BGI planning supported by mapping the environmental quality of university campus areas is therefore an appropriate part of the sustainable development of universities. Quality BGI brings benefits not only for the university environment itself, but also for the development of the urban landscape in a wider territorial context. The development of university campuses as hot spots of modern trends in urban and suburban landscape transformation is a crucial contribution of universities to the sustainable development of regions.

Acknowledgements This research was supported by the Technology Agency of the Czech Republic, Grant SS03010080 "Interdisciplinary approaches to efficient rainwater management on development sites of urban areas in the economic, social and environmental context".

References

Alshuwaikhat HM, Abubakar IR, Aina YA, Adenle YA, Umair M (2017) The development of a GIS-based model for campus environmental sustainability assessment. Sustainability 9(3):439

Alves A, Vojinovic Z, Kapelan Z, Sanchez A, Gersonius B (2020) Exploring trade-offs among the multiple benefits of green-blue-grey infrastructure for urban flood mitigation. Sci Total Environ 703:134980

Antoszewski P, Świerk D, Krzyżaniak M (2020) Statistical review of quality parameters of blue-green infrastructure elements important in mitigating the effect of the urban heat island in the temperate climate (C) zone. Int J Environ Res Public Health 17:7093

Aznarez C, Svenning J-C, Taveira G, Baró F, Pascual U (2022) Wildness and habitat quality drive spatial patterns of urban biodiversity. Landsc Urban Plan 228:104570

Bacchin T, Ashley R, Blecken G, Viklander M, Gersonius B (2016) Green-blue infrastructure for sustainable cities: innovative socio-technical solutions bringing multifunctional value. Novatech, Lyon, France 28 June–1 July 2016, pp 1–4

Baiamonte G (2020) A rational runoff coefficient for a revisited rational formula. Hydrol Sci J 65(1):112–126

Baynes TM, Wiedmann T (2012) General approaches for assessing urban environmental sustainability. Curr Opin Environ Sustain 4:458–464

Becker W et al (1990) The biotope area factor as an ecological parameter. principles for its determination and identification of the target. Becker Giseke Mohren Richard, Landschaft Planen & Bauen, Berlin, p 24

Carrus G, Scopelliti M, Lafortezza R, Colangelo G, Ferrini F, Salbitano F, Agrimi M, Portoghesi L, Semenzato P, Sanesi G (2015) Go greener, feel better? The positive effects of biodiversity on the well-being of individuals visiting urban and peri-urban green areas. Landsc Urban Plan 134:221–228

City of Helsinki (2014) Climate-Proof City—the planner's workbook. City of Helsinki Environment Centre, Helsinki. https://ilmastotyokalut.fi/en/. Accessed 5 Sept 2022

City of Helsinki (2016) Developing the city of Helsinki green factor method. Report summary. iWater—Integrated Storm Water Management. Interreg Central Baltic. City of Helsinki Environment Centre, Helsinki. https://www.integratedstormwater.eu. Accessed 5 Sept 2022

CpKP (2018) Kampus snů? Závěrečná zpráva. Zapojení studentů a zaměstnanců do zpracování návrhu koncepčního řešení areálu ZČU na Borských polích s důrazem na veřejná prostranství

[Dream Campus? Final Report. Involvement of students and employees in drafting the conceptual solution of the UWB complex in Borské pole with an emphasis on public spaces]. Centrum pro komunitní práci západní Čechy (Centre for Community Work West Bohemia), Pilsen, Czech Republic, p 132

Čibik M, Back Prochnow S, Stiles R, Štěpánková R (2020) Recognising green infrastructure as a part of the fourth nature concept through university campuses. Acta Hortic Regiotecturae 23(2):71–75

De Jong M, Joss S, Schraven D, Ch, Zhan, Weijnen M (2015) Sustainable–smart–resilient–low carbon–eco–knowledge cities; making sense of a multitude of concepts promoting sustainable urbanization. J Clean Prod 109:25–38

Depietri Y, McPhearson T (2017) Integrating the grey, green, and blue in cities: nature-based solutions for climate change adaptation and risk reduction. In: Kabisch N et al (eds) Nature-based solutions to climate change adaptation in urban areas. Linkages between science, policy and practice. Springer, Cham, pp 91–109

Derkzen ML, van Teeffelen AJA, Verburg PH (2015) Quantifying urban ecosystem services based on high-resolution data of urban green space: an assessment for Rotterdam, the Netherlands. J Appl Ecol 52:1020–1032

Ferguson BC, Fratzeskaki N, Brown R (2013) A strategic program for transitioning to a water sensitive city. Landsc Urban Plan 117(1):32–45

Fránek Z, Horňasová T, Šenekel L, Hamrová M, Vlková K, Železný J, Kučírková V (2022) Územní studie ZČU [Regional study of UWB]. University of West Bohemia, Pilsen, Czech Republic, Fránek Architects

Göçer Ö, Göçer K, Başol AM, Kıraç MF, Özbil A, Bakovic M, Siddiqui FP, Özcan B (2018) Introduction of a spatio-temporal mapping based POE method for outdoor spaces: Suburban university campus as a case study. Build Environ 145:125–139

Gulwadi GB, Scholl KG (2017) Campus infrastructure and sustainable resource management practices: mapping campus DNA for human resiliency. In: Leal Filho W, Brandli L, Castro P, Newman J (eds) Handbook of theory and practice of sustainable development in higher education. Springer, Cham, pp 103–118

Gunasekara C (2006) Reframing the role of universities in the development of regional innovation systems. J Technol Transf 31(1):101–113

Hansen R, Rall E, Chapman E, Rolf W, Pauleit S (2017) Urban GreenInfrastructure planning: a guide for practitioners. Green Surge. Freising, Munich. http://greensurge.eu/working-packages/wp5/. Accessed 10 Sept 2022

Herzner A, Hommerová D (2022) Multi-dimensional assessment of a Bavarian and Czech University: a case study of sustainability implementation. Eur J Sustain Develop 11(2):93–113

Hoang L, Fenner RA (2016) System interactions of stormwater management using sustainable urban drainage systems and green infrastructure. Urban Water J 13(7):739–758

Iojă CI, Grădinaru SR, Onose DA, Vânău GO, Tudor AC (2014) The potential of school green areas to improve urban green connectivity and multifunctionality. Urban for Urban Green 13(4):704–713

Juhola S (2018) Planning for a green city: the green factor tool. Urban for Urban Green 34:254–258

Keeley M (2011) The green area ratio: an urban site sustainability Metric. J Environ Plan Manag 54(7):937–958

Kestin T, van den Belt M, Denby L, Ross K, Thwaites J, Hawkes M (2017) Getting started with the SDGs in universities: a guide for universities, higher education institutions, and the academic sector. Sustainable Development Solutions Network, Australia/Pacific, Melbourne, p 52

Kohout M, Kopp J (2020) Green space ideas and practices in European cities. J Environ Plan Manag 63(14):2464–2483

Kopp J, Frajer J, Lehnert M, Kohout M, Ježek J (2021a) Integrating concepts of blue-green infrastructure to support multidisciplinary planning of sustainable cities. Problemy Ekorozwoju 16(2):137–146

Kopp J, Frajer J, Novotná M, Preis J, Dolejš M (2021b) Comparison of ecohydrological and climatological zoning of the cities: case study of the city of Pilsen. ISPRS Int J Geoinf 10(5):1–21

Kopp J, Preis J (2019) The potential implementation of stormwater retention ponds into the blue-green infrastructure of the suburban landscape of Pilsen, Czechia. Appl Ecol Environ Res 17(6):15055–15072

Kopp J, Raška P, Vysoudil M, Ježek J, Dolejš M, Veith T, Frajer J, Novotná M, Hašová E (2017) Ekohydrologický management mikrostruktur městské krajiny. Czech Republic, University of West Bohemia, Pilsen, p 166

Kruuse A (2011) GRaBS Expert paper 6: the green space factor and the green points system. Town and Country Planning Association, Malmö, p 12

Leal Filho W (2011) About the role of universities and their contributions to sustainable development. High Educ Policy 24:427–438

Lehnert M, Geletič J, Kopp J, Brabec M, Jurek M, Pánek J (2021a) Comparison between mental mapping and land surface temperature in two Czech cities: a new perspective on indication of locations prone to heat stress. Build Environ 203:108090

Lehnert M, Tokar V, Jurek M, Geletič J (2021b) Summer thermal comfort in Czech cities: measured effects of blue and green features in city centres. Int J Biometeorol 65(8):1277–1289

Lerer SM, Righetti F, Rozario T, Mikkelsen PS (2017) Integrated hydrological model-based assessment of stormwater management scenarios in Copenhagen's first climate resilient neighbourhood using the three point approach. Water 9(11):883

Liu Y, Huang TT, Zheng X (2022) A method of linking functional and structural connectivity analysis in urban green infrastructure network construction. Urban Ecosyst 25:909–925

Macháč J, Louda J (2019) Urban wetlands restoration in floodplains: a case of the city of Pilsen, Czech Republic. In: Hartmann T, Slavíková L, McCarthy S (eds) Nature-based flood risk management on private land. Springer, Cham, pp 111–126

Mattanovich E, Bürger G, Fischer M, Neubauer U, Stebegg K (2018) Handlungsziele für Stadtgrün und Deren Empirische Evidenz. Indikatoren, Kenn-und Orientierungswerte. Bundesinstitut für Bau-, Stadt-und Raumforschung, Bonn, p 141

McCulloch L, Robertson M (2015) Southampton city council green space factor guidance notes. Southampton City Council, Southampton, p 7

Melo EFRQ, Melo RHRQ, Libera LTD, Melo RHRQ, Melo JHQ (2020) Recognizing sustainability in a university campus through a green trail. In: Leal Filho W, Tortato U, Frankenberger F (eds) Universities and sustainable communities: meeting the goals of the agenda 2030. World Sustainability Series. Springer, Cham, pp 767–776

Morison PJ, Brown RR (2011) Understanding the nature of publics and local policy commitment to Water Sensitive Urban Design. Landsc Urban Plan 99(2):83–92

Palliwoda J, Priess JA (2021) What do people value in urban green? Linking characteristics of urban green spaces to users' perceptions of nature benefits, disturbances, and disservices. Ecol Soc 26(1):28

Pauleit S, Duhme F (2000) Assessing the environmental performance of land cover types for urban planning. Landsc Urban Plan 52:1–20

Peroni F, Pristeri G, Codato D, Pappalardo SE, De Marchi M (2020) Biotope area factor: an ecological urban index to geovisualize soil sealing in Padua, Italy. Sustainability 12:150

Prague Institute of Planning and Development (2002) Metodický pokyn k Územnímu plánu sídelního útvaru hlavního města Prahy [Methodological instruction for the Spatial Plan of the residential unit of the capital city of Prague]. Prague Institute of Planning and Development, Prague, p 37

Qiu L, Lindberg S, Nielsen AB (2013) Is biodiversity attractive?—On-site perception of recreational and biodiversity values in urban green space. Landsc Urban Plan 119:136–146

Simperler L, Himmelbauer P, Stöglehner G, Ertl T (2018) Siedlungswasserwirtschaftliche Strukturtypen und ihre Potenziale für die dezentrale Bewirtschaftung von Niederschlagswasser. Oesterr Wasser- Abfallwirtsch 70:595–603

Smets V, Verbeiren B, Hermy M, Somers B (2019) Urban spatial configuration and functional runoff connectivity: influence of drainage grid density and landscape metrics. Water 11(12):2661

Stránský D, Kabelková I (2015) Review of the implementation process od sustainable stormwater management in the Czech Republic. In: Hlavínek P, Zeleňáková M (eds) Storm water management, examples from Czech Republic, Slovakia and Poland. Springer, Cham, pp 13–26
Sulzer W (2016) Reale Abbildung der Stadtentwicklung Graz –Landnutzungskartierung 1945–2015. Eigenverlag der Stadt Graz, Graz, Austria, p 260
Szulczewska B, Giedych R, Borowski J, Kuchcik M, Sikorski P, Mazurkiewicz A, Stańczyk T (2014) How much green is needed for a vital neighbourhood? in search for empirical evidence. Land Use Policy 38:330–345
Thorne C (ed) (2016) Delivering and evaluating multiple flood risk benefits in blue-green cities. Key project outputs. EPSRC Project EP/K013661/1. University of Nottingham, Nottingham
UPDICP (2020) Koncepce odtokových poměrů města Plzně. Urban Planning and Development Institute of the City of Pilsen, Pilsen, Czech Republic. https://ukr.plzen.eu/zivotni-prostredi/koncepce-odtokovych-pomeru-mesta-plzne/koncepce-odtokovych-pomeru-mesta-plzne.aspx. Accessed 3 June 2022
UWB (2022) Strategy 2021–2025. University of West Bohemia, Pilsen, Czech Republic. https://www.zcu.cz/en/University/About-us/strategy-2025.html. Accessed 10 June 2022
Vítek J, Stránský D, Kabelková I, Bareš V, Vítek R (2015) Hospodaření s dešťovou vodou v ČR [Rainwater management in the Czech Republic]. ZO ČSOP Koniklec, Prague, p 127
Voskamp IM, van de Ven FHM (2015) Planning support system for climate adaptation: composing effective sets of blue-green measures to reduce urban vulnerability to extreme weather events. Build Environ 83:159–167
WEF (2014) Green infrastructure implementation: a special publication. WEF Special Publication, Water Environment Federation, Alexandria, Virginia, p 491
Wojciechowska E, Gajewska M, Nawrot N, Kilanowska M, Obarska-Pempkowiak H (2021) Combination of architectual, environmental, and social aspects in urban stormwater management: a case study of the university campus. Int J Conserv Sci 12(1)681–700
Woods-Ballard B et al (2015) The SUDS manual (C753). CIRIA, London, p 937
Ziersen J, Clauson-Kaas J, Rasmussen J (2017) The role of greater Copenhagen utility in implementing the city's cloudburst management plan. Water Pract Technol 12(2):338–343
Žák P (2021) Potenciál rozvoje modro-zelené infrastruktury kampusu Západočeské univerzity v Plzni [The development potential of the blue-green infrastructure of the campus of the University of West Bohemia in Pilsen] (Diploma thesis). Czech Republic, University of West Bohemia, Pilsen, p 83

The Effect of Architectural Parameters on the Risk of Infection by Airborne Diseases in Classrooms

Pedro Carmo e Souza and Joyce Correna Carlo

LATECAE—n/n P.H. Rolfs Ave., Campus UFV, Viçosa-MG.
36570-900, Brazil. Website: http://www.latecae.ufv.br.

Abstract The COVID-19 pandemic brought increased concern for the indoor air quality (IAQ) of spaces like classrooms from late 2019 to late 2022, prompting the production and consumption of over 8 millions of tonnes of use plastics (Peng et al. 2021) in the form of face-masks, HEPA filters, gloves and more. As we move forward, it is important to prevent the spread of such diseases as much as we can, without compromising our sustainable development goals. This paper underlines the importance and effectiveness of natural ventilation in reducing airborne infection risk, through the case study of an educational building in Brazil. It demonstrates natural ventilation strategies can keep infection risk to a statistically negligible number, even when the occupants eschew facemasks.

Keywords Ventilation · Covid-19 · Sustainable development · Indoor air quality · Classrooms

1 Introduction

COVID-19 is a viral respiratory disease that can be transmitted through the air via droplets of mucus and saliva (Zhang et al. 2020; Morawska and Milton 2020). Understandably, there are concerns about the spread of COVID in indoor environments, especially in indoor environments such as classrooms, which offer a greater risk of infection than others due to their usage (Zhang et al. 2020). In the past, similar

P. C. e. Souza · J. C. Carlo (✉)
Architecture and Urbanism at Federal University of Viçosa, Viçosa, Brazil
e-mail: joycecarlo@ufv.br

P. C. e. Souza
e-mail: pedro.carmo@ufv.br

W. Leal Filho et al. (eds.), *The Contribution of Universities Towards Education for Sustainable Development*, World Sustainability Series,
https://doi.org/10.1007/978-3-031-49853-4_14

risks have prompted the adoption of face masks and air filters (Riley et al. 1978; Gammaitoni and Nucci 1997; WHO 2020), but current understanding of sustainability and of the world's over-reliance on energy consumption and disposable plastics stimulates the search for alternative strategies. This paper quantifies the risk of airborne disease transmission in two naturally ventilated university classrooms, which rely on thermal chimneys and electric fans to regulate indoor temperature, and are located in a tropical climate. This paper suggests safe usage protocols for similar classrooms. This study focuses on transmission in classrooms because the typical classroom is a densely occupied, potentially poorly ventilated space, where people spend hours at a time exhaling significant amounts of air by talking, sometimes loudly (Morawska et al. 2009; Zivelonghi and Lai 2021). These factors all contribute to poor air quality and increased infection risk, Riley et al. (1978) but it becomes evident that the full estimation of risks relies heavily on understanding the behaviour of air in the room.

2 Education for Sustainable Development

A building's users are frequently pointed out as the most important component in its energy performance (Chong et al. 2021). The operation of windows, doors, light switches and equipment, plays a major role in determining not only a building's energy consumption, but also the effectiveness of the sustainability strategies employed in its design. Educating users about the importance of natural ventilation in buildings and its benefits can promote a more sustainable and healthy environment. Showing the effectiveness of natural ventilation in raising indoor air quality might prompt users to rely more on these strategies than on artificial conditioning, maybe even reverting the trend of "AC Dependency" (Lamberts 2008; IEA 2018).

More than educate users, universities also teach architects and engineers, which shape the space we live in. According to the International Energy Association, buildings account for 40% of the world's energy consumption (IEA 2018). Raising university students' awareness about passive ventilation strategies will surely prompt them to employ those strategies in their professional lives, even more if they are acquainted with how effective those strategies are, how they work, and when to employ them.

3 Mechanics of Infection

Studies have calculated the risk of COVID infection by simulating airflow in a room with Computational Fluid Dynamics (CFD) software (Bhagat et al. 2020). This type of software can be very accurate in predicting specific airflow pathways in a room, but it also requires a high computational cost and has a steep learning curve (Foster and Kinzel 2021). Alternatively, ventilation data about a room can be obtained using software based on thermal- and mass-balance equations, such as EnergyPlus (United

States Department of Energy 2022). Coupling this ventilation data with a mathematical model for airborne infection risk (which has been called the GN-Riley model (Zivelonghi and Lai 2021)) requires less computational cost, has a more forgiving learning curve, and still produces results within 6% error (Foster and Kinzel 2021). This study bases itself on that assumption, and a 2021 study by Zivelonghi and Lai (2021), for classroom infection risk calculation purposes.

As classrooms differ not only in size and configuration but also in usage, it is beneficial to calculate risks for as many cases as feasible. Feasibility also dictated, in this case, that only two classrooms should be analysed, so the authors chose classrooms in the campus' main lecture hall, which also were equipped with thermal chimneys. These thermal chimneys were built to naturally permit air exchanges with the outside environment, allowing the air in the classroom to escape upwards as it heats up and becomes less dense, the so-called Chimney Effect. Even when the outside and indoor temperature are similar, the aforementioned chimney effect should be present, if delayed, since it is mainly driven by a pressure-temperature gradient (Maciel 2016). Maciel's results also clearly show the contribution of the chimney height to ventilation surpasses the contribution of the temperature of the absorbing wall alone.

To delve further into infection mechanics, it must first be emphasised that this study treats infection as a physical transport problem, rather than a physiological susceptibility problem. That is to say, the authors are more concerned with the physical trajectory of infective particles, rather than with the likelihood of a specific person or group contracting the disease. By basing itself on the Wells-Riley model, this study makes a series of assumptions, chief among them the notion of Quanta. A Quantum, as defined by Wells (1934), is a dose of infective particles big enough to infect a susceptible subject. By abstracting the exact number of viruses or bacteria that a patient needs to come in contact with to be infected, this model needs not concern itself with the individual susceptibility of individuals, and can rather focus on the probability of coming in contact with one or more "doses", however big or small that dose may be. The model in question also assumes that the doses instantly and homogeneously disperse into the room once emitted (which also justifies the usage of thermal and mass-balance algorithms, rather than flow pathway-specific CFD software) and that the infective agent is introduced and removed from the room at a constant rate. It has been said that this level of simplification comes at the expense of accuracy (Foster and Kinzel 2021) (Gammaitoni and Nucci 1997), so recent interpretations of this model also took into account the time required for the pathogen to fully disperse into the room, the lifespan of viral particles outside a host (known as viral inactivation), and the time required for airborne particles to depose onto surfaces when the room air stays undisturbed (Gammaitoni and Nucci 1997; Zivelonghi and Lai 2021).

The actual equation was proposed by Riley (1978), and then improved by Gammaitoni and Nucci (1997) was described by Buonanno et al. (2020) as Eqs. (1) and (2)

$$R = 1 - e^{-IR\int_0^T n(t)dt} \quad (1)$$

$$n(t) = \frac{ER_q * I}{IVRR * V} + \left(n_0 + \frac{ER_q * I}{IVRR}\right) * \frac{e^{-IVRR*t}}{V} \quad (2)$$

where:

- *R* is the risk of infection (%)
- *e* is Euler's Number
- *IR* is the Inhalation Rate of the exposed subject (a function of the subject's metabolism)
- *T* is the total time of contact with infected air, in hours
- *n(t)* is the quanta (units of infection) concentration in the air
- *ERq* is the disease's quanta emission rate
- *I* is the number of infected people
- *IVRR* is the Infective Virus Removal Rate, that is, the rate at which viruses depose onto surfaces, get ventilated out, or die without infecting a host
- *V* is the volume of the room
- n_0 is the initial quanta in the room

By analysing Eqs. (1) and (2) one can see that the risk of infection by an airborne disease such as COVID-19 is directly proportional to the time spent in contact with infected air, to the number of infected, to the specifics of the disease, and to the amount of air changes per hour in the room. Some of these factors are non-controllable or unknowable a priori, such as the quanta emission rate, the number of infected, and the average pulmonary volume of a person (which Gammaitoni assumes as 0.01 m^3/min (Gammaitoni and Nucci 1997)), but it is possible to control for exposure time and room ventilation. These represent the two variables this work shall concern itself with.

4 Methodology

Two classrooms equipped with thermal chimneys were selected and compared as to their safety relative to airborne infection. The classrooms are located in Pavilhão de Aulas II (PVB), a high-traffic educational building on the main campus of Universidade Federal de Viçosa, in southeastern Brazil. The campus is a low-rise area, with large open spaces between buildings, and is located in a tropical climate (Classified as "Humid Subtropical" or "*Cwa*" by Köppen-Geiger, and as Highland Tropical by IBGE (Guimarães and Carlo 2011)). The classrooms are located directly on top of each other, separated by one floor, in such a way that the main differences between them are the height of their thermal chimneys, and their own height from the ground. Data was collected on the classrooms, then used to build a simulation model of them.

After gauging the model for accuracy, the simulation was run and outputted ventilation data for both classrooms for a year. Finally, using the data acquired and Eqs. (1) and (2) it was possible to calculate the mean risk of infection by airborne COVID-19 in each classroom. This process is further detailed in this section.

4.1 Data Collection

All on-site data collection took place on the same day, April 13th 2022, from 14:00 to 16:45. Weather data was sourced from a weather file by Guimarães and Carlo (2011). The two classrooms' physical dimensions were gathered on-site, as well as the effective ventilation area of their openings, building orientation, surface temperatures, and air temperature. Numbered 106 and 306, the rooms are rectangular, measuring 7.32 m by 9.30 m (68.07 m^2), with a ceiling height of 3.25 m. They are bordered by two balconies: one on the southwest façade of the building, and one which serves as a corridor between all classrooms and also to the staircase. The rooms' exterior walls are 20 cm cellular concrete block, mortar, plaster and white paint; while the interior walls are 10 cm drywall with glass wool insulation for the walls between classrooms. The ceiling and floor are concrete topped steel decks, leaving the girders and lower steel deck surface apparent. The rooms also feature a set of wooden doors topped with a transparent glass pane; two groupings of four single steel doors each, and maxim-air windows all through the length of the wall to the balcony. Each classroom also featured a thermal chimney: a roofed vertical shaft between the classroom ceiling and the top of the building, built to aid ventilation via thermal gradient (Asadi et al. 2016).

The key difference between both classrooms was the height of their chimneys, which measured 9.24 m for room 106 and 2.24 m for room 306. As noted before, Room 106 is located on the ground floor, while room 306 is located two floors directly above, so the last one's chimney is much shorter. Each room's chimney was adjacent to the same wall, but offset from the chimneys of the other rooms above and below.

Four data loggers (Onset HOBO model U10-003, accurate to ±0.53 °C), were used to measure the air temperature in and around the empty classrooms, logging the air temperature every five minutes. Two of the data loggers were placed inside the classroom (one at the orthogonal centre of the classroom and one directly below the thermal chimney opening, on top of desks), one on the balcony away and one in the hallway. The data loggers were started up immediately before being placed, their time of placement was recorded, and the temperatures recorded up to 5 minutes after placement were discarded to avoid distortions in the data. After measuring for two hours, the data loggers were moved from room 106 to room 306, and continued measuring. The temperatures logged during the move and up to 5 minutes after placement were also discarded. Additionally, before placing the data loggers, the room's external and internal surfaces were photographed with a thermal imaging camera (Flir camera, model TG165-X, accurate to ±2.5 °C), and an anemometer (Kimo, model VT100, accurate to ±0.05 m/s and ±0.3 °C) was used to measure the

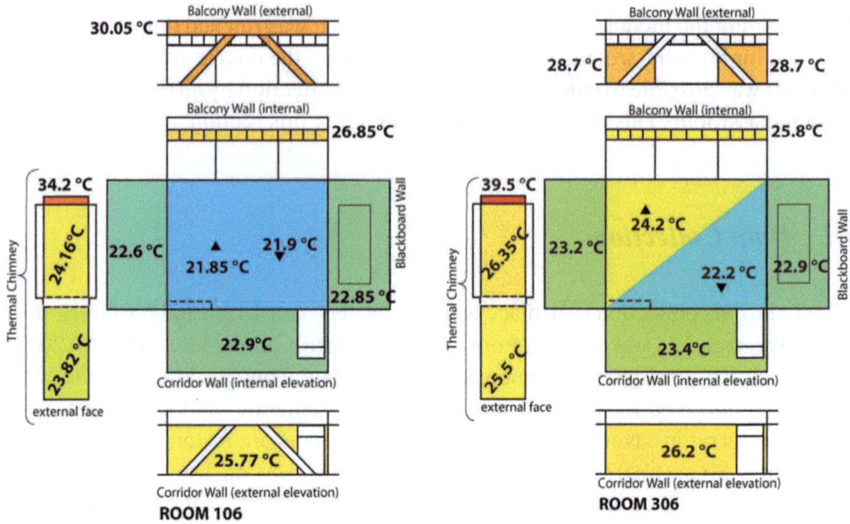

Fig. 1 Surface temperatures of both classrooms, as measured on-site with a thermal camera. These temperatures were used to gauge the model. **a** Room 106. **b** Room 306

air temperature and velocity through each thermal chimney's lower opening. The data points gathered (Fig. 1) allowed the creation and gauging of a virtual model of the room on EnergyPlus 22.1.0.

4.2 Simulation Model

Based on the data obtained, a virtual model of the classrooms was constructed in EnergyPlus (United States Department of Energy 2022). As seen on Fig. 2, the model represents only classrooms 106 and 306, the thermal chimneys, and the upper corridor floors, which function as shading for this simulation. While the building does feature classrooms on either side of the rooms that were modelled, the authors chose not to model the adjacent rooms for the sake of simplicity. Instead, shared walls were modelled as adiabatic. The model was then tested for accuracy by running the simulation and comparing the surface and air temperatures generated with the surface and air temperatures gathered on-site under similar weather conditions. For the model gauging, the authors considered the rooms to be unoccupied and with all possible fenestrations closed, since the building had been out of typical use for the previous two years due to the pandemic (Note that even with fenestrations closed, ventilation was still present by infiltration, which was simulated with the Airflow Network module) These simulated temperatures were compared with the temperatures measured on site. If a discrepancy between the two sets of data was found, the virtual model was updated accordingly, the simulation parameters were adjusted, and the simulation

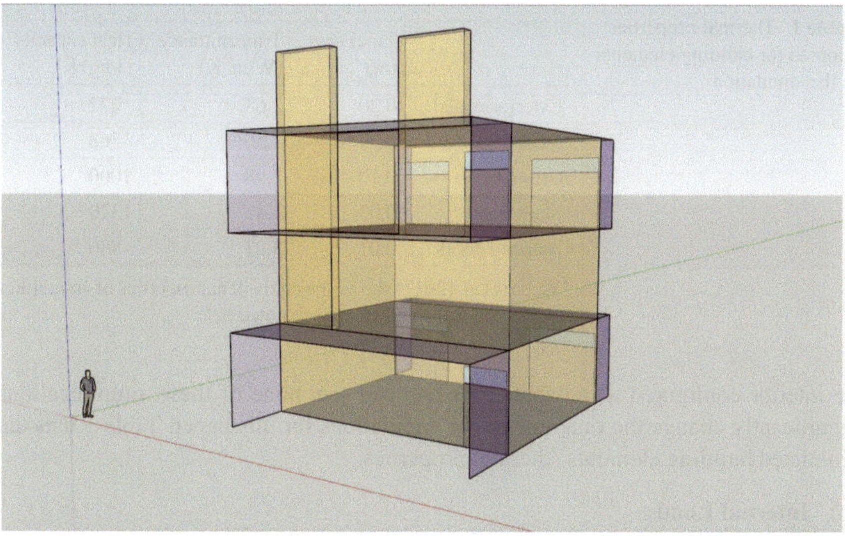

Fig. 2 Model Visualization as seen on the 3D modelling application *Sketchup*, by way of the *OpenStudio* plug-in. (Trimble Inc. 2019; OpenStudio Coalition 2021)

was run again, until the surface and air temperatures differed by less than 2 degrees Celsius. Only after the model was deemed accurate were the air changes per hour extracted. Any configurations not mentioned in this item followed specification on item 2.3.

4.3 Simulation

The settings for the model and the simulation are as follows:

Version: 22.1.0

External Façade Orientation: 39 degrees from North

Terrain: Suburbs

Solar Distribution Calculation Model: FullInterior And Exterior

Site Location: Viçosa (−20.753889, −42.881944, 648 m. Köppen-Geiger Cwa)

Run Period: (1/1/2022 to 31/12/2022). No holidays, special days, daylight saving, or holiday rule

Undisturbed Ground Temperature Model: Xing

As in all simulations, some assumptions and simplifications were employed. As an example, only a section of the whole building was modelled, with walls that would

Table 1 Thermal properties adopted for building elements in the simulation

	Thickness (m)	Transmittance (W/m^2K)	Heat capacity (J/kg*K)
Exterior walls[1]	0.20	2.68	272
Steel deck slab[1]	0.39	2.56	368
Ground slab	0.10	2.98	1000
Main door	0.03	1.41	1340
Balcony doors	0.03	3.23	880

[1] Weber et al. (2017) shows the equivalent properties of assemblies such as the walls and steel deck noted here

be interior configured as adiabatic. As one can see, none of these simplifications significantly change the outcome of the simulations for this paper. Table 1 lists the simulated building elements' thermal properties.

(a) Internal Loads

After model gauging, internal loads were introduced: The fluorescent light models emitted 9 W per m^2 of floor area, the computer and projector models emitted a total of 1300 W, and the room was considered to be occupied by 30 people, all at a metabolism rate of 100 W. These loads were applied from 8:00 to 22:00 on weekdays for the whole year.

(b) Materials and Constructions

This model uses a Typical Meteorological Year 3 (TMY3) weather file for Viçosa, as developed by (Guimaraes and Carlo 2011). For ground temperature, and therefore ground-interacting concrete slabs, an undisturbed ground temperature calculation model by Xing (2014) was employed. The corridor-classroom and balcony-classroom walls were modelled as 20 cm thick mortared solid cellular concrete, yielding an equivalent thermal transmittance of 2.68 W/m^2K, and a heat capacity of 272 kJ/m^2K, as calculated according to (Weber et al. 2017). The walls between classrooms on the same floor were modelled as adiabatic. The ground slabs were considered to be 10 cm thick, made of mesh concrete, and the hallway slabs were modelled as shading groups, as were the lateral walls separating each classroom's balcony from the adjacent ones. For ventilation, the Airflow Network module was used. The doors were considered to be 3 cm thick, and they appeared to be semi-solid wood panel doors on site.

(c) Ventilation

Each room was modelled as a thermal zone, as were each of the ventilation chimneys, following guidelines established by Oliveira and Carlo (2018). The zones that represented the chimneys were declared in EnergyPlus as Thermal Chimneys in the "*ZoneThermalChimney*" object, outside the Airflow Network Module. Since the rooms were naturally ventilated, the AirflowNetwork

Table 2 Effective ventilation area for modelled openings

Opening	Area (m^2)	Discharge coefficient
Main door	1.68	0.78
Balcony windows	0.28	0.78
Balcony doors	6.12	0.6
Thermal chimney	1.17	0.6

"*MultizoneWithoutDistribution*" ventilation mode was used. The discharge coefficient for windows (including the thermal chimney horizontal opening) was considered to be 0.6 when fully open, which happened, as detailed in item a), from 08:00 to 22:00 on weekdays on the gauged simulation. For doors, that coefficient was 0.78. Each opening's effective ventilation area can be seen in Table 2.

After the simulation was gauged and working correctly, the Output Variable "*Zone Air Changes per Hour (ach)*" was also requested for the Run Period of 3900 h a year, corresponding to the hours of operation for the educational building being studied: from 08:00 to 22:00 on each day of the week, and closed on weekends.

(d) **Risk**

The risk of infection by COVID-19 in each classroom was calculated using the Gammaitoni-Nucci-Riley model, using Eqs. (1) and (2). The risk was calculated for a maximum total exposure time of 4 hours, which exceeds the maximum class duration in the campus by 40 minutes. The rate of quanta emission ERq by the 1 infective source was considered to be 32 q/hour, following findings by (Buonanno et al. 2020), while the Infective Virus Removal Rate (IVRR) was calculated by simple addition of the deposition factor $k = 0.24$, the viral inactivation factor $\lambda = 0.63$, and the corresponding air change rate for that hour, in accordance to (Zivelonghi and Lai 2021; Buonanno et al. 2020). Also according to these works, the inhalation rate (IR) of the susceptibles was considered to be 0.96 m^3/hour, which corresponds to the mean value between stationary standing inhalation and light exercise inhalation rates.

5 Results

The Air Changes per Hour (ACH) were requested for the period the classrooms are typically occupied in a year, that is, from 08:00 to 22:00 on weekdays, resulting in 3900 data points. Figure 3 summarises the air renovation rates in each classroom.

Air renovation rates for both classrooms differed in line with predictions: Room 106, on the ground floor, exhibited greater overall air change rates than room 306. These discrepancies were expected, since room 106 has a much taller thermal chimney than room 306, which contributed to a greater airflow. Additionally, overall wind speeds in the area have ranged most commonly from 0.5 to 4.0 m/s in the weather file of the 10 years series, according to (Guimaraes and Carlo 2011), which

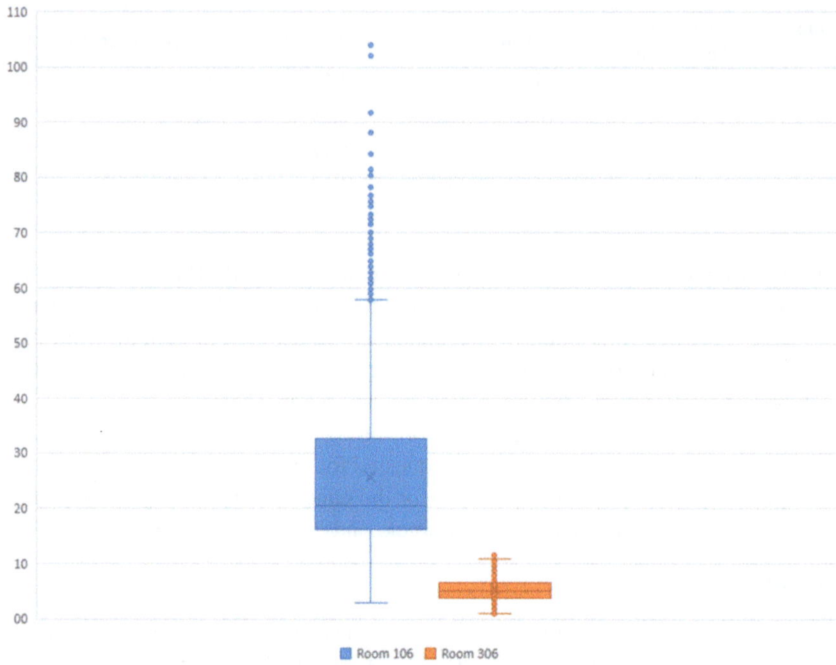

Fig. 3 Frequency distribution of air changes per hour values obtained in the simulation, for both rooms

explains why Room 306 did not exhibit better ventilation despite being higher off the ground. Room 106 also showed a broader range of air changes per hour (from 2.92 to 103.96 air changes per hour), but 90% of values fell within 15 and 51 ACH. Meanwhile in room 306, the values obeyed a tighter margin (0.9 ACH to 12.3 ACH), but were more distributed within that margin: Room 306's most common range is 3 to 4.34 ACH, with 342 data points (around 8.7% of all simulated hours), but it is closely followed by the 3.48 to 3.91 ACH category, with 341 hours. The majority (>50%) of simulated hours is spread over the 6 most common categories, while room 106 only needs the 3 most common categories to surpass 50%. The Mode air change rate range across all rooms was 15.32 to 18.42, as shown in Fig. 4 a result of the tighter variation in room 106. The results highlighted the fickle nature of ventilation, and the necessity of running infection risk calculations for each case separately, since even rooms with so much in common can yield vastly different ventilation data. Lastly, one can see that the air changes per hour (which, once again, ranged from 0.9 to 103.97) contributed overwhelmingly more than viral inactivation and deposition factors (respectively 0.63 h^{-1} and 0.24 h^{-1}) for the Infective Virus Removal Rate (IVRR), and thus to the risk of infection, as seen in Eq. (1).

As shown by that same equation, the risk of infection is mainly driven by the IVRR and exposure time. As such, Fig. 5 shows the risk of infection for every simulated IVRR, categorized by total time spent in contact with the infected air. This graph

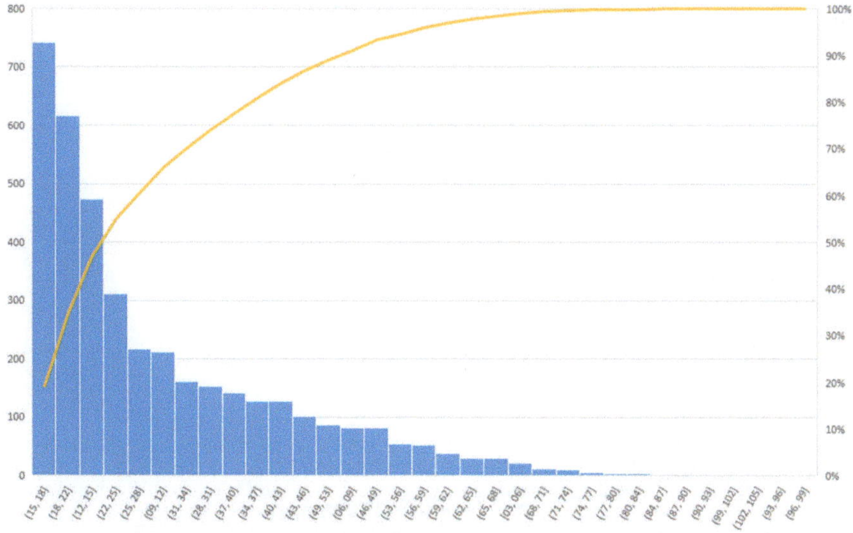

Fig. 4 Frequency distribution for all simulated hourly air changes. The horizontal axis shows the ACH ranges from most to least common

assumes the IVRR stays the same for the entire period of contact, which is unlikely, but doing so gives us an idea of the order of magnitude for the risks. The lowest possible risk (excluding R(0), which is always 0 by definition), was the risk for 50 minutes of total exposure time at 103.97 IVRR, which was 1.13\%, and occurred in Room 106. The maximum risk was associated with Room 306, the longest time of exposure (240 min) and lowest IVRR (0.9), as expected, and showed a 23.65\% chance of infection. The lowest risk for that same IVRR in the same classroom (306) was 3.07\% for 50 minutes of exposure. This shows that however low the risk of infection it can never be 0 for these circumstances, and more so, that even at low exposure times, there is a risk of infection if without proper ventilation.

If one considers the risk of infection as a metric for new infections, as done by (Zivelonghi and Lai 2021), the "one-infection threshold" for this classroom starts at 3,33% or 1 in 30 subjects. By that metric, with an IVRR of just 2.97 and exposure time of 50 minutes, the room can be occupied by 47 people before an infection arises, which is interesting, considering there are 72 seats present in each classroom. Maximum safe occupancies for each room were calculated in Table 3 using that room's mean risk for every time category. The room's maximum capacity of 72 seats was considered, so numbers above that were ignored.

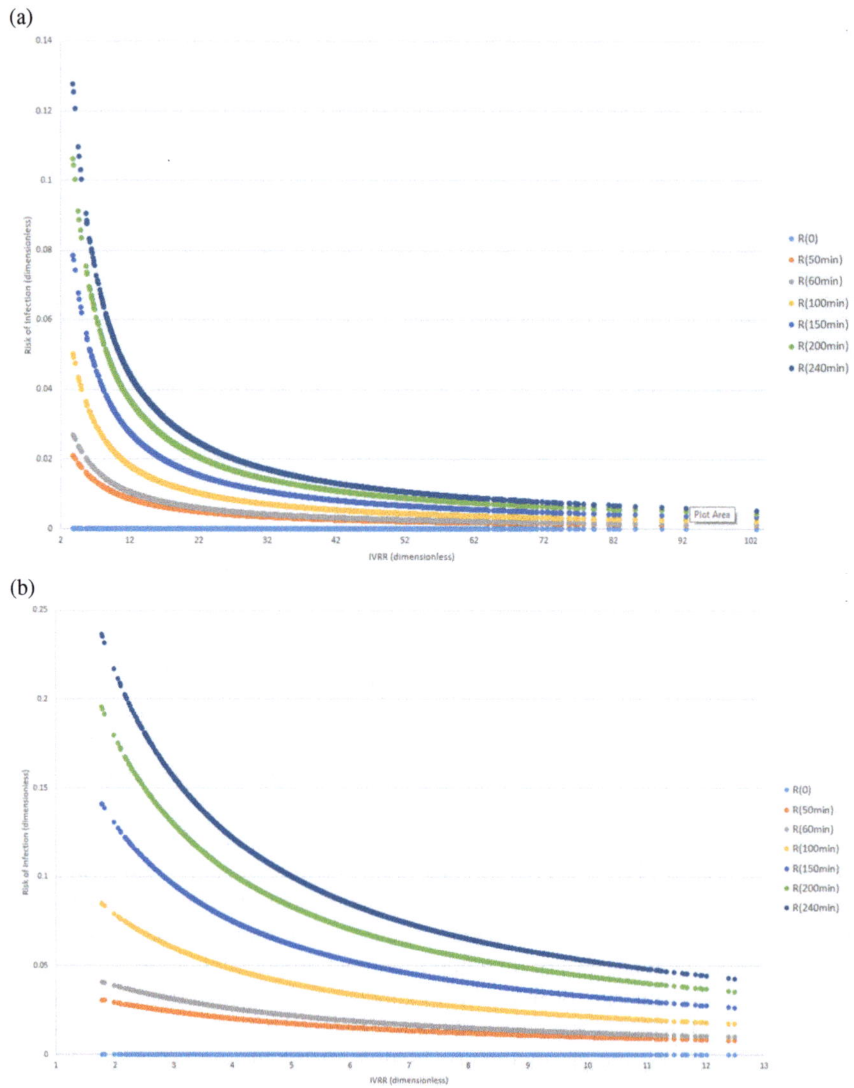

Fig. 5 Risk of infection for t = 0 to t = 240 min, for every IVRR found. Note that the bounds for horizontal and vertical axes change with the ACH range for each classroom. **a** Room 106. **b** Room 306

5.1 Sustainable Development

One final result catches the attention when one considers the implications of air safety being accomplishable by natural ventilation alone: In the case of tropical climates, the alternative to natural ventilation would be exhaust fans. Considering each classroom

Table 3 Maximum occupancy resulting in <1 new infections for classes lasting 50 to 240 min. Risks over the one-infection threshold have been highlighted, meaning the assumed occupancy of 30 people needs to be reduced

Room 306						
	R (50 min)	R (60 min)	R (100 min)	R (150 min)	R (200 min)	R (240 min)
Mean risk	0.01606	0.0215	0.03657	0.05676	0.07648	0.09
New infections	0.48	0.60	1.10	1.70	2.29	2.76
Maximum safe occupancy	62.28	49.62	27.34	17.62	13.07	10.88
Room 106						
	R (50 min)	R (60 min)	R (100 min)	R (150 min)	R (200 min)	R (240 min)
Mean risk	0.00509	0.00621	0.01057	0.001600	0.02139	0.0257
New infections	0.15	0.19	0.32	0.48	0.64	0.77
Maximum safe occupancy	72	72	72	62.51	46.76	38.96

would need three exhaust fans, of about 200W each, running for 12 hours a day every weekday for 30 weeks a year, the authors estimate the use of natural ventilation for air safety in classrooms, as detailed in this chapter, directly saves as much as 27.000 kWh every year for that building alone, in electricity not consumed by exhaust fans.

6 Conclusions

This paper studies the influence of passive ventilation strategies of classrooms on a university campus, as well as their usage, on the probability of infection by COVID-19 and on indoor air quality. The methods consisted in creating and gauging an *EnergyPlus* (United States Department of Energy 2022) simulation model based on the dimensions, air temperatures and surface temperatures of two classrooms in the same building on an university campus in Viçosa, southeastern Brazil. After certifying the model was accurate by comparing its temperatures to the temperatures gathered on site, a simulation was run on the same software, which outputted the air renovation rates for each of the two classrooms for each hour of the year they could possibly be occupied. This ventilation data was then used to calculate the risk of infection by COVID-19, and compared between classrooms to determine which passive ventilation strategy yielded the smallest infection risk, and by how much. A comparison is also drawn between passive ventilation strategies and forced ventilation, to highlight the sustainability of the former in face of the use of materials and energy of the latter.

This work highlighted the vastly different nature of ventilation (and therefore, the risk of airborne infection) for classrooms in a tropical climate, and classrooms in temperate climates such as the ones studied by Buonanno et al., Morawska et al. and Zivelonghi et al. (Buonanno et al. 2020; Zivelonghi and Lai 2021). Passive ventilation strategies were shown here to guarantee a degree of safety vastly superior to the air filters in the referenced works, at a much higher sustainability level. These passive ventilation strategies have zero cost, zero carbon emission, zero plastic waste, and less than zero energy expenditure, if one considers the energy saving brought about by them. It also showed that a constant, static Air Change Rate is not reflective of the actual nature of ventilation in these classrooms and that the contribution of thermal chimney height to ventilation outstripped that of windowsill height for these classrooms.

In the studied classrooms, the air renovation varied from 0 to 100 ACH in the room with the taller thermal chimney, while the one that contained the shorter chimney did not reach 15 ACH. In consequence, it has been found that safe occupation rates of these two classrooms (safe here being defined not as zero-risk, but as risk below the one-infection threshold) are possible. Speaking practically, this means that if all windows and doors can be kept open, and class times are kept under 100 minutes, the taller thermal chimney safely enables full occupation of the classroom, while the shorter one enables over 50%. It is important to underline that these calculations did not consider the use of masks or face coverings of any kind, and that safe occupancy rates could be further improved by them. These findings are of great importance because university campuses and classrooms are highly susceptible to infectious disease outbreaks, airborne or otherwise, due to the high amount of contact between people both within and outside campus; and preventing such outbreaks can become costly if passive strategies like the ones outlined in this paper are not used.

The limitations of this article include the relatively quick data collection period, of only two and a half hours on a single day. Future works on this subject should and will include a lengthier data collection period, which would guarantee a more accurate gauging of the simulation. Another limitation was that the classrooms studied were facing the same orientation, which affects air change rates since wind direction significantly affects façade pressure coefficients, which are known drivers of ventilation. Future research will further investigate the influence of window orientation on these results.

References

Asadi S, Fakari M, Fayaz R, Mahdaviparsa A (2016) The effect of solar chimney layout on ventilation rate in buildings. Energy Build 71–78

Bhagat RK, Wykes MS, Dalziel BS, Linden PF (2020) Effects of ventilation on the indoor spread of COVID-19. J Fluid Mech

Big Ladder Software, apud Xing L (2014) Estimations of undisturbed ground temperatures using numerical and analytical modeling. PhD Thesis. Stillwater, OK, United States of America, Oklahoma State University

Buonanno G, Stabile L, Morawska L (2020) Estimation of airborne viral emission: quanta emission rate of SARS-CoV-2 for infection risk assessment. Environ Int 10574

Chong A, Gu Y, Jia H (2021) Calibrating building energy simulation models: a review of the basics to guide future work. Energy Build 253:111533. https://doi.org/10.1016/j.enbuild.2021.111533

Foster A, Kinzel M (2021) Estimating COVID-19 exposure in a classroom setting: a comparison between mathematical and numerical models. Phys Fluids

Gammaitoni L, Nucci MC (1997) Using a mathematical model to evaluate the efficacy of TB control measures. Emerg Infect Dis 335–342

Gu L (2007) Airflow network modelling in EnergyPlus. In 10th international building performance simulation association (pp 964–971). IBPSA, Beijing

Guimaraes I, Carlo J (2011) Caracterização bioclimática da cidade de Viçosa-MG [Bioclimactic characterization of the town of Viçosa-MG]. ENCAC, Armação de Búzios

International Energy Association (2018) https://www.iea.org/data-and-statistics/charts/percentage-of-households-equiped-with-ac-in-selected-countries-2018

Lamberts R (2008) Trends in Brazilian building ventilation market and drivers for change. Ventilation Information Paper, vol 23. Air Infiltration and Ventilation Centre

Maciel LF (2016) Desenvolvimento de dispositivo de proteção solar com ênfase na ventilação natural: Um método de avaliação baseado em CFD. [Designing a shading device focused on natural ventilation] Dissertation (Architecture and Urbanism MSc). UFV-Universidade Federal de Viçosa, Viçosa, MG, Brazil

Morawska L, Milton DK (2020) It is time to address the airborne transmission of COVID-19. Clin Infect Dis 2311–2313

Morawska L, Johnson GR, Ristovski ZD, Hargreaves M, Mengersen K, Corbett S, Katoshevski D (2009) Size distribution and sites of origin of droplets expelled from the human respiratory tract during expiratory activities. J Aerosol Sci 256–269

Oliveira MM, Carlo JC (2018) Análise da Qualidade da Simulação do Fluxo de Ar de Chaminés Solares com o EnergyPlus [Quality Analysis of Aiflow Simulation in Solar Chimneys with EnergyPlus]. PARC-Pesquisa em Arquitetura e Construção.

OpenStudio Coalition (2021) OpenStudio plug-in. https://github.com/openstudiocoalition/openstudio-sketchup-plugin/releases

Peng, Y., Wu, P., Schartup, A. T., Zhang, Y. (2021). Plastic waste release caused by covid-19 and its fate in the Global Ocean. Proceedings of the National Academy of Sciences, 118(47). https://doi.org/10.1073/pnas.2111530118

Riley EC, Murphy G, Riley RL (1978) Airborne spread of measles in a suburban secondary school. Am J Epidemiol 421–432

Trimble Inc (2019) Sketchup. Sunnyvale, California, United States of America. https://www.sketchup.com/

United States Department of Energy (2022) EnergyPlus web-based documentation. Big Ladder Software. https://bigladdersoftware.com/epx/docs/

Weber F, Melo AP, Marinoski DL, Guths S, Lamberts R (2017) Desenvolvimento de um modelo equivalente de avalição de propriedades térmicas para a elaboração de uma biblioteca de componentes construtivos brasileiros para o uso no programa EnergyPlus. [Designing an equivalent model for thermal property evaluation, to develop a library of Brazilian construction elements for EnergyPlus]. LabEEE. https://labeee.ufsc.br/sites/default/files/publicacoes/relatorios_pesquisa/Biblioteca_ComponentesConstrutivos_0.pdf

Wells WF (1934) On air-borne infection. Study II. Droplets and droplet nuclei. Am J Epidemiol 611–618

WHO (2020) Mask use in the context of COVID-19. World Health Organization; Geneva (https://www.who. int/publications/i/item/advice-on-the-use-of-masks-in-the-community-during-home-care-and-in-healthcaresettings- in-the-context-of-the-novel-coronavirus-(2019-ncov)-outbreak, accessed 12 February 2021).

Xie X, Li Y, Chwang A, Ho P, Seto W (2007) How far droplets can move in indoor environments: revisiting the Wells evaporation-falling curve. Indoor Air 211–225. https://doi.org/10.1111/j.1600-0668.2007.00469.x

Zhang R, Li Y, Zhang AL, Wang Y, Molina ML (2020) Identifying airborne transmission as the dominant route for the spread of COVID-19. PNAS. https://doi.org/10.1073/pnas.2009637117.

Zivelonghi A, Lai M (2021) Mitigating aerosol infection risk in school buildings: the role of natural ventilation, volume, occupancy and CO2 monitoring. Build Environ 108139. https://doi.org/10.1016/j.buildenv.2021.108139

Implementation of a Transversal and Multidisciplinary Sustainability Project for the Development of Student's Final Projects at the Engineering School of Bilbao, UPV/EHU (Spain)

Esther Acha, Miren Martínez-Santos, Mireia Irazola, Mikel Oregui, Alexander Lopez-Urionabarrenechea, Borja B. Perez-Martinez, Miryam Gil-Calvo, and Carmelo Zabala

Abstract A crosscutting project between teaching and research has been developed by a multidisciplinary team to involve students, teaching and research staff, and administration and services staff in research with application in sustainability and circular economy. The topic to investigate has been the production of activated carbon for the elimination of antibiotics from wastewater using waste biomass from the university campus. The students have been able to apply previously studied technical knowledge and have been trained in new competencies related to sustainability. The knowledge acquired by the teaching and research staff has allowed them to reorient the teaching content of the subjects they teach in terms of sustainability, especially in circular economy. In addition, the participation of the administration and services staff increased their awareness of the need to recover and valorise the waste generated at the university. The research result of the collaboration between these three groups was the production of carbonaceous adsorbents capable of eliminating more than 60% of the antibiotics present in water.

Keywords Sustainability · Engineering university degrees · Multidisciplinary · Biomass waste

E. Acha (✉) · M. Martínez-Santos · M. Oregui · A. Lopez-Urionabarrenechea · B. B. Perez-Martinez · M. Gil-Calvo · C. Zabala
Chemical and Environmental Engineering Department, School of Engineering of Bilbao, University of the Basque Country (UPV/EHU), Plaza Ingeniero Torres Quevedo, 1, 48013 Bilbao, Bizkaia, Spain
e-mail: esther.acha@ehu.eus

M. Irazola
Analytical Chemistry Department, Faculty of Science and Technology, University of the Basque Country (UPV/EHU), Barrio Sarriena S/n, 48940 Leioa, Bizkaia, Spain

W. Leal Filho et al. (eds.), *The Contribution of Universities Towards Education for Sustainable Development*, World Sustainability Series,
https://doi.org/10.1007/978-3-031-49853-4_15

1 Introduction

At the beginning of the twenty-first century, modern societies still face a large amount of major challenges, many of which are not evident in the short term, which are grouped under the term of sustainability. In 2015, the United Nations (UN) established the 2030 Agenda for Sustainable Development, which resulted in 17 Sustainable Development Goals (SDGs), the major problematic nodes that, if accomplished, will hypothetically ensure a real sustainable future. The 2030 Agenda is, to date, the most ambitious document drafted by the UN in terms of equality between people and the protection of the planet, from a joint economic, social and environmental perspective (United Nations 2015).

The SDGs can be a strategic roadmap for universities by means of the alignment of teaching, research, innovation and social initiatives with sustainability principles. The University of the Basque Country (UPV/EHU) is an example of this reality, as it took advantage of the publication of the UN's 2030 agenda to carry out an internal analysis of those sustainability global goals. This process resulted in the creation of its own guideline based on its own reality: the so-called *EHUagenda 2030 for Sustainable Development (2019–2025)* (UPV/EHU 2019a). This EHUagenda 2030 included the UPV/EHU's contribution to 12 of the 17 SDGs of the 2030 Agenda together with another goal about its commitment to linguistic and cultural diversity, the so-called SDG 17 + 1. This agenda was built around the education model of the UPV/EHU (IKD the acronym in Basque for "cooperative and dynamic learning"), which was redefined as IKD i^3 (teaching x research x sustainability, 3 words beginning with "i" in Basque). Besides, three sectoral plans were also defined as building blocks of the EHUagenda (Campus Equality, Campus Inclusion and Campus Planet). The integration of these models for sustainability in Higher Education are complex, and need to be done in a systematic way to allow the implementation, diagnosis, and evaluation. The implementation of this agenda at the UPV/EHU can be found in the work published by Saez de Cámara and collaborators (Saez de Cámara et al. 2021).

In its attempt to establish and implement the EHUagenda 2030 for sustainability among the university community, the UPV/EHU promotes the *Campus Bizia Lab* (CBL) programme, which could be translated into English as "Campus living lab". The aim of this programme is to trigger a collaborative process between students, teaching and research staff (TRS), and administration and services staff (ASS) in order to advance the sustainability of the university. Consequently, this programme offers to the university community the possibility to work collaboratively on projects that address real issues related with the sustainable development of the UPV/EHU, contributing to the achievement of the objectives of the three sectoral plans of the EHUagenda 2030—Inclusion, Equality and Planet—and, especially, to the UPV/EHU's own education model, IKD i^3. Using campuses as learning laboratories, the projects based on the CBL programme promote the development of a high impact practice on students, such as the transdisciplinary learning based on challenges related to sustainability, with curricular recognition through the development of their Final Degree Project (FDP) and Final Master's Project (FMP). The 2020/21 course was

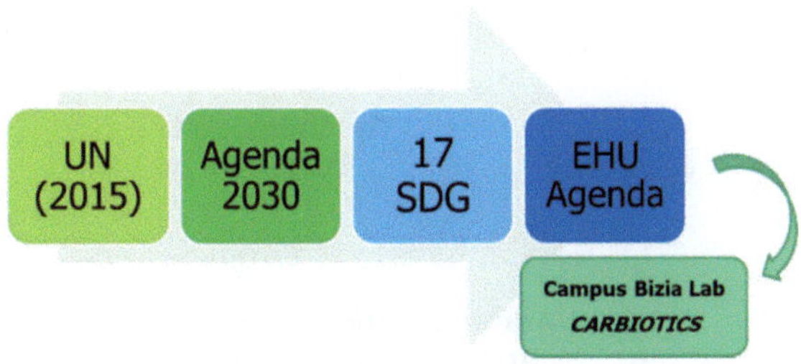

Fig. 1 Diagram of contextualization of SDGs by the University of the Basque Country and the Campus Bizia Lab programme

the fifth edition of the CBL, with 26 projects approved (https://www.ehu.eus/es/web/iraunkortasuna/campus-bizia-lab-2021-2022).

Among these approved projects was the so-called CARBIOTICS project, which was developed over a period of two consecutive academic courses (from 2020 to 2022). Two objectives were pursued in this project. The first one was of a scientific nature: the transformation of pruning and gardening waste generated on the Bizkaia Campus into a value-added product (carbonaceous adsorbent). The second objective was related to the promotion of collaborative work between different university bodies (TRS, ASS and students) in the area of sustainable development, specifically, on circular economy actions focused on the university itself. The contextualisation of this project is shown in Fig. 1.

This article is divided into four main sections: (1) the first of them, the introduction, aims to contextualise the reader in the trajectory developed by the UPV/EHU in sustainability issues, taking into consideration the European guidelines; (2) in the second section, an example of the actions of this university in sustainability policies is set out in detail, describing the specific case of the CARBIOTICS project carried out at the UPV/EHU; (3) in the results section, not only the most experimental results obtained are presented, but especially those results that have involved the personal and professional development of the members of the working team in aspects related to sustainability; and finally, (4) the main conclusions derived from this work are summarised.

2 The CARBIOTICS Project

CARBIOTICS is the acronym of the CBL project titled "Production of carbonaceous adsorbents from biowaste generated at the UPV/EHU for the removal of antibiotics from wastewater". The project was running from October 2020 to September 2022. As said before, the three bodies of the university took part in this transversal and multidisciplinary sustainability project, where different disciplines were merged, as summarised in Fig. 2.

The project was led by six members of the TRS body from two different departments: Chemical and Environmental Engineering Department and Analytical Chemistry Department. This group of people belong to different research and teaching lines

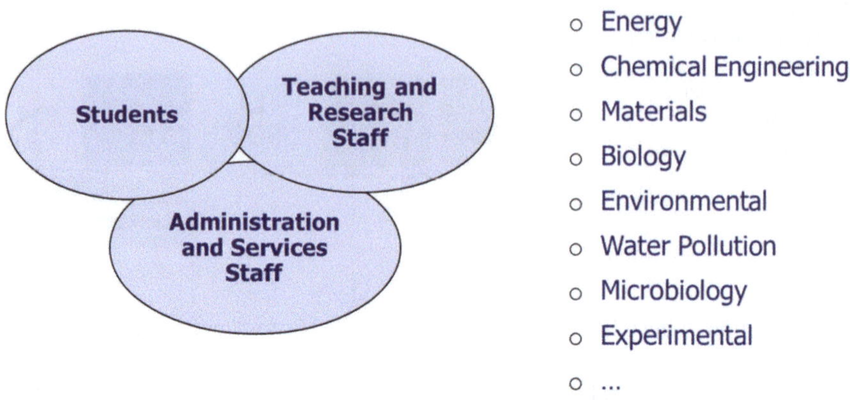

Fig. 2 University people and disciplines involved in the CARBIOTICS project

and have stable research groups behind them. The later fact was critical for the development of the project, since the CBL funding for CARBIOTICS was relatively small. Having the support of stable research groups helped in the financing of the experimental work, and allowed the project members to use devices and equipment already available. In this way, the funding obtained for the project was mainly employed in buying laboratory consumables and paying for characterization techniques not available at the laboratories of these research groups. Apart from them, a member of the Administration and Services Staff took part: the laboratory technician of the Chemical and Environmental Engineering Department, whose main objective in the project was to advise and supervise the more experimental tasks, but, at the same time, he participated in the discussion of the results obtained. Finally, four students were involved in the project, belonging to the following academic programmes: (i) Degree in Environmental Engineering; (ii) Degree in Chemistry; (iii) Master's Degree in Sustainable Energy Engineering; (iv) Master's Degree in Microbiology and Health. For all the four students involved in this project, it was the first time being part of such a multidisciplinary and experimental research work. The participation of students in the project allowed them to work on several SDGs: SDG5 (gender equality), SDG6 (clean water and sanitation), SDG7 (affordable and clean energy), SDG9 (industry, innovation and infrastructure), SDG12 (responsible consumption and production), and SDG18 (linguistic and cultural diversity, from the EHUagenda 2030, UPV/EHU). The students that participated in the project had different academic backgrounds, and therefore, their scientific knowledge differ considerably. Furthermore, their studies were really focused on technical and scientific knowledge and, generally, they had lack of a transversal social, economic and environmental point of view. Thanks to the collaboration in this project, the students have also worked with sustainable development goals, gaining a more general, social and transversal point of view of the technical work they do.

Concerning gender equity, it was also guaranteed in this project, as a key factor for sustainable development (being described in the SDG 5). On the one hand, 50%

of the total participants were women and, on the other hand, a woman carried out the coordination of the project. This promotes the empowerment of women scientists through the direction and execution of projects and constitute a step forward in breaking down the barriers for women in science, where they still make up less than 30% of the workforce (Unesco 2019). Even so, women continue to face obstacles in this field.

As it has been previously mentioned, the research objective of this project was to transform a waste into a value-added product. Specifically, pruning and gardening waste was valorised to carbonaceous adsorbents with the potential of eliminating antibiotics from wastewater. Thus, the project aimed not only to improve the water quality of our environment but also to reduce the generated waste through a responsible management.

The scientific objectives of the project were three:

1. Valorisation of biowaste into selective carbonaceous adsorbents for the elimination of antibiotics, following the principles of circular economy.
2. Thermophysical and/or chemical activation of the sustainable carbonaceous adsorbents, in order to improve their efficiency and selectivity for antibiotic removal.
3. Laboratory scale adsorption tests of antibiotics present in wastewater and quantification of the efficiency in antibiotic removal.

The methodology to carry out the three main tasks and their sub-tasks are briefly described below:

- Task 1: Selection and characterization of residual biomass to produce carbonaceous adsorbents. The companies contracted for this service in the UPV/EHU were contacted and two types of biowaste of interest were selected: pruning and gardening waste. The criteria for the selection were the proximity to the laboratories (Bizkaia better than Gipuzkoa better than Araba), the stationarity of the waste produced and the availability of the waste needed to proceed with the experimental work. Later, the collected biomass was characterised to determine its quality by means of elemental analysis (C, H, N, O, S content), analysis of the moisture and volatile matter, thermogravimetric analysis, and analysis of the metal content.
- Task 2. Production and activation of carbonaceous adsorbents by pyrolysis, and thermophysical and/or chemical activation to improve their efficiency and selectivity for antibiotic adsorption. The biowaste was pyrolysed under different operational conditions. The materials obtained in this way were non-active carbons with low porosity and surface area. A lab-scale pyrolysis plant (3.5 L capacity for waste) was employed and experimental design methodology was applied to evaluate the effect the pyrolysis conditions: pyrolysis temperature, reaction time (time of the isothermal step) and particle size of the waste. The solid biochar remaining after pyrolysis was collected and chemical activation with KOH was carried out in order to increase the porosity, adsorption capacity and selectivity). In parallel, the biowaste was also thermo-physically activated with CO_2 after being pyrolysed. Finally, the activated and non-activated carbonaceous adsorbents were

characterised to establish relationships between the efficiency and selectivity of the adsorbents and their physicochemical properties. In the solid materials, i.e. adsorbents, elemental analysis, superficial area, porosity, and functional groups were analysed, while the liquids and gases produced in the pyrolysis were analysed in order to determine their calorific value and chemical composition.

- Task 3. Laboratory-scale adsorption tests were carried out at different working regimes: in a stirred tank (batch) and in a fixed bed (fixed column). The remaining antibiotic in the water after being in contact with the adsorbent was determined by LC–MS/MS.

The project was defined as transdisciplinar, but the transdisciplinarity of a project does not emerge until its participants interact in a critical and open discussion of the results, challenges and problems. This was carried out through the coordination of the project members directly involved in each task, and through the organization of periodic meetings to address those challenges and problems that arose in the course of the project. These meetings were planned to be the final activity of each task, which were completed once the results of the task were presented and discussed among all team members (teachers, laboratory technician and students). Collaborative and close work between all the participants was necessary to make the proposal succeed. The students mentioned above where involved in the consecution of the different tasks of the project through their FDP or FMP. The methodology of the project, together with the students' collaboration and FDP or FMP defence is summarised in Fig. 3.

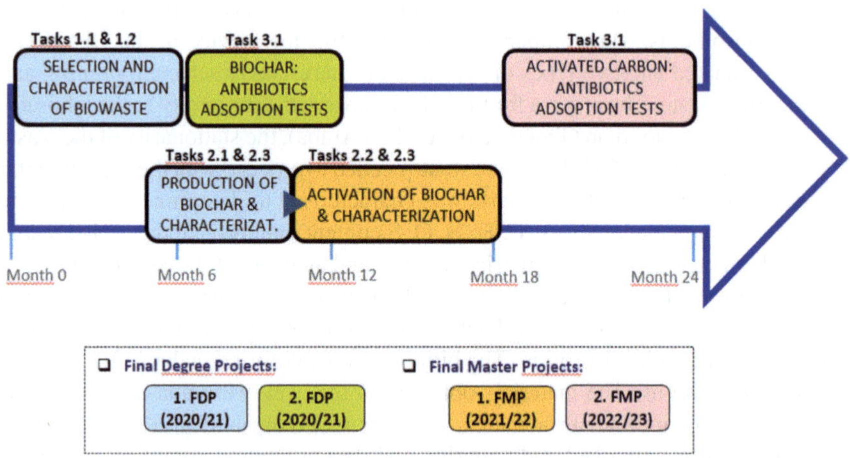

Fig. 3 Summary of the CARBIOTICS Project indicating the students' participation and their Final Degree or Master's Project defence

3 Scientific and Educational Results

Due to the idiosyncrasy of the CARBIOTICS project, results of different nature were achieved during the completion of the project. Some of the results are more scientific-driven, such as the advances in the activated adsorbents production from biowaste thermal treatment; but others can be accounted as educational achievements, since they were implemented with a pedagogic view. The main limitations of the performed work were time, as previously described the Project duration was 1 + 1 years and the few students involved, because the project was planned to be implemented in research and not on regular subjects. In the following sections, these results are categorised and analysed one by one.

3.1 *Experimental Results*

As a result of the experimental work carried out, 13 different biochars were produced at three different temperatures, reaction times and particle sizes. Besides, the chemical activation process applied to the biochars was very effective for those that initially had a low specific surface area (<50 m^2/g) and pore volume, increasing its surface area up to 358 m^2/g. However, the activation was not so effective for those biochars with a high initial surface area (>400 m^2/g). In the case of thermo-physical activation with CO_2, high surface area adsorbents were obtained in most of the cases (around 670 m^2/g). In the antibiotic adsorption tests, the percentages of antibiotic adsorption obtained by stirred tank were higher than those obtained by fixed bed regime, reaching adsorption percentages up to 100% for the biochar produced at 800 °C.

3.2 *Learning Competences Acquired During the Development of the Project*

The three university bodies involved in the development of this project did a cross-cutting work and reached a series of competences and skills in the field of sustainability, specifically in those aspects related to waste management, valorisation and the circular economy strategies.

In addition to the aforementioned competences, the development of this project allowed them to achieve a series of more transversal competences (UPV/EHU 2019b), always from a sustainability perspective. Some of those competences are summarised below:

- Social commitment: the three bodies developed a critical and empathetic contextualization of knowledge in the face of social, economic and, especially, environmental challenges at local (UPV/EHU) and global (Planet) levels.

- Innovation and entrepreneurship: this refers to the generation, transformation and implementation of an idea, a procedure or an attitude to provide a satisfactory response to different needs of problems.
- Critical thinking and teamwork: within the transdisciplinary team, this competence was worked by means of active collaboration among team members to achieve the project objectives, and through an open, reflective and constructive attitude. This was the main source that allowed them to take decisions and solve problems that arose during the implementation of the project.

3.3 Training Students in Sustainability

One of the main and most transversal result obtained in this project was to improve the sustainability training of the students. The transversality between education and research trained students in the following competences:

- Basic knowledge and application of environmental technologies and sustainability.
- Sustainable use of resources and the prevention of negative impacts on the natural and social environment.
- Application of ethical principles related to the values of sustainability in personal and professional behaviours.

Having such a real and well-structured context helped the students improving their learning outcomes, increasing their motivation and allowing them to benefit from a positive experience. All this could be a reason to follow their career taking the sustainability as inherent part of any future job they have. The students own experience can be materialised in the following experience summary:

- Thanks to the collaboration in this project, she became interested in the real problem the planet faces with waste, and now she works for a waste management company.
- It was such a good and interesting experience, that he is now pursuing a PhD thesis on a topic related to sustainability (wind energy use).
- The breadth of context, the wider application to society and the improvement of water quality were positive aspects for her. This together with having a global framework in which to apply the specific 6-months laboratory work were the reasons why this student chose this project.

3.4 Knowledge Acquired by Teaching and Research Staff

The participation in this project improved our ability to reorient the teaching content and the way of teaching, with the aim of continuously training students in sustainability-related competencies. The teachers taking part in the project are in charge of subjects that directly address aspects related to waste management from the perspective of sustainability, such as "Sampling and Analysis of Waste" and "Environmental Science and Technology". Therefore, their participation allowed the incorporation of Inquiry-based Learning into these or other related subjects. The knowledge acquired will also be implemented in research, through the submission of proposals in Basque and National project calls.

3.5 Knowledge Acquired by Administration and Services Staff

The participation of the laboratory technician in the different tasks of the project increased his knowledge of SDGs and of the recovery and valorisation of waste generated at the UPV/EHU, for which he is responsible in the Chemical and Environmental Department laboratories. In addition, his work in teaching, supporting and supervising the students during the laboratory experiments has been crucial for the development of the project and the achievement of reliable experimental results.

3.6 Dissemination

Together with doing and learning, another important part is the dissemination of a project. Effective dissemination raises awareness and fosters collaboration, as well as ensures the efficient distribution of new knowledge. It is important to guarantee the visibility of a project by promoting it and its activities. In the CBL projects application itself, the FDP/FMP proposals have to be already indicated and these proposals are disseminated through the university corporate website.

The students use the experimental results obtained to present their FDP/FMP but, in addition, those results are also being prepared for publication in high impact research journals. Moreover, the results were also presented in three research/teaching conferences organised by the UPV/EHU, which are specially focused in students, such as the annual CBL conference, the Student Congress "Our Final Degree Projects (FDP) serve to transform the world" and "IKD i^3 JENDARTEA 2022 conference: the development of the IKD i^3 strategy to face new challenges".

4 Conclusions

Based on our experience and as the main conclusions of this project, we will summarise the difficulties and positive aspects encountered during its implementation. Starting with the difficulties, sometimes we found it costly to work together with people from different university bodies (TRS, ASS and students), as we are not too used to do it. Moreover, it can also sometimes be critical to find interested students, as the offer of FDP/FMP inside and outside the university is very varied and broad, which leads many students to opt for other type of project or research lines. Finally, the duration of the CBL projects is short (1 year with the possibility of renewal for another one), which leads to a large offer of new projects every year, but which turns out to be too short period to find interested students, to develop the planned tasks, to get results and to disseminate them. Everything had to be very well thought out and planned before starting, which can be difficult to do.

Finally, we would like to highlight the most positive aspects of taking part in the project. Applying for a project of this type helps to clarify ideas, structure them, search for possible partners in teaching courses, and to establish contacts with teachers who are also working in the field of sustainability. The framework of this project and the official certification of participation is a bonus for all the participants, and therefore allows to develop transversal and stable multidisciplinary research groups that otherwise would not interact; an idea could remain just that, if there was not this call. In addition, the students involved in these type of projects see more clearly the interest and the need for what they do, they are officially part of the teamwork, it appears to them as a merit, the participation in conferences also goes to their curricula, and the support of the university for the project increases their motivation. Overall, the experience of the research team has been very positive. Although given the usual workload (research plus teaching) it never seems to be a good time to carry out this kind of "extra work". We think that it is very interesting to take advantage of these small internal calls of the university to generate these links teachers-administration-students, which give us a great option to develop transversal competencies and knowledge, labour relations, etc., which we can later use to improve our teaching, our research and the sustainability knowledge of the university graduates.

Therefore, as a general conclusion, it should be noted that all participants have extrapolated the knowledge acquired in terms of sustainability, during their participation in the CARBIOTICS project, to other personal or professional fields, being more aware of the need to manage waste in a more economically and environmentally efficient way. Finally, future prospects include the application of circular economy principles in the management of other types of waste generated at our university, and to continue working with students in the deepening of sustainability in their training at the university.

Acknowledgements The authors wish to thank the Basque Government (Consolidated Groups of Sustainable Engineering Processes IT1554-22, and Hydrogeology and Environment IT1678-22) and the University of the Basque Country (UPV-EHU, 20MART and 21MARTI) for supporting this research.

References

Saez de Cámara E, Fernández I, Castillo-Eguskitza N (2021) A holistic approach to integrate and evaluate sustainable development in higher education. The case study of the university of the Basque Country. Sustainability (Switzerland) 13(1):1–19. https://doi.org/10.3390/su13010392

Unesco Institute for Statistics (2019) Women in science

United Nations (2015) Transformar nuestro mundo: la Agenda 2030 para el Desarrollo Sostenible. Resolution approved by General Assembly on 25th of September 2025

UPV/EHU (2019a) EHUagenda 2030 for sustainable development

UPV/EHU (2019b) Catálogo de competencias transversales de la UPV/EHU, 42, Ed. Universidad del País Vasco. Accessed March 2023. https://www.ehu.eus/documents/1432750/12757375/Cat%C3%A1logo+de+Competencias+trasnversales_cas.pdf

European Universities' Discourse on SDGs in Social Networks

Salvador Ruiz de Maya and Elvira Ferrer-Bernal

Abstract The 2030 Agenda for Sustainable Development has become an increasing focus of activities in universities, particularly affecting research and education and integrating the SDGs into the curricular schedule. Universities can contribute to the SDGs not only through education and research but also by boosting public participation in addressing SDGs and promoting cross-sectoral dialogue through social networks, since they are now considered essential communication tools to interact and build relationships with stakeholders. In this chapter, we apply big data techniques to analyze universities' messages related to SDGs on Twitter to demonstrate how the content of these messages influences stakeholders' engagement. Using a large sample of recent tweets from European universities, we draw conclusions that shed light on universities' impact when disseminating information about SDGs.

Keywords SDGs · Universities · Sustainability · 2030 agenda

1 Introduction

The concept of "Sustainable Development" started gaining popularity in 1987 aft the United Nations published the Brundtland Report (Du Pisani 2006). Since then, it has evolved into a precept connecting the environment, economy, and society (De Iorio et al. 2022). In the 2000s, the Millennium Development Goals (MDGs) were introduced as a further step toward attaining Sustainable Development. The MDGs were eight objectives, with the principal focus of fighting poverty and improving global health over the 15-year span from 2000–2015 (Griebeler et al. 2021). These goals were a precedent to the Sustainable Development Goals (SDGs) approved in

S. R. de Maya (✉) · E. Ferrer-Bernal
Consumer Research Group, Faculty of Economics and Business, University of Murcia, Campus de Espijnardo, 30100 Murcia, Spain
e-mail: salvruiz@um.es

E. Ferrer-Bernal
e-mail: elvira.ferrerb@um.es

W. Leal Filho et al. (eds.), *The Contribution of Universities Towards Education for Sustainable Development*, World Sustainability Series,
https://doi.org/10.1007/978-3-031-49853-4_16

September 2015 by the United Nations General Assembly. The SDGs were then included in the 2030 Agenda and are comprised of 17 goals, 169 targets, and 232 indicators aimed at guiding countries towards a sustainable path and equitable society (Cotaffava et al. 2022; Serafini et al. 2022). To achieve the SDGs, all countries and industries, including both private and public sector organizations, must take responsibility and take them into consideration when designing their strategies and business models (De Iorio et al. 2022; De la Poza et al. 2021).

To implement the SDG framework as an effective scheme for leading the transformation of society, a creative, persistent, and holistic approach must be followed. In other words, it is essential that goals are tackled jointly since they are interconnected in nature. Consequently, a specific goal can only be achievable if the others are accomplished. Therefore, because any development in one area typically depends on progress in the others, trying to reach one goal independently could result in adverse outcomes (Alcántara-Rubio et al. 2022; Leal Filho et al. 2019). An example of this can be found in Agusdinata's (2022) work, in which challenges associated with fighting poverty and upgrading a community's income are identified (SDGs 1 and 2). By addressing these challenges, there is progress in other fields; for instance, solar energy is proposed as a possible solution to improve the community's income, which boosts energy efficiency (SDG 7) and occasionally, the upgraded income could be used to support education (SDG 4).

Education plays a crucial role in the realization of the SDGs and the 2030 Agenda for Sustainable Development. Although "Quality Education" is a standalone goal (SDG 4), its connections with all the other global goals make its contribution fundamental for attaining sustainable development. This means that when "SDG 4: Quality Education" is pursued, additional development outcomes also tend to improve (Alcántara-Rubio et al. 2022; De Iorio et al. 2022). Ensuring inclusive and equitable high-quality education enhances citizens' lives by boosting access to employment, nutrition, and health, among others (SDSN Australia/Pacific 2017). Therefore, Higher Education Institutions (HEIs) have officially been recognized as an essential driving force in implementing the 2030 Agenda due to their position in society as creators and disseminators of knowledge (Alcántara-Rubio et al. 2022). In fact, the education sector has traditionally been identified as a catalyst for leading the societal transformation (Caiado et al. 2018). Consequently, the concept of Education for Sustainable Development (ESD) has emerged to reflect the importance of education and universities in integrating the various SDGs (Kopnina and Meijers 2014). Nonetheless, there has been no research to date that has focused on how universities use social media to communicate SDGs and engage their stakeholders through these messages.

In this chapter, we apply big data techniques to analyze universities' tweets related to SDGs to demonstrate how the content of these messages influences stakeholders' engagement. Using a large sample of recent tweets from European universities, we draw conclusions that shed light on universities' impact when disseminating information about SDGs.

In order to achieve this objective, we first review how universities are dealing with SDGs and communicating their actions related to these 17 goals. In the next section, we then review how organizations are using the social platform, Twitter, to communicate existing knowledge and the progress made in the context of the SDGs. This is followed by a description of the methodology used, including data collection and processing of the tweets. The findings of the analysis allow us to draw specific conclusions and provide recommendations for universities.

2 SDGs in Universities

Historically, the role of universities in implementing sustainable development has been extensively discussed. The Talloires Declaration, created in 1990, can be considered the official starting point of this process, aiming to become an action plan for universities to achieve sustainability (Griebeler et al. 2021). Since then, HEIs have been encouraged to include sustainable development in their missions, core functions, study plans, and other activities (Serafini et al. 2022; Soini et al. 2018). Following the launch of the 2030 Agenda in 2015, there has been a sharp increase both in scientific publications related to sustainability and in universities with sustainability centres (Alcántara-Rubio et al. 2022; Soini et al. 2018). However, as it is likely that the SDGs may not be accomplished by 2030, HEIs need to urgently adopt sustainability as a core value and increase their efforts to implement measures to achieve the SDGs (Leal Filho 2020). Although there is no standard way to embrace the SDGs, the SDSN Australia/Pacific (2017) specifies five steps that universities could follow to strengthen engagement with the SDGs. These include (1) being aware of what universities are doing, (2) acquiring SDG ownership and capacity, (3) identifying opportunities and gaps, (4) including SDGs into the university's strategies and policies, and (5) monitoring and communicating these activities to stakeholders.

As universities are responsible for preparing the future generation of professionals and decision-makers (Caiado et al. 2018), these institutions play a critical role in societal transformation (Amaral et al. 2015). Universities pledge to raise awareness of the values of responsible corporate citizenship; thus, they must empower learners with the necessary skills, knowledge, abilities, and motivation to direct actions responsibly to preserve the environment and the economy, and ensure an equitable society (Adams 2018; García-Feijoo et al. 2020). Ideally, a creative, collaborative, and practical approach should be adopted to achieve the SDGs so that learners are ready for the global context (Manolis and Manoli 2021). Technological and social innovation, leadership, education, and research will be key to overcoming these challenges and speeding up the transition towards a better world (Griebeler et al. 2021). In this sense, the absence of involvement of these institutions, which act as change agents, could result in failure to address these sets of social, economic, and environmental challenges (Alcántara-Rubio et al. 2022).

To achieve sustainable development, universities must embed SDGs across their core four dimensions, namely, learning and teaching, research, community outreach, and governance (Owens 2017). Moreover, universities are expected to address sustainability-related challenges holistically since these key areas are interconnected

(Leal Filho et al. 2019). From the learning and teaching perspective, universities can provide learners with sufficient knowledge to comprehend the most compelling global challenges and understand how they can contribute to addressing them (Leal Filho et al. 2021). In the research area, SDGs are a great opportunity to stimulate sustainable development research and find innovative responses to global problems (De Iorio et al. 2022; SDSN Australia/Pacific 2017). By enhancing interdisciplinary collaborations, solutions to contemporary challenges can be identified (Cottafava et al. 2022). In the context of governance and management activities, these institutions can contribute to the accomplishment of the SDGs by including the associated principles in their missions, strategic plans, and reports (Serafini et al. 2022). In addition, universities can develop a structure to coordinate the implementation of the SDGs and integrate the decisions derived from spaces for dialogue with HEIs, corporations, governments, and society (Alcántara-Rubio et al. 2022). Finally, universities must raise awareness and empower civic engagement through outreach activities and training in the community (Serafini et al. 2022). For example, the project "ODSesiones" conducted at the University of Murcia aimed to disseminate knowledge about the SDGs and promote sustainable actions in the region (Marín et al. 2019). Universities across the globe are implementing distinctive strategies to address the sustainability challenge (Fia et al. 2022).

It is important to note that ensuring synergy between HEIs and SDGs benefits not only society but also the HEIs themselves (SDSN Australia/Pacific 2017). The implementation of the SDGs could offer new opportunities to universities in terms of potential multidisciplinary partnerships, collaborations, and networks (Leal Filho et al. 2021). Thus, enhancing their role as change agents in tackling real-life problems might increase the institutions' relevance to stakeholders (Leal Filho et al. 2021) and give them access to new funding schemes (SDSN Australia/Pacific 2017). Further benefits may include an improvement in the institutions' corporate image and reputation (De Iorio et al. 2022). Consequently, future students and employers may choose the university due to the appealing educational offerings (Leal Filho et al. 2019).

To do this, universities can adopt an array of practices, including the systematization of the execution of the SDGs and the promotion of awareness (Serafini et al. 2022). However, even though integrating SDGs into universities' plans is a unique opportunity for them to showcase their crucial role in society, certain HEIs are not yet enhancing their efforts to tackle the challenge (Leal Filho et al. 2018). This may be because pursuing sustainability is a complex task due to the various barriers and constraints that universities may face (Leal Filho et al. 2021). These can be related to a lack of resources, trained staff, and sustainable institutional culture, among others (Ávila et al. 2017; Leal Filho et al. 2018; Serafini et al. 2022). As a result, implementing integrated actions to achieve the SDGs becomes challenging and these universities end up performing isolated and uncoordinated actions (Serafini et al. 2022). Consequently, their educational offerings may eventually become insufficient since they do not train learners to face sustainability issues in the global context (Mader and Rammel 2015).

3 SDG Communication in Universities

Communication has been identified as a factor that is likely to play an important role in furthering the 2030 Agenda due to the burgeoning expectations of accountability from educational institutions (Mulholland 2019). Stakeholders expect universities to disclose their initiatives aimed at achieving sustainability to enhance their transparent image and reputation (Ceulemans et al. 2015; Garde Sánchez et al. 2021; Secundo et al. 2016). As each country faces a different context and has varying necessities, there are no specific requirements or guidelines on how to disclose SDG-related information (Mulholland 2019). The lack of a standard communication strategy and trained staff on SDGs complicates the dissemination of these efforts in universities (Leal Filho et al. 2021). In fact, recent literature claims that universities are failing to provide information about SDGs and their contributions in the context of sustainability (De la Poza et al. 2021). Moreover, students have highlighted that they are receiving insufficient information about sustainability across the different media channels in their universities (Zamora-Polo et al. 2019). In terms of online communication, prior research emphasizes the absence of information from universities in search engine results for queries related to the 2030 Agenda and SDGs (Vállez et al. 2022).

Therefore, to achieve the goal of sustainable development in a timely manner, universities must implement effective communication on SDGs (Mulholland 2019; Findler et al. 2019). Universities around the globe must publish reports outlining their practices focused on social, economic, and environmental transformation (Fonseca et al. 2011). Based on content analysis of universities' sustainability reports, Weybrecht (2022) and Caputo et al. (2021) identify the SDGs that are being disclosed more frequently. Although sustainability reports have long been considered a major and fundamental element in the communication of universities' commitment towards sustainability, institutions have started to use social media to engage more easily with stakeholders and raise awareness (Di Tullio et al. 2021). By fostering dialogues and linking SDGs to people's lives, sustainable concepts can become more attractive and gain more attention.

4 SDG-Related Communication on Twitter

Research on SDG disclosure on social media is still scarce (ElAlfy et al. 2020). However, organizations are increasingly turning to social media to enhance their relationships with stakeholders through direct, two-way dialogues (De Luca et al. 2022; Wang 2016). Through these social media campaigns, citizens can engage more easily with actions and initiatives (Mulholland 2019). In fact, in the university field, evidence suggests that institutions that have historically reported on sustainability are using social media to involve stakeholders and strengthen their sustainability practices (Di Tullio et al. 2021). Among the social media platforms being used, Twitter has been specifically recognized as a valuable tool for communication, considering its high accessibility. University staff, learners and stakeholders are taking advantage of this platform to stay updated and involved in the community (Malik et al. 2019).

Recent literature has investigated how organizations communicate on Twitter in the context of SDGs. De Luca et al. (2022) identify the SDGs that are drawing the most engagement across different industries and countries. Furthermore, the authors determine the characteristics of tweets that influence stakeholder engagement the most (e.g., content type, hashtag presence, number of characters, and industry type). At the same time, ElAlfy et al. (2020) suggest that companies are more likely to disclose SDGs related to their core business while Salvatore et al. (2022) demonstrate the interconnection between the different SDGs.

5 Methodology

The activities and their sequence implemented in this study to collect and analyze the data are described in Fig. 1.

5.1 Data Collection

We collected a total of 122,495 tweets through searches from September 30 to December 23, 2022. Tweets that mention at least one of the Twitter accounts associated with the 35 European Universities included in the top 100 list of the Times Higher Education (THE) World University Ranking 2022. First, we identified the primary Twitter accounts of the universities and collected data using the R (R Core Team 2022) package 'rtweet' (Kearney 2019). We ran our query twice a week (Tuesdays and Fridays at 1 p.m.) during the period of data collection (from 22 September 2022 to 23 December 2022), excluding retweets and discarding duplicate tweets after the integration of the search results into a single database.

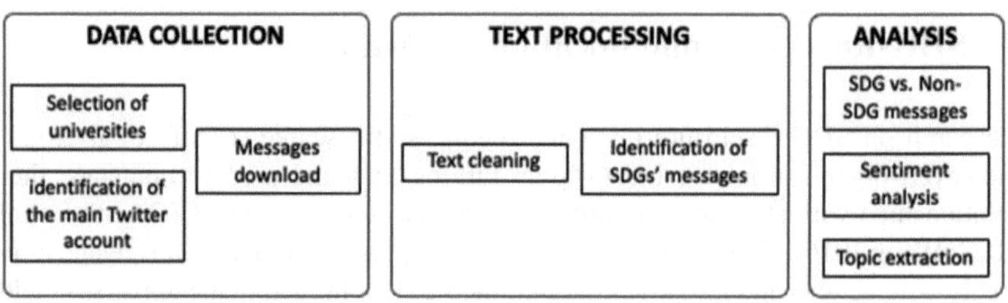

Fig. 1 Research design

5.2 *Data Processing, Sentiment Extraction and Topic Identification*

With the RStudio IDE (Posit team 2022), we processed the tweets, analyzed the data, and visualized the results using other R packages, such as "texclean" (Rinker 2018), "psych" (Revelle 2022), "MOTE" (Buchanan et al. 2019), "sentimentr" (Rinker 2021), "quanteda" (Benoit et al. 2018), "keyATM" (Eshima et al. 2020), "ldatuning" (Nikita 2020), "seededlda" (Watanabe and Xuan-Hieu 2022), and "ggplot2" (Wickham 2016).

The text of the tweets was cleaned using the "textclean" package. We removed URLs, mentions, numbers, single letters, unnecessary spaces, and special characters. In addition, the text of each tweet was transformed to lowercase because this data was subsequently used for topic modelling and sentiment analysis. Punctuations were not removed because the package "sentiment" takes sentences into consideration to calculate text sentiment and then aggregates it by row.

Then we searched the database for tweets related to the SDGs. We ran a very strict search with single words ("poverty", "hunger", "wellbeing", "equality", "sanitation", "inequality", "infrastructure", "sustainable", "peace", and "justice") and word expressions (e.g., "zero hunger") related only to the names of the SDGs and the terms "sdg" and "sdgs". With this information, a new 0–1 column was added to the database identifying the relevant tweets.

Additionally, sentiment analysis was performed to better understand Twitter users' opinions about SDGs. Among the available algorithms to run sentiment analysis, we used "sentiment" (Rinker 2022). This R package shows higher accuracy because, in addition to considering a sentiment lexicon, it also takes valence shifters into account (e.g., negators and amplifiers). With this package, we assigned a score from –2 to + 2 to each sentence in each tweet, with the value denoting a positive (score > 0) or negative sentiment (score < 0).

Finally, we adopted a semi-supervised approach to extract topics from the subsample of SDG-related tweets. We applied the keyword-assisted topic model (keyATM) after tokenizing the corpus of tweets and removing stop words using the package "quanteda". keyATM allows improved interpretability of the topics by conducting guided topic extraction. In this extraction procedure, a set of seed words that describe each potential topic is incorporated and the topic-word distribution is used to assign tweets to topics. In order to develop the keyword catalogue, we restricted ourselves to the terms included in the names of the SDGs ("poverty", "hunger", "health", "wellbeing", "quality", "education", "gender", "equality", "clean", "water", "sanitation", "affordable", "clean energy", "decent", "work", "economic", "growth", "innovation", "infrastructure", "inequality", "sustainable", "city", "community", "responsible", "consumption", "production", "climate", "change", "life", "water", "life", "land", "institutions", "peace", "justice", and "partnership") as multiword expressions were not present in the selected tweets. Using single-word terms, we obtained better values of log-likelihood and perplexity, which are the two main indicators in keyATM.

We fitted the keyATM model with 1500 iterations, allowing only one topic without prespecified keywords because the topics were clearly defined with the names of the SDGs. According to the top words identified by keyATM for each topic in the output, and taking into consideration only those topics with a pi value (probability of the words associated with the topics were drawn from the set of keywords for the topic) greater than 3%, the best-represented topics in the subsample of SDG-related tweets were: "no poverty", "good health and wellbeing", "quality education", "gender equality", "affordable and clean energy", "decent work and economic growth", "climate action", and "peace, justice and strong institutions". A ninth topic, "other", was also presented to account for tweets related to the remaining topics. This result was corroborated by the R package "ldatuning", a useful tool for identifying the optimal number of topics in Latent Dirichlet Allocation (LDA) models.

6 Results

Out of the complete database of downloaded tweets, representing the Twitter communication of the 35 European universities, only 1.5% of the tweets were clearly related to the SDGs. A simple comparison of the two types of messages in terms of their extension (Fig. 2a) revealed that SDG-related tweets were longer than those that were unrelated. The average number of characters in SDG-related tweets was 260.03, while the remaining tweets showed an average of 190.5 characters.

The impact of SDGs on stakeholders' engagement was of high interest. As shown in Fig. 2b and c, SDG tweets received a higher number of likes and retweets, which are two indicators of engagement on this social network. The average number of likes for SDG tweets was 7.98, while other tweets received an average of 6.75 likes.

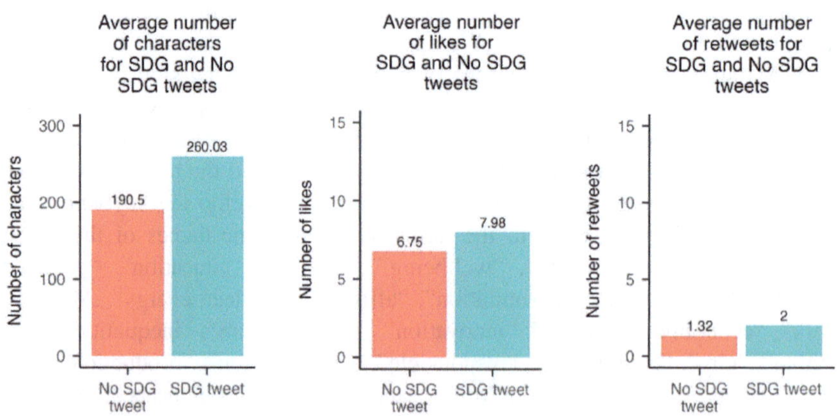

(a) (b) (c)

Fig. 2 Extension and engagement of SDG and Non-SDG tweets

The number of retweets is usually smaller than the number of likes; thus, the average number of retweets was 2 for SDG messages and only 1.32 for the remaining tweets.

The sentiment analysis revealed a significant difference between the two types of messages. SDG-related tweets were written with a more positive sentiment than non-SDG tweets (Fig. 3). This difference in sentiment may explain the stronger engagement of these messages.

The distribution of the SDG-related tweets among the nine identified topics is shown in Fig. 4. It can be observed that the most represented topics in universities' communication about SDGs were "no poverty", "good health and wellbeing", "quality education", "gender equality", "affordable and clean energy", "decent work and economic growth", "climate action", and "peace, justice and strong institutions". The proportion of these eight topics in the subsample of SDG tweets varied between 10 and 16%.

We also analyzed the number of likes and the sentiment of the tweets for each of the most represented SDGs in the sample. Figures 5 and 6 reveal that there was a certain

Fig. 3 Average sentiment of SDG and Non- SDG tweets

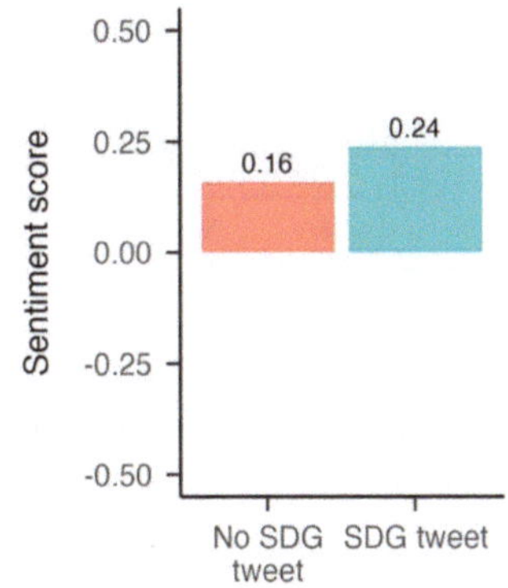

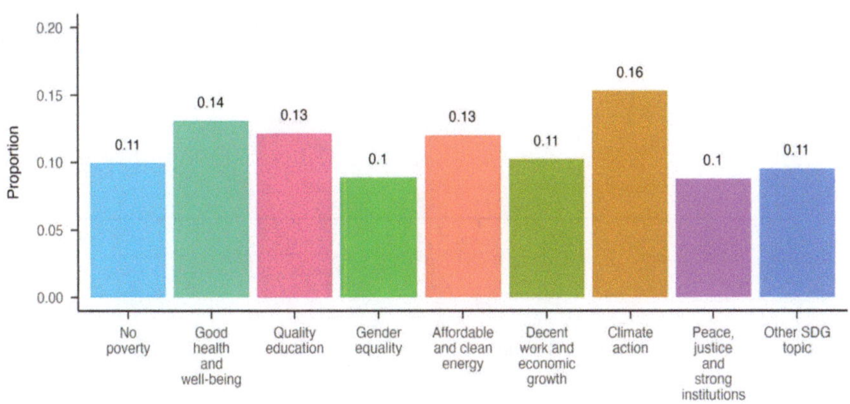

Fig. 4 Distribution of SDG tweets among the 17 SDGs

consistency between the likes and sentiment for topics such as "no poverty", "good health and well-being", "quality education", "decent work and economic growth", and "peace, justice and strong institutions". However, the goals "affordable and clean energy" and "climate action" received a lower number of likes although they had a sentiment score similar to that of "quality education". This indicates that universities are communicating these two goals in a similar proportion to other goals and with a similar sentiment, but they are not engaging with stakeholders at the same rate as they do with tweets related to other goals. The opposite was observed in the case of gender equality. A lower sentiment score in these messages garnered a higher number of likes.

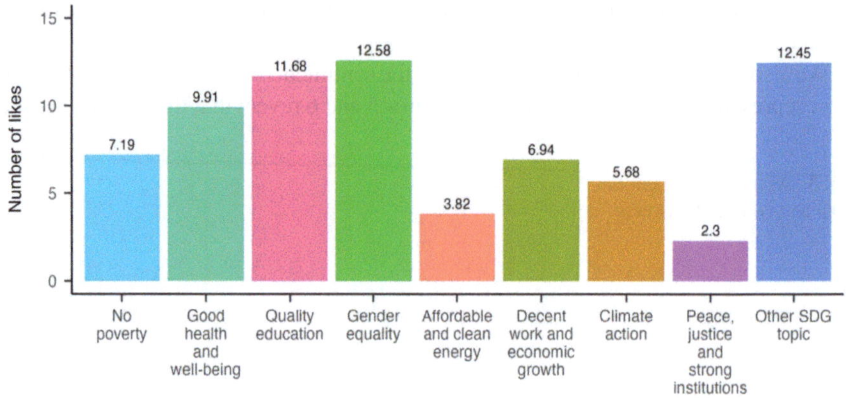

Fig. 5 Average number of likes per SDG topic

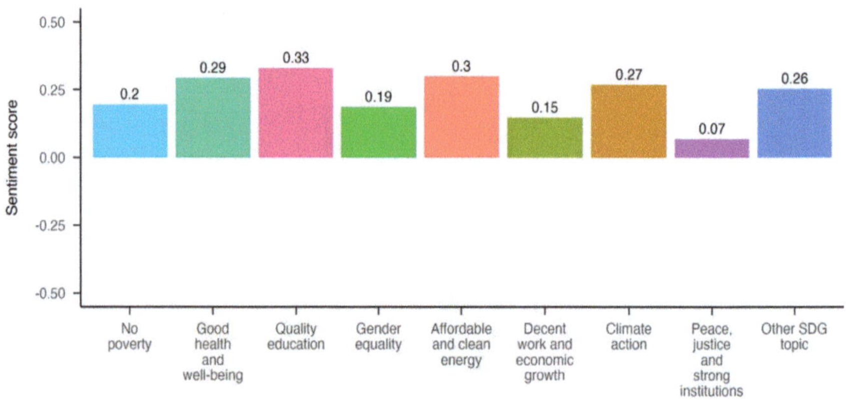

Fig. 6 Average sentiment per SDG topic

7 Conclusions

The 2030 Agenda has become an increasing focus of activities in universities, which are responsible for adopting and engaging with the SDGs. These institutions contribute to the achievement of the SDGs by communicating and favouring stakeholders' participation in related activities, in addition to education and research. Thus, it is important to understand their involvement in SDGs by studying how they are addressing the task of communication. In this study, we analyzed universities' messages on SDGs and their capacity to engage their stakeholders.

Our findings allow us to conclude that SDGs are not among the major priorities of universities. SDG-related tweets represented a low proportion of their Twitter activity. Moreover, this activity only covered half of the SDGs. However, the relevant tweets were composed differently in comparison to other tweets. For instance, they were longer, i.e., they included more content, and were composed to express more positive sentiments. These two significant differences may be responsible for the greater engagement of stakeholders through these messages. Therefore, a good strategy for universities to increase their stakeholders' involvement with their activities would be to relate their communication of their teaching and research with the SDGs.

Finally, we also have to acknowledge that further research is needed to gain insight into how the characteristics of the universities and their activities influence their communication of the SDGs. For example, future studies may analyze why universities are less interested, in terms of communication on social networks, in SDGs such as "zero hunger", "clean water and sanitation", or "reduced inequalities", among others.

This research is not without limitations. First, we selected a limited number of universities that were restricted to a single continent and held very high positions in a raking, which clearly limits the generalization of the conclusions. In addition, we used messages only from Twitter; citizens' behaviour may be different for messages published on other social networks. Finally, we only used messages published through the main account of the universities, but it is possible that they may be using other accounts through which they may channel SDG-related messages.

Acknowledgements This research was supported by the grant PID2020-116247GB-I00 from the Spanish Agencia Estatal de Investigación (MCIN/ AEI/ https://doi.org/10.13039/501100011033).

References

Adams CA (2018) Debate: integrated reporting and accounting for sustainable development across generations by universities. Public Money Manag 38(5):332–334

Agusdinata DB (2022) The role of universities in SDGs solution co-creation and implementation: a human-centered design and shared-action learning process. Sustain Sci 17(4):1589–1604

Alcántara-Rubio L, Valderrama-Hernández R, Solís-Espallargas C, Ruiz-Morales J (2022) The implementation of the SDGs in universities: a systematic review. Environ Educ Res 28(11):1585–1615

Amaral LP, Martins N, Gouveia JB (2015) Quest for a sustainable university: a review. Int J Sustain High Educ 16(2):155–172

Ávila LV, Leal Filho W, Brandli L, Macgregor CJ, Molthan-Hill P, Özuyar PG, Moreira RM (2017) Barriers to innovation and sustainability at universities around the world. J Clean Prod 164:1268–1278

Benoit K, Watanabe K, Wang H, Nulty P, Obeng A, Müller S, Matsuo A (2018) quanteda: an r package for the quantitative analysis of textual data () [Computer software]. The Comprehensive R Archive Network. https://quanteda.io

Buchanan EM, Gillenwaters A, Scofield JE, Valentine K (2019) {MOTE: Measure of the effect}: package to assist in effect size calculations and their confidence intervals (R package version 1.0.2) [Computer software]. The Comprehensive R Archive Network. http://github.com/doomlab/MOTE

Caiado RGG, Leal Filho W, Quelhas OLG, de Mattos Nascimento DL, Ávila LV (2018) A literature-based review on potentials and constraints in the implementation of the sustainable development goals. J Clean Prod 198:1276–1288

Caputo F, Ligorio L, Pizzi S (2021) The contribution of higher education institutions to the SDGs—an evaluation of sustainability reporting practices. Admin Sci 11(3):97

Ceulemans K, Molderez I, Liedekerke LV (2015) Sustainability reporting in higher education: a comprehensive review of the recent literature and paths for further research. J Clean Prod 106:127–143

Cottafava D, Ascione GS, Corazza L, Dhir A (2022) Sustainable development goals research in higher education institutions: an interdisciplinarity assessment through an entropy-based indicator. J Bus Res 151:138–155

De Iorio S, Zampone G, Piccolo A (2022) Determinant factors of SDG disclosure in the university context. Adm Sci 12(1):21

De Luca F, Iaia L, Mehmood A, Vrontis D (2022) Can social media improve stakeholder engagement and communication of sustainable development goals? A cross-country analysis. Technol Forecast Soc Change 177:121525

De la Poza E, Merello P, Barberá A, Celani A (2021) Universities' reporting on SDGs: using the impact rankings to model and measure their contribution to sustainability. Sustain 13:2038

Di Tullio P, La Torre M, Rea MA (2021) Social media for engaging and educating: from universities' sustainability reporting to dialogic communication. Adm Sci 11(4):151

Du Pisani JA (2006) Sustainable development–historical roots of the concept. J Environ Sci 3(2):83–96

ElAlfy A, Darwish KM, Weber O (2020) Corporations and sustainable development goals communication on social media: corporate social responsibility or just another buzzword? Sustain Develop 28(5):1418–1430

Eshima S, Imai K, Sasaki T (2020) Keyword assisted topic models () [Computer software]. The Comprehensive R Archive Network. https://arxiv.org/abs/2004.05964

Fia M, Ghasemzadeh K, Paletta A (2022) How higher education institutions walk their talk on the 2030 agenda: a systematic literature review. High Educ Policy 1–34

Findler F, Schönherr N, Lozano R, Reider D, Martinuzzi A (2019) The impacts of higher education institutions on sustainable development: a review and conceptualisation. Int J Sustain High Educ 20(1):23–38

Fonseca A, Macdonald A, Dandy E, Valenti P (2011) The state of sustainability reporting at Canadian universities. Int J Sustain High Educ 12(1):22–40

García-Feijoo M, Eizaguirre A, Rica-Aspiunza A (2020) Systematic review of sustainable-development-goal deployment in business schools. Sustain 12(1):440

Garde Sánchez R, Rodríguez Bolívar MP, López Hernandez AM (2021) Which are the main factors influencing corporate social responsibility information disclosures on universities' websites. Int J Environ Res Public Health 18(2):524

Griebeler JS, Brandli LL, Salvia AL, Leal Filho W, Reginatto G (2021) Sustainable development goals: a framework for deploying indicators for higher education institutions. Int J Sustain High Educ 23(4):887–914

Kearney MW (2019) rtweet: collecting and analyzing twitter data (R package version 0.7.0) [Computer software]. The comprehensive R archive network. https://joss.theoj.org/papers/ https://doi.org/10.21105/joss.01829

Kopnina H, Meijers F (2014) Education for sustainable development (ESD): exploring theoretical and practical challenges. Int J Sustain High Educ 15(2):188–207

Leal Filho W (2020) Accelerating the implementation of the SDGs. Int J Sustain High Educ 21(3):507–511

Leal Filho W, Pallant E, Enete A, Richter B, Brandli LL (2018) Planning and implementing sustainability in higher education institutions: an overview of the difficulties and potentials. Int J Sustain Dev World Ecol 25(8):713–721

Leal Filho W, Shiel C, Paço A, Mifsud M, Veiga Ávila LV, Brandli LL, Molthan-Hill P, Pace P, Azeteiro UM, Ruiz Vargas V, Caeiro S (2019) Sustainable development goals and sustainability teaching at universities: falling behind or getting ahead of the pack? J Clean Prod 232:285–294

Leal Filho W, Frankenberger F, Salvia AL, Azeteiro U, Alves F, Castro P, Will M, Platje J, Orlovic Lovren V, Brandli L, Price E, Doni F, Mifsud M, Veiga Avila L (2021) A framework for the implementation of the sustainable development goals in university programmes. J Clean Prod 299:126915

Mader C, Rammel C (2015) Brief for GSDR 2015 transforming higher education for sustainable development. UN Sustain Dev Knowl Platf 22(01)

Malik A, Heyman-Schrum C, Johri A (2019) Use of Twitter across educational settings: a review of the literature. Int J Educ Technol High Educ 16(1):1–22

Manolis EN, Manoli EN (2021) Raising awareness of the sustainable development goals through ecological projects in higher education. J Clean Prod 279:123614

Marín L, Rubio A, López I, Bote M (2019) The role of the university in the development of SDGs practices. Proc Int Conf Sustain Develop 151:10–17. https://burjcdigital.urjc.es/bitstream/handle/10115/20648/ACTAS%20ICSF22.pdf?sequence=1&isAllowed=y

Mulholland E (2019) Communicating sustainable development and the SDGs in Europe: good practice examples from policy, academia, NGOs and media. ESDN Q Rep 51:1–22

Nikita M (2020) ldatuning: tuning of the latent dirichlet allocation models parameters (R package version 1.0.2) [Computer software]. The Comprehensive R Archive Network. https://CRAN.R-project.org/package=ldatuning

Owens TL (2017) Higher education in the sustainable development goals framework. Eur J Educ 52(4):414–420

Posit team (2022) RStudio: integrated development Environment for R. Posit Software, PBC, Boston, MA. http://www.posit.co/.

R Core Team (2022) R: A language and environment for statistical computing. R Foundation for Statistical Computing, Vienna, Austria. https://www.R-project.org/.

Revelle W (2022) psych: procedures for psychological, psychometric, and personality research (R package version 2.2.9) [Computer software]. The Comprehensive R Archive Network. https://CRAN.R-project.org/package=psych

Rinker TW (2018) {textclean}: text cleaning tools (version 0.9.3) [Computer software]. The Comprehensive R Archive Network. https://github.com/trinker/textclean

Rinker TW (2022) {sentimentr}: calculate text polarity sentiment (version 2.9.0) [Computer software]. The Comprehensive R Archive Network. https://github.com/trinker/sentimentr

Salvatore C, Biffignandi S, Bianchi A (2022) Corporate social responsibility activities through twitter: from topic model analysis to indexes measuring communication characteristics. Soc Indic Res 164(3):1217–1248

SDSN Australia/Pacific (2017) Getting started with the SDGs in universities: a guide for universities, higher education institutions, and the academic sector. Australia, New Zealand and Pacific Edition. Sustainable Development Solutions Network–Australia/Pacific, Melbourne

Secundo G, Dumay J, Schiuma G, Passiante G (2016) Managing intellectual capital through a collective intelligence approach: an integrated framework for universities. J Intellect Cap 17(2):298–319

Serafini PG, de Moura JM, de Almeida MR, de Rezende JFD (2022) Sustainable development goals in higher education institutions: a systematic literature review. J Clean Prod 370:133473

Soini K, Jurgilevich A, Pietikäinen J, Korhonen-Kurki K (2018) Universities responding to the call for sustainability: a typology of sustainability centres. J Clean Prod 170:1423–1432

Vállez M, Lopezosa C, Pedraza-Jiménez R (2022) Data: web visibility of the SDGs and the 2030 agenda on university websites. Int J Sustain High Educ 23(8):41–59

Wang Y (2016) US State education agencies' use of Twitter: mission accomplished? SAGE Open 6(1):2158244015626492

Watanabe K, Xuan-Hieu P (2022) seededlda: Seeded-lda for topic modeling (R package version 0.8.2) [Computer software]. The Comprehensive R Archive Network. https://CRAN.R-project.org/package=seededlda

Weybrecht G (2022) Business schools are embracing the SDGs–But is it enough?–How business schools are reporting on their engagement in the SDGs. Int J Manag Educ 20(1):100589

Wickham H (2016) ggplot2: elegant graphics for data analysis () [Computer software]. The Comprehensive R Archive Network. https://ggplot2.tidyverse.org

Zamora-Polo F, Sánchez-Martín J, Corrales-Serrano M, Espejo-Antúnez L (2019) What do university students know about sustainable development goals? A realistic approach to the reception of this UN program amongst the youth population. Sustain 11(13):3533

Transformative Marketing Services to Fight Poverty at the Bottom of the Pyramid

Mercedes Marcela Pincay Pilay, Inés López López, and Salvador Ruiz de Maya

Abstract This chapter explores the role of transformative marketing services as a new framework to address poverty. In contrast to traditional economic conceptualisations of poverty, more recent approaches have acknowledged that different dimensions may play a role, thereby highlighting it as a complex phenomenon. Poor individuals are consumers with consumption needs and aspirations but limited opportunities to satisfy them. Companies are partly responsible for this unfulfillment, as they have ignored poor individuals for a long time under the assumption that they are not profitable or worthy. In addition to product adaptation and provision in vulnerable areas, transformative marketing services focus on creating uplifting opportunities that contribute to well-being and are an effective solution for achieving the first of the United Nations' Sustainable Development Goals (i.e. no poverty).

Keywords Sustainable development goals · No poverty · Transformative marketing services · Bottom of the pyramid · Felt deprivation · Powerlessness

1 Introduction

Poverty is undoubtedly one of the greatest challenges that society currently faces. The United Nations, in their summit in September 2015 where the 2030 Agenda was agreed upon, called for an end to poverty as the first Sustainable Development Goal.

M. M. Pincay Pilay (✉) · I. López López · S. Ruiz de Maya
University of Murcia, Murcia, Spain
e-mail: marcela.pincay@unesum.edu.ec

I. López López
e-mail: ineslop@um.es

S. Ruiz de Maya
e-mail: salvruiz@um.es

M. M. Pincay Pilay
Universidad Estatal del Sur de Manabí, Jipijapa, Ecuador

W. Leal Filho et al. (eds.), *The Contribution of Universities Towards Education for Sustainable Development*, World Sustainability Series,
https://doi.org/10.1007/978-3-031-49853-4_17

Poverty rates increased in 2020, reaching 9.2% of the world population, compared to 8.3% registered in the previous year. In absolute terms, the number of people living in global extreme poverty is roughly 800 million; these people live below the international poverty line, set at $1.90 per day (United Nations 2022). Indeed, the COVID-19 pandemic has made this situation even worse.

The population living under these extreme conditions is frequently referred to as the bottom of the pyramid (BoP); they are often the targets of development objectives (London et al. 2014) and consist of four billion people living in poverty or extreme poverty (Prahalad 2012). Similarly, according to the World Resources Institute (2006) the BoP includes people with less than $3,000 in annual purchasing power. BoP individuals are distributed in various regions across the globe, such as Africa, Asia, Eastern Europe, Latin America, and the Caribbean (World Resources Institute 2006).

Poverty is a material disadvantage frequently linked to economic inequality, which can lead to political, economic, and social imbalance. However, it is important to highlight that, in addition to economics, poverty is linked to a number of other dimensions, such as the deprivation of capacities, marginalisation, discrimination, social exclusion, and poor health (Amorós and Cristi 2011). Indeed, BoP people live under difficult conditions, and they aspire for a higher social status and better well-being (Martin and Hill 2015).

The goal of this chapter is to show the role of transformative marketing services as a new framework to deal with poverty at the BoP. In terms of structure, first, the features of poor individuals are analysed. Then, their situation as consumers is described and the role of marketing in providing them with products and services is discussed. Subsequently, transformative marketing services are introduced as an interesting opportunity to offer not only access to an adapted market but also the skills and knowledge required to lift people out of poverty. Finally, conclusions are drawn and limitations acknowledged.

2 Poverty at the Bottom of the Pyramid: Consumption Patterns

2.1 *Poverty: Concept Evolution*

The concept of poverty has changed over the years in light of new empirical evidence. From the 1900s to the 1960s, the concept of poverty mainly consisted of economic indicators and was measured via per capita income (Vollmer 2010). In the late 1960s, experts began to question whether economic indicators were sufficient to capture poverty and whether social and political issues could also play a role. Consequently, the definition of poverty evolved to focus on the satisfaction of basic needs (e.g. income and employment) and physical needs for a basic standard of living (e.g. housing, property, and food). Despite these efforts, the concept of poverty remained essentially economic, as highlighted by Seers (1969).

In the 1970s, as countries began to focus on economic growth and modernisation (Webster 1990), the concept of poverty shifted to include material aspects and the inability to satisfy basic requirements for a decent life, such as adequate nutrition, housing, and clothing (Kanbur and Squire 2002; Sumner 2003). However, other factors such as education and health care (ILO 1976; The World Bank 1978) were also included.

Subsequently, in the 1980s, the concept of poverty went beyond tangible drivers to satisfy basic needs and incorporated intangible non-monetary factors such as vulnerability, empowerment, isolation, powerlessness, and well-being (Chambers 1983).

The World Bank (2001) stated that, in the twenty-first century, 'poverty is pronounced deprivation in well-being', where well-being can be measured through an individual's income, sanitary conditions, nutrition, educational level, possession of property or housing, and certain rights such as freedom of expression.

The concept of poverty has evolved from being purely economic in nature to include a multidimensional perspective. As such, poverty involves a myriad of dimensions and refers to a capacity failure that substantially reduces the options available to an individual to pursue the life they want to live (Ali-Akpajiak and Pyke 2003; Bourguignon and Chakravarty 2003; Chambers et al. 1981; Chambers 1983, 1995; Devereux et al. 2012). These dimensions include the following:

- *Financial*: the financial dimension involves low level or a lack of income; it is measured as income below the minimum national wage, having limited access to loans, and a lack of savings and indebtedness (Banerjee 2016; Wong 2012).
- *Economic*: the economic dimension involves a lack of resources needed for individuals to move past the poverty line, to live a decent life, or to satisfy basic needs (Topsoe-Jensen and Qusaj-Mustafa 2021).
- *Material*: the material dimension involves the living conditions of households or individuals (Terraneo 2017); it also includes material deprivation and the possession of low-quality consumer goods and services in terms of, say, television, radio, transportation means, clothing, food, housing, and public services (Chambers 2006; Gordon 2010; Kus et al. 2016; Townsend 1979).
- *Social*: the social dimension includes a general feeling of being excluded from society (Bartkus and Davis 2009; Brand 2002; Ostrom 2009), which can result in limited participation in social activities or roles or no participation at all (Gordon 2010; Raphael 2011).
- *Environmental dimension*: the environmental dimension focuses on areas where poor people live. For example, they frequently reside in regions with high seismic activity, scarcity of infrastructure and communication systems, vulnerability to disasters (such as floods, droughts, and landslides), or places affected by crime and drug use (Chambers 1995; Narayan et al. 2000).
- *Health-related dimension*: the health-related dimension suggests that poor people live with bad health and, more importantly, do not have access to health care (Chen and Pan 2019).

- *Seasonal dimension*: the seasonal dimension suggests that the experience of being poor changes with the seasons (Devereux et al. 2012; Chambers 2012). For example, in the wet seasons, the poor may experience a lack of food and money, high food prices, debt, and debilitating diseases such as diarrhoea, dengue fever, and malaria.

In this vein, Sachs (2005, p. 20) summarised the living experience of people under extreme poverty as follows:

> … households cannot meet basic needs for survival. They are chronically hungry, unable to access health care, lack the amenities of safe drinking water and sanitation, cannot afford education for some or all of the children, and perhaps lack rudimentary shelter and basic articles of clothing, such as shoes.

A significant share of the world's population live under these extreme conditions, and thus the fight against poverty is vital. In this regard, Thin (2004) identified three main routes to reducing poverty:

a. *Poverty alleviation*: this involves alleviating the symptoms of poverty and/or reducing the severity of poverty without taking people out of poverty, that is, not producing the shift from poor to non-poor. In this case, public policies can be created along with companies engaging in specific programs to alleviate poverty at the BoP. As brand products are not free or even affordable for the poor, brands can offer subproducts to those below the poverty line (London 2008). However, this is purely a band-aid-type solution.
b. *Reduce the number of poor people*: human capital and education are essential for economic growth and, ultimately, for poverty reduction. Education and poverty are closely related: the higher the level of education of the population, the fewer the number of poor. The underlying mechanism builds on the idea that education transfers knowledge and skills, putting individuals in a better position to get better jobs with higher wages. This is beneficial with respect to increasing income, satisfying basic needs, and, finally, ensuring a better quality of life such that individuals can get out of poverty (Awan et al. 2011).
c. *Poverty prevention*: poverty can be avoided by reducing vulnerability. One route to this prevention comes from academic training, which is a guarantee for job stability (Minujin and Comas 2009).

2.2 Consumption at the Bottom of the Pyramid

The term BoP was first introduced 20 years ago by Prahalad and Hart (2002). The BoP framework has induced governments and non-governmental organisations to address poverty through policies and participation (Gardetti 2006; Halme et al. 2016; Simanis and Milstein 2012). However, after an extensive search for benefits through programs designed with the intention to fight poverty, many studies have suggested that spending large amounts of money on such programs has no great impact on the lives of the poor (Banae and Yandell 2006; Riddell 2007; London et al. 2014; Sachs 2005).

Among other factors, individuals at the BoP are consumers. However, they generally lack resources, suffer from low literacy rates, have limited access to education, lack basic services, and experience poor quality of life (Chakravarti 2006; Anderson et al. 2010). Despite this, they usually believe in a brighter future, where both their status and well-being can be improved (Jaikumar et al. 2018).

Consumers at the BoP represent cognitive and social vulnerabilities (Arnold and Valentin 2013; Garrette and Karnami 2010; Karnami 2007). On the one hand, cognitive vulnerability refers to illiteracy or limited education, as poverty hinders access to education and children tend to quit school to support their families. Consistently, low-income consumers do not have the adequate knowledge and skills to make efficient decisions due to ignorance of their rights as consumers (Gupta and Raj Sharma 2021). Often, they are unable to process product information, so they trust the salesperson and follow their recommendations. On the other hand, social vulnerability is due to poverty itself and appears in the form of exclusion and marginalisation (Arnold and Valentin 2013).

Additionally, a large proportion of the BoP live in highly populated communities and are often disconnected from the rest of the world, with a higher concentration in rural areas compared to urban areas in some cases (Lanjouw and Murgai 2009). For example, rural markets dominate most of the BoP in Asia and Africa, whereas urban areas are prevalent in Eastern Europe and Latin America (Subrahmanyan and Gómez-Arias 2008). In India, which is a paradigm of poverty, 70% of the BoP lives in rural areas; however, in Brazil, the majority of the poor live in urban slums (Prahalad 2012). In rural areas, other factors can aggravate this situation, such as a lack of culture, difficulty in accessing products, and deficiencies in communications due to low government investment (Bird et al. 2002). Moreover, in rural areas, consumers at the BoP are not homogeneous because they live in different types of settlements and have different levels of income and expenditure (Shukla and Bairiganjan 2011).

In light of this, it is evident that the BoP is typified by limited availability of goods and services suitable for consumption (Martin and Hill 2012). Additionally, these consumers can hardly plan their consumption because they live in constant uncertainty. Therefore, products are purchased when they are needed, when the product is available, or when the household can afford it, not because of a planned distribution of expenditure (Chikweche et al. 2012). These individuals do not plan their shopping list, ask about prices and product quality, or change stores to improve their shopping experience. They simply live from one day to the next.

This idea of acting on the spot and apart from regular markets leads to the emergence of informal economies, that is, unregulated but well-organised business activities with access to poor quality products, high prices, and poor distribution (Acemoglu and Robinson 2012; Subrahmanyan and Gómez-Arias 2008). These informal economies replace the markets available for regular consumers, as they are perceived as a convenient alternative to obtaining necessary products. In the context of informal economy, purchasing habits are influenced by cultural norms and traditions along with the price and attributes of the product (Webb et al. 2009).

According to D'Andrea et al. (2004), consumers at the BoP spend most of their income on consumer goods (between 50 and 70%), compared to the highest income segments (around 35%). These authors also identified daily purchases as a pattern for low income. BoP people are reluctant to shop in places that are far from home due to a lack of transportation, which, in turn, nurtures informal economies (Gupta and Srivastav 2016).

The purchase decisions and behaviour of BoP consumers are complex, progressive, and divergent; they are also influenced by the opinions of community members, their ethical stance, and psychological, conditional, and cognitive variables (Sheth et al. 1991). Poor people may aspire to have the same products as wealthy individuals, but they risk their own subsistence if they ignore their real budget (Karnami 2007). Accordingly, BoP people are associated with conspicuous consumption, as they try to keep up with their neighbours or with society in general in terms of social class, accumulating material goods, and living a consumer lifestyle (Christen and Morgan 2005; Dreze and Nunes 2009). Jaikumar and Sarin (2015) found that conspicuous consumption improves the social status of poor people by increasing their confidence, happiness, and satisfaction, which is considered a positive effect. However, these long-term changes affect their standard of living and, in the long term, can reduce their ability to fulfil basic needs such as food, education, and health. Unfortunately, this aspiration to conspicuous consumption has a negative long-term impact on financial sustainability and its priority areas (Matos 2019).

3 Transformative Marketing Services

3.1 The Role of Marketing in Alleviating Poverty

Individuals at the BoP are consumers with consumption needs and aspirations but also with important limitations in terms of income. Their condition frequently prevents companies from addressing them as a target, under the assumption that they do not have the financial capacity to afford the products and brands offered to the 'regular' market.

Consequently, smaller businesses are in charge of the provision of goods for poor people. In rural areas, retailers serving the BoP are mostly small independent family businesses with considerable local market power that sell groceries and, to a lesser extent, consumer durables (Lenartowicz and Balasubramanian 2009). Sometimes, rural BoP retailers are the only source of market information. For example, when a consumer orders laundry soap, the retailer decides which brand to offer. In this environment, the store owner is empowered, as they are the only link with the market (Alur and Schoormans 2013). Urban retailers, however, are more organised, offer a greater variety of products, and are closer to general markets (Alur and Schoormans 2013).

Small companies struggle to serve the poor, and the private sector pays little attention to them. This makes BoP people feel ignored due to a lack of access to desired goods and services. Even when needs are met, BoP people often pay more for the same product than wealthier consumers due to mobility constraints and distribution inefficiencies, which are caused, at least partly, by the geographical dispersion of this population across mainly rural areas (Anderson and Markides 2007). However, Prahalad and Hart (2002) claimed that international leading companies could increase their wealth and alleviate poverty for millions of BoP people by conducting business with them, urging companies to engage with this well-known segment of the world's population. According to Prahalad (2005), the growth of large- and medium-sized companies is not based on the small segment of high-income people but, rather, on the massive number of low-income people who enter the market for the first time. Prahalad (2005) noted that attending to these consumers may require the application of adapted business strategies. The peculiarities of the BoP make it a complex and difficult market to serve, but, at the same time, it represents a business opportunity that few companies take advantage of. In essence, consumers at the BoP have relatively low incomes compared to those in developed countries but have significant purchasing power at the aggregate level (Prahalad and Hammond 2002).

The success of these markets requires adaptations in the accessibility, availability, and knowledge of the products to overcome economic, institutional, and cultural barriers (Prahalad 2009; Johanson and Vahlne 2003). In other words, knowledge from traditional markets cannot be easily and directly transferred to the BoP (Ezequiel and Patricia 2012). Consumers at the BoP are price sensitive and vulnerable to inflation compared to those in the middle and upper classes (Jacobs and Smit 2010; Duvenage et al. 2010). As a result, they prefer small transactions rather than large ones, which can lead to expensive operations costs for manufacturers (Karnami 2007). In response, manufacturers are required to restructure their budgets, business models, and commercial practices (Nicole 2003). Making low-income markets profitable requires a different business model (Chesbrough et al. 2006) that emphasises low costs, high quality, sustainability, and profitability (Prahalad and Hart 2002).

Arnold and Valentin (2013) posited that companies that target the BoP must be affordable and profitable while empowering local players in the development of local products. Prahalad (2005) outlined the following four key elements for companies to thrive in the BoP market:

1. Create purchasing power;
2. Meet aspirations through innovation and consumer education;
3. Improve distribution channels and communication systems;
4. Solve problems based on local needs.

At the same time, firms should keep in contact with other agents, such as local and central governments, financial institutions, and non-government organisations, to offer coordinated interventions. Governments alone cannot solve the problem of poverty through economic policies and income redistribution. Instead, the business world must become more sensitive to social responsibility issues and the concerns of the poor in general (Mohsin and Leon 2010).

3.2 *Transformative Marketing Services: A Step Further*

Companies can radically improve the lives of billions of people not only by selling affordable products and services to low-income customers, as described in the previous section, but also by creating employment and business opportunities (Simanis and Hart 2008). In doing so, these companies may incorporate social goals along with traditional profit-maximisation objectives (London and Hart 2004).

The concept of transformative marketing services originated in the early 2000s with the aforementioned view on fighting poverty. Transformative marketing services seek to 'create uplifting changes' in the well-being of individuals, groups, and ecosystems via interactions between service providers and consumers (Anderson et al. 2013) and by focusing on solving the problems that affect poverty in both developed and developing countries (Ostrom et al. 2010). They enable services by promoting tools for the development of intangible resources to generate income and create opportunities that contribute to well-being (Vargo and Lush 2016). These transformative services are aimed at improving the well-being of consumers and society, which is to say they allow not only individual but also collective well-being to be achieved (Kuppelwieser and Finsterwalder 2016). These services also look at well-being from a wide perspective, including, among other things, the dimensions of physical health (objective and subjective impressions), mental health (resilience, stress, and exhaustion), financial well-being, discrimination, marginalisation, lack of education, inclusion, and life satisfaction (Anderson et al. 2013; Rosenbaum et al. 2011).

Transformative marketing services go a step beyond the conceptualisation of social marketing. They are not simply about providing products and services to people at the BoP or engaging in social initiatives. Rather, they contribute to reducing suffering and improving well-being (Russell-Bennett et al. 2019). In this sense, individuals at the BoP can feel less deprived and more empowered than ever.

4 Conclusions

Nowadays, an international consensus has been reached with regards to poverty as one of the main challenges facing the world. This was confirmed by the United Nations, where, in the 2030 Agenda, an end to poverty was confirmed as the first Sustainable Development Goal.

Millions of people around the world live with less than $2 a day; new avenues to deal with this alarming situation must be explored. The traditional economic view of poverty has been abandoned in favour of concepts that typify it as not only an economic construct but also a complex phenomenon comprised of a diverse array of dimensions.

From a marketing perspective, this paradigm shift was a necessary way of addressing consumers at the BoP. They need products and services; they aspire to obtain items they can hardly afford, and they cannot be attended to following the rules of regular markets. Their defining features make them a complex segment. However, they also represent a great opportunity for firms in terms of constituting a large consumer base in need of attention. The most integrative way to address them is through so-called transformative marketing services, which aim at improving the well-being of poor people by offering them uplifting solutions, resources, skills, and, in sum, a path towards a brighter future.

This chapter focused on poverty as a major societal challenge, as pointed out by the United Nations in its 2030 Agenda. Organisations should not avoid their responsibility to help alleviate poverty. This can be achieved via transformative marketing services, which tackle poverty at the root and attempt to improve the lives of BoP individuals by offering them the opportunity to gain skills. In doing so, they feel less deprived and more empowered. In addition, public policies should promote the application of these services to confer well-being and personal growth to those in vulnerable situations.

In terms of limitations, evidence is lacking from empirical studies demonstrating the true impact of transformative marketing services on BoP people, which offers an interesting avenue for future research. Additionally, we did not analyse developing countries compared to developed countries or urban areas compared to rural ones as targets for organisations engaged in transformative services.

Acknowledgements This research was supported by the grant PID2020-116247GB-I00 from the Spanish Agencia Estatal de Investigación (AEI / https://doi.org/10.13039/501100011033).

References

Acemoglu D, Robinson J (2012) Why nations fail: the origins of power, prosperity, and poverty. Crown Publishers, New York

Ali-Akpajiak SC, Pyke T (2003) Measuring poverty in Nigeria. Oxafam GB, Hendon

Alur S, Schoormans JP (2013) Retailers and new product acceptance in India's base of pyramid (BoP) markets. Int J Retail Distrib Manag 1(3):1898–2200

Amorós JE, Cristi O (2011) Poverty and entrepreneurship in developing countries. In: Minniti M (ed) The dynamics of entrepreneurship: evidence from global entrepreneurship monitor data. Oxford University Press, Oxford, pp 209–230

Anderson J, Markides C (2007) Strategic innovation at the base of the pyramid. MIT Sloan Manag Rev 49(1):83

Anderson JL, Markides C, Kupp M (2010) The last frontier: market creation in conflict zone, deep rural areas, and urban slums. Calif Manage Rev 52(4):6–28

Anderson L, Ostrom AL, Corus C, Fisk RP, Gallan AS, Giraldo M, Williams JD (2013) Transformative service research: an agenda for the future. J Bus Res 1203–1210

Arnold DG, Valentin A (2013) Corporate social responsibility at the base of the pyramid. J Bus Res 1904–1914

Awan MS, Malik N, Sarwar H, Waqas M (2011) Impact of education on poverty reduction. Int J Acad Res 3(1):659–664

Banae ME, Yandell D (2006) Development strategies and opporunities: the case of Africa. Rev Human Factor Stud 12(1):114–133

Banerjee MM (2016) We routinely borrow to survive: exploring the financial capability of income-poor people in India. Soc Work 61(4):349–358
Bartkus VO, Davis JH (2009) Introduction: the concept yet undiscovered value of social capital. In: Barkus VO, Davis JH (eds) Social capital: reaching out, reaching. Edward Elgar, Cheltenham, UK, pp 1–6
Bird K, Hulme D, Moore K, Shepherd A (2002) Chronic poverty and remote rural areas. Chronic Poverty Res Centre 1–57
Bourguignon F, Chakravarty SR (2003) The measurement of multidimensional poverty. J Econ Inequal 1(1):29–49
Brand C (2002) Making the livelihoods framework work (RUP No.20). Department of Rural and Urban Planning, University of Zimbabwe
Chakravarti D (2006) Voices unheard: the psychology of consumption in poverty and development. J Consum Psychol 363–376
Chambers R (1995) Poverty and livelihoods: whose reality counts? Environ Urban 7(1):173–204
Chambers R, Longhurst R, Pacey A (1981) Seasonal dimensions to rural poverty. Frances Printer Ltd
Chambers R (1983) Rural development: putting the last first. Longman
Chambers R (2006) What is poverty? Concepts and meaures in poverty in focus. Internacional Poverty Center
Chambers R (2012) Provocations for development. Practical Action Publishing
Chen C, Pan J (2019) The effect of the health poverty alleviation project on financial risk protection for rural residents: evidence from Chishui city, China. Int J Equity Health 2–16
Chesbrough H, Ahern S, Finn M, Guerraz S (2006) Business models for technology in the developing world: the role of non-governmental organizations. Calif Manage Rev 48(3):47–62
Chikweche T, Stanton J, Fletcher R (2012) Family purchase decision making at the bottom of the pyramid. J Consum Mark 29(3):202–213
Christen M, Morgan M (2005) Keeping up with the Joneses: analysing the effect of income inequality on consumer borrowing. Quant Mark Econ 3(2):145–173
Devereux S, Sabates-Wheelers R, Longhurst R (2012) Seasonality, rural livelihoods and development. Earthscan
D'Andrea G, Stengel EA, Goebel-Krstelj A (2004) Six truths about emerging-market consumers. Strategy Bus 2–12
Dreze X, Nunes JC (2009) Feeling superior: the impact of loyalty program structure on consumers' perception of status. J Consum Res 35(6):890–905
Duvenage S, Schonfeldt H, Kruger R (2010) Food product attributes guiding purchasing choice of maize meal by low-income South African consumers. Dev Sout Africa 27(3)
Ezequiel R, Patricia M (2012) Inclusive networks for building BOP markets. Bus Soc J 512–554
Fellowes M (2006) From poverty, opportunity: putting the market to work for lower income families. The Brookings Institute, Washington, DC
Gardetti MA (2006) Base-of-the-pyramid approach in Argentina: preliminary findings for a BOP Lab. Greener Manag Int 65–77
Garrette B, Karnami AG (2010) Challenges in marketing socially useful goods to the poor. Ross School of Business, Working Paper N° 135. https://deepblue.lib.umich.edu/bitstream/handle/2027.42/64347/1135_r10_karnani.pdf?sequence=4
Gordon D (2010) Poverty. En P. Townsend. The peter townsend reader, pp 129–268. Policy Press
Gupta N, Raj Sharma B (2021) A qualitative investigation of shopping behavior of customers at bottom of the pyramid. J Global Market
Gupta S, Srivastav P (2016) An exploratory investigation of aspirational consumption at the bottom of the pyramid. J Int Consum Mark 1:2–15
Halme M, Kourula A, Lindeman S, Kallio G, Lima-Toivanen M, Korsunova A (2016) Sustainability innovation at the base of the pyramid through multi-sited rapid ethnography. Cor-Porate Soc Responsib Environ Manag 23:113–128
ILO (1976) Employment, growth and basic needs: a one-world problem. ILO, Geneva

Jacobs G, Smit E (2010) Materialism and indebtedness of low income consumers: evidence from South Africa's largest credit granting catalogue retailer. South Afr J Bus Manag 41(4):11–33
Jaikumar S, Sarin A (2015) Conspicuous consumption and income inequality in an emerging economy: evidence from India. Mark Lett 26:279–292
Jaikumar S, Singh R, Sarin A (2018) I show off, so I am well off': subjective economic well-being and conspicuous consumption in an emerging economy. J Bus Res 86:386–393
Johanson J, Vahlne JE (2003) Business relationship learning and commitment in the internationalization process. J Int Entrep 1(1):83–101
Kanbur R, Squire L (2002) The evolution of thinking about poverty: exploring the interactions. In: Meier GM, Stiglitz JE (eds) Fontiers of development economics-the future in perspective. Oxoford University Press Inc., New York, pp 183–226
Karnami A (2007) The mirage of marketing to the bottom of the pyramid: how the private sector can help alleviate poverty. Calif Manag Rev 90–111
Kuppelwieser VG, Finsterwalder J (2016) Transformative service research and service dominant logic: quo vaditis? J Retail Consum Serv 28:91–98
Kus B, Nolan B, Whelan CT (2016) Material deprivation and consumption. In: Braddy D, Burton LM (eds) The oxford handbook of the social science of poverty. Oxford University Press, pp 577–601
Lanjouw P, Murgai R (2009) Poverty decline, agricultural wages, and nonfarm employment in rural India: 1983–2004. Agric Econ 40(2):243–263
Lenartowicz T, Balasubramanian S (2009) Practices and performance of small retail tores in developing economies. J Int Mark 17(1):58–90
London T, Hart SL (2004) Reinventing strategies for emerging markets: beyond the transantional model. J Int Bus Stud 35(5):350–370
London T, Esper H, Grogan-Kaylor A, Kistruck GM (2014) Connecting poverty to purchase in informal markets. Strateg Entrep J 8(1):37–55
London T (2008) The base of the pyramid perspective: a new approach to poverty, vol 2008. Academy of Management, Briarcliff Manor, NY 10510
Martin KD, Hill R (2012) Life satisfaction, self-determination, and consumption adequacy at the bottom-of- the- pyramid. J Consum Res 38(6):1155–1168
Martin KD, Hill RP (2015) Saving and well-being at the base od the pyramid: implications for transfromative financial service delivery. J Serv Res 18(3):405–421
Matos C (2019) Inequality and crisis: conspicuous consumption as the missing link in the Portuguese case. J Econ Issues 53:26–38
Minujin A, Comas G (2009) Vulnerability and resilience of the middle class in Latin America. International Affairs Working Paper, 2
Mohsin H, Leon Z (2010) The bottom of the pyramid: key roles for business. J Bus Econ Res 8(5)
Narayan D, Patel R, Schafft K, Redemacher A, Koch Schulte S (2000) Voices of the poor: can Anyone hear us? Oxford University Press, New York
Nicole B (2003) The base of the pyramid (BOP): reperceiving business from the bottom up. Global Business Network (GBN) Working Paper
Ostrom A, Bitner MJ, Brown S, Burkhard K (2010) Moving forward and making a difference: research priorities for the science of service. J Serv Res 1–33
Ostrom E (2009) What is social capital? In: Barkus O, Davis JH (eds) Social capital: reaching out, reaching in. Edward, Elgar, Cheltenham, UK, pp 17–38
Peredo AM, McLean M (2006) Social entrepreneurship: a critical review of the concept. J World Bus 41(1):56–65
Prahalad CK (2005) The fortune at the bottom of the pyramid: eradicating poverty through profits. Wharton School Publishing, Upper Saddle River, NJ
Prahalad CK (2009) The fortune at the bottom of the pyramid revised and updated 5th anniversary edition: eradicating poverty through profits. PearsonEducation Inc., Upper Saddle River, NJ
Prahalad CK (2012) Bottom of the pyramid as a source of breakthrough innovations. J Prod Innov Manag 29(1):6–12

Prahalad CK, Hammond A (2002) Serving the world's poor, profitably. Harv Bus Rev 80(9):48–57
Prahalad CK, Hart SL (2002) The fortune at the bottom of the pyramid. Strategy Bus 1–14
Raphael, D. (2011). *Poverty in Canada: Implications for health and quality of life.* Canadian Scholars' Press Inc.
Riddell RC (2007) Does foreign aid really work? Oxford University Press, Oxford
Rosenbaum MS, Canan C, Ostrom AL, Anderson L, Fisk RP, Gallan AS, Williams JD (2011) Conceptualization and aspirations of transformative service research. J Res Consum 19:1–6
Russell-Bennett R, Fisk R, Rosenbaum MS, Zainuddin N (2019) Transformative service research and social marketing – converging pathways to social change. J Serv Mark 33(6)
Sachs JD (2005) The end of poverty. The Penguin Press, New York
Seers D (1969) The meaning of development. Int Dev Rev 11(4):2–6
Sheth JN, Newman BI, Gross BL (1991) Why we buy what we buy: a theory of consumption values. J Bus Res 22(2):159–170
Shukla S, Bairiganjan S (2011) The base of pyramid distribution challenge: evaluating alternate distribution models of energy product for rural. Center for Development Finance, India
Simanis E, Hart SL (2008) The base of the pyramid protocol: toward next generation BoP strategy. Cornell University, NY
Simanis E, Milstein M (2012) Back to business fundamentals: making "bottom of the pyramid" relevant to core business. J Field Actions Ield Actions Sci Rep Spec Issue 4:82–88
Subrahmanyan S, Gómez-Arias JT (2008) Integrated approach to understanding consumer behavior at bottom of pyramid. J Consum Mark 402–412
Sumner A (2003) Economic and non-economic wellbeing: a review of the progress on the meaning and measurement of poverty. Paper presented at the WIDER conference 'Inequality, Poverty and Human Well-Being'
Terraneo M (2017) Material deprivation, health and well-being in old age: a comparative analysis of 14 European Countries. In: White R (eds) Measuring multidimensional poverty and deprivation: Incidence and determinants in developed countries. Palgrave Macmillan, Switzerland, pp 113–138
The World Bank (2001) World development report 2000/2001: attacking poverty. Oxford University Press, Oxford
The World Bank (1978) World development report. Prospects for growth and alleviation of poverty. The World Bank, Washington DC
Thin N (2004) Post-basic education and training, enabling environments, and pathways to poverty reduction. University of Edinburgh, Edinburgh
Topsoe-Jensen B, Qusaj-Mustafa A (2021) Sida Decentralised Evaluation. Sida, Kosovo. https://cdn.sida.se/publications/files/sida62385en-final-evaluation-of-the-framework-agreement-regarding-direct-grant-support-and-consultancy-to-the-agency-for-gender-equality-kosovo.pdf
Townsend P (1979) Poverty in the United Kingdom. Penguin Books, Harmondsworth
United Nations (2022) Sustainable development goals. https://www.un.org/sustainabledevelopment/poverty/
Vargo SL, Lush RF (2016) Institutions and axioms: an extension and update of service-dominant logic. J Acad Mark Sci 44(1):5–23
Vollmer F (2010) Reviewing 'poverty' as an object of study: Seeking a conceptual match of well-being with the inter-subjective understanding of ill-being. In-Spire J Law Polit Soc 67–85
Webb JW, Tihanyi L, Ireland RD, Sirmon DG (2009) You say illegal, I say legitimate: entrepreneurship in the informal economy. Acad Manag Rev 34(3):492–510
Webster A (1990) Introduction to the sociology of development, 2nd edn. The Macmillan Press Ltd., London
Wong SY (2012) Understanding poverty: comparing basic needs approach and capability approach. Available at SSRN 2066179.
World Resources Institute (2006) The market of the majority: the BOP opportunity map of Latin America and the Caribbean. Inter-Am Development Bank, Washington, DC

Energy-Efficient Renovation of University Buildings and Their Use as Living Labs—The Technical Sciences Building at the University of Innsbruck as a Best Practice Example

Lorenzo Rieg, Bertram Knoflach, Hanspeter Mühlegger, Rainer Pfluger, and Wolfgang Streicher

Abstract CO_2 emissions from the heating, cooling, and ventilation of university buildings are a large part of many universities' total greenhouse gas emissions. Modern, energy-efficient buildings have a huge potential to reduce energy consumption and CO_2 emissions and help universities in reaching carbon neutrality. However, universities are often restricted by various factors in their ability to build and finance new, highly energy-efficient buildings, which would reduce emissions significantly. This makes refurbishment of existing university buildings a high priority for many universities aiming at a reduction of their emissions. At the University of Innsbruck, the Technical Sciences Building, an 8-story reinforced concrete building built in 1971 with a 19.399 m^2 gross floor area, typical for many university buildings in Central Europe, has been extensively refurbished with a strong focus on energy efficiency and reduction of renovation and operating costs. The presented renovation can be seen as a demonstration project for similar buildings awaiting refurbishment, with the building being the first office tower that achieved the EnerPHit certification for energy efficiency. In addition, the building has been equipped with a variety of sensors and is used as a living lab to collect learnings from the refurbishment, the management of the building and to improve the comfort and well-being of the over 700 employees working in the building. The following article introduces the process as well as the technical parameters of the building and its refurbishment, which is exceptionally well-documented and measured, as a best practice and shows how the building has been and still is used as a living lab for increasing energy efficiency as well as comfort.

L. Rieg (✉)
University of Innsbruck, Innsbruck, Austria
e-mail: lorenzo.rieg@uibk.ac.at

B. Knoflach · H. Mühlegger
Austrian Federal Property Association, Innsbruck, Austria

R. Pfluger · W. Streicher
Unit for Energy Efficient Building, University of Innsbruck, Innsbruck, Austria

W. Leal Filho et al. (eds.), *The Contribution of Universities Towards Education for Sustainable Development*, World Sustainability Series,
https://doi.org/10.1007/978-3-031-49853-4_18

Keywords Energy-efficient buildings · University buildings · Building refurbishment · Sustainable construction · Sustainable refurbishment

1 Introduction

The goal of limiting global warming to well below 2 °C of the Paris Agreement (UNFCCC 2015) requires the achievement of net-zero emissions by the mid of the twenty-first century. The IPCC special report on 1.5 °C (IPCC 2018) emphasizes that this entails using the large energy efficiency potential of the building stock. Due to this potential, renovations for improving energy efficiency are considered pivotal to reaching carbon neutrality (He et al. 2019; Bertoldi et al. 2020).

Universities are well-placed to play a leading role in driving change towards a more sustainable future since they educate future leaders and citizens. In addition to research and teaching as their main areas of impact, universities are often large institutions with thousands of students and employees as well as numerous buildings. With buildings accounting for roughly 28% of total global energy consumption (International Energy Agency 2022), it is clear that significant greenhouse gas emissions are also associated with university buildings. Especially heating and cooling of buildings are often responsible for a large part of the total emissions of a university, with Eskander and Nitsche (2021) report, that a majority of the CO_2 emissions of UK Universities originate from their use of natural gas and grid energy. Energy-efficient buildings, therefore, offer a huge potential for reductions in energy consumption and greenhouse gas emissions. In addition, saving energy leads to cost savings which often offset additional costs for energy-efficient construction or materials. In addition, the bold demonstration of novel technologies and practices encourages and accelerates the wider adoption of sustainable and energy-efficient planning and building practices. Especially the development of universally understood and accepted tools and performance criteria can foster their wider uptake (Amaral and Martins 2013). At the University of Innsbruck, the energy demand of buildings (heating, cooling, and electricity) is responsible for about 56% of the total CO_2 emissions of the institution.

New buildings are generally becoming more and more energy-efficient over time, even though building standards are often below the ones which would be required to reach the Paris Goals of the UNFCCC. Nevertheless, many universities operate in buildings from the 1960ies and 1970ies, which demand large amounts of energy for cooling and heating. Refurbishing those buildings is often complicated and costly, leading to various challenges for universities that want to reduce their energy demand and greenhouse gas emissions. The research on challenges regarding campus operations and sustainability has been summarized by Amaral et al. (2020) who often mention lack of (monetary) resources as well as missing or reluctant leadership support as the main issues. Especially regarding renovation and refurbishment projects, possibilities for energy efficiency are often under-used. This is partly due to cost issues, but also risk adversity and gaps in knowledge on possibilities, costs, and actual risks.

Therefore, demonstration projects in the classical sense can show the possibilities of energy-efficient, sustainable buildings and contribute to the wider uptake of existing technologies. The Technical Sciences Building of the University of Innsbruck has been extensively refurbished in order to improve energy efficiency and to create an attractive location for research and teaching based on the 1970ies building.

The Technical Sciences building of the University of Innsbruck was planned and built in the course of the research funding program of the Austrian Research Promotion Agency and the Federal Ministry for Climate Protection. The project was scientifically accompanied and documented from 2010—2016. The project is a demonstration building of the Austrian Federal Real Estate Agency (BIG), as owner of the building. The main objective of the research program "BIGMODERN" was to develop criteria that can be applied to future renovations with a view to ensuring that energy-efficient renovated buildings are also economically viable. BIG has implemented many criteria from its experiences as a building standard for nearly zero-energy buildings. Today, BIGMODERN criteria have become a mandatory standard for all new buildings and modernizations of the Group.

For the renovation of the Technical Sciences building of the University of Innsbruck, the following goals had been defined in order to ensure the transferability of the developed innovation:

- The demonstration of the feasibility of renovating the building to achieve very high energy-efficiency
- The formulation of additional criteria for energy efficiency and sustainability as orientation for the design and execution planning phases
- Implementation of an integrative building standard, which enables significant energy savings and the use of a high percentage of renewable energy, with financially feasible costs.

The Technical Sciences Building of the University of Innsbruck has been extensively refurbished within such a demonstration project, and respective research was carried out using a living lab approach, from 2012 to 2014. According to the European Network of Living Labs (2023), Living Labs are open innovation ecosystems in real-life environments using iterative feedback processes throughout a lifecycle approach of an innovation to create sustainable impact.

In the presented case, offices used by university staff for standard office use are equipped with special monitoring devices. This allows innovative solutions in terms of control, daylight, and artificial light as well as HVAC to be checked and further developed for comfort, health and energy efficiency in an open innovation process. In this way, the university building works as a living lab, the results are used directly in research projects as well as teaching (demonstrations and internship exercises). While the learnings might be mainly applicable to university buildings in Austria, the demonstration effect will go beyond them and can be useful for other office buildings in Austria and beyond. In addition, the refurbishment was accompanied by extensive user consultations and research, and the building is still used as a living lab to investigate the refurbishment results, as well as technical details during operation and comfort for users. This follows the contemporary focus on involving stakeholders and conducting research in real environments (Evans et al. 2015).

2 The "Technical Sciences" Building at the University of Innsbruck

The Technical Sciences Building of the University of Innsbruck can be considered a relatively typical non-residential or university building of the post-war period. It has been built in 1971 and is located centrally at the "Campus Technik". In addition to the 9-story flat roof building with a net floor area (before modernization) of 11.800 m^2, two adjacent wings with lecture halls are part of the building complex. The building is a point-supported flat slab construction, with the reinforced-concrete core containing the stairway as well as the elevator and technical shafts.

Before the modernization, the horizontal carrying elements protruded from the façade and were used as external escape routes consisting of surrounding balconies connected with stairs, which were acting as thermal bridges (see Fig. 1). The building can be considered a rather typical one for its era, with hundreds of similar buildings being used by educational institutions such as universities or schools throughout Europe, with most of them still in dire need of refurbishment (Achammer et al. 2015).

Fig. 1 The Technical Sciences Building before the refurbishment. The clearly visible but hardly used balconies were removed during the refurbishment (photograph University of Innsbruck)

3 The Modernization Project

The refurbishment of the Technical Sciences Building was initially not planned with such a focus on energy efficiency. However, the general planning contract (ATP for architecture, HAVAC and electrical planning, and Passive House Institute for building physics) allowed the development of an integrated concept that enabled the high energy efficiency standard EnerPHit[1] and made the refurbishment project suitable to demonstrate the potential of energy-efficient refurbishment, especially for publicly owned post-war buildings. The presented modernization project has been carried out as a sub-project of the flagship project "BIGMODERN—Sustainable modernization standards for buildings owned by the Federal Republic of Austria of the period from the 1950s to the 1980s".

To reach the aim of sustainable refurbishing standards in practice, extensive changes in different technologies are needed. Many refurbishment projects are focusing on the technical, architectural, and monetary optimization of single components. Innovative technologies in the context of modernization mean additional efforts in planning and coordination, which are hardly possible in standard planning processes.

Since changes in the standard planning process in technologies lead to risks for the building owner like strongly increasing costs, lower than planned savings in the operation of the building, or even failure, risks have to be minimized to innovate modernization standards. Therefore, the implementation of demonstration projects with above-average energy efficiency and sustainability standards while complying with requirements for industrial management aid in breaking new ground toward innovative yet cost-effective renovations.

The Technical Sciences Building at the University of Innsbruck was used as such a demonstration project, and its modernization represents a shining example of energetic standards in modernization. The project aimed to demonstrate the feasibility of modernizing the building to achieve high energy efficiency while respecting sustainability criteria, define additional quality standards regarding energy efficiency and sustainability and implement an integrated building standard for significant energy savings without significant cost increases.

According to Jäger et al. (2015), The main innovations and results of the implemented demonstration project are:

- Reduction of primary energy demand (−75%) and CO2 emissions (−80%)
- The newly developed facade (wood-aluminum facade element), which was mounted at the same time on a rail system from two sides of the building (see Fig. 2)
- The innovative windows with 4 glasses, a weather-protected shading system, and automatic control (radiation-dependent with manual override)

[1] https://passipedia.org/certification/enerphits=enerphit.

Fig. 2 The Technical Sciences building during the refurbishment of the façade. The new façade and windows are already clearly visible above the scaffolding (photograph Rainer Pfluger)

- The innovative ventilation concept with automatic night ventilation and comfort ventilation of the offices, which allows a combination of free ventilation and mechanical ventilation (see Fig. 4)
- Cascade ventilation with flow into the offices, overflow to the central rooms via the corridor, and exhaust from the central rooms to reduce the length of air ducts and therefore costs.
- Newly designed, transparent overflow over the office doors with very good acoustic characteristics (see Fig. 5)
- Innovative control concept (DDC Automation System) for the individual control of the shading system, ventilation, and windows incl. night ventilation
- Installation of "light traps" (transparent areas for lighting the core areas of the building)
- and transparent fields above the doors
- Cooling with groundwater; 100% cooling energy via groundwater cooling. This cooling backup was never needed in the last years.

To optimize energy efficiency measures and determine the best context for investments, parameter studies were conducted using the Passive House Planning Package (PHPP, see also Kah 2009). For example, thermal targets were used to determine the permitted investment costs within the scope of the savings achieved by the corresponding measure, and suitable implementation concepts that fit within this cost framework were compared. In order to determine the heat produced inside the building through IT equipment and other electrical equipment, information on all electrical consumers, their power load, and times of use have been collected. Due to the high power load from IT equipment, lighting, and the high people density, a zonal rating of the building in summer conditions has been established using a dynamic simulation (Schnieders 2003).

The passive cooling principle determined for the building is based on night ventilation using the windows. This type of cooling is powerful enough for sufficient cooling of the Technical Sciences Building, as the climate in the Tyrolean Inn valley offers relatively cool nights even during hot summer days. However, certain requirements are to be fulfilled to ensure the required airflow volume for the night ventilation (see sections "Windows and façade" and "Ventilation").

After the presented analyses, the focus of the refurbishment of the Technical Sciences building was on the interaction of the façade (thermal protection in summer and winter), sun protection, use of daylight, lighting, ventilation, heating, and cooling. The optimization of those interactions through intelligent coordination of the individual components has been important for reaching the high post-refurbishment energy standard. Especially the interaction between the building envelope and lighting as well as heating and cooling is an important part, which should be tackled in an integrative way in any refurbishment project. In the presented case, the glazing of the façade has been designed with the optimal compromise between the highest possible use of daylight and the lowest possible demand for heating and cooling. The building was therefore greatly improved regarding its energy demand, air quality, and lighting.

3.1 Windows and Façade

The façade of the building has proven to be especially challenging during the project. In order to reach the ambitious goals regarding energy efficiency, the insulation of the façade and the roof have been increased to 24 cm. However, the concrete balustrades did not provide sufficient bearing strength, which is why the windows were mounted onto the steel skeleton of the façade elements via wooden crossbeams. With the low thermal bridge façade concept, a very well-insulated opaque building envelope was achieved.

The top-hinged, outward-tilting windows (see Figs. 3 and 4) are another core element of the refurbishment project and have been specifically developed for the building. A standard three-pane thermal glazing with $Ug = 0.6\ W/m^2K$ with increased solar heat gain coefficient SHGC ≤ 0.4 and "warm edge" in combination with an external fourth pane was installed. The frame profiles were designed with a high construction depth but narrow face width, in order to achieve better thermal behavior (fewer transmission losses) and higher solar gains during the heating period plus a higher proportion of daylight in the offices at the same time. The motorized sun

Fig. 3 View of the façade and windows (*source* ATP Architekten ingenieure for BIG)

protection in the form of movable horizontal lamellas is located in the air space between the outer pane and the inner triple glazing and is thus protected from wind damage. The outer pane together with the aluminum frame can be opened for maintenance of the motorized sunshade and cleaning. The installation of the window in front of the insulation level was due to the architecturally desired façade design (flawless outer surface of glass and curtain wall). Nevertheless, due to the successful development efforts, a thermal bridge loss coefficient of the window installation of only $\Psi = 0.049$ W/mK could be achieved (Malzer and Pfluger 2015).

The windows are automatically controlled based on temperature and wind conditions but provide the possibility of manual override. Despite the difficult conditions of a high-rise building in an area affected by strong winds, they allow the users to individually open their windows and also enable highly efficient cooling with outside air during the night. The windows and blinds are primarily controlled based on the felt room temperature, which is measured with 7 sensors on each floor. In addition, solarimeters have been installed on each main façade to regulate the shading.

The natural cooling via the top-hinged, outward-tilting windows (see Fig. 4) is regulated based on the temperature difference between the rooms and the outside. The windows open or stay open if the room temperature is $\geq$23 °C and the outside temperature is at least 0,5 °C lower. They close or stay closed if the room temperature is lower than 20 °C or the difference between room temperature and outside temperature is lower than 0,5 °C. In extremely hot periods, the concrete core can be

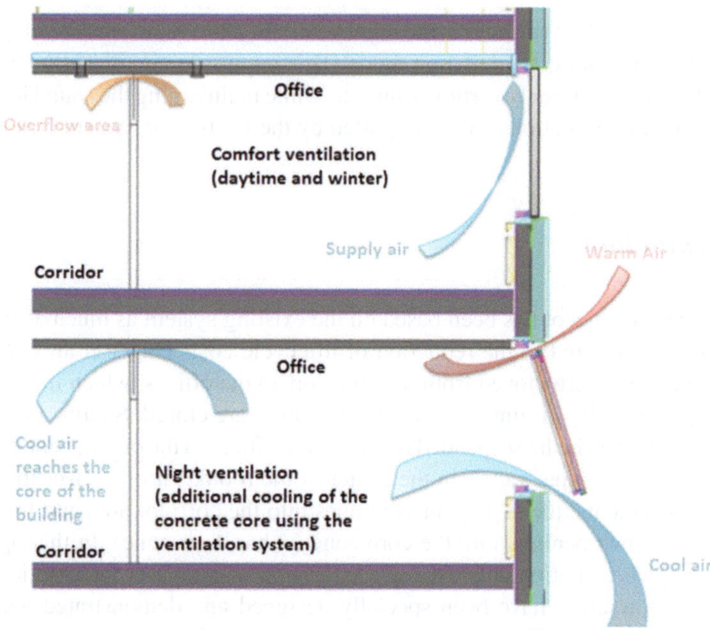

Fig. 4 Schematic view of the ventilation system. Night ventilation is used to cool down the building if the temperature difference between the interior of the building and the outside is 4 K or higher (adapted from Malzer and Pfluger 2015)

Fig. 5 Overhead lights with overflow openings enable the exchange of air between the offices and corridors (adapted from Malzer and Pfluger 2015)

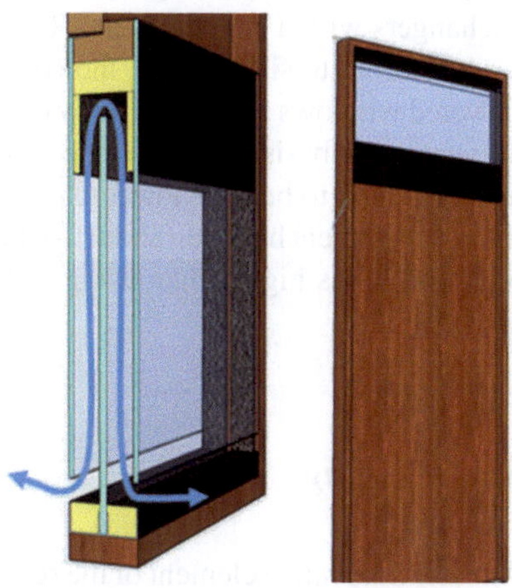

additionally cooled by starting the ventilation system in exhaust-only mode. Wind and rain sensors on the outside are used as safety measures.

The night ventilation is activated from the beginning of April until the end of September and works similarly by automatically opening the windows based on the room temperature and its difference from the outside temperature (see also Schnieders

2009). The required air volumes in the night ventilation mode are considerably higher compared to the comfort ventilation mode. This requires a significantly larger free opening than conventional overflow solutions while maintaining the sound insulation level. The night ventilation is also supported by the ventilation system (see Fig. 4).

3.2 Ventilation

The ventilation concept has been based on the existing system as much as possible, which also contributed to the reduction of life cycle costs (Jäger et al. 2013). The outflow openings were moved from the corridors to the offices, which improved the air quality, especially in summer when the windows are closed. Supply air is routed from the main ducts in the suspended ceiling to the offices in the outdoor zones, where it is admitted via poppet valves in the center of the room. Through overflow valves in the door elements, the supply air continues into the corridor area and from there through overflow openings into the core zone (exhaust air zone). In this operating mode, window ventilation only takes place through manual user intervention.

The overflow areas have been specially designed and demonstrated within the refurbishment project. They improve the ventilation of the offices and allow the cooling of the core of the building and are supported by the extraction of air in the corridors. The overflow areas are integrated into the existing doors and have been extensively tested to achieve a low flow resistance while providing sufficient acoustic insulation (see Fig. 5). The ventilation system is equipped with rotary heat exchangers with a heat recovery of $\geq$75% and a moisture recovery of ca. 75%.

According to simulations, the cooling and ventilation of the building via motor-powered windows are significantly cheaper than using classic air conditioning. Using 40 years as a basis, the life cycle costs for ventilation and cooling of the building are calculated to be 60% lower using the presented innovative solution, even though initial investment has been about five times higher and maintenance costs are assumed to be ten times higher than using a standard air conditioning solution (Achammer et al. 2015).

3.3 Cooling

Another innovative element of the refurbishment project is the cooling using groundwater while focusing on reducing energy demand for cooling. The pre-cooling of the air for the ventilation system is achieved by using water from an existing well, which is an effective and low-cost method. Nevertheless, this option was never needed so far, as night ventilation by automatic window control was sufficient up to now. However, pre-cooling could also become necessary as a backup in Innsbruck in the future as global warming progresses.

3.4 Lighting

It has been shown by different authors, that the use of light control systems and adapted control can save about 50% of the energy demand for artificial light (Leslie et al. 2005; Roisin et al. 2008; Hernández 2012). This requires a distinction between control based on room-related sensor data and control based only on data from an outdoor sensor and simulation calculations for the respective room under specified boundary conditions. In the refurbishment of the Technical Sciences Building, a respective control approach was implemented and investigated through a living lab approach.

On the roof of the building, radiation sensors are aligned with the surface normal of the main facades. These sensors are used to maximize the use of daylight and automatically control the blinds; however, the automatic control can be overridden manually if required. The artificial lighting in the offices of the refurbished building is switched on manually. The process of switching on artificial lighting was deliberately not automated but left to the users, because it has been found in practice that this leads to significantly shorter switch-on times compared to sensor automation. However, the switch-off is automated with a time delay via IR presence sensors on the ceiling in case the user forgets to switch off the light. The dim factor of the luminaire row near the window is reduced by a constant 10% compared to the luminaire row further from the window to compensate for the reduced daylight input in the depth of the room. The daylight-dependent dimming of the artificial light takes place via look-down sensors on the ceiling and has been calibrated room by the room due to the different degrees of reflection (Malzer and Pfluger 2015).

4 Building Modernization Using a Living Lab Approach

The modernization was accompanied by specific research to support decisions and to better understand the impact of decisions made during planning and construction. The research encompassed conducting a life cycle cost analysis of different approaches to determine the most cost-effective approach over the life cycle of the building. This also helped to use the total costs, meaning net-rent and operating costs as a basis to show the advantages of reducing operating costs. Feasibility analyses have been conducted on the use of innovative and sustainable technologies to reduce perceived and actual risks on the sides of the planer and building owner. For example, the overflow areas which have been integrated into the existing doors have been intensively tested regarding their impact on the acoustic insulation of the doors as well as regarding their airflow resistance.

As mentioned, the main focus of the refurbishment has been on energy efficiency, with a predicted decrease in specific primary energy demand of ca. 75% (see Fig. 6). The predicted values could be achieved in most areas, for example, the specific demand for energy for heating decreased to less than 20 kWh/m^2 (16.58 kWh/m^2

in 2020 and 19.28 kWh/m^2 in 2021, measured values), which shows the significant impact of the refurbishment.

In addition to the decrease in energy demand for heating, the building nearly never needs energy for cooling. The innovative groundwater-based cooling system is not used and room temperatures usually stay below 27 °C with few peaks reaching up to 30 °C. The duration of temperatures in Fig. 7 shows that comfortable temperatures dominate and temperatures above the comfort range are rare. The refurbishment, therefore, did not only significantly increase the energy efficiency of the building, but also the comfort for the users.

The additional costs only for the energetic high performance, including the high insulation, could be kept at 3.8% compared to a "normal" refurbishment according to the owner. Energy efficiency in the refurbishment of buildings does not lead to much higher investment costs, these costs are compensated by reduced energy costs within a few years.

But not only energy demand and respective reductions were part of monitoring the building refurbishment. In the VisErgyControl research project, an integral, simulation-based daylight and artificial lighting control system has been developed and compared with the state of the art. The control approach, which was implemented in the renovated university building, is examined and evaluated by monitoring the switching states in addition to recording energy consumption, temperatures, and indoor air quality.

Even several years after the completion of the refurbishment project, research is still carried out while using the building or parts of it as a living lab. For example, the OptiFas project, which is conducted by the Unit for Energy Efficient Building at the Department of Structural Engineering and Material Sciences of the University of

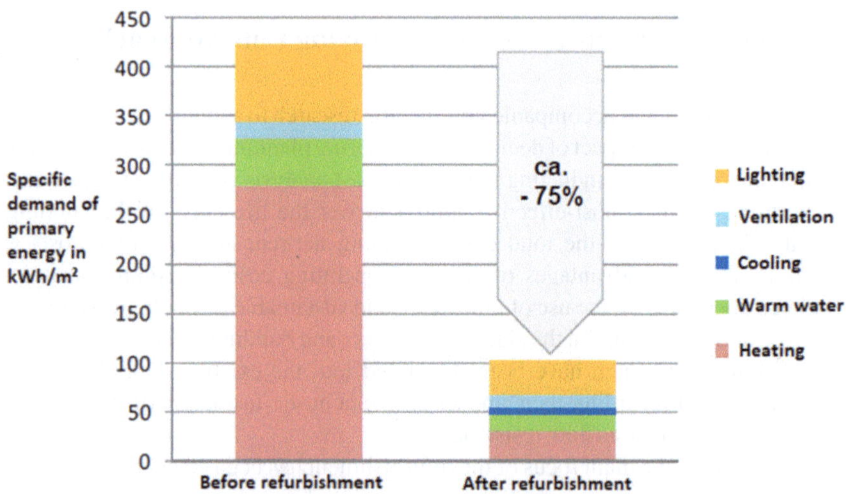

Fig. 6 Predicted changes in specific primary energy demand after the refurbishment (adapted from Jäger et al. 2015)

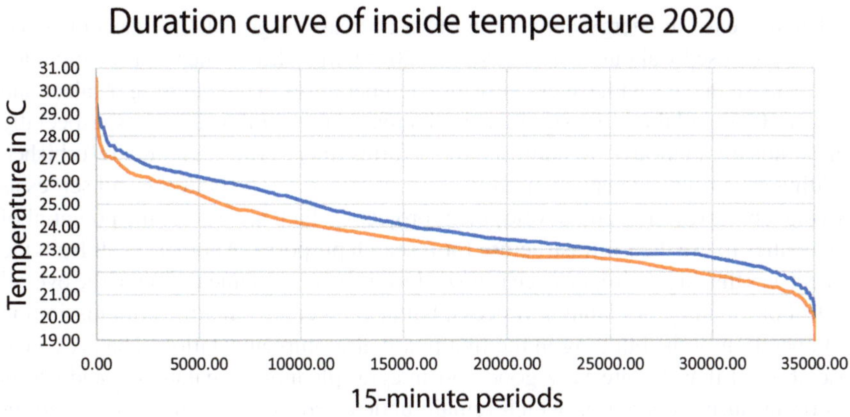

Fig. 7 15-min duration curve of temperature for two offices after refurbishment. The blue line shows the values for a south-west facing office, and the red line for an east-facing office

Innsbruck. The project aims the implementation of an innovative concept of specific daylight and artificial lighting control in a real office setting and to validate it in terms of efficiency and user acceptance by employing a long-term test with office users. For this purpose, a representative office room in the Faculty of Technical Sciences building will be equipped with an innovative control approach via the existing building management system. This includes individual control of the two louvers (height position and inclination of the louvers) as well as individual control of the individual rows of luminaires in the office (on/off and dimming function).

5 Conclusions

Overall, the refurbishment of the Technical Sciences Building of the University of Innsbruck, which has been presented in this article, can be seen as an example of massive improvements in energy efficiency through a renovation with very low additional investment costs (3.8%). There was no need to install an air conditioning system with high operating and maintenance costs since the night ventilation using approximately 300 automated windows (opening and sun protection) is controlled via the integrated building control system. The energy demand could be reduced by ca. 75% (see Fig. 6), with an especially significant reduction in the energy demand from heating and no heat demand for cooling. Even after reduction, the still very high internal heat loads due to IT, high person density, and lighting help in winter, but certainly put the concept of window night ventilation for summer to the test. However, the high internal heat loads also compensate for the relatively large ventilation losses due to the high density of people and the resulting large air volumes in return (Malzer and Pfluger 2015).

The decrease in energy demand led to a decrease in greenhouse gas emissions from energy use by about 80% (Jäger et al. 2015). In addition, the refurbishment led to a decrease in days with excessively high temperatures in the building (Malzer and Pfluger 2015) and therefore significantly increased the comfort for users, especially during hot summer days. In addition, the refurbishment project and the refurbished building have been extensively used to conduct research. The building still serves as a living lab for research and development projects, which underlines the potential of universities to use their infrastructure for research purposes. Another result from the project was that higher importance is now placed on the implementation of the overall control of the integral planning process, but also on controlling the commissioning and operation of the building in the first phase after implementation. The presented case showed that despite very good and integral planning, weaknesses and errors can occur during operation, which remain undetected in the absence of operation-management responsibility and can therefore lead to not achieving possible savings targets (Jäger et al. 2015).

After the refurbishment, the building has been certified by several institutions regarding its energy efficiency. It reached 917 of 1000 points in the TQB certification of the Austrian Sustainable Building Council, which is amongst the highest ratings ever achieved by a refurbishment project (Austrian Sustainable Business Council 2015). The building has also been certified according to the EnerPhit standard of the Passive House Institute, which has been developed for refurbishment and renovation projects (Passivehouse Database 2014). The building was the largest building that reached the EnerPhit certification at that time. In addition, the building reached the maximum amount of 1000 points in the Austrian klimaaktiv Standard (klimaaktiv 2017) and was awarded an energy globe (UIBK 2016).

Even though the refurbishment project has been highly successful, there are some potential limitations that should be kept in mind. For example, the relatively high expenditure for the temporary relocation of employees is seen as limiting the general reproducibility of this renovation. This is obviously the case with most refurbishment projects on this scale and would be highly difficult to avoid. Even when using extensive prefabrication as part of a so-called serial renovation (assembly of the prefabricated façade elements), which would allow for minimization of construction times, the extensive scope of work inside the building would still require a, potentially shorter, relocation.

In addition, the refurbishment project was part of a research project, which did not only help in covering the additional costs, but also by involving highly accomplished experts in the area of energy-efficient buildings during the planning and execution of the refurbishment. This was of course very helpful in determining optimal solutions and reducing operating costs.

Despite those limitations, the refurbishment project has served as an important example for the Austrian Federal Real Estate Agency. It demonstrates the possibilities of high-quality refurbishments in the area of energy efficiency and can aid in convincing stakeholders to invest in similar projects. It has also been proven that a very low additional investment can lead to significant savings over the building life cycle, even without taking the reduction in greenhouse gas emissions into account.

If greenhouse gas emissions were to be considered as suggested by authors such as Bookhart (2008), return on investment would increase significantly, and simple payback would be achieved even faster.

References

Achammer C, Mark M, Kelca R, Hessenberger K, Winkler C, Ohnmacht P, Stolz L, Neuhauser H, Oberhuber P, Gebhart K (2015) Refurbishment oft wo faculty buildings in Innsbruck. DETAIL green–Review of Sustainable Architecture and Energy-Efficient Refurbishment, 02, 15, pp 44–51

Amaral LP, Martins N (2013) Quest for a sustainable university: a review. Int J Sustaina Higher Educ 16(2):155–172. https://doi.org/10.1108/IJSHE-02-2013-0017

Amaral AR, Rodrigues E, Rodrigues Gaspar A, Gomes Á (2020) A review of empirical data of sustainability initiatives in university campus operations. J Clean Prod 250.https://doi.org/10.1016/j.jclepro.2019.119558

Austrian Sustainable Building Council (2015) Die Projekte 2015 [The 2015 projects]. https://www.oegnb.net/upload/file/2015_BauZ_Projekte.pdf

Bertoldi P, Economidou M, Palermo V, Boza-Kiss B, Todeschi V (2020) How to finance energy renovation of residential buildings: review of current and emerging financing instruments in the EU. Wires Energy Environ 10:1. https://doi.org/10.1002/wene.384

Bookhart D (2008) Strategies for carbon neutrality. Sustainability 1(1):34–40. https://doi.org/10.1089/SUS.2008.9991

Eskander SMSU, Nitschke J (2021) Energy use and CO2 emissions in the UK universities: an extended Kaya identity analysis. J Clean Prod 309:1. https://doi.org/10.1016/j.jclepro.2021.127199

Evans J, Jones R, Karvonen A, Millard L, Wendler J (2015) Living labs and co-production: university campuses as platforms for sustainability science. Curr Opin Environ Sustain 16:1–6. https://doi.org/10.1016/j.cosust.2015.06.005

European Network of Livin Labs (2023) What are living labs? What are Living Labs-European Network of Living Labs European Network of Living Labs (enoll.org). Accessed 7 Feb 2023

He Y, Liao N, Bi J, Guo L (2019) Investment decision-making optimization of energy efficiency retrofit measures in multiple buildings under financing budgetary restraint. J Clean Prod 215:1078–1094

Hernández JL (2012) Studie über den Einfluss eines Lichtsteuerungs- und Sonnenschutz-Systems auf den Energieverbrauch eines Bürogebäudes [Study on the influence of a lighting control and solar shading system on the energy consumption of an office building]

International Energy Agency (2020) Greenhouse gas emissions from energy: overview-emissions by sector. International energy agency report greenhouse gas emissions from energy. Emissions by sector–Greenhouse Gas Emissions from Energy: Overview–Analysis–IEA. Accessed 30 June 2022

IPCC (2018) Global warming of 1.5°C. An IPCC special report on the impacts of global warming of 1.5°C above pre-industrial levels and related global greenhouse gas emission pathways. In: Masson-Delmotte V, Zhai P, Pörtner H-O, Roberts D, Skea J, Shukla PR, Pirani A, Moufouma-Okia W, Péan C, Pidcock R, Connors C, Matthews JBR, Chen Y, Zhou X, Gomis MI, Lonnoy E, Maycock T, Tignor M, Waterfield T (eds) The context of strengthening the global response to the threat of climate change, sustainable development, and efforts to eradicate poverty

Jäger D, Hofer G, Leutgöb K, Grim M, Kuh C, Bucar G (2013) BIGMODERN Subprojekt 3: Demonstrationsgebäude Universität Innsbruck - Hauptgebäude der Fakultät für technische

Wissenschaften Planungsprozess [BIGMODERN Subproject 3: Demonstration building University of Innsbruck-Main building of the Faculty of Technical Sciences Planning process]. Nachhaltig Wirtschaften–Berichte aus Energie- und Umweltforschung, 15

Jäger D, Leutgöb K, Bucar G (2015) BIGMODERN Subprojekt 9: Demonstrationsgebäude Universität Innsbruck – Umsetzung [BIGMODERN Subproject 9: Demonstration building University of Innsbruck - Implementation]. Nachhaltig Wirtschaften – Berichte aus Energie- und Umweltforschung, 30

Kah O (2009) Wirtschaftlichkeit von Energiesparmaßnahmen bei der Gebäudemodernisierung, Protokollband Nr.39 des Arbeitskreises kostengünstige Passivhäuser - Phase IV; Passivhaus Institut; Darmstadt 2009 [Economic efficiency of energy saving measures in building modernization, Protocol Volume No. 39 of the Working group on cost-effective passive houses-phase iV; Passive House Institute; Darmstadt 2009]

klimaaktiv (2017) UNI IBK Bauingenieurgebäude, Sanierung Fakultät für Bauingenieurwissenschaften der LFU Innsbruck. Database for certified non-residential buildings [UNI IBK Civil Engineering Building, Renovation Faculty of Civil Engineering of the LFU Innsbruck. Database for certified non-residential buildings]. https://www.klimaaktiv-gebaut.at/gebaut/objekte/klimaaktiv/dienstleistung/uni-ibk-bauingenieurgebaude-sanierung-fakultat-fur-bauingenieurwissenschaften-der-lfu-innsbruck/. Accessed 1 July 2022

Leslie RP, Raghavan R, Howlett O, Eaton C (2005) The potential of simplified concepts for daylight harvesting. Light Res Technol 37(1):21–40

Malzer H, Pfluger R (2015) Erstes EnerPHit-zertifiziertes Bürohochhaus- Modernisierung mit Passivhaus-Komponenten [First EnerPHit-certified high-rise office building-modernization with passive house components]. In: Feist, Wolfgang: 19. Passivhauskonferenz [Passive House Conference]. 17–18. April 2015, Leipzig. Darmstadt: Passivhaus Institut Darmstadt., ISBN 978–3–00–048604–3 , S. 325–332

Passivehouse Database (2014) Fakultät für Technische Wissenschaften Universität Innsbruck. Database for EnerPhit certified buildings [Faculty of Technical Sciences University of Innsbruck. Database for EnerPhit certified buildings]. https://passivehouse-database.org/#d_4200. Accessed 1 July 2022

Roisin B, Bodart M, Deneyer A, Dherdt P (2008) Lighting energy savings in offices using different control systems and their real consumption. Energy Build 40(4):514–523

Schnieders J (2003) Der Einfluss verschiedener Lüftungsstrategien auf das Sommerklima – vergleichende Untersuchung mittels dynamischer Gebäudesimulation, Protokollband Nr.22 des Arbeitskreises kostengünstige Passivhäuser - Phase III; Passivhaus Institut; Darmstadt 2003 [The influence of different ventilation strategies on the summer climate-comparative study using dynamic building simulation, Protocol Volume No. 22 of the Working Group on Cost-effective Passive Houses-Phase III; Passive House Institute; Darmstadt 2003]

Schnieders J (2009) Nachtlüftung mit natürlichem Auftrieb, Protokollband Nr.15 des Arbeitskreises kostengünstige Passivhäuser - Phase II; Passivhaus Institut; 4. Auflage Darmstadt 2009 [Night ventilation with natural buoyancy, Protocol Volume No. 15 of the Working Group on Cost-effective Passive Houses - Phase II; Passive House Institute; 4th edition Darmstadt 2009]

UIBK (2016) Technik-Gebäude erneut ausgezeichnet [Technology building honored again]. https://www.uibk.ac.at/de/newsroom/2016/technik-gebaeude-erneut-ausgezeichnet/. looked up 21st Oct 2022

UNFCCC (2015) The Paris Agreement. FCCC/CP/2015/10/Add.1

Lorenzo Rieg is the Sustainability Coordinator of the University of Innsbruck and has been working on sustainability and climate change matters in research and practice.

Bertram Knoflach is Project Manager at the Austrian Federal Property Association and is not only working on maintaining properties but also focuses on improving their energy efficiency.

Hanspeter Mühlegger is Architect and Asset Manager at the Austrian Federal Property Association,with a long-standing experience in managing real estate.

Rainer Pfluger is associate Professor at the Unit for Energy Efficient Building at the University of Innsbruck. He is an expert for energy efficient buildings.

Wolfgang Streicher Wolfgang Streicher is Professor for Energy Efficient Building with Focus on Renewable Energy at the University of Innsbruck. He is head of the Unit for Energy Efficient Buildings, and therefore has not only long-standing scientific expertise in the area of buildings and their energy-efficiency, but is also dealing with the management of the university buildings on a daily basis.

An Exploration of Early Childhood and Primary School Pre-service Teachers' Knowledge of the SDGs: Proposals for Its Implementation in Initial Teacher Training

María Calero, José Cantó, Olga Mayoral, Tatiana Pina, M. Àngels Ull, and Amparo Vilches

Abstract This research focuses on the Sustainable Development Goals (SDG) knowledge of teachers in initial training in Early Childhood Education and Primary Education. It also includes teaching–learning proposals, to contribute to overcoming the shortcomings detected, as well as the first results of their implementation.

Keywords Education for sustainable development (ESD) · Sustainable development goals (SDGs) · Sustainability science (SS) · Teacher training · Pre-service teachers

---

M. Calero (✉) · J. Cantó · O. Mayoral · T. Pina · A. Vilches
Department of Experimental and Social Science Education. Faculty of Teacher Training, University of Valencia, Valencia, Spain
e-mail: Maria.Calero@uv.es

J. Cantó
e-mail: Jose.Canto@uv.es

O. Mayoral
e-mail: Olga.Mayoral@uv.es

T. Pina
e-mail: Tatiana.Pina@uv.es

A. Vilches
e-mail: Amparo.Vilches@uv.es

O. Mayoral
Botanical Garden, University of Valencia, Valencia, Spain

M. Àngels Ull
ERI Sustainability Studies. University of Valencia, Valencia, Spain
e-mail: Angels.Ull@uv.es

W. Leal Filho et al. (eds.), *The Contribution of Universities Towards Education for Sustainable Development*, World Sustainability Series,
https://doi.org/10.1007/978-3-031-49853-4_19

1 Introduction

The early years of this century saw the emergence of a new scientific field: Sustainability Science (SS), announced in 2001 by twenty-three researchers from different fields in an article published in the journal Science: *A new field of sustainability science is emerging that seeks to understand the fundamental character of interactions between nature and society* (Kates et al. 2001, p. 641). Another article published in Proceedings of the National Academy of Sciences specified: *The emergence of "sustainability science" builds toward an understanding of the human-environment condition with the dual objectives of meeting the needs of society while sustaining the life support systems of the planet* (Turner et al. 2003, p. 8074). Sustainability Science has arisen in response to the understanding of the gravity of the interconnected and mutually reinforcing problems that humanity must face, which cannot be addressed separately.

No single problem—be it resource depletion, climate change, extreme poverty, or the energy crisis—can be solved without taking into account its linkages with the others. This leads to an awareness of the need to tackle globally the increasingly complex system of human societies and the natural systems with which they interact and of which they are a part. *The problems that sustainability science confronts are not only complex but also interconnected. If we are to find solutions to them, we must first clarify the relationships among them* (Komiyama and Takeuchi 2006, p. 4).

In addition to its interdisciplinary nature, SS is also transdisciplinary by promoting citizen participation from the origins of the studies, breaking the isolation of the academic world, and multiplying the effectiveness of collaborative work. Its strategies must respond to a broad, holistic vision, both spatially and temporally. However, integrating social development with the processes of the so-called natural world, to understand the interactions between nature and society, cannot be achieved only with the creation of a new area, but must constitute a new paradigm, a new orientation that must permeate all fields of knowledge and human activities. The work of professionals in any area and of the whole of society, as well as the teaching of the different disciplines and citizenship education in general, must be guided by the requirements for inter- and transdisciplinarity and *global* approaches in a broad temporal perspective (Vilches and Gil-Pérez 2015).

At the same time, since the beginning of the twenty-first century, different events have been taking place that point out the way towards favouring and promoting attention to Sustainability and Education for Sustainability (ESD). These include, among others, the institution by the United Nations of the Decade of Education for Sustainable Development (DESD 2005–2014) and, more recently, the approval in 2015, after a process of broad participation, of the Sustainable Development Goals (SDGs) and the 2030 Agenda for their implementation. The 17 major universal goals proposed, with 169 specific targets, aim to respond to the set of interconnected problems that constitute the socio-environmental crisis we are facing, ensuring there is no omission of any of them that could prevent real progress in the Agenda as a whole. The convergence of the SDGs with the characteristics of SS can be clearly seen here. We are facing a confluence of several crises, comprising a global crisis that threatens the survival of humanity; and all these crises, and the ways to prevent and overcome them, are comprehensively addressed in the SDGs.

The broad consensus reached and the way in which SDGs were developed have been decisive for the recognition of their urgency and the involvement of different social, academic, economic, political and citizen groups, entailing specific actions in all areas of our lives. They are not proposed as ideal goals to which we aspire, but as urgent obligations to which all people in all spheres must contribute. This formulation highlights the key role that education must play directly through SDG 4 and in the achievement of the other SDGs for the necessary transition to Sustainability (Stevens and Kanie 2016; Leal Filho 2020).

Convergently, there have been numerous proposals, in particular from the DESD, to contribute to the advancement of ESD, especially in universities. In this regard, we can point out the importance of incorporating competences for Sustainability into different programmes, as well as the numerous proposals for ESD in different areas and countries and, more recently, actions in the field of the SDGs to advance citizen education and their active involvement in the transition to Sustainability (Aznar et al. 2011; Wiek et al. 2011; Rieckmann 2012; Albareda et al. 2019; Calero et al. 2019; Chuliá et al. 2022).

This is precisely the central objective of the study we present, which aims to contribute to the training of future Early Childhood Education and Primary School Education teachers for the transition to Sustainability. To engage teachers in training on Sustainability and the SDGs, this paper is divided into three sections following an introduction on the importance of ESD. In the first section, we explore the pre-service teachers' conceptions about SDGs. Based on the findings, we design and implement teaching strategies about the SDGs and their importance, which are assessed in two subjects: *Science Teaching: Matter, Energy and Machines* and *Multidisciplinary Workshop on the Physical, Natural, Social and Cultural Environment*. Finally, we summarize the main conclusions and limitations of this chapter.

2 Knowledge of the SDGs Among Pre-service Teachers. Results and Implications

Various studies have pointed out that despite progress in addressing Sustainability and ESD in education in general, and particularly in Higher Education, it remains insufficient, especially considering the worsening of socio-environmental issues in recent decades (Aznar et al. 2018; Leal Filho et al. 2018; Cheeseman et al. 2019). Moreover, the latest data and estimates reveal that the 2030 Agenda is in serious jeopardy due to the various interrelated crises such as pandemics, climate change and conflicts. Each of these problems and their complex interactions impact on all the SDGs and generate serious consequences at different levels.

This leads us to expect that, although there is a clear involvement of political and academic institutions in the different countries, and in particular in Europe, we would still find similar problems with regard to the insufficient attention paid to the SDGs in different university degrees, especially in initial teacher training. This is particularly relevant considering teacher´s influence and the multiplier effect they have on the training of citizens in Sustainability from the very beginning of education.

This highlights the need to increase efforts as education for the SDGs (ESDGs) must contribute to providing all people, in all areas and at all levels, with the knowledge, skills and values needed to address the challenges posed by the SDGs and contribute to the transformations required in our societies to move towards more sustainable societies.

This is especially important considering the scope of the role of universities in different SDGs, such as those related to education and research. This is evident in SDG 4: *Ensure inclusive and equitable quality education and promote lifelong learning opportunities for all* and when referring to ESD in its Target 4.7: *By 2030, ensure that all learners acquire the knowledge and skills necessary to promote sustainable development, including through education for sustainable development and sustainable lifestyles, human rights, gender equality, promotion of a culture of peace and non-violence, global citizenship and appreciation of cultural diversity and the contribution of culture to sustainable development* (UN 2015, p. 17). This is a goal that, as we have already mentioned, will contribute to advancing the achievement of the SGDs as a whole.

In order to test our hypothesis about the lack of preparation acquired by pre-service teachers of Early Childhood Education and Primary School Education on the SDGs and the need for interventions and didactic training proposals in ESDGs, a comprehensive study has been carried out in these degrees at the University of Valencia over three academic years.

In this chapter we first analyse the knowledge of the SDGs among future Early Childhood Education and Primary School Education teachers, the results of which will guide the design of the evidence-based intervention proposals that we develop in the following section.

2.1 Methodology

The study was conducted during the academic years 2019–20, 2020–21, 2021–22 in the Degrees in Early Childhood Education and Primary School Education at the Faculty of Teacher Training of the University of Valencia (UV) (in the UV, the Early Childhood Education stage, from 0 to 6 years of age, is officially called Preschool).

To determine their basic knowledge about the SDGs and the origin of this knowledge, we designed a questionnaire with three open-ended questions and one multiple-choice question (Table 1) to be answered by future teachers. The sample consisted of a total of 853 UV students, 144 of whom were enrolled in the Degree in Early

Table 1 Questionnaire on the knowledge of the SDGs used with students of the Degrees in Early Childhood Education and in Primary School Education

Question	Statement/formulation
Q1	Do you know what the SDGs are? Explain them briefly
Q2	If you do so, please specify how many there are. If you don´t, write "I don't know"
Q3	Describe what you remember of their contents
Q4	Where have you heard about them? Select all the options that apply and write a comment: (a) At school (b) On TV (c) At home (d) Elsewhere (specify where)

Childhood Education (DECE) and 709 in the Degree in Primary School Education (DPSE).

The questionnaire was used in the first session of compulsory subjects related to the scientific education of future teachers: *Natural Sciences for Teachers* (9 European Credit Transfer System, ECTS); *Teaching Natural Sciences at Nursery School* (6 ECTS); *Science Teaching: Matter, Energy and Machines* (4.5 ECTS) and *Teaching Science: Environment, Biodiversity and Health* (4.5 ECTS).

2.2 Results

Throughout the period analysed, the results of the responses of the 853 participants indicate that:

- In relation to the first question, 95.8% of the DECE students and 83.4% of the DPSE students show a high lack of knowledge or very insufficient knowledge about the SDGs. Specifically, 79.2% of the DECE students and 53.0% of the DPSE students did not answer the question or gave an incorrect answer. No significant differences are observed, in general, throughout the period, although a slight improvement is perceived in the last year in the case of DPSE.
- Regarding the second question on how many SDGs there are, the highest percentage corresponds to "I don't know/not answer" responses, in coherence with the previous answer, with 88.9% of the DECE students and 73.2% of the DPSE students not answering.
- Regarding the content of the SDGs referred to in the third question, the result is really low in both degrees, with SDG 13 on climate action being the most cited. The average values obtained in terms of the number of SDGs to which they refer are really low in all cases (less than one in the case of DECE students–0.68–and less than two in the case of DPSE students–1.23 -), which, together with the other results, justifies the need to raise awareness of the SDGs and to address them in detail.
- In the last question, which aims to find out where their knowledge about the SDGs comes from, 47.1% of DPSE students cited the educational centre as the main

source of information about the SDGs, while in the case of the DECE students, the percentage does not reach 22%.

We can state that, in general, most of the pre-service teachers in Early Childhood and Primary School Education participating in the study have little knowledge of the SDGs and their meaning, with the lack of knowledge among future Early Childhood Education teachers being particularly relevant. The last question also reveals that few of them consider that the SDGs are addressed in the studies they carry out.

In conclusion, these results confirm our hypotheses and converge with those of other studies carried out at other educational levels in our country, as shown in Master's Theses and different research in the field of Secondary Education (Ochando et al. 2021).

And, above all, these results point to the need to strengthen the focus on Sustainability and the SDGs within ESD in Higher Education and, in particular, to continue contributing more deeply to the incorporation of the SDGs in teacher training, promoting specific actions and intervention proposals to promote the implementation of the ESDG in initial teacher training. It is necessary to make future teachers raise awareness of the seriousness of the global crisis we are facing and to contribute to their involvement in the necessary measures to advance in the construction of fairer and more sustainable societies. This will also encourage their commitment to contribute in their professional future to the formation of a responsible citizenry prepared to make decisions and implement the actions required by the global crisis.

3 Proposals for Addressing the SDGs Within the Framework of ESD

One of the main difficulties of policies promoted by universities is that they are not effectively engaging teachers, particularly in the need to effectively train students in competences for Sustainability (Collazo and Geli 2018; Calero et al. 2019). It is necessary to recognise the existence of a gap between what is proposed by institutions and what is put into practice in the classrooms. There is also a mismatch between the attitudes, knowledge, and skills that are generally provided through Higher Education and those required to address the SDGs, which are considered crucial for progress in the transition to Sustainability (UNESCO 2017).

We have already referred to the growing number of Higher Education institutions that are concerned with developing educational processes to contribute to a more sustainable future. And we have also pointed out studies that suggest that the curricular changes made so far have been insufficient and, above all, have not reached the classroom, i.e., are not contributing to citizen engagement (Aznar et al. 2011; Albareda et al. 2019). This is something we have noted in the previous section regarding the knowledge and attention to the SDGs in teacher training.

It is therefore necessary to reorient teacher training in the framework of ESDG through the design of activities and interventions that foster their involvement in the promotion of a responsible citizenship regarding social and environmental problems. To address the SDGs, ESDG requires methodologies and a set of transformative teaching and learning practices that are not common practice in most universities.

It is essential, among other things, to promote action-oriented learning and citizen participation involving multiple stakeholders. The interconnectedness of the SDGs implies that successfully addressing a particular goal requires understanding and managing the interactions and consequences for other goals, that is to say, it requires a polyhedral approach (Montero-Pau et al. 2020). Sustainability must be a cross-cutting dimension that permeates the curriculum in teacher training (Vilches and Gil-Pérez 2012), addressing the problem of the current socio-environmental crisis through a process of research and immersion in the culture of Sustainability that can be applied with their students in their future teaching practice.

Therefore, with a view to promoting the achievement of the SDGs and being aware of the limited knowledge of university students about them, we wondered how to promote their inclusion, from a cross-cutting approach, in the initial training of Early Childhood and Primary School Education teachers. In other words, what intervention proposals could be implemented with pre-service teachers to encourage their involvement in the actions needed to achieve the SDGs and facilitate the incorporation of ESDG into their future teaching practice. Taking into account previous research cited above, we believe that the participation of pre-service teachers in a didactic intervention proposal in which the SDGs are incorporated under a cross-cutting approach contributes to facilitating the treatment of competences for Sustainability in their future work as teachers.

To this end, we present below some examples and first results of the cross-cutting incorporation of the SDGs in the Faculty of Teacher Training of the UV, within the framework of an extensive research on Education for Sustainability.

3.1 Methodology

During the 2020–21 academic year, taking into account the initial results of the study on students' knowledge of the SDGs (Sect. 2. Knowledge of the SDGs among pre-service teachers), we designed and implemented didactic proposals in different subjects, among others, *Science Teaching: Matter, Energy and Machines* in the third year of the DPSE and *Multidisciplinary Workshop on the Physical, Natural, Social and Cultural Environment* in the fourth year of the DECE. Table 2 shows some data on the sample used.

In the teaching and learning process in both subjects, the methodology used during the course is designed to involve pre-service teachers in the study of the problems and the construction of viable solutions posed by science education, based on an abundance of literature. For this reason, the participation of future teachers is encouraged in the construction of knowledge, which can then be compared with the proposals of the scientific community (Fraser et al. 2012).

Table 2 Characteristics of the sample in which the didactic proposals were implemented

Degree	Subject	Year course	Description (ECTS)**	Participants
Degree in primary school education	Science teaching: matter, energy and machines	3rd	Compulsory subject quaterly (4.5 ECTS)	86
Degree in early childhood education	Multidisciplinary workshop on the physical, natural, social and cultural environment	4th	Optative quaterly (6 ECTS)	25

**ECTS: European Credit Transfer System

The main purpose is to analyse, through collective reflection, key issues in the process of teaching and learning science, with well-founded proposals to contribute to improving science education from the outset. These proposals address a set of knowledge and competences that should contribute to providing a rich and complex vision of the teaching activity that calls for ongoing research and innovation. The aim is to guide teacher training as a collective work of innovation, enquiry, and lifelong learning, i.e., as a process of immersion in scientific culture such as the one they are intended to use later with their students. And in this immersion in scientific culture, among these key issues, attention to Education for Sustainability and the SDGs, in the framework of Sustainability Science, as we propose, play an essential role.

Thus, after an initial reflection on the first topics, a transversal treatment is conducted, permeating the rest of the syllabus with the measures that need to be adopted, within the framework of the ESD. The aim is to encourage student involvement through reflection on the SDGs, showing their multiple interactions and universal nature. This contributes to the acquisition of a global vision that highlights the close links between the problems (which include the climate, economic, food, energy, health, etc. crises, as well as the conflicts that affect these crises) by proposing actions from different areas, and in particular from education. By way of example, we will briefly comment below on the guiding line of the interventions carried out in two subjects and the assessment of the participants.

3.2 Proposal for the Subject: *Science Teaching: Matter, Energy and Machines*

The intervention implemented in this third-year subject of the DPSE begins with an initial activity on the SDGs in Lesson 1, "Science Education as an essential component of citizenship education", which aims to address the need for scientific literacy for all people. Initially, students are asked "To what extent can science and technology literacy help to enable citizens to participate in decision-making on the challenges facing humanity?".

The analysis of some examples, which highlight the necessity and possibility of citizens participation in decision-making, easily leads to the conclusion that this need is becoming increasingly urgent in view of the global crisis we are currently experiencing. Subsequently, reference is made to the numerous calls from worldwide institutions (UN Conferences, Earth Summits, etc.) for education to help citizens to acquire a correct perception of the serious problems and the necessary measures.

After reflecting on the need to consider these measures collectively, the students, divided into small groups, are asked to consult the information about the 2030 Agenda and the SDGs as well as the history and origin of the 17 SDGs and their targets, on the website dedicated to them by the United Nations and UNESCO. Later, each group focuses on the study of one SDG, analysing what it aims to achieve and proposing some ideas of what it could involve in our context (region, country, etc.). Each SDG is then presented and discussed by the groups to the whole class.

The aim of this activity is to initiate the reflection on the SDGs, to introduce pre-service teachers to what they involve, by analysing their initial conceptions and studying whether they link each SDG with others, whether they highlight the local or global dimension or both, whether they relate them to the impact of the pandemic, whether they propose actions based on education, etc. Framed in the relevance of SDG 4 for their work, this is a first step that helps to understand where to work next, but it is insufficient to fully comprehend the meaning and importance of the SDGs. Therefore, it is necessary to address them more deeply, with a cross-cutting treatment that is conducted in the rest of the course, incorporating them into the different topics.

Thus, in Lesson 2, "Teaching and learning science. What to do in Primary School Education?", which addresses how to teach science in this level, future teachers are encouraged to incorporate activities and links to the different SDGs in their proposals, and in specific cases to those related to central issues in the Primary Education curriculum, such as the study of energy, responsible consumption, and climate action (SDGs 7, 12, and 13, respectively), while recognising their close relations with the rest of the SDGs.

Lesson 3, "First steps in the acquisition of technoscientific culture", dedicated to early tools and machines, deals with the importance and evolution of technology, and the final part of the lesson focuses on addressing where to direct technology to contribute to Sustainability. Thus, reference is again made to the SDGs (in particular SDGs 8, 9, 10, 11, 14, and 15) and their close links.

The focus on the SDGs is reinforced in the recapitulation carried out in Lesson 4, "Didactic analysis of school science content in Primary School education", by studying materials from different levels of Primary School education, analysing the occasions when different SDGs can be incorporated and proposing examples of specific activities.

3.3 Proposal for the Subject: *Multidisciplinary Workshop on the Physical, Natural, Social and Cultural Environment*

Another example of the transversal incorporation of the SDGs implemented by our research group is carried out in the subject *Multidisciplinary Workshop on the Physical, Natural, Social and Cultural Environment* in the fourth year of the DECE.

Initially, a session of Lesson 1 is devoted to reflecting with students on the global problems that humanity must face, and the SDGs are introduced through the websites dedicated to them by the UN or UNESCO. Later, in Lesson 2, "Analysis, research, and evaluation of curricular materials related to the physical, natural, social, and cultural environment", students are asked to analyse different materials in order to establish a possible link between the topics presented and the different SDGs, and to what extent they could incorporate ESD in Early Childhood Education.

This activity allows students to become aware of how the SDGs could be addressed in Early Childhood Education, despite the fact that the materials used may not explicitly refer to them. A template such as the one shown in Fig. 1 is used for the analysis.

And finally, in Lesson 3, "Design and development of curricular materials related to the Physical, Natural, Social, and Cultural Environment area", students are asked

Fig. 1 Activity linking curricular materials to the SDGs in Lesson 2 of the subject *Multidisciplinary Workshop on the Physical, Natural, Social and Cultural Environment*. (*Source* Own development based on the United Nations website (UN 2015). Translated from Spanish)

The SDGs are a list of humanity's pending tasks for a sustainable planet, a clear roadmap for a better future. They concern us all, as we are all responsible for being part of the change, which is why it is essential to address them in the classroom from the earliest levels of education.

We propose the design of an educational proposal to raise awareness of the SDGs in Early Childhood Education, addressing the following questions:

- At what level of Early Childhood Education could it be carried out?
- With which objectives of the curriculum in the area of The physical, natural, social and cultural environment could it be related?
- What curricular content could it be linked to?
- What would be its timing?
- Briefly describe the proposed activity
- Would it be convenient to make any adaptation? Could it be complemented with any other activity, resource, workshop…?
- How could it be evaluated?
- Indicate the bibliography consulted

Fig. 2 Design of a teaching proposal for the treatment of the SDGs in Early Childhood Education in Lesson 3 of the subject *Multidisciplinary Workshop on the Physical, Natural, Social, and Cultural Environment* (*Source* Own development. Translated from Spanish)

to design a teaching proposal to introduce the SDGs, following the instructions shown in Fig. 2.

3.4 Results

In the case of the subject *Science Teaching: Matter, Energy, and Machines*, the results of the collective work developed by the students on a specific SDG allow us to confirm the initial hypotheses, to detect which aspects have been highlighted, whether they have *contextualised* it at a global or local level, whether they link it to other SDGs and/or to the pandemic, whether they make proposals, etc. We can highlight, among others, the results shown in Table 3.

Table 3 Results in the subject *Science Teaching: Matter, Energy and Machines*

Category of analysis (N = 86)	% students (standard deviation)
Summarises the central idea	94.2 (2.5)
Contextualises it at a global level (reference to a planetary problem, it always appears in the SDGs)	76.7 (4.5)
Contextualises it at a local level (reference to what is happening in Spain)	65.1 (5.1)
Contextualises it at a *glocal* level (both global and local; in general, how the global part appears in the SDGs)	58.1 (5.3)
Links the SDGs to some (other SDGs without explicitly mentioning them)	16.3 (3.9)
Shows holistic view of the SDGs (own content of the SDG)	2.3 (1.6)
Refers to the pandemic (connection to nearby and affecting problems)	40.7 (5.3)
Proposes educational measures in general	30.2 (4.9)
Proposes more explicit educational measures with activities for Primary School Education	18.0 (4.1)

These results allowed us to focus and reinforce the study of the SDGs throughout the course, paying particular attention to the acquisition of a global vision of the problems and the links between the SDGs. It is noteworthy that in the final evaluation, most participants (80%) cited the activities carried out on the SDGs throughout the course as among the activities they value most positively.

In the case of the subject *Multidisciplinary Workshop on the Physical, Natural, Social, and Cultural Environment*, at the end of the course, students were asked to rate from 0 to 10 the extent to which they considered that the activities carried out could help them to work on ESD with Early Childhood Education students and/or what other instruments or resources they would need. Some of the results obtained are shown in Table 4.

Table 4 Results in the subject *Multidisciplinary Workshop on the Physical, Natural, Social and Cultural Environment*

Evaluation of didactic proposal (N = 25)	Mean value (Standard deviation)
Please rate to what extent the work done on the SDGs helped you to better understand the planetary emergency we are facing and, above all, the need and possibility of progressing towards Sustainability	8.8 (1.4)
Assess the interest of the activities carried out	8.4 (1.6)
Please rate the extent to which you think this work can enhance your commitment to the challenge of achieving the SDGs	8.5 (1.4)
Assess whether you consider that the activities carried out can help you to work on Education for Sustainability in Early Childhood Education	8.8 (1.3)

4 Conclusions

In the first part of the study, focused on the analysis of the knowledge about the SDGs over three academic years, the results show that the majority of the 853 participants, teachers in initial training for Early Childhood and Primary School Education, have limited knowledge of the SDGs and their meaning, with a significant lack of knowledge among future Early Childhood Education teachers.

The initial results obtained point to the need to reinforce attention to Sustainability and the SDGs within ESD in Higher Education and, therefore, the urgency of promoting specific actions and intervention proposals to promote the implementation of ESDGs in initial teacher training.

To this end, proposals for the mainstreaming of the SDGs have been designed, implemented and evaluated, two examples of which are shown in this chapter. It should be noted, as a limitation of this work, that it is a study carried out in the Degrees in Early Childhood Education and Primary Education at the University of Valencia—as pointed out in the previous sections—, and that in the case of the Degree in Primary Education, the didactic proposal was implemented with the group of Major in Science and Mathematics Specialist.

The results of the activities carried out on the SDGs indicate that they have contributed to the acquisition of a global vision of the current socio-environmental problems among pre-service teachers. They were able to identify the close links between the problems, proposing actions from different areas, and in particular from education, in convergence with our initial hypotheses and with other research carried out by our research group and other authors who have conducted similar studies (Murga-Menoyo and Novo 2015; Aznar et al. 2018; Collazo and Geli 2018). The proposals implemented have helped future teachers to better understand the meaning and importance of the SDGs as a new global framework for redirecting humanity towards building fairer and sustainable societies (UNESCO 2017).

The set of results supports the initial hypothesis about the need and possibility of implementing well-founded intervention proposals that promote the reflection of future Early Childhood and Primary School Education teachers and the (re)construction of a global vision of the planetary emergency situation and the possible solutions related to the achievement of the SDGs, so that pre-service teachers acquire an adequate perception of the problems addressed, as well as favourable attitudes and behaviours in their future work.

The serious situation we are facing requires Sustainability to be a cross-cutting dimension that permeates the curriculum in the training of future teachers, favouring a process of research and immersion in the culture of ESD and the SDGs; in this way, we will contribute to empower pre-service teachers as essential agents of change in the transition towards Sustainability.

Acknowledgements This research was partially funded by Climate-KIC, a Knowledge and Innovation Community, supported by the EIT, a body of the European Union through the Young Innovators Project 2019—2022 (EIT_4.1.2_220147_P165_1B) and by the Spanish Ministry through the project "Study on the teaching of science in Early Childhood and Primary School Education. Improvement proposals (formaprofe3)—PID2019-105320RB-I00".

References

Albareda S, Azcárate P, Muñoz JM, Valderrama R, Ruiz J (2019) Evaluar competencias en sostenibilidad en los grados y posgrados de educación: propuesta de un instrumento [Assessing sustainability competencies in undergraduate and graduate education: proposal of an instrument]. Ensen Cienc 37(3):11–29. https://doi.org/10.5565/rev/ensciencias.2670

Aznar P, Martinez MP, Palacios B, Piñero A, Ull A (2011) Introducing sustainability into university curricula: an indicator and baseline survey of the views of university teachers at the University of Valencia. Environ Educ Res 17(2):145–166. https://doi.org/10.1080/13504622.2010.502590

Aznar P, Calero M, Martínez MP, Mayoral O, Ull A, Vázquez V, Vilches A (2018) Training secondary education teachers through the prism of sustainability: the case of the Universitat de València. Sustainability 10(11):4170. https://doi.org/10.3390/su10114170

Calero M, Mayoral O, Ull A, Vilches A (2019) La educación para la sostenibilidad en la formación del profesorado de ciencias experimentales en Secundaria [Education for sustainability in the training of experimental science teachers at the secondary level]. Ensen Cienc 37(1):157–175. https://doi.org/10.5565/rev/ensciencias.2605

Cheeseman A, Wright TSA, Murray J, McKenzie M (2019) Taking stock of sustainability in higher education: a review of the policy literature. Environ Educ Res 25(12):1697–1712. https://doi.org/10.1080/13504622.2019.1616164

Chuliá R, Vilches A, Calero M (2022) The press as a resource for promoting sustainability competencies in teacher training: the case of SDG 7. Sustainability 14(2):857. https://doi.org/10.3390/su14020857

Collazo LM, Geli AM (2018) Avanzar en la educación para la sostenibilidad. Combinación de metodologías para trabajar el pensamiento crítico y autónomo, la reflexión y la capacidad de transformación del sistema [Advancing education for sustainability. Combination of methodologies to work on critical and autonomous thinking, reflection and the ability to transform the system]. Revista Iberoamericana de Educación (RIE) 73:131–154. https://doi.org/10.35362/rie730295

Fraser BJ, Tobin K, McRobbie CJ (2012) Second international handbook of science education. Springer

Kates RW, Clark WC, Corell R, Hall JM, Jaeger CC, Lowe I, Mccarthy JJ, Schellnhuber HJ, Bolin B, Dickson NM, Faucheux S, Gallopin GC, Grübler A, Huntley B, Jäger J, Jodha NS, Kasperson RE, Mabogunje A, Matson P, Mooney H, Moore BI, O'Riordan T, Svedin U (2001) Sustainability science. Science 27 April 2001 292(5517):641–642. https://doi.org/10.1126/science.1059386

Komiyama H, Takeuchi K (2006) Sustainability science: building a new discipline sustainability. Science 1(1):1–6. https://doi.org/10.1007/s11625-006-0007-4

Leal Filho W (2020) Viewpoint: accelerating the implementation of the SDGs. Int J Sustain High Educ 21(3):507–511. https://doi.org/10.1108/IJSHE-01-2020-0011

Leal Filho W, Brandli L, Becker D, Skanavis C, Kounani A, Sardi C, Papaioannidou D, Paço A, Azeiteiro U, de Sousa L, Raath S, Pretorius R, Shiel C, Vargas V, Trencher G, Marans R (2018) Sustainable development policies as indicators and pre-conditions for sustainability efforts at universities: fact or fiction? Int J Sustain High Educ 19(1):85–113. https://doi.org/10.1108/IJSHE-01-2017-0002

Montero-Pau J, Álvaro N, Gavidia V, Mayoral O (2020) Development of environmental health competencies through compulsory education. A polyhedral approach based on the SDGs. Sustainability 12(8):3215. https://doi.org/10.3390/su12083215

Murga-Menoyo MA, Novo M (2015) The processes of integrating sustainability in higher education curricula: a theoretical-practical experience regarding key competences and their cross-curricular incorporation into degree courses. In: Leal Filho W (ed) Transformative approaches to sustainable development at Universities, pp 119–135. Springer. https://doi.org/10.1007/978-3-319-08837-2

Ochando J, Vilches A, García-Ferrandis I (2021) Conocimiento de los objetivos de Desarrollo Sostenible y su importancia para la transición a la sostenibilidad por alumnado de Educación Secundaria Obligatoria [Knowledge of the Sustainable Development Goals and their importance for the transition to sustainability by students in Compulsory Secondary Education]. In: Lucena H et al (eds) Innovación e investigación educativa para la formación docente [Educational innovation and research for teacher training]. Dykinson, pp 688–701

Rieckmann M (2012) Future-oriented higher education: which key competencies should be fostered through university teaching and learning? Futures 44:127–135. https://doi.org/10.1016/j.futures.2011.09.005

Stevens C, Kanie N (2016) The transformative potential of the sustainable development goals (SDGs). Int Environ Agreem-Polit Law Econom 16(3):393–396. https://doi.org/10.1007/s10784-016-9324-y

Turner BL, Kaspersonb R, Matsone P, McCarthy J, Corell R, Christensen L, Eckley N, Kasperson J, Luers A, Martello M, Polsky C, Pulsipher A, Schiller A (2003) A framework for vulnerability analysis in sustainability science. Proc Natl Acad Sci USA 100:8074–8079. https://doi.org/10.1073/pnas.1231335100

UNESCO (2017) Education for sustainable development goals. Learning objectives. UNESCO. https://unesdoc.unesco.org/ark:/48223/pf0000247444

UNITED NATIONS (UN) (2015) Transforming our world: the 2030 agenda for sustainable development, resolution adopted by the General Assembly on 25 September 2015

Vilches A, Gil-Pérez D (2015) Ciencia de la Sostenibilidad: ¿Una nueva disciplina o un nuevo enfoque para todas las disciplinas? [Sustainability science: a new discipline or a new approach for all disciplines?] Revista Iberoamericana de Educación (RIE) 69(9):39–60. https://doi.org/10.35362/rie691152

Vilches A, Gil-Pérez D (2012) La educación para la sostenibilidad en la Universidad: el reto de la formación del profesorado [Education for sustainability at the University: the challenge of teacher training]. Profesorado 16(2):25–43

Wiek A, Withycombe L, Redman CL (2011) Key competencies in sustainability: a reference framework for academic program development. Sustain Sci 6(2):203–218. https://doi.org/10.1007/s11625-011-0132-6

[illegible]

Carbon Balance in Educational Institutions: Greenhouse Gases Emission and Carbon Removal Inventory

Lauana Blenda Silva, Thaís de Almeida Rocha, Thaynara Pereira Albuquerque, Valéria de Fatima Silva, Mariany Filipini de Freitas, and Laércio Antônio Gonçalves Jacovine

Abstract Greenhouse gases (GHG) emissions are escalating at a rapid pace and the consequences related to climate change are the focus of major global discussions. Educational Institutions (EIs) are increasingly active in promoting social changes toward sustainable development by elaborating and implementing alternatives that consider environmental concerns in their principles and infrastructure. Climate change is an environmental problem that has been challenging organizations, including EIs, in the search for solutions to minimize it, such as carbon neutrality. The GHG Emission and Carbon Removal Inventory is the first step toward a carbon–neutral organization. Therefore, the aim of this study is to present a proposal for standardizing guidelines to allow EIs to prepare their GHG inventory, and if necessary, to neutralize their GHG emissions. The work was based on theoretical information about the GHG inventory elaboration methodologies and practical experiences lived by the authors. The main contribution was a guide for the management of GHG emissions, in addition to showing what is done at the EI, especially at the Federal

L. B. Silva (✉) · T. de Almeida Rocha · T. P. Albuquerque · V. de Fatima Silva · M. F. de Freitas · L. A. G. Jacovine
Department of Forest Engineering, Federal University of Viçosa, Peter Henry Rolfs Avenue, Viçosa, Minas Gerais 36570-900, Brazil
e-mail: lauana.silva@ufv.br

T. de Almeida Rocha
e-mail: thais.a.rocha@ufv.br

T. P. Albuquerque
e-mail: thaynara.albuquerque@ufv.br

V. de Fatima Silva
e-mail: valeria.fatima@ufv.br

M. F. de Freitas
e-mail: mariany.freitas@ufv.br

L. A. G. Jacovine
e-mail: jacovine@ufv.br

W. Leal Filho et al. (eds.), *The Contribution of Universities Towards Education for Sustainable Development*, World Sustainability Series,
https://doi.org/10.1007/978-3-031-49853-4_20

University of Viçosa (UFV), through the "*Carbono Zero UFV*" Program. The initiatives adopted by some EIs already indicate an advance in terms of the sustainability of the academic community. GHG inventories and any sustainable practices adopted by EIs can help publicize the subject, promoting changes in the academic community and society in general.

Keywords Academic community · Carbon neutrality · Climate change · Sustainability

1 Introduction

One of the greatest common concerns of humanity is climate change. The increase in the global average temperature, acidity, salinity, and level of the oceans, changes in the precipitation regime, the retreat of glaciers, and the greater frequency and intensity of extreme events are some consequences of the increase in the greenhouse gases (GHG) concentration in the atmosphere (IPCC 2021). Human influence significantly contributes to the increase in the GHG concentration in the atmosphere and worsening of this scenario (IPCC 2021). The global surface temperature increased by 1.09 °C between 1850–1900 and 2011–2020, and probably 1.07 °C of this increase was caused by human actions (IPCC 2021).

Discussions related to the causes, consequences, and ways of adapting to climate change involve the political, public, private, and academic environment. The Conference of the Parties (COP)—the supreme body of the United Nations Framework Convention on Climate Change (UNFCCC)—annually brings together National Parties, with one of the objectives being to establish global agreements, joining efforts to mitigate GHG emissions and their consequences.

The Paris Agreement signed at COP21 in 2015 has the goal (among others) to keep the increase in the global average temperature well below 2 °C above pre-industrial levels and to seek efforts to limit this increase to 1.5 °C (UNFCCC 2015). National Parties at COP26 in 2021 recognized the urgent need for more ambitious actions concerning mitigation, adaptation, and financing to contain climate change. In this sense, the Parties signed the agreement to seek efforts to limit the increase in temperature to 1.5 °C, recognizing that rapid and deep sustained reductions in emissions and cooperative and multilevel action between the various actors in society are necessary for this to occur (UNFCCC 2021).

The Paris Agreement, together with the 2030 Agenda for Sustainable Development, which brings 17 Sustainable Development Goals (SDGs), define a framework of ambitious global commitments for transformative changes in a world with serious social, economic, political and environmental challenges (Fuso Nerini et al. 2019). Much evidence indicates that these commitments are normatively and empirically interconnected (Sachs et al. 2019), highlighting how the impacts of climate change will make it difficult to achieve development goals, especially those linked to physical and material well-being, provision of food, energy, water availability and health

(Fuso Nerini et al. 2019). Since the creation of the SDGs, their effects have been mainly discursive, being used as a point of reference in the pronouncement of international policies, but without the necessary effectiveness in normative and institutional reforms which occur in isolation, indicating that they are still not a transforming force by themselves alone (Biermann et al. 2022).

Educational Institutions (EIs) are increasingly active in promoting social changes toward sustainable development. Discussions about these actions have been integrated into the EI cores, mainly because they play an important role in shaping the character of future professionals and world leaders (Guerra et al. 2018). They have been involved in activities ranging from participation in regional development (Dlouhá et al. 2013) and reducing GHG emissions (Klein-Banai and Theis 2013), to committing to include sustainable development in their mission and vision statements (Lee et al. 2013).

The GHG inventory is the first step towards elaborating climate policy at any level, be it organizational, municipal, or national. This tool can be comprised of two parts: (i) GHG Emission Inventory; and (ii) Carbon Removal Inventory.

Anthropogenic activities emit GHGs at some stage to a greater or lesser extent. The burning of fossil fuels, the enteric fermentation process of ruminant animals, and industrial processes are among the main activities which contribute to worsening climate change worldwide (WRI 2022). Conversely, carbon sinks perform the important ecosystem service of removing and storing carbon. A relevant example is forest ecosystems, which have the potential to remove carbon from the atmosphere and store it in their various components, namely: vegetation, soils, litter, and necromass (Gogoi et al. 2022; Harris et al. 2021).

It is possible to map the emission and removal sources through the GHG inventory, and verify the contribution of each to the emissions profile. This provides a fundamental basis for managing anthropogenic GHG emissions, including mitigation forms, monitoring impacts (IPCC 2019), and neutralization.

A possible result of GHG emissions and carbon removal inventory is the carbon balance (or net emissions). The carbon balance is the index which enables direct verification of an organization's participation in global emissions and its contribution to climate change. Many organizations are already interested in obtaining the title "Net zero emissions", indicating that all their GHG emissions have been neutralized through carbon removal by some sink under their responsibility.

Educational Institutions (EIs), as opinion makers and environmental awareness makers, can use their GHG inventories for transformations, not only on their campuses, but in training professionals. Thus, the aim of this work was to present a proposal for standardizing guidelines for EIs to prepare their GHG emission and carbon removal inventory to help promote this change, and if necessary, to neutralize their GHG emissions.

2 The Role of Educational Institutions

Educational Institutions (EIs) are important trainers of trained professionals to contribute to the development of peaceful, inclusive, fair and sustainable societies, from basic to applied research, transforming scientific knowledge into practical actions (Chankseliani et al. 2021). The innovative character of EIs is meeting the main global challenges, such as social problems involving food security, supporting women as agents of change and future leaders and sustainable use of ecosystems (Chankseliani and McCowan 2021). In addition, the faculty and researchers of these institutions are included in sustainability committees aimed at implementing programs to improve the quality of life and sustainability due to their potential to serve as a model for society, for example by demonstrating how to make their operations more sustainable (Santovito and Abiko 2018; Rappaport and Creighton 2007).

Climate change stands out among the great challenges of the twenty-first century, which challenges all EIs to draw up comprehensive plans towards climate neutrality in increasingly shorter terms as a way to popularize the topic in their operating regions (Klein-Banai et al. 2010). The basis for a climate neutrality plan is a GHG emission inventory, which can provide a first-degree quantitative measure of each EI's participation in global climate change, as well as its progress towards sustainability (Klein-Banai et al. 2010).

The preliminary step for the emission inventory is to understand, identify and quantify the activities which contribute to GHG emissions, but constitutes a task that can be difficult due to the type and size of the EI, as well as the lack of a specific standardized methodology for EIs (Valls-Val and Bovea 2021). Several American and European institutions rely on the Campus Carbon Calculator (CCC) and the Greenhouse Gases Protocol for Communities (GPC) to prepare their inventories, but both are incomplete when the object of study is teaching units (Santovito and Abiko 2018). Methodologies, especially in relation to Scope 3 emissions, are less developed, with inconsistencies and are complex, demanding more research to improve and simplify inventories (Klein-Banai and Theis 2013). EIs can draw the baseline from the GHG emission inventory to assess the effect of future mitigation efforts (Letete et al. 2011), which reflects in more effective climate actions or a sustainability plan with targets and target dates for emission reductions (Klein-Banai and Theis 2013).

3 The GHG Emission and Carbon Removal Inventory

Greenhouse Gas (GHG) inventories must be prepared following recognized guidelines, methodologies, and supported by scientific findings to ensure the reliability of the results. The standardization proposed in this work was based on the guidelines of the Intergovernmental Panel on Climate Change (IPCC), the GHG Protocol Program of the World Resources Institute (WRI), and the *Associação Brasileira de Normas Técnicas* (*ABNT*).

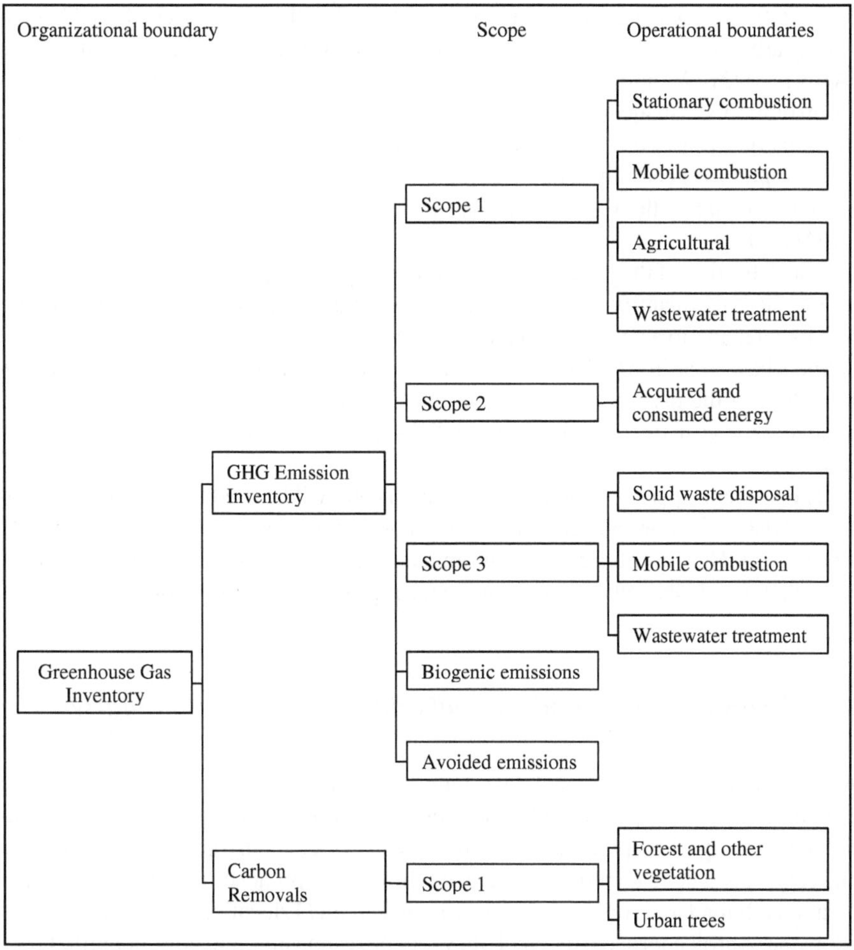

Fig. 1 Structure of greenhouse gas (GHG) inventories. *Source* Elaborated by the authors

The structure of GHG inventories along with possible emission and removal sources, are shown in Fig. 1, and the steps for their preparation are described underneath it.

3.1 Definition of the Base year and Reference Period

The first step in carrying out the Greenhouse Gas (GHG) emission and carbon removal inventory is to define the base year and reference period of the inventory. The base year generally consists of a period of 12 months and has the function of allowing to make future comparisons of the results obtained in the inventories of GHG emissions, carbon removals, or any other related information to GHG, in different periods (ABNT 2022). Without such definition, it is impossible to assess the evolution of an institution's issuances and removals over time. If there is not enough historical information to carry out an inventory of emissions and removals for a base year, the first inventory is generally used to define this parameter. It should be noted that institutions can change their base year, but they must explain the reason for the change (ABNT 2022).

The definition of the base year must follow some precepts, and therefore the institution must: (i) quantify emissions and removals using the largest possible number of data referring to a year, or through averages, weighted or not, of several years; (ii) define a base year in which data are available and verifiable; (iii) explain the reason for selecting the base year in question; and (iv) carry out a GHG inventory for the defined base year using available guidelines, such as the ABNT NBR ISO 14064 (ABNT 2022).

The reference period is the period in which data will be used to carry out inventories of GHG emissions and carbon removals, which may or may not be the base year. Normally, GHG inventories have an annual cut, with a reference period between January 1st and December 31st of the reporting year (GHG Protocol Brasil 2017).

3.2 Definition of Organizational and Operational Boundaries

After defining the base year and the reference period, the organization must limit the organizational and operational boundaries covered by the inventory. The organizational boundary corresponds to the facilities where greenhouse gas (GHG) emissions and removals will be quantified. There are different approaches to defining the organizational boundary (ABNT 2022). It is indicated that Educational Institutions (EIs) adopt the control approach, in which they will be responsible and must quantify GHG emissions and removals from facilities in which they have operational or financial control.

Operational boundary is the source of GHG emission and carbon removal associated with the operations of the organization carrying out the GHG inventory. Each EI must identify its operational boundary and categorize it into direct or indirect, within scopes 1, 2, and 3 (ABNT 2022; WRI 2004).

"Scope 1—Direct GHG emissions" covers GHG emissions from own sources controlled by the institution (WRI 2004). Some common sources of direct GHG emissions by EIs are the consumption of liquefied petroleum gas (LPG) and firewood for energy generation within the institution's limits (stationary combustion) and fuel consumption by the institution's vehicle fleet (mobile combustion). However, each educational institution must map its activities to define its direct sources of emissions. An educational institution that has animals, cultivated areas, and treatment of solid waste and effluents within its organizational boundary must consider the emissions generated by these sources in Scope 1.

Scope 1 also covers EI's responsible carbon sinks. Normally, all carbon removal sources will be considered in Scope 1, since if they are present in its organizational boundary, they will be controlled by the institution. It is indicated that the EIs consider all tree components as carbon removal sources, such as forest fragments, urban trees, and other plant formations.

"Scope 2—Indirect GHG emissions from energy" includes GHG emissions from energy generation (electricity, heat, steam, and cooling) acquired and consumed by the institution. In this scope, the emissions are considered to occur outside the institution's limits, where energy is generated (WRI 2015). However, the emission is the indirect responsibility of each educational institution, since it would not exist if the energy were not generated to be consumed.

"Scope 3—Other indirect GHG emissions" is an optional but strongly recommended reporting category. It also accounts for indirect emissions, which are consequences of the institution's activities, but are not owned or controlled by it (WRI 2004). In this scope, the educational institution must decide which GHG emission sources it wants to consider, according to its activities.

3.3 Biogenic Emissions

Carbon dioxide (CO_2) emissions from burning biomass are called biogenic emissions. These emissions can be considered neutral in terms of climate impact, as the CO_2 emitted is equal to the CO_2 removed from the atmosphere by the photosynthesis process. In other words, CO_2 is generated through a biological and not a geological cycle, as is the case with CO_2 of fossil origin (Assad et al. 2020). Firewood, biodiesel, and ethanol are some examples of biogenic emission sources.

Biogenic emissions must be reported separately in greenhouse gases (GHG) inventories, outside of Scopes 1, 2, and 3 (Table 1). Methane (CH_4) and nitrous oxide (N_2O) emissions from biomass burning are not considered neutral, as these gases are not removed from the atmosphere during biomass growth (GHG Protocol Brasil 2010).

Table 1 Greenhouse gases (GHG) emission allocation according to the origin of the emission source

Emission source	GHG emission		
	CO_2	CH_4	N_2O
Biomass	Biogenic emission	Scope 1 or 3	Scope 1 or 3
Fossil	Scope 1 or 3	Scope 1 or 3	Scope 1 or 3

Source ABNT (2022), WRI (2004)

3.4 Avoided Emissions

Some activities, products, and services cause so-called "avoided emissions". This term corresponds to emission reductions that occurred outside the life cycle of a product or value chain, but which are a result of the use of that product (GHG Protocol 2014). The consumption of photovoltaic solar energy can be understood as one of these activities, as it avoids emissions from energy generation through burning mineral coal, for example. Recycling and reuse of solid waste also avoid the emission from the decomposition of this waste.

The adoption of measures, activities, and use of products that promote avoided emissions are highly recommended. All these initiatives would generally contribute to the greater environmental sustainability of the EI.

3.5 Methodologies for Calculating GHG Emissions and Carbon Removals

3.5.1 GHG Emissions

There are dozens of greenhouse gases (GHG) between natural and synthetic (IPCC 2001). It is recommended that Educational Institutions (EIs) at least consider the following gases in their GHG inventory: CO_2, CH_4, and N_2O. This recommendation is due to the greater contribution that these three gases have to climate change (IPCC 2021). However, each EI must include the gases emitted with greater frequency and/or intensity in its GHG inventory according to its activities.

GHG emission is generally estimated by the product of the activity data (which is a measure that expresses the intensity, work, or consumption of a given emitting source), and the emission factor (which indicates the emission associated with a unit of the activity data in question) (Eq. 1) (IPCC 2006b). The emission of each greenhouse gas must be converted to the standard unit CO_2e (carbon dioxide equivalent) by multiplying it by its respective Global Warming Potential (GWP) (Table 2). The GWP is a metric that makes it possible to compare the potential climate impact of different GHG compared to CO_2 (IPCC 2013).

$$E_i = \sum \left(AD_i \times EF_{ij} \times GWP_j\right) \tag{1}$$

In which:

E_i = total GHG emission by source *i*, in $MgCO_2e\ year^{-1}$.

AD_i = activity data by source *i*, in units.

EF_{ij} = emission factor of gas *j* by source *i*, in MgGHG (activity data unit)$^{-1}$.

GWP_j = global warming potential of gas *j*.

The specific methodologies for quantifying GHG emissions from each emitting source (tier, equations, emission factors, and other parameters) must be verified in

Table 2 Global warming potential (GWP) of greenhouse gases (GHG) that must be included in GHG inventories of Educational Institutions

GHG	GWP*
Carbon dioxide (CO_2)	1
Methane (CH_4)–fossil	29.8
Methane (CH_4)–non fossil	27.0
Nitrous oxide (N_2O)	273

Source IPCC (2021)
* There are other metrics (IPCC 2021)

the IPCC publications, especially in the documents: (i) 2006 IPCC Guidelines for National Greenhouse Gas Inventories (IPCC 2006a) and (ii) 2019 Refinement to the 2006 IPCC Guidelines for National Greenhouse Gas Inventories (IPCC 2019). It is important to remember that the IPCC, even indicating default values, recommends the use of emission factors and other parameters defined to the most similar reality to the organization that is carrying out the GHG inventory (IPCC 2006a). Thus, it is recommended that EIs look for emission factors and other parameters in national inventories and other related documents.

Obtaining the results of the GHG emission inventory makes it possible to calculate emission indices, which express the emission related to a unit of interest to the institution. Some examples of these indices are emission per person in the university community, per total campus area, and per built-up area. The emission indices can help the institution to prepare future GHG inventories, and especially to monitor the evolution of its emissions over the years. Other institutions can also use the emission indices generated to estimate their emissions if they do not have the resources available for the detailed preparation of their GHG inventory. In addition, issuance indices allow comparisons between institutions with different characteristics.

When detailed, the emission index per person of the university community, for example, can elucidate individual responsibility regarding GHG emissions. The EI can then use this index to encourage each member of the university community to neutralize their emissions. This could happen through institutional programs and campaigns that promote the donation of seedlings and community neutralization plantations.

3.5.2 Carbon Removal

Carbon removal takes place from incorporating CO_2 in the biomass of the plants through the photosynthetic process. The carbon increment must be considered in greenhouse gas (GHG) inventories, which expresses how much carbon was removed by the vegetation during the reference period, and not the total carbon stock present in the tree individuals. This indicator is obtained from dividing the total carbon stock by the age of the forest and/or arboreal individuals.

There are different ways of estimating the carbon removal by forests, urban trees, and/or perennial crops existing within the limits of the Educational Institutions (EIs).

The definition of which methodology to use will depend on the existing equations and factors for the region and forest typology present in the EI. The EI must choose the methodology that is most specific to its reality.

The first way to estimate the carbon removal by forests is through volume, basic wood density, and tree age. The basic wood density of tree species can be found in the literature, as in Zanne et al. (2009). The tree volume can be obtained in different ways, such as by measuring (destructive and non-destructive) and by forest inventory. Non-destructive cubing consists of measuring all sections of an individual tree to obtain its volume, without cutting it down (Soares et al. 2012). The inventory is carried out through primary measurements of the trees, such as diameter at breast height (dbh) and height, and the use of existing allometric equations (Soares et al. 2012). Carbon removal can be calculated by Eq. 2, if only one forest inventory has been carried out in the area, and by Eq. 3, if at least two forest inventories have been carried out in the area.

$$R_i = \sum \frac{(V_i \times d_i \times 0.47)/1000}{a_i} \tag{2}$$

$$R_i = \sum \frac{\left[\left(V_{i(n)} - V_{i(n-1)}\right) \times d_i \times 0.47\right]/1000}{In} \tag{3}$$

In which:
R_i = total carbon removal by the tree species *i*, in $MgCO_2e$ year^{-1}.
V_i = total volume of the tree species *i*, in m^3.
d_i = basic density of wood from the tree species *i*, in kg m^{-3}.
a_j = individual age of tree species *i*.
0.47 = Biomass to Carbon Conversion Factor (IPCC 2006c).
N = year of the last forest inventory.
n – 1 = year of the penultimate forest inventory.
In = interval between the two forest inventories, in years.

Like the allometric equations to estimate the tree volume, some equations directly estimate the carbon stock of individuals through primary measurements (Albuquerque 2021; Morais Junior et al. 2020). In this case, simply divide the total carbon stock by the age of the forest and/or tree to obtain the carbon increment.

Another method for estimating carbon removal is the use of secondary carbon increment factors. Carbon increments for forest fragments are generally given in units of carbon per area (i.e. MgC ha^{-1} year^{-1}). Silva (2017) presented a compilation of secondary carbon increment factors for all Brazilian biomes and several tree species at different ages and forest conditions.

Carbon increments of urban trees, planted singly or with wide spacings, are generally given in units of carbon per tree (e.g. MgC tree^{-1} year^{-1}). Carbon increments in this category can be found for different species and planting conditions (Brianezi et al. 2014; Silva 2017, 2019).

The use of carbon increment factors to estimate carbon removal requires that the EI only define the forest area and/or the number of urban trees and apply the specific increment factors. The increment factors found in the literature can only be used if they are for the same region and typology of the forest in question; otherwise, the factors do not apply.

3.6 Carbon Balance

Carbon balance is the difference between carbon removal and Greenhouse Gas (GHG) emission in terms of carbon dioxide equivalent (CO_2e). The result of the carbon balance can be positive when removals are greater than carbon emissions (which is the desired result), or negative when the opposite occurs.

Given the climate change scenario and its consequences, it is important that Educational Institutions (EIs) strive to achieve a positive carbon balance and to make it increasingly greater. This would indicate that the removal sources present in the EI are absorbing more carbon from the atmosphere than the quantity of GHGs being emitted by the activities carried out by the institution. Many actions can be taken to achieve this goal.

Installing photovoltaic plants for solar energy generation is an example of an action that can reduce effective emissions. Electricity consumption has been one of the main discussions at the Conference of the Parties (COP) because most countries still do not have low-emission energy systems (UNFCCC 2021). Brazil stands out in the generation of electricity from renewable sources, being responsible for 84.8% of the domestic electricity supply in the country (EPE 2021). The water source accounts for 65.2% of the domestic supply (EPE 2021) and Brazil still has great potential for exploring other renewable energy sources, such as wind, solar, and biomass (Garcez 2017; Lima et al. 2020). The use of energy generated in photovoltaic plants promotes avoided emissions (Lira et al. 2019), as it reduces the consumption of energy generated by other sources such as that of hydraulic origin, which although is a clean energy generation source, it still emits certain amounts of GHGs (Dos Santos et al. 2017).

Biomass use is another way to reduce effective emissions according to the concept of biogenic emissions. Therefore, when it comes to mobile combustion, it is advisable that EIs use ethanol (a biofuel), instead of fuels of non-renewable origin such as gasoline, which is mostly composed (73%) of fossil fuel (MAPA 2015). Likewise, it is recommended to use firewood in stationary combustion instead of fossil fuels such as liquefied petroleum gas (LPG) and natural gas, which are widely used in food preparation in EIs.

Waste recycling and reuse in EIs are also alternatives to reduce GHG emissions. These actions increase the useful life of materials, preventing their decomposition and consequent GHG emissions. In addition, carrying out these activities contributes to the inherent problem of the disposal and accumulation of solid waste, which causes devastating impacts on the environment, such as water and soil contamination, among others (Onel and Mukherjee 2017). Composting is a type of treatment of organic waste which allows its reuse (Boldrin et al. 2009), which enables reducing GHG emissions, as the amount emitted in this process may be lower than the amount emitted in other types of waste treatment. In addition, composting represents a promising alternative because it provides valuable products which can be used as organic fertilizers, replacing chemicals, and in turn causing lower GHG emissions and promoting positive impacts on soils and plants (Milinković et al. 2019).

4 GHG Emissions Neutralization

The preparation of the greenhouse gases (GHG) inventory allows Educational Institutions (EI) to know their GHG emissions and carbon removals and enables better management of emissions to achieve a positive carbon balance. If an educational institution has a negative carbon balance, it can neutralize its emissions as a further step toward the institution's environmental sustainability.

Carbon neutralization can be done in two ways: by purchasing forest carbon credits or other activities and by planting trees. Purchasing carbon credits is a simpler alternative, however, when not coming from forestry projects, it brings fewer environmental benefits when compared to planting. In addition to contributing to carbon offsetting, planting tree seedlings provides several benefits promoted by forests, such as maintaining water and biodiversity (Cardoso 2021).

Seedling plantations related to climate concern associated with forest restoration of degraded areas are called "carbon neutralization plantations" (Morais Junior et al. 2020). These plantations promote carbon neutralization and contribute to recovering the floristic structure and ecological functions of the ecosystem as close as possible to their pre-degradation condition, which is usually human-caused (Silva et al. 2021).

Priority in carbon–neutral plantations should be given to the use of regional native species to conserve the biodiversity of human-dominated landscapes (Janishevski et al. 2015; Possingham et al. 2015). In addition, it is necessary to stick to monitoring and maintenance activities of plantations, such as controlling leaf-cutting ants and weed competition. The purpose of these activities is to reduce the mortality rate of seedlings, creating a more favorable environment for their growth.

Another important point in carrying out the plantations is to involve the university community in the activities. Involvement can take place through promoting lectures and events (such as "Field Day"), and conducting research in planting areas. Thus, EIs combine neutralizing their emissions with environmental education by promoting greater environmental awareness in the community, and with research, thereby contributing to science.

5 Current Scenario of Educational Institutions (EIs) in the Face of Climate Change

With each passing day, more Educational Institutions (EIs) are seeking actions to reduce their impact on climate change. Some EIs have already carried out their greenhouse gases (GHG) inventories as the first step toward achieving the desired transformation. Some examples are Clemson University, located in the United States (Clabeaux et al. 2020), Montfort University—United Kingdom (Ozawa-Meida et al. 2013), University of Cape Town—South Africa (Letete et al. 2011), Federal Technological University of Paraná—Brazil (Mesquita 2015; Yabushita 2013), and the Federal University of Viçosa—Brazil (Brianezi et al. 2014; Rocha 2020), among others (De Carvalho et al. 2017; Gonzaga 2016; Güereca et al. 2013; Martins 2015).

The Federal University of Viçosa (UFV) in Brazil is recognized worldwide for its sustainable initiatives. The institution stands out in the UI GreenMetric World University Ranking, which ranks EIs in terms of their current conditions and their sustainability-related policies. One of the criteria evaluated is GHG emissions and the extent to which the institution is committed to the issue of climate change. The ranking has the participation of more than 900 EIs, 40 of which are Brazilian, and UFV has maintained its position among the 10 most sustainable EIs in the country (UI Green Metric 2022; UFV 2020).

One of UFV's initiatives was the creation of the "*Carbono Zero UFV*" Program, intending to quantify and neutralize GHG emissions generated by the "*Semana do Fazendeiro*", the largest rural extension event in Brazil, which takes place on the institution's campus (Alves 2014; Carbono Zero 2022). Through the actions of the "*Carbono Zero UFV*" Program, more than 4,770 seedlings have already been planted to neutralize the GHG emissions of this event.

The institution also implemented the "*Diálogos UFV + Sustentável*" project, which holds meetings in which University employees or guests from other institutions discuss sustainable actions that can be replicated at UFV (UFV 2021). This project also aims to advance actions to become a "Zero Waste University", from which a compost bin was installed where the organic waste generated in the institution is destined.

UFV also installed a Voluntary Delivery Point (VDP) for the disposal of paper, plastic, glass, and metal, which are destined for the "Association of Workers of the Viçosa Sorting and Recycling Plant" (*Associação dos Trabalhadores da Usina de Triagem e Reciclagem de Viçosa - Acamare*). There is also a specific waste destination for fluorescent lamps, batteries, and used batteries, which are destined for specialized companies that transform them back into raw material and/or energy. Used cooking oil is also collected at the same point and sold to a company in Viçosa to reuse the residue. The resources obtained through these projects are used in student assistance actions by the University's Dean of Community Affairs (UFV 2022).

Like the UFV, other EIs must commit to carrying out actions aimed at environmental improvement. The trend is for the reporting of GHG emissions to be mandatory, as is already the case in commitments signed by some EIs in several countries (Clabeaux et al. 2020). Thus, the EIs that are already taking initiatives in this direction will be ahead of possible agreements for carbon neutrality.

6 Conclusion

Educational Institutions are opinion makers and environmental sensitizers and can generate various impacts on the economy, society, culture, politics, public services, health, environment and quality of life that go beyond their organizational or academic limits. Accounting for these impacts is not simple, as there are several indirect and long-term impacts, but the direct ones, such as greenhouse gas emissions and contribution to the local economy can be observed, described and quantified with greater assertiveness. By accounting for these direct effects, institutions are playing an essential role in promoting sustainable socio-environmental transformations, in addition to transmitting skills for sustainable development to students and the community.

The sustainability actions developed in EIs, in particular performing an inventory of greenhouse gas emissions, are essential to become a sustainable development model for the community and provide guidelines so that municipal managers can also develop them in the cities. However, preparing these inventories is a time-consuming and expensive process, especially with specific and detailed guidelines on how to perform calculations at the campus level, which are very similar to neighborhood coverage.

Different calculators and tools for carrying out emission inventories are freely available and can be used by any university, but many were developed to characterize a specific moment and particular conditions of the institution being inventoried, without enabling consistency and comparability studies, both internally to assess the changes in emissions, such as between different institutions. Thus, the guidelines presented in this article allow the EI to take the first steps in preparing its GHG inventory; in addition to defining basic issues such as the base year and reference period and organizational and operational limits, it allows planning the management of mitigation actions and as a tool for managing the sustainability of the EI. In addition, reference documents were presented, indicating more robust methodologies for calculating GHG emissions and carbon removals.

A limitation of this article is that it does not provide more detailed guidelines for preparing GHG inventories. However, this is justified, as the specificities of each EI must be considered in the GHG inventory. These specificities will influence the definition of GHG emission and removal sources, and the choice of calculation methodology to be followed, including emission and removal factors. It is important to emphasize that each EI must choose the methodologies and factors that most apply to their reality in order to achieve more accurate results. Another point of attention is the constant updating of methodologies and calculation factors. Each educational institution must prioritize the use of these most up-to-date methodologies every year, always seeking to improve the quality of their inventories.

References

Albuquerque TP (2021) Sobrevivência e estoque de carbono de espécies florestais em plantio de neutralização, aos 9 e 10 anos, em Viçosa, Minas Gerais [Survival and carbon stock of forest species in a neutralization plantation, at 9 and 10 years, in Viçosa, Minas Gerais]. Universidade Federal de Viçosa, Thesis

Alves EBBM (2014) Mudanças Climáticas: Percepção do produtor, Balanço de Carbono em Propriedades Rurais e Neutralização de evento da Universidade Federal de Viçosa [Climate change: producer perception, carbon balance in rural properties and neutralization event at the Federal University of Viçosa.]. Thesis, Universidade Federal de Viçosa

Assad ED, Martins SC, Pavão ED, Gusson E, Silva JP, Prado A, Pontes C, Biderman R, Esturba T, Romeiro V (2020) Greenhouse gas protocol calculation tool for forestry in Brazil. WRI Brasil: Technical Note. https://wribrasil.org.br/sites/default/files/ghg_protocolo-florestas-technicalnote.pdf. Accessed 29 Apr 2022

[ABNT] Associação Brasileira de Normas Técnicas (2022) ABNT NBR ISO 14064-1: Gases de efeito estufa: Parte 1: Especificação com orientação no nível da organização para quantificação e notificação de emissões e remoções de gases de efeito estufa [ABNT NBR ISO 14064-1: greenhouse gases: part 1: specification with guidance at the organization level for quantification and reporting of greenhouse gas emissions and removals]. ABNT, Rio de Janeiro

Biermann F, Hickmann T, Sénit CA, Beisheim M, Bernstein S, Chasek P, Grob L, Kim RE, Kotzé LJ, Nilsson M, Llanos AO, Okereke C, Pradhan P, Raven R, Sun Y, Vijge MJ, van Vuuren D, Wicke B (2022) Scientific evidence on the political impact of the sustainable development goals. Nat Sustain 5:795–800. https://doi.org/10.1038/s41893-022-00909-5

Brianezi D, Jacovine LAG, Soares CPB, Gonçalves W, Rocha SJSSD (2014) Balance of emissions and removals of greenhouse gases at campus of Universidade Federal de Viçosa. Floresta e Ambiente 21:182–191. https://doi.org/10.4322/floram.2014.017

Boldrin A, Andersen JK, Møller J, Christensen TH, Favoino E (2009) Composting and compost utilization: accounting of greenhouse gases and global warming contributions. Waste Manage Res 27:800–812. https://doi.org/10.1177/0734242X09345275

[Carbono Zero] Programa Carbono Zero UFV (2022) O Programa [The Programme]. http://www.carbonozero.ufv.br/?page_id=167. Acessed 04 May 2022

Cardoso EJBN (org) (2021) A sustentabilidade ambiental da agricultura e de florestas tropicais: uma visão científica, ecológica, política e social [The environmental sustainability of agriculture and tropical forests: a scientific, ecological, political and social vision]. Appris, Curitiba

Chankseliani M, McCowan T (2021) Higher education and the sustainable development goals. High Educ 81:1–8. https://doi.org/10.1007/s10734-020-00652-w

Chankseliani M, Qoraboyev I, Gimranova D (2021) Higher education contributing to local, national, and global development: new empirical and conceptual insights. High Educ 81:109–127. https://doi.org/10.1007/s10734-020-00565-8

Clabeaux R, Carbajales-Dale M, Ladner D, Walker T (2020) Assessing the carbon footprint of a university campus using a life cycle assessment approach. J Clean Prod 273:122600. https://doi.org/10.1016/j.jclepro.2020.122600

De Carvalho JPAF, Van Elk AGHP, Romanel C (2017) Inventário de emissões de gases de efeito estufa no campus gávea da PUC-Rio [Inventory of greenhouse gas emissions at PUC-Rio's Gávea campus]. Engenharia Sanitária e Ambiental 22:591–595. https://doi.org/10.1590/S1413-41522017155865

Dos Santos MA, Damazio JM, Rogerio JP, Amorim MA, Medeiros AM, Abreu JLS, Maceira MEP, Melo AC, Rosa LP (2017) Estimates of GHG emissions by hydroelectric reservoirs: the Brazilian case. Energy 133:99–107. https://doi.org/10.1016/j.energy.2017.05.082

Dlouhá J, Huisingh D, Barton A (2013) Learning networks in higher education: universities in search of making effective regional impacts. J Clean Prod 49:5–10. https://doi.org/10.1016/j.jclepro.2013.01.034

[EPE] Empresa de Pesquisa Energética (2021) Brazilian Energy Balance 2021 Year 2020. EPE, Rio de Janeiro. https://www.epe.gov.br/sites-pt/publicacoes-dados-abertos/publicacoes/PublicacoesArquivos/publicacao-601/topico-596/BEN2021.pdf. Accessed 26 May 2022

Fuso Nerini F, Sovacool B, Hughes N, Cozzi L, Cosgrave E, Howells M, Tavoni M, Tomei J, Zerriffi H, Milligan B (2019) Connecting climate action with other sustainable development goals. Nat Sustain 2:674–680. https://doi.org/10.1038/s41893-019-0334-y

Garcez CG (2017) Distributed electricity generation in Brazil: an analysis of policy context, design and impact. Utilities Policy 49:104–115. https://doi.org/10.1016/j.jup.2017.06.005

Gogoi A, Ahirwal J, Sahoo UK (2022) Evaluation of ecosystem carbon storage in major forest types of Eastern Himalaya: implications for carbon sink management. J Environ Manage 302:113972. https://doi.org/10.1016/j.jenvman.2021.113972

Gonzaga EAR (2016) Neutralização de carbono por meio de inventário das emissões de gases de efeito estufa a partir do planejamento institucional da UFU [Carbon neutralization through an inventory of greenhouse gas emissions based on UFU's institutional planning]. Universidade Federal de Uberlândia, Thesis

Guerra JBSOA, Garcia J, Lima MA, Barbosa SB, Heerdt ML, Berchin II (2018) A proposal of a balanced scorecard for an environmental education program at universities. J Clean Prod 172:1674–1690. https://doi.org/10.1016/j.jclepro.2016.11.179

[GHG Protocol] Greenhouse Gas Protocol (2014) GHG protocol standard on quantifying and avoided emissions: summary of online survey results. https://ghgprotocol.org/sites/default/files/ghgp/Avoided%20emissions%20survey%20report_final%20draft.pdf. Accessed 27 Apr 2022

[GHG Protocol Brasil] Programa Brasileiro GHG Protocol (2010) Especificações do Programa Brasileiro GHG Protocol: Contabilização, Quantificação e Publicação de Inventários Corporativos de Emissões de Gases de Efeito Estufa [Specifications of the Brazilian GHG protocol program: accounting, quantification and publication of corporate greenhouse gas emissions inventories]: 2ed. https://s3-sa-east-1.amazonaws.com/arquivos.gvces.com.br/arquivos_ghg/152/especificacoes_pb_ghgprotocol.pdf. Accessed 29 Apr 2022

[GHG Protocol Brasil] Programa Brasileiro GHG Protocol (2017) Nota técnica: Período de relato de inventário de gases de efeito estufa [Technical note: greenhouse gas inventory reporting period]: 1ed. http://mediadrawer.gvces.com.br/ghg/original/ghg-protocol_nota-tecnica_ano_calendario_v1.pdf. Accessed 17 May 2022

Güereca LP, Torres N, Noyola A (2013) Carbon footprint as a basis for a cleaner research institute in Mexico. J Clean Prod 47:396–403. https://doi.org/10.1016/j.jclepro.2013.01.030

Harris NL, Gibbs DA, Baccini A, Birdsey RA, Bruin S, Farina M, Fatoyinbo L, Hansen MC, Herold M, Houghton RA, Potapov PV, Suarez DR, Roman-Cuesta RM, Saatchi SS, Slay CM, Turubanova SA, Tyukavina A (2021) Global maps of twenty-first century forest carbon fluxes. Nat Clim Chang 11:234–240. https://doi.org/10.1038/s41558-020-00976-6

[IPCC] Intergovernmental Panel on Climate Change (2001) TAR climate change 2001: the scientific basis: atmospheric chemistry and greenhouse gases. https://www.ipcc.ch/site/assets/uploads/2018/03/TAR-04.pdf. Accessed 29 Apr 2022

[IPCC] Intergovernmental Panel on Climate Change (2006a) 2006 IPCC guidelines for national greenhouse gas inventories. https://www.ipcc-nggip.iges.or.jp/public/2006gl/. Accessed 29 Apr 2022

[IPCC] Intergovernmental Panel on Climate Change (2006b) 2006 IPCC guidelines for national greenhouse gas inventories: general guidance and reporting: introduction to the 2006 Guidelines. https://www.ipcc-nggip.iges.or.jp/public/2006gl/pdf/1_Volume1/V1_1_Ch1_Introduction.pdf. Accessed 29 Apr 2022

[IPCC] Intergovernmental Panel on Climate Change (2006c) 2006 IPCC guidelines for national greenhouse gas inventories: agriculture, forestry and other land use. https://www.ipcc-nggip.iges.or.jp/public/2006gl/vol4.html. Accessed 19 May 2022

[IPCC] Intergovernmental Panel on Climate Change (2013) Climate change 2013: the physical science basis. Contribution of working group I to the fifth assessment report of the intergovernmental panel on climate change. https://www.ipcc.ch/report/ar5/wg1/. Accessed 02 May 2022

[IPCC] Intergovernmental Panel on Climate Change (2019) 2019 refinement to the 2006 IPCC guidelines for national greenhouse gas inventories. https://www.ipcc-nggip.iges.or.jp/public/2019rf/index.html. Accessed 29 Apr 2022

[IPCC] Intergovernmental Panel on Climate Change (2021) Climate change 2021: the physical science basis: summary for policymakers. https://www.ipcc.ch/report/ar6/wg1/downloads/report/IPCC_AR6_WGI_SPM_final.pdf. Accessed 18 Apr 2022

Janishevski L, Santamaria C, Gidda SB, Cooper HD, Brancalion PHS (2015) Ecosystem restoration, protected areas and biodiversity conservation. Unasylva 245:19–28. http://www.fao.org/3/a-i5212e.pdf

Klein-Banai C, Theis TL, Brecheisen TA, Banai A (2010) A greenhouse gas inventory as a measure of sustainability for an urban public research university. Environ Pract 12:35–47. https://doi.org/10.1017/S1466046609990524

Klein-Banai C, Theis TL (2013) Quantitative analysis of factors affecting greenhouse gas emissions at institutions of higher education. J Clean Prod 48:29–38. https://doi.org/10.1016/j.jclepro.2011.06.004

Lee KH, Barker M, Mouasher A (2013) Is it even espoused? An exploratory study of commitment to sustainability as evidenced in vision, mission, and graduate attribute statements in Australian universities. J Clean Prod 48:20–28. https://doi.org/10.1016/j.jclepro.2013.01.007

Letete T, Mungwe NW, Guma M, Marquard A (2011) Carbon footprint of the University of Cape Town. J Energy South Afr 22:2–12

Lima MA, Mendes LFR, Mothé GA, Linhares FG, de Castro MPP, da Silva MG, Sthel MS (2020) Renewable energy in reducing greenhouse gas emissions: reaching the goals of the Paris agreement in Brazil. Environ Dev 33:100504. https://doi.org/10.1016/j.envdev.2020.100504

Lira MAT, Melo MLS, Rodrigues LM, de Souza TRM (2019) Contribution of photovoltaic systems connected to the electrical network for reducing CO2 in the state of Ceará. Revista Brasileira De Meteorologia 34:389–397. https://doi.org/10.1590/0102-7786343046

[MAPA] Ministério da Agricultura, Pecuária e Abastecimento (2015) Portaria nº 75, de 5 de março de 2015 [Ordinance No. 75 of March 5, 2015]. https://pesquisa.in.gov.br/imprensa/jsp/visualiza/index.jsp?jornal=1&pagina=17&data=06/03/2015. Accessed 27 May 2022

Martins FRR (2015) Balanço de gases de efeito estufa da Universidade Federal Rural do Rio de Janeiro, no campus Seropédica [Greenhouse gas balance at the Federal Rural University of Rio de Janeiro, Seropédica campus]. Universidade Federal Rural do Rio de Janeiro, Thesis

Mesquita GS (2015) Estimativa de emissão e absorção de carbono na Universidade Tecnológica Federal do Paraná - Câmpus Dois Vizinhos [Estimating carbon emissions and absorption at the Federal Technological University of Paraná - Dois Vizinhos Campus]. Universidade Tecnológica Federal do Paraná, Thesis

Milinković M, Lalević B, Jovičić-Petrović J, Golubović-Ćurguz V, Kljujev I, Raičević V (2019) Biopotential of compost and compost products derived from horticultural waste—effect on plant growth and plant pathogens' suppression. Process Saf Environ Prot 121:299–306. https://doi.org/10.1016/j.psep.2018.09.024

Morais Junior VTM, Jacovine LAG, Alves EBBM, Torres CMME, Faustino IS, França LCJ, Rocha SJSS, Simiqueli GF, Silva LB, Alcántara-De La Cruz R (2020) Growth and survival of potential tree species for carbon-offset in degraded areas from southeast Brazil. Ecol Ind 117:106514. https://doi.org/10.1016/j.ecolind.2020.106514

Onel N, Mukherjee A (2017) Why do consumers recycle? A holistic perspective encompassing moral considerations, affective responses, and self-interest motives. Psychol Mark 34:956–971. https://doi.org/10.1002/mar.21035

Ozawa-Meida L, Brockway P, Letten K, Davies J, Fleming P (2013) Measuring carbon performance in a UK University through a consumption-based carbon footprint: De Montfort University case study. J Clean Prod 56:185–198. https://doi.org/10.1016/j.jclepro.2011.09.028

Possingham HP, Bode M, Klein CJ (2015) Optimal conservation outcomes require both restoration and protection. PLoS Biol 13:e1002052. https://doi.org/10.1371/journal.pbio.1002052

Rappaport A, Creighton SH (2007) Degrees that matter, climate change and the University. MIT Press. https://doi.org/10.7551/mitpress/2322.001.0001

Rocha TA (2020) Balanço de carbono do campus-sede da Universidade Federal de Viçosa: inventário das emissões de gases de efeito estufa e remoção de carbono pelas florestas no ano-base de 2019 [Carbon balance of the headquarters campus of the Federal University of Viçosa: inventory of greenhouse gas emissions and carbon removal by forests in the base year 2019]. Thesis, Universidade Federal de Viçosa

Sachs JD, Schmidt-Traub G, Mazzucato M, Messner D, Nakicenovic N, Rockström J (2019) Six transformations to achieve the sustainable development goals. Nat Sustain 2:805–814. https://doi.org/10.1038/s41893-019-0352-9

Santovito RF, Abiko AK (2018) Recommendations for preparation of anthropogenic greenhouse gases emission inventory for University Campuses. Towards green campus operations: energy. Clim Sustain Dev Initiat Univ 297–313.https://doi.org/10.1007/978-3-319-76885-4_19

Silva LB (2017) Aprimoramento da ferramenta de cálculo de emissões e remoções de gases de efeito estufa em propriedades rurais [Improving the tool for calculating greenhouse gas emissions and removals on rural properties]. Universidade Federal de Viçosa, Thesis

Silva TM, Caldeira CRT, Caldeira MCO, Balieiro CPP, Ferreira ME (2021) Classificação de uso e cobertura da terra e o monitoramento de áreas em restauração florestal por RPAS [Land use and cover classification and monitoring of forest restoration areas using RPAS]. Revista Cerrados 190:84–112. https://doi.org/10.46551/rc24482692202104%20

Silva VF (2019) Neutralização de carbono: adaptabilidade e desenvolvimento de espécies florestais no ambiente urbano [Carbon neutralization: adaptability and development of forest species in the urban environment]. Universidade Federal de Viçosa, Thesis

Soares CPB, Neto FP, Souza ALS (2012) Dendometria e Inventário Florestal [Dendometry and forest inventory]. http://www.mensuracaoflorestal.com.br/inventario-florestal. Accessed 27 May 2022

[UFV] Universidade Federal de Viçosa (2020) UFV é a sexta mais sustentável do Brasil, aponta ranking [UFV is the sixth most sustainable in Brazil, according to ranking]. https://www.cpa.ufv.br/sem-categoria/ufv-e-a-sexta-mais-sustentavel-do-brasil-aponta-ranking/. Accessed 24 May 2022

[UFV] Universidade Federal de Viçosa (2021) Lançamento do Projeto Diálogos UFV + Sustentável apresenta novo programa para gerenciamento e consumo de água na Universidade [Launch of the UFV + Sustainable dialogues project presents new program for water management and consumption at the University]. https://www2.dti.ufv.br/noticias/scripts/exibeNoticiaMulti.php?codNot=34918&link=corpo. Accessed 24 May 2022

[UFV] Universidade Federal de Viçosa (2022) Resíduo orgânico do campus Viçosa ganha novo destino [Organic waste from the Viçosa campus gets a new destination]. https://www2.dti.ufv.br/noticias/scripts/exibeNoticiaMulti.php?codNot=36755. Accessed 24 May 2022

[UI Green Metric] UI GreenMetric World University Rankings (2022) Home. https://greenmetric.ui.ac.id/. Accessed 27 May 2022

[UNFCCC] United Nations Framework Convention on Climate Change (2015) Adoption of the Paris agreement: decision -/CP.21. https://unfccc.int/sites/default/files/resource/docs/2015/cop21/eng/l09r01.pdf. Accessed 18 Apr 2022

[UNFCCC] United Nations Framework Convention on Climate Change (2021) Glasgow climate pact: decision -/CP.26 (advance unedited version). https://unfccc.int/sites/default/files/resource/cop26_auv_2f_cover_decision.pdf. Accessed 18 Apr 2022

Valls-Val K, Bovea MD (2021) Carbon footprint in Higher Education Institutions: a literature review and prospects for future research. Clean Technol Environ Policy 23:2523–2542. https://doi.org/10.1007/s10098-021-02180-2

[WRI] World Resources Institute (2004) The greenhouse gas protocol: a corporate accounting and reporting standard: Revised Edition. https://ghgprotocol.org/sites/default/files/standards/ghg-protocol-revised.pdf. Accessed 25 Apr 2022

[WRI] World Resources Institute (2015) GHG protocol: scope 2 guidance: an amendment to the GHG protocol corporate standard. https://ghgprotocol.org/sites/default/files/standards/Scope%202%20Guidance_Final_Sept26.pdf. Accessed 25 Apr 2022

[WRI] World Resources Institute (2022) Climate watch platform. https://www.climatewatchdata.org/. Accessed 26 Apr 2022

Yabushita EER (2013) Inventário e proposta de gerenciamento de gases de efeito estufa (GEE) na UTFPR: estudo de caso do Câmpus Campo Mourão [Greenhouse gas (GHG) inventory and management proposal at UTFPR: a case study of the Campo Mourão Campus]. Universidade Tecnológica Federal do Paraná, Thesis

Zanne AE, Lopez-Gonzalez G, Coomes DA, Ilic J, Jansen S, Lewis SL, Miller RB, Swenson NG, Wiemann MC, Chave J, Lopez-Gonzalez G (2009) Global wood density database. https://datadryad.org/stash/dataset/doi:10.5061/dryad.234. Accessed 26 Apr 2022

Lauana Blenda Silva Forest Engineer (2017), graduated from the Federal University of Viçosa (UFV), MSc (2020) and PhD (2024) in Forest Science at the same University. She works in the areas of environmental economics and climate change. She focuses on the study of methodologies and the development of tools for greenhouse gas emissions and carbon removals inventories, especially in farms and rural activities.

Thaís Almeida Rocha Forestry Technician (2014) from the Federal Institute of Rondônia, Forest Engineer (2021) and MSc (2023) in Forest Science from the Federal University of Viçosa (UFV). Since graduating, she has been developing work related to climate change, specifically with methodologies and elaboration of greenhouse gas emission inventories.

Thaynara Pereira Albuquerque Forestry Technique (2014) by the Federal Institute of the Southeast of Minas Gerais, Rio Pomba campus, Forest Engineer (2021) and MSc (2024) in Forest Science by the Federal University of Viçosa (UFV). She works on environmental economics and climate change, focusing on greenhouse gas inventory and carbon quantification in coffee plantations.

Valéria de Fatima Silva Forest Engineer (2016), graduated from the Federal University of Viçosa (UFV), MSc (2019) and PhD student in Forest Science, and is working in the Forest Management concentration area at the same University. She focuses on studying the carbon market, management and compensation of greenhouse gas emissions and municipal public policies with forecast payment for environmental services.

Mariany Filipini de Freitas Forestry Engineer (2021) from the Federal University of Viçosa (UFV). She is currently studying a Master's degree in Forest Sciences at the same institution in the area of Forest Management. She works in environmental economics and climate change, with a focus on greenhouse gas inventory, management and offsetting.

Laércio Antônio Gonçalves Jacovine Forestry Engineer (1993) from the Federal University of Viçosa (UFV), MSc (1996) and PhD (2000) in Forestry Science from the same University. He is currently an associate professor at the Federal University of Viçosa. He has experience in Forest Resources and Forestry Engineering, with emphasis on Environmental Economics, mainly working on the following topics: carbon credits, carbon sequestration, forest certification, sustainable development, and environmental management.

Sustainable Futures in UK Higher Education Through Multi-disciplinary Innovation and Advanced Practice: A Northumbria University Case Study

Marco Zilvetti, Matteo Conti, and Richard Kotter

Abstract The following chapter intends to critically reflect on multi-disciplinary collaboration and co-creation practices as educational design strategies for near-term impact within higher education, involving external organisations. As part of transformative learning applied to the area of sustainability, three distinct case studies from Northumbria University are presented to illustrate how collaborative projects with external partners around sustainable challenges are tackled by teams of university students supervised by experienced academics. This practice represents not only a testing and professionally rewarding learning and delivery experience for students and academics respectively, but also an effective and leading-edge educational delivery model to reflect on. Its associated pedagogic benefits and challenges will be discussed in order to provide a strategic overview of one of Northumbria University's strands of 'future-ready' approaches to Higher Education which strongly supports students' proactive attitude towards both innovation and an understanding of how to respond to 'real-world' sustainability-related challenges.

Keywords Higher education · Education for sustainable development · Multi-disciplinary design · Innovation practice · Advanced practice · Sustainable futures

M. Zilvetti (✉)
Lecturer and Researcher in Interior Design and Industrial Design, Research Focus On Smart Urban Mobility, Smart Cities, and Product/Service Design, School of Design, Northumbria University, Newcastle upon Tyne NE1 8ST, UK
e-mail: marco.zilvetti@northumbria.ac.uk

M. Conti
Design Innovation/MA/MSc Multi-disciplinary Innovation Programme Leader, Specialist in Industrial and Transportation Design Practice With a Research Focus On Smart Mobility, e-Mobility, and Low Carbon Vehicles, School of Design, Northumbria University, Newcastle upon Tyne NE1 8ST, UK
e-mail: matteo.conti@northumbria.ac.uk

R. Kotter
Department of Geography and Environmental Sciences, Engineering and Environment, Northumbria University, Newcastle upon Tyne NE1 8ST, UK
e-mail: richard.kotter@northumbria.ac.uk

W. Leal Filho et al. (eds.), *The Contribution of Universities Towards Education for Sustainable Development*, World Sustainability Series,
https://doi.org/10.1007/978-3-031-49853-4_21

1 Introduction

Sustainability is a recurring ambition that tends to gain lacklustre meanings if not supported by sound visions and comprehensive strategies. The United Nations' (UN) 2030 Agenda for Sustainable Development represents an authoritative and internationally leading agenda for embodying what society should really be focussing on. It follows that academic establishments are no exception in the dissemination and co-creation of sustainability through teaching and collaborative practices to the next generations of graduates entering the professional arena. Design Thinking and Design as a practice clearly can make substantial contributions towards enabling the implementation of the UN Sustainable Development Goals (DMI Review Staff 2018).

One of the emerging UK universities' educational strategies currently adopts a student-centred and research-informed model (Jenkins et al. 2007). This blended teaching approach is also applied to design disciplines to enable co-creation activities with students involved at all levels of higher education (HE). This method plays a key role in shaping and supporting both the outlook and the skills formation of the future professionals (which can be seen ideally as 'intrapreneurs'), implementers, and decisions-makers.

Here, a multi-disciplinary team of authors who have previously collaborated on this canvas critically set out their higher education experiences, strategies and lessons learnt by pooling their reflections across two different academic Faculties (Arts, Design and Social Sciences, and Engineering and Environment) and Departments (School of Design, and Department of Geography and Environmental Sciences respectively). Three distinct case studies are introduced—alongside an account of wider underpinning designed teaching and learning systems in two areas of the university—to provide a broader understanding of how the discussed teaching methods when combined with experiential and authentic learning offer students a dynamic, critical, and yet applied learning experience towards a sustainable future. The common denominator is Northumbria's university-wide sustainable strategy alongside its supportive, nurturing academic environment, designed to enhance learning and individual skills development.

The first case study is based on the 'Creative Cities' Design module that has been recently introduced to enable undergraduate students from diverse backgrounds to work in multi-disciplinary teams and envision sustainable future scenarios through 'live' innovation projects. This practice-based learning involves stakeholders to provide a more inclusive and realistic project environment to tangibly address key challenges as outlined in United Nations (UN) Sustainable Development Goal (SDG) 4 (Quality Education), SDG 9 (Industry, Innovation, and Infrastructure), SDG 10 (Reduced Inequalities), SDG 11 (Sustainable Cities and Communities), and SDG 12 (Responsible Consumption and Production).

The second focusses on the MA/MSc in Multi-Disciplinary Innovation (MDI), a distinctive postgraduate taught course with innovation practice, teamwork, and project-led learning at the core of its operating model. The programme has for a long time focussed, inter alia, on sustainability issues, with a mixed-design disciplinary student-team centred problem challenge set through engaged clients, including external organisations. A reflective account of two recent projects with Northumbria University's central corporate sustainability team/manager is presented.

The third revolves around the "Subject Exploration" module on the MSc Disaster Management and Sustainable Development (a social science course, but with a very wide intake across all disciplines and classed as a 'conversion' or career change one by the university) based on an individual secondary-data based critical exploration in the wider fields of disaster management, disaster risk reduction, humanitarian development/logistics and sustainable development (and into environmental and humanities topics). This is a steppingstone for (even more) challenging modules students must take in their programme, both in the one-year programme and the 'with Advanced Practice' prolonged version.

2 Relevant Pedagogical Theories

Design is a discipline of making where theoretical and practical knowledge are intertwined. Key learning theories presented here provide one of the conceptual and pedagogical frameworks that inform the curricula delivery at Northumbria University. Learning within the classroom or studio is intended to be student-centred and accomplished through active discovery where the role of the lecturer is to facilitate this learning process through one-to-one and group tutorials, alongside lectures and seminars.

Independent learning and peer support are also regarded as key drivers in the learning process (Donald and Ford 2022). They can be enhanced by action learning (Northumbria University 2015) when a real-world problem or scenario are used as a vehicle to set outcome-focussed learning in a tangible and applied manner.

As part of the learning process, the four learning cycle stages introduced by Kolb or similar models are applied and reviewed by postgraduate taught students with academic staff acting as facilitator throughout the academic year, and particularly at the end of each project. The so-called 'unpacking session' is delivered in the form of a critical and reflective tool which allows the class/cohort/student team to critically review their project work, the dynamics around teamwork, to identify good practice and any issues or areas for improvement (Fig. 1).

This framework also directly relates to Morgan and Jaspersen's (2022: 333–334) four-step learning model (Research—Doing—Reflecting—Further Research) and the seven skills areas (Innovation and Design Thinking; Collaboration and teaming; Critical thinking and problem solving; Commercial awareness; Empathy and communication; Resilience and managing change; Initiative and active learning).

Several pedagogical and design theories inform and influence the teaching delivery within the School of Design and on other multi-disciplinary programmes across the university, including the 'with Advanced Practice' disciplinary-based Masters programme provision in the Faculty of Engineering and Environment. It can

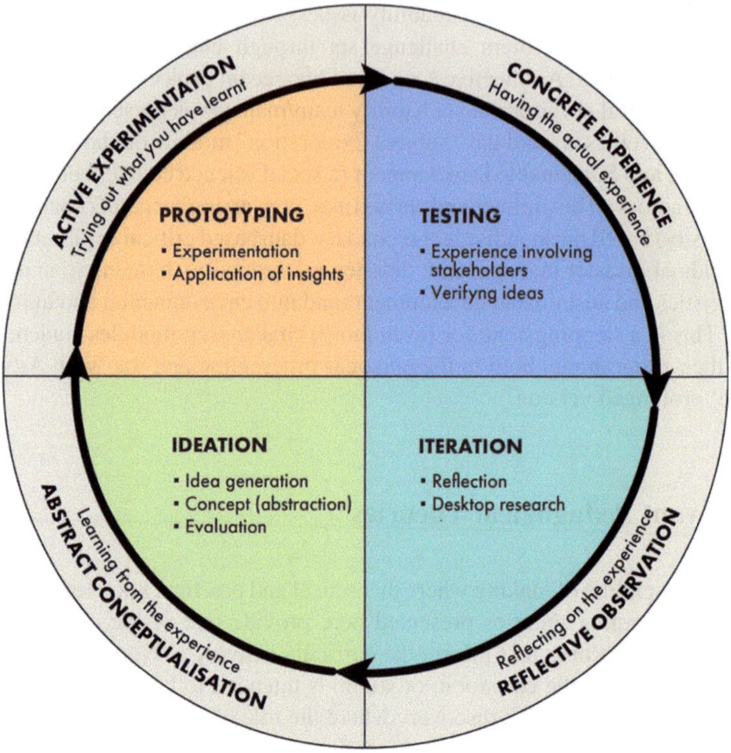

Fig. 1 Design disciple learning cycle (adapted from Kolb's reflective cycle, Mcleod 2023a)

be argued that this approach, widely applicable to academia and other professional sectors, embodies some of main underlying principles of sustainability as it induces objective introspection, critical and analytical thinking, and the ability to propose constructive recommendations and ideas to implement as part of the process (Biggs and Tang 2011). These skills, combined with resolve, are required more than ever to question, challenge and respond to the widespread status quo around sustainability, circular economy and related systemic issues the world is currently facing.

Cognitive Constructivism

This perspective about cognitive development has been developed by the psychologist Piaget and underlines the key role of the individual who is the centre of the knowledge-creation and acquisition process. The focus of Piaget's theory is the series of long-term developments, which represent a sequence of thinking patterns that occur at different ages between birth and the end of adolescence, as children gradually become able to think logically and scientifically. As Thompson (2019) pointed out, this Cognitive Development Theory identifies four major stages of personal growth creating a 'staircase' model of children's development (Fig. 2).

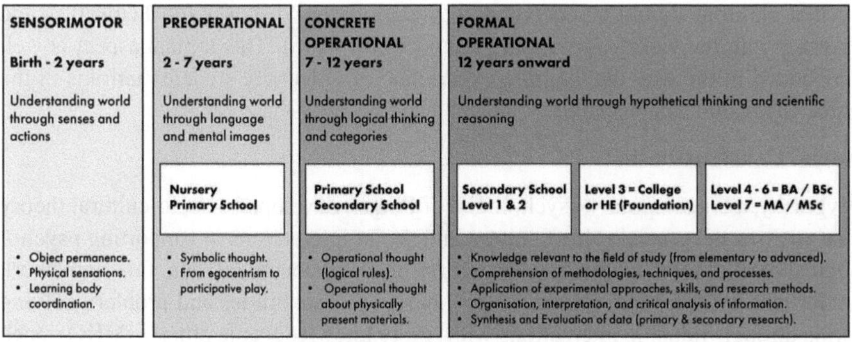

Fig. 2 Piaget's theory of cognitive development (informed by McLeod 2023b, and Northumbria University module specs informed inter alia by Bloom's taxonomy)

– Each stage represents a 'sequence of thinking patterns' with key features:
– The four stages always happen in the same chronological order.
– No stage can be skipped.
– Significant transformations take place between each stage and the previous one.
– Each stage incorporates the earlier stages within itself.

The last stage, which occurs approximately from the age of twelve onwards, is the formal operational stage where the individual can handle abstract and hypothetical problems, and deal with 'forms'/representations. This allows children to think about abstract concepts, logically test hypothesis and formulate broader scientific thinking. Such abilities are then reflected in the delivery of taught contents, as students can be asked to manipulate ideas in several ways (Mcleod 2023b).

Piaget claimed that knowledge cannot emerge solely from sensorial experience and that supporting structures must be created. The notion of 'schema' emerges as "a cohesive, repeatable action sequence possessing component actions that are tightly interconnected and governed by a core meaning" (Piaget and Cook 1952:7). Two aspects become fundamental in this regard:

– Assimilation, which is the process of fitting new information into existing cognitive schemes, perceptions and understanding.
– Accommodation, which occurs when schemes need to radically change as the new experience is quite different from what encountered before.

The notion of 'readiness' (referring to when information or concepts should be taught) is of utmost importance in this framework, as contents should be taught only at appropriate stages of cognitive development (Bond 2008). More specifically, Smith et al. (1998) point out that some children develop earlier than Piaget predicted through group work by learning to appreciate the views of others in preparation for the concrete operational stage. In consideration of that, curricula should be flexible to allow for variations in ability of different students of the same age. Therefore, education must be stage-specific, instead of being given ready-made knowledge, so that children should be encouraged to explore and discover for themselves and interact with the world surrounding them (Smith 1996). This topical aspect is well embedded in the way the learning experience of university students unfolds in the three cases studies presented.

Social Constructivism

Vygotsky, a contemporary psychologist of Piaget, developed a socio-cultural theory that stresses the crucial role of culture and social interactions in supporting psychological development in children. As a form of social constructivism, this framework underlines how children acquire values, language capabilities and problem-solving skills through valuable interactions with 'more knowledgeable others' (MKO), such as peers and adults (Mcleod 2023c). In this respect both Piaget and Vygotsky view children as actively constructing their knowledge of the world and going through phases in their learning process. However, the latter emphasises how development does not follow universal stages, as it depends on cultural diversity instead (Vygotsky 1978).

Cultural and social context deeply influence cognitive development, which stems from guided learning activities in the so-called 'zone of proximal development' (ZPD). This theory accentuates the vital role of culture in the development of mental abilities and the relevance of adults in teaching children the meaning of objects and experiences (Chaiklin 2003). Here communities play a crucial part, as the cognitively prepared student needs guidance and social interaction through challenging tasks to ensure full development of their higher mental functions and capabilities. Once again, this type of educational model is one of the core principles considered in the project work outlined through the case studies.

The concept of 'scaffoldings' defines the difference between what a student can achieve independently and what can be achieved by receiving support and encouragements from MKOs. In the ZPD, teachers or tutors alongside more experienced peers provide learners with instruction and guidance. As language and social interactions play a vital role in making meanings and shaping thought, the delivery of contents should be calibrated not only according to learner's age but also how social factors entail qualitative changes in thinking to support co-construction of meaning and development of advanced skills (Verenikina 2003). This aspect is self-evident at Northumbria University in the design of module learning contents and project briefs to ensure that learning is aligned to evolving project stages which correspond to the students' enhanced ability to deal with increased complexity in a team-based setting.

Design Thinking

This framework of knowledge identifies key principles and approaches to problem solving that underpin professionals within the Design practice, as they go through stages of ideation according to which they identify issues and creatively turn ideas into solutions. As a theoretical and practice-based approach it identifies different forms of intellectual competences spanning from linguistic, logical-mathematical, spatial, musical, bodily-kinaesthetic, and personal, to cognitive skills. Such abilities can be applied at various levels throughout the thinking process and support designers in their approach to problem solving (Cross 2011).

There is no unique definition of Design Thinking (DT) as it can be seen in the form of a strategy, a method, a framework, a toolkit, and mindset (IDEO LLC 2012). It is widely recognised as an abductive process in which the designer/innovator produces a composition, a product, a service, or an intervention to respond to a given problem or opportunity. Unlike the classic scientific method of thinking, the Design process cannot derive a solution directly from the problem: cognition is not merely 'intra-mental' but rather a practical, interactive, and cohesive activity that involves a combination of both thinking and acting (Gedenryd 1983). Consequently, a range of techniques as sketching, drawing, model making, prototyping etc. become part of a project strategy that allow designers to outline and interrogate future scenarios, deriving strategies for effective solutions.

The Design process is formed by a series of both planned and unplanned activities, in which elaboration of information, exploration of alternative ideas, development of solution concepts and reiterations are intertwined factors. In this setting teamwork represents a core aspect within design practice, as social/professional interactions and communication have massive impact on the ideation process. Not surprisingly, the DT perspective emphasises the relevance of collaboration as a key strategic factor for work practice to optimise the workflow and the range of cognitive tasks (e.g., inquiring, evaluating, formulating, visualising, modifying, etc.) that each professional elaborates and produces (Goldschmidt 1995). When designers work within a team, successful collaboration is achieved when group roles are well balanced (Bereton et al. 1996) and both communication and negotiation are managed so that they become constructive elements that support team members' contributions and effective discussions towards coherent idea development (Dong 2005). This is a fundamental premise applied to teamwork through DT alongside design and business tools which Northumbria University students from different background are exposed to, as described in the case studies.

While Design Thinking is inherent in the way practitioners approach design and envision solutions, innate talent is 'relatively unimportant' compared to the accumulation of experience. Dedication and motivation become pivotal in the development of expertise from novice to expert, and education has a key role in providing 'novice' designers and innovators with knowledge, technical insights and experiences that would eventually support the changes in behaviour, approaches to problem exploration and design ideation (Orr and Shreeve 2019; Cross 2011). As Brown (2008) stated: "*Design thinking is a human-centred approach to innovation that draws from the designer's toolkit to integrate the needs of people, the possibilities of technology, and the requirements for business success.*" This refers to the key dimensions of design innovation, e.g., the human, technical and economical perspectives, as DT lies at the intersection of three key domains: desirability, feasibility, and viability. It follows that inspiration, ideation and implementation are the three core activities of DT.

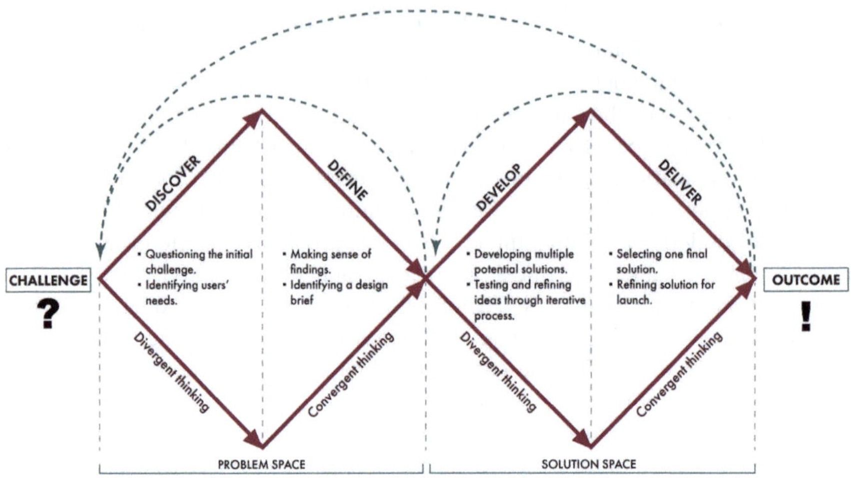

Fig. 3 The double diamond design thinking process (adapted from Design Council 2019 and informed by Ball 2019)

The Double Diamond Framework

Divergence and convergence are two sides of contextualised Design Thinking as they refer to the dynamics of the ideation process (IDEO LLC 2023). In this regard, the British Design Council in 2003 developed a visual representation of the design and innovation process called 'Double Diamond' to provide a sort of navigation tool to practitioners, academics, and students as they plan and develop projects. This framework effectively indicates the fundamental steps taken in any design and innovation project, and identifies four key phases (Ball 2019) as (Fig. 3):

- **Discover**—in which the initial challenge is questioned to identify users' needs.
- **Define**—in which designers make sense of findings, to understand how users' needs and problems align, and identify a design brief that defines the challenge based on such insights.
- **Develop**—in which multiple potential solutions are developed, tested, and refined through iterative processes.
- **Deliver**—in which one final solution is selected and prepared for launch.

Human-Centred Design

Human-centred Design (HCD) is an approach to problem-solving which is commonly utilised within product, service, and systems design, as well as engineering. The HCD technique allows designers to adopt the perspective of the users, with their needs, desires, and pain points at the core of the design process to lead to solutions tailored around the actual needs of the audience (Landry 2020). Townson (2017) argues that HCD should consider not just the users but also all human beings that the design idea has an impact on. Good design should work for all people it encounters and this includes stakeholders that designers may not be fully aware of at the beginning of the ideation process. For this reason, designers should undertake research to observe the environment and engage with users to gain knowledge and derive primary insights that would facilitate the ideation process.

Informed by the Design Thinking framework, the HCD approach identifies five phases within the problem-solving process, which are (see also the representation in Fig. 4).

- **Empathise**—This initial phase expands on the early assumptions by undertaking research and considering customers' perspectives (emotions, needs, expectations etc.) to clarify the problem and outline pathways for effective solutions. At this stage it is important to discover customer's pain points, which can be either explicit (e.g., the ones the user is aware of) or latent (e.g., the ones the user is totally or partially unaware of). The role of researchers is to leverage their expertise and utilise primary research (interviews, observations etc.) to derive findings and observations that lead to transformative innovations (Datar and Bowler 2015).
- **Define**—The designer task is to make sense of the insights gathered in the initial stage and define a problem statement(s) from the users' perspective which will then inform the ideation process.
- **Ideate**—Ideas are explored, evaluated, and then combined to produce potential solutions to the previously identified problems. At this point it is critical to embed desirability, feasibility, and viability in the evaluation process. Several DT tools

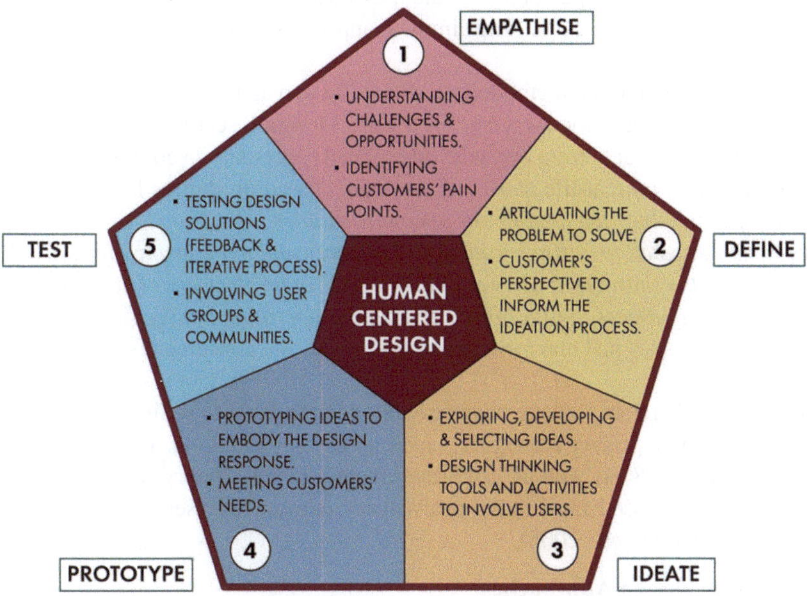

Fig. 4 Diagram of HCD process (adapted from Tamarack Institute 2022)

(e.g., brainstorming) and activities that involve the end-users, can be utilised to overcome cognitive bias and functional fixedness (Brown 2008).
- **Prototype**—As ideas are evaluated, the ones that meet users' needs are selected, and prototyped to embody what the design response (product, service, policy etc.) could look like.
- **Test**—In this final stage, design solutions and prototypes are tested with user groups and communities the design ideas are meant for.

Following this procedure HCD blends divergent and convergent thinking throughout the evolutionary stages, in line with the Double Diamond approach to co-design. As this process is mostly iterative, ideas may be combined, evaluated, and revised multiple times, as user groups are involved at different stages of completion. Testing proposals and prototypes is very likely to generate insights as feedback from users and wider groups is collected. This in turn informs the design process and leads to improved solutions (Nous Group 2023).

Microsoft Research authors (Harper et al., eds, 2008:58–59) had proposed an extension to the research and design cycle in Human–Computer-Interaction (HCI) by adding a further stage of conceptual analysis which considers, in a deeper and systematic manner, human values rather than just focusing on understanding the examined problem. This involves running experiments and building scenarios through discourse with professionals and designers from different disciplines and backgrounds to capture insights on human values. The proposed approach revolves around the quest to answer the why, what, and how to design in a more meaningful manner through prototypes being created, evaluated by selected participants, and improved in an iterative cycle. The aim is to apply this methodology at all stages of the process.

In consideration to the methodological side, Bueno and Ramckers (2003:2) have described an approach for carrying out Philips Design research that makes use of rich and illustrative data from everyday life, namely 'personas in context' or fictional characters based on real-life stories. They argue, based on Philips Design's (2003) 'Pulse' research that '*the creation of personas enables [consumer] research results to be communicated to designers in a much clearer and more inspiration way. This approach is also characterized by a greater degree of involvement of designers early in the research process, and a higher level of co-operation between designers and researchers.'* This is just one form of linking design and consume research (Veryzer 2000).

Of course, there are a range of methodologies available to designers (as well as innovation and/or project managers), such as Design Thinking, Lean Six Sigma, Agile (Asante 2018) as well as LogFrame Analysis and 'Theory' of Change. And teaching user-focussed innovation is also very relevant for (service and product) design innovation (Read 2005).

Synthesising Design Approaches

Nowadays, the multitude of ready-to-use design methods and tools allow for a wide range of project approaches which are typically discussed and planned before or from the start of any project. Usually, experienced practitioners, researchers and academics are in an advantageous position not only to think strategically about what methodologies, theories, and tools to adopt in the planning of a project, but also how to blend and/or adapt those based on practical experience and expertise. Even if there are a few unknowns to contend with at the beginning of the project, including how students individually and in teams may work together and respond to the set brief, this is part of the typical scenario associated with the proposed case studies. The dynamics between students, tutors, project partners and other stakeholders are normally seen as the core melting pot where different capabilities and mindsets are put to the test when working together towards the realisation of the 'live' project aims.

Often a revised and improved project approach is considered or even implemented at the key stages of the project, as design remains a fluid activity which cannot be entirely pre-planned and executed in a very defined, chronological, and rigid manner. Within this context what is often suggested and discussed is the ability to work collectively to map out in the form of diagrams, info graphics or flow charts, a process which simulates how project tasks and outputs relate to each other and will be executed. To illustrate this Design Thinking planning task the above-mentioned Double Diamond and HCD processes can be combined as in Fig. 5, to provide students more clarity and confidence in initial and recurring project development discussions.

Application of Design Thinking and Learning Within Higher Education

In the fast-pacing digitalised world we live in customers' needs and expectations evolve and lead to an ever-changing demand for new services and products (Droga and Shah 2022; Budge 2012)). Therefore, in the quest to innovate, designers need to be proactive in their research efforts by observing local and global dynamics to meet people's desires and expectations (IDEO 2015). In this regard, Higher Education plays a vital role in providing future professionals with HCD insights, DT skills, and training opportunities that would allow them to gain expertise in the multifaceted areas of design and innovation. This aim aligns to the universities' mission to generate

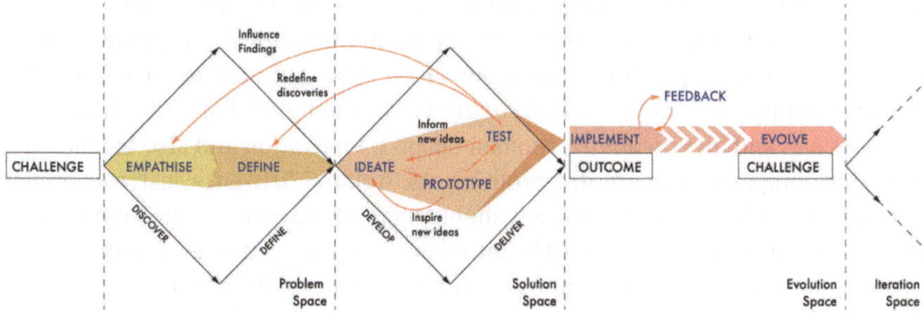

Fig. 5 Diagram of double diamond and HCD (*source* authors)

knowledge and create socio-cultural and economic value through its contribution to the advancement of society at large (Luka 2019).

Empirical studies confirm the value of DT beyond the original fields of design, architecture, and art (Dorst 2011), and, when adopted by a broader audience, such an empathetic, iterative, and flexible approach to problem-solving promotes a holistic view of 'real-world' challenges. Embedding DT methods in HE teaching activities is a successful strategy to engage students in relevant group work that supports the ideation of desirable, feasible and multi-disciplinary solutions to complex problems (McLaughlin et al. 2022). Integration, critical observation, self-reflection, flexibility, and creative collaboration become crucial aspects of the DT-informed research and ideation processes (Val et al. 2017). With regards to pedagogy, and relevant to the purpose of the case studies analysed here, Lake et al. (2021) underline the relevance of such aspects as:

- Situatedness, intended as the relationship between students' subjective experiences and their response to the surrounding world.
- Iterative exploration alongside stakeholders.
- Divergent idea-generation processes to explore alternative possibilities.
- Prototyping and testing to evaluate potential solutions.
- Commitment to change and sustainability.

In this regard, team dynamics, studio-based activities, multi-disciplinary research, and curated approach to 'Live Projects' play a crucial role in supporting students' commitment to collaborative work and foster meaningful design outcomes (Blair et al. 2008). Embedding various Sustainable Development Goals into, for instance, business or business-related subjects through an experiential learning approach (Fang and O'Toole 2023) is gaining momentum at both under-graduate and Masters degree levels. Fang and O'Toole (2023) state that students' learning, understanding and application of SDGs during their business schools' programmes have enabled them to develop a skills-set and a new mindset which make them more employable. However, the big challenge remains for academic institutions to devise successful strategies which enable SDGs to be well integrated into their curriculum and students' learning.

In line with the "Cox Review of Creativity in Business: building on the UK's strengths of 2005", which concerned smaller business and in particularly manufacturing, it is also important not just to look at the multi-national companies/ corporations of the world operating in and out of the UK, but also to look for value added and quality (and 'clean'/ NetZero) growth beyond the FTSE 100, 'blue-chip' companies etc. and into creativity, (human-shaped) innovation and also technology/ design-rich companies and non-for-profit actors. Acklin (2010), for instance, argues that small and medium sized businesses should incorporate both design methodology and design management based on DT into their company activities and innovation processes in an integrated and in-house fashion. As Julier (2017) argues, beyond the way design has historically been promoted for its capacity to add values to products and services, in contemporary capitalism it assumes a much more central and complex role which is influenced by, and in turn actively shapes, economic systems around us. The same can be argued for applied environmental and social science, especially in interaction with business, design, and societal issues (Davidson et al. 2016). For instance, amongst the approaches and tools promoted on all the modules/ courses which we have chosen here as case studies, (group) concept mapping applies for (future) planning and evaluation can be taught (Kane and Trochim 2007), nurtured and supported in a curated educational 'real-world' environment. Here, academic tutors and professional staff (with internal moderators and external examiners input) set regular post-and mid-term delivery analytical reviews of how well the (experiential) learning provision, the student engagement and the learning outcomes process have been working.

Educators and higher education institutions aim to embed into (post-)graduates specific skills for the sake of employability, career journeys, life-long learning, social mobility and impact. They relate to individuals that once graduated find themselves working in organisations (whether for or non-for-profit), social contexts, and in most cases (at least temporary, 'project') teams. As Dawson and Andriopoulus (2021:414) argue, change or flux, adaptation and anticipation are important; cognitive factors and personality traits that relate to individual creativity. This includes the domains of the 'literal', the 'abstraction', and the 'association', but furthermore judgment (discernment) and originality of thinking. All of these have implications for creativity, innovation, and change management in the workplace. However, one should not overlook that knowledge is also needed to be truly creative, in the domain of both formal/ explicit and informal/ tacit knowledge. The ultimate difference is then made by motivation—both extrinsic and intrinsic (Dawson and Andriopoulus 2021:415–430).

As Andersson (2009:15) pointed out, the (industrial, product) '*designer is usually seen as a specialist iform, colour, graphics, visualisation and aesthetics'. However, what is perhaps less well known is that the designer has [acquired] knowledge about the whole chain, from the users' behaviour to material engineering and manufacturing process, to market conditions.'* This really should be said about applied environmental scientists, specialists and managing/policy 'generalists'/professionals in disaster and humanitarian and socio-economic development/ integrated emergency management/ disaster risk reduction. In this vein, writing about the design on intelligent systems, Hummels and Vinke (2009:24) argue that industrial design students need to take an 'active position' about the role of design, considering the social, moral, aspects of it, and factoring in such aspects as the role of technology (as a mediator) and the societal transformations. *'Designers have to put this [ethical and impact] topic of social and societal transformation on the table despite the fact that it is not their sole responsibility, because designers often work in multi-disciplinary teams and because technology is context-dependent and multi-stable'.* Considering digital/virtual and service design, he underlines that in the past designers worked on the appearance of tangible products and their application, whereas today they often work on the experience of intangible products and their 'cognitive application'. He underlines the evolving role of designers as design is increasingly influenced by information from computer-based products/ services (for learning, leisure, and decision-making). *'The design process employed by architects, designers and others is of great interest today to industry as well as public authorities and organisations.*

*There are many indications that the design process can become an alternative way of handling complex problems connected with progress. [...] Aesthetic aspects are always very much present. For the designer they are the core of, and expression of, the concern for the humanistic aspect which is the foundation of all design activity. This is also true of ethical considerations. Both ecological and social sustainability are ethical questions which occupy the designer. [...] The designer has to be familiar with many different areas. In fact, it is typical for a designer to know a little about a great many things' (*Andersson 2009:15).

For the UK, Crisp (2010:24) outlines the journey of design research in initially different forms of higher education until the Further and Higher Education Act of 1992 eliminated the formal divide between polytechnics and universities and joined them (in terms of provision) into a "unified" system. Essentially, in the 1970s technological design was embedded within other subjects as a component of engineering disciplines and blended with physics, chemistry, and mathematics-related subjects. It was also taught in Art Colleges as a key driver for theatre, ceramic and furniture design. It was only in the 1990s that design became a discipline on its own, recognised in its value by a political landscape that was keen to adopt the 'creative industries' as a key factor for producing wealth in the United Kingdom (Wilson 2022).

Blended Approaches to Learning

During the COVID-19 period Northumbria University swiftly adapted to the imposed government restrictions in terms of isolation and the requirement for individuals to study or work from home. Consequently, on-campus teaching delivery was replaced by online delivery through a dedicated Virtual Learning Environment (VLE) called Blackboard (Northumbria University 2023a). This platform allowed teaching and learning for all courses to be delivered using digital tools and resources. In addition to that, the School of Design created a website named X Studio (Northumbria University 2020a) to share online internal resources for team collaboration, online sessions, research, and DT tools, as well as a series of guides. It is worth noting that the sudden switch to online teaching had been challenging for both students and staff. It has taken some time for students to get used to this new way of being taught, as well as working and collaborating on Live Projects. Staff had been trained and managed to adapt to this new way of teaching.

The School of Design operates an extended studio approach (X-Studio) that embraces our physical studios and makes workshops as well as our virtual spaces to create a flexible and rich environment that is responsive to our learners' needs. This dedicated approach has been developed over many years of continuous teaching innovation, research, and industry collaboration. It has created a supportive extended studio culture that encourages our students to learn flexibly, dynamically blending their learning and University experience between physical and digital interactions both in real-time and online at their own pace. This extended studio culture not only helps our students to become more confident autonomous learners but also prepares them for a dynamic design industry in which an extended studio philosophy is current practice.

More broadly, Northumbria University encourages all students to attend taught studios sessions, seminars and lectures. The course timetable is carefully planned to maximise staff availability, make best use of students' availability through academic and/or technical contact time, provide access to internal technical resources to support directed and independent learning. However, it is the student's responsibility to maintain a professional approach to taught sessions, undertake independent study and manage their time. Students should be prepared to work on a project in the timetabled sessions and be pro-active with their work. From the end of January 2021, the university has implemented a return to face-to-face teaching on campus which is once again the main and preferred teaching delivery method although some reduced online delivery is still applied to some courses where appropriate. The challenges and limitations of teaching entirely online are well known today as students and staff have experienced for an extended period of the pandemic.

Currently, as academic and research practitioners, working again in normal operating conditions, we are reflecting on how to better implement:

- Sustainability and the United Nations (UN) Sustainable Development Goals (SDGs).
- Multi-disciplinary approaches to human-centred design (HCD).
- The Programme Framework for Northumbria Awards (PFNA) in a way that encompasses both the SDGs and HCD.

3 Sustainable Development Goals and (UK-Based) Higher Education

Three case studies are critically explored in this chapter to illustrate how higher education at Northumbria University applies multi-disciplinary approaches to teaching and explores internationalisation opportunities in collaboration with partner institutions. Overall, they sit within a university-wide strategy to supports its teaching and learning approach, contribute to research (including by bringing in students to this at all levels), as well as (including social and mobility) impact. Northumbria University's evolving narrative, having used "*Research-Intensive Modern University*" as a shorthand to describe it is as a community and organisation (which has won the UK title of the "Times Higher Education (THE) University of the Year 2022" is currently moving to describing ourselves as a "*research-intensive university that unlocks potential for all, changing lives regionally, nationally and internationally.*" [Vice-Chancellor's staff briefing by e-mail, 31/03/2023].

One way of zooming into sustainability is via the UN Sustainable Development Goals (SDGs), which are the blueprint for achieving more virtuous and futures that respect both humans and the environment. Many are relevant for our university and its stakeholder impacts (Northumbria University 2022a); however, the focus here is on:

- SDG4 Quality Education—developing knowledge and skills to promote sustainable development.
- SDG8 Decent Work and Economic Growth.
- SDG9 Industry, Innovation and Infrastructure.
- SDG11 Sustainable Cities and Communities.
- SDG13 Climate Action.
- SDG17 Partnership for the Goals.

To confirm the University's support to the UN SDGs, a corporate organisational statement clarified that: "*Northumbria (University) is committed to supporting the Sustainable Development Goals. We have made some excellent achievements to date including ranking 4th in the UK (People and Planet) and as 27th in the world (THE Impact League) for activity in support of sustainability, as well as reducing our carbon emissions by 48% in the past 5 years. There is however more to be done and many complexities to consider and we need your support!*" (Northumbria University 2021a) (Fig. 6).

According to the UK's Quality Assurance Agency (QAA)'s as one of the regulators also for the higher educator sector, the Higher Education Academy (HEA)'s 2014 "Education for Sustainable Development Group" (2014), Education for Sustainable Development (ESD) is '*the process of equipping students with the knowledge and understanding, skills and attributes needed to work and live in a way that safeguards environmental, social and economic wellbeing, both in the present and for future generations'*. Consequently, Northumbria University's strategy aims at shaping teaching activities so that they support the development of knowledge and skills with clear focus on raising sustainability awareness. In this regard, the institutional staff and student community is constantly encouraged to engage and get involved in sustainability-related initiatives (Northumbria University 2022b), as their actions through teaching and research are considered extremely important.

The University's dedicated platform invites those involved in teaching to consider how to use their activity to further deliver and advance ESD. In this regard, the strategy:

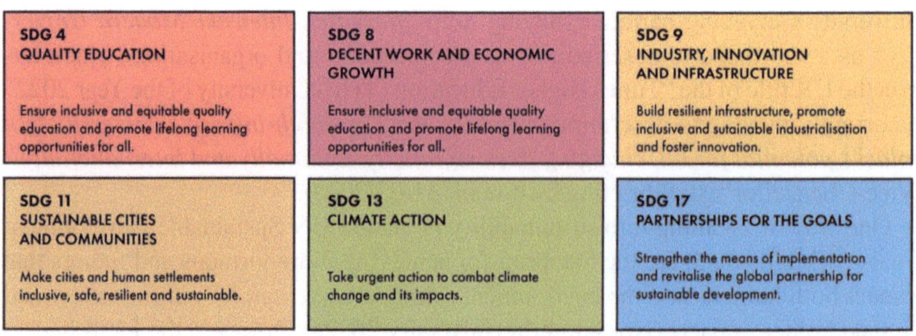

Fig. 6 UN sustainable development goals considered in this study (adapted from United Nations 2022)

- Indicates the focus on using operations, research and education activity to support its contribution to the UN SDGs (Northumbria University 2021b).
- Outlines NU's Sustainable Living Lab Projects (Northumbria University 2020a)
- Provides a framework for Environmental Management System (EMS) for carbon reduction on campus, and this includes the installation of solar photovoltaic (PV) and air source heat pumps, as well as improved lighting, heating, and building controls.
- Outlines the Environmental Sustainability Policy (ESP), a strategy—reviewed on an annual basis—that considers the impact of activities and strives to meet and exceed (where possible) the current legislation, industrial standards, and environmental requirements. It also outlines plans to support the institutional activities towards the UN SDGs in several areas.

As can be appreciated from Fig. 7, all SDGs are addressed in the ESP, especially SDG 11, SDG 12, SDG 13, and SDG 15. Northumbria University is also a signatory of the SDG Accord and the United Nations Academic Impact, in recognition of the positive impact it makes as part of its portfolio of sustainability initiatives (Northumbria University 2023d). Moreover, the university boasts a series of key institutional achievements in the field of sustainability to acknowledge its commitments. In 2022 Northumbria University was ranked in the top tier (15th) in the People and Planet's University League table, which is annually compiled by the UK's largest student campaigning network. Northumbria was also placed in the top tier of Uswitch's Green University Report 2022 for its commitment to sustainability, acknowledging the university's institutional engagement towards sustainability. This follows on from Northumbria's National Green Gown Award for 'Benefitting Society' (2019) and its ISO14001 certified Environmental Management System and Platinum EcoCampus Award (2022). The university was placed 50th in the world for sustainability in the 2021 Times Higher Education (THE) Impact Rankings (a global league table which measure universities' success in delivering the UN Sustainable Development Goals) after having come 27th in the 2020 global ranking (and 8th in the UK).

4 Programme Framework for Northumbria Awards and Continuous Programme Performance Review

As a widespread practice in academia, Northumbria University periodically re-design its programmes to reflect both the real-world trends and the feedback from students and staff. As part of this re-validation process, a strategy has been outlined to respond to the challenges of the constantly evolving world as well as to the UN SDGs. In this regard, the guidelines include the following objectives:

- Module and Programme validation.
- Involving students in the process.
- Embedding UN SDGs within research-informed teaching across the programmes.

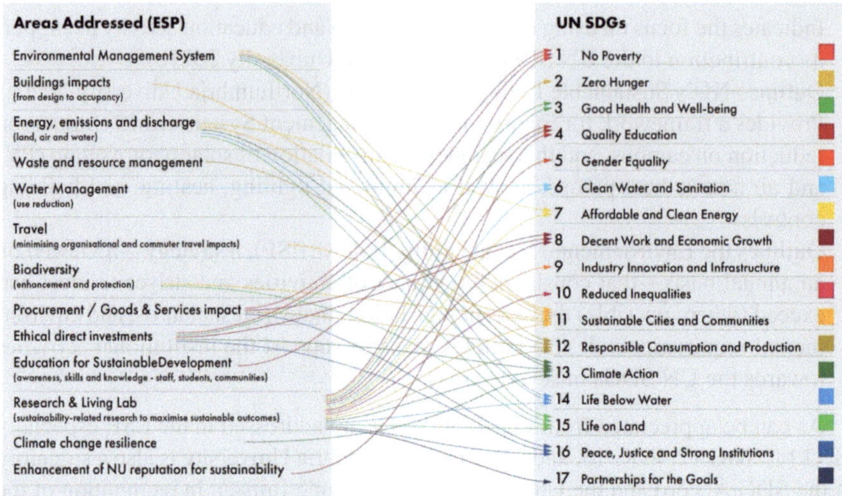

Fig. 7 Environmental Sustainability Policy (ESP) and relevant plans aligned to the UN SDGs (Northumbria University 2022c)

The current basis for structuring both programmes and modules within the institution is the Programme Framework for Northumbria Awards (PFNA), which has been developed after a university-wide re-validation process of all programmes and modules within the institution. It maps outputs against operational considerations and four proposed design pillars (Northumbria University 2023b):

1. Technology Enhanced Learning.
2. Assessment.
3. Research-rich learning embedded into the curriculum (programme and module specs).
4. Employability.

The PFNA represents the starting point for the development of both modules of study and programmes, as it identifies specific Learning Outcomes (LOs) for that define the Graduate Characteristics and the expected achievements across programmes and modules. In turn, these LOs inform the project briefs and the assessment criteria according to which students work is marked (Holgate 2016) (Figs. 8 and 9).

One other element that must be pointed out in the University's central approach is the—for some years now centralised and digital format-based—surveys of student. This is beyond the University-independently organised surveys of the National Student Survey (NSS) [except for Oxford and Cambridge universities] for the final under-graduate degree level year [typically their 3rd on a Bachelor programme, unless they do a 'sandwich' study format with a year in either or half/half "industry" or study abroad, in which case it would be their 4th] and the on–off (with many and increasing number of UK universities opting out) running of the 'national' Postgraduate [Taught]

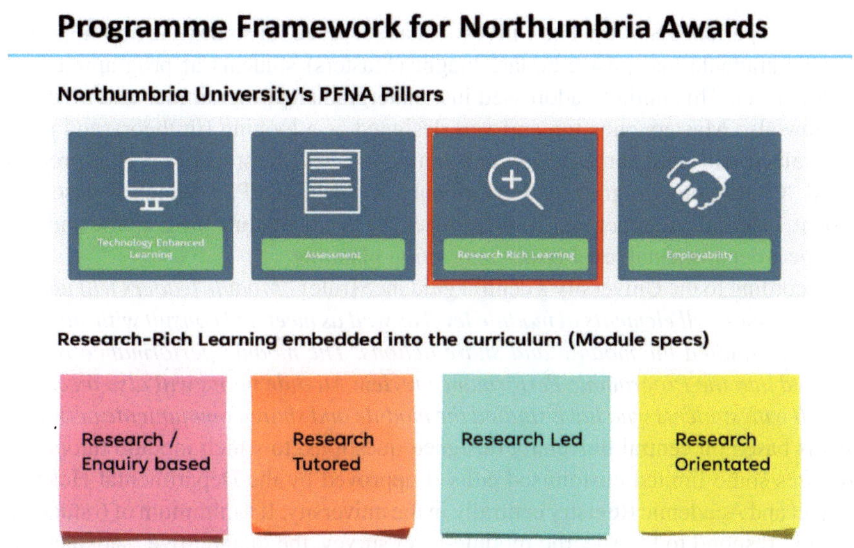

Fig. 8 Northumbria University's PFNA pillars (Northumbria University 2023c)

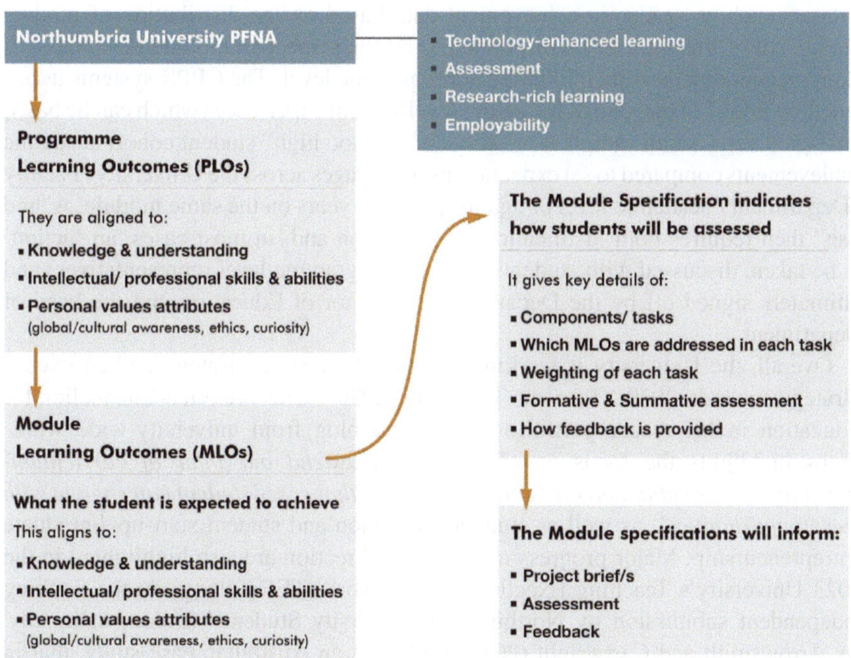

Fig. 9 PFNA programme and module learning outcomes diagram (*source* authors)

Student Experience Survey. The University has surveyed for some time now all levels of undergraduate and post-graduate taught (Masters) students at programme and module level. This initially addressed just undergraduate programmes and modules but now also Masters ones too and was designed as a looping (in theory and principle, albeit still with some design and timing "issues" in practice) digital process called "Continuous Programme Performance Review" (CPPR), which is based on student, external examiner and staff feedback, to tackle module programme and—if need be—departmental level issues and action planning.

According to the University's central guidance [rule] *"Module leaders will undertake a review of all elements at module level as well as meet and consult with students who have studied the module and share actions. The module performance review will feed into the Programme Performance review. Module tutors will also meet and consult with students who have studied the module and share/communicate act*ions." This is based on central university designed questions, to which module tutors can proposes some limited customised edits if approved by the Departmental Head of Subject and Academic Registry centrally in the university. If a minimum of 6 students digitally respond to the specific module-level survey, the quantitative 'satisfaction' data (on scale of 1–5) will be produced alongside collated [and anonymised] qualitative comments the students may have submitted. It is also informed by background datasets such as academic achievement data based on the distribution of module marks. All of these are to be responded to by the respective module teaching staff team on a module, and then likewise at programme level. The CPPR systems uses a "methodology" of thresholds for below-level student experience (which can be based on—even very- small samples) or "too low" or "too high" student cohort academic achievements compared to set expectations or averages across the University / Faculty / Department / academic level or against previous years on the same module. A "red flag" then requires both justification / explanation and, in most cases, an 'action' to be taken, discussed with students (or their programme level representatives) and ultimately signed off by the Departmental Director of Education and the head of Department.

Overall, the University is working to a new Education Strategy and an overall Strategy towards 2030 "Take on Tomorrow 2030". The Pro-Vice-Chancellor for Education in his recent portfolio update and blog from university-wide workshops highlights the focus on *'How can we extend our offer of curriculum-embedded experiential and authentic learning so that every student can engage with these opportun*ities', as well as student innovation and student start-ups/ graduate entrepreneurship. Major progress made in this direction are also highlighted in the 2023 University's Teaching Excellence Framework (TEF), alongside the flanking independent submission by Northumbria University Students' Union, nationally. As Arrowsmith and Cartwright (2019) find from an Australian case study, that in universities with a close industry, technical and vocational focus, the engagement of academics and students (at various levels) with 'industry' has been valued in terms of solving 'real-world' problems. This often corresponds to active participation of industry and professional organisations in program development, which is the long-established operating model of the School of Design at Northumbria University.

5 Multi-disciplinary Approaches to Human-Centred Design (Thinking, Student-Centred and Research-Informed Practice) Across Academic Subjects

As previously introduced, multi-disciplinary innovation and human-centred design are fundamental methods applied at Northumbria University in connection with experiential learning within a student-centric environment which usually derives from the learning by doing approach. This is achieved through project briefs which ask students to apply theoretical insights and design thinking methodology to real-world challenges and opportunities as reflected in both the programme and module specifications.

Conscious of the importance of developing students' creative confidence, intended as *"the ability to come up with new ideas and the courage to try them out"* (Kelley and Kelley 2013) attained through proactive engagement and based on planned learning—activities within the Design programmes are informed by the above described PFNA pillars and structured in their purpose on the following key drivers:

- Nurturing multi-disciplinary teams to support inclusive teamwork and project management. This will, in turn, ensure thorough cross-disciplinary contribution to students' curricula.
- Tutor-led co-creation educational design strategies to support students throughout the design process.
- Instilling proactive and 'can do' attitude using advanced practice, specific language, and tools to face real world challenges aligned to the professional sector.
- Embedding digital tools and prototyping within the design process.
- Assignments and assessment for learning which includes both formative and summative feedback to inform students' progression of studies.
- Education for Sustainable Development, aligned to the institutional commitments towards the SDGs and to the societal need for inclusive innovation.

'Live Project' Provision and Experiential Learning

One counterpart to the provision around 'industry'-connected group projects is Northumbria University's School of Design and the 'Clinics' in the Business and Law Schools is the "Advanced Practice" (AP) approach in the Faculty of Engineering and Environment across a range of Masters courses in its four different Departments. At core, there are 'curated' 12 week, and typically group, 'live projects' which are 'industry' (for and non-for-profit businesses/ organisations, the public sector, charities/voluntary sector) connected with a problem/ challenge. This takes place in the form of a curated/ edited/ negotiated brief at the university end, as students are supported by a small Professional Practice central Faculty team as well as AP/

module leads and academic supervisors (and programme leaders also, where necessary and appropriate). At the time of writing [April 2023], this has spanned over 7 semesters so far.

According to the "Draft AP Handbook for Engineering and Environment (E&E) Advanced Practice provision" (Arthur et al. 2022), Advanced Practice is: *"A content managed, deadline dependent, timetabled system of management which guides a multi-disciplinary team of internal and external professionals. These professionals engage simultaneous student cohorts along a critical path of recruitment, assessment, training, experience, supervision and evaluation centred on enhancing their real-world employability."* AP is a supervised semester (60 credits module), and in the AP governance model there are four levels of supervision which enhance student experience, oversee student wellbeing, and promote student success. There are four overlapping "executive", expert and organisational roles:

1. Cohort supervision—by the (Faculty) Professional Practice Co-ordinator.
2. Client supervision—by the (Faculty) Professional Practice Director of Clients and external Clients.
3. Subject Matter supervision (and academic assessment)—by the respective Departmental team.
4. Industrial supervision—by the Director of Professional Practice.

The first two roles are (since Autumn 2022) a sub-set of the two Faculty-level professional support staff in E&E for 'Experiential Learning' provision in a hub that is branching out now to cater for demand from under-graduate programmes (with one more professional staff member to be added to the team).

As agreed by all E&E AP colleagues, academic subject matter supervision (and final assessment) is essential to the success of any AP 'Live Project' (or a chosen legitimate alternative, such as an applied research project). The ideal approach to this comprises of expert input, performance communications, and final formal/ summative academic assessment (a group poster/presentation and an individual reflective essay). Academic supervisors must be regularly available to the Student Team to guide their subject matter choices—e.g. areas to research, solution directions to pursue, theory or methodology to apply etc.—and be able to interact with student requests for commentary and guidance (blending in-person and online modes).

A particular input effort is required around three critical touchpoints in the AP semester timetable:

The *Kick-off Process*—The academic supervisor of any AP Student Team is to meet them as soon as possible in the first week(s) for a detailed discussion around the Project Brief. Student Teams must have spent several days' work studying the brief and conducted initial research; they are required to come to that first Academic Supervision armed with questions. If the students are not prepared, the Academic Supervisor could withhold the first supervision until this criterion is met, as the result of the first academic supervision must be an energetic understanding of, and an application plan for either meeting, developing, or challenging the Project Brief. This is expressed in the delivery of the Client Kick-off Presentation created by the student teams.

The *Mid-term Review*—This is a critical stage gate indicator of student progress. Speed, quantity, and quality of their work to that point will all be revealed through the Mid-Term Review Presentation. Student Teams are required to submit this prior to seeing their client, and it must detail the anticipated final deliverable of the project (in draft). The Mid-term Review meeting can indicate two roads for the respective Student Team (mediated by the Academic Supervisor collaborating with the 'Industrial' Supervisor). The first road is negative: if the presentation demonstrates that the students have done little work and/or made no inroads to addressing the content of the brief, they will not be permitted to see the Client. They will be invited to a remedial session with the Professional Practice Team and the Academic Supervisor instead. They will be issued with a formal 'Notice to Improve' and the Academic Supervisor has a role in shaping the requirements of that. The second road is positive: If the work is good/ exemplary, the Academic Supervisor can provide coaching notes to the Student Team and support their communication strategy towards their client.

The *Close-out Process*—This is monitored by the 'Industrial' Supervisor. During the 'Coming-in-to-Land' meeting, the final deliverable that was tabled at the mid-term is critically assessed for its emergent feasibility. A new statement of the final deliverable for the Client is agreed with both the Student Team and the Professional Practice team. This is what the Student Team is required to produce to close-out and hand-over the 'Live Project' deliverables. The Academic Supervisor does have two key roles to play in supporting the Student Team to successfully arrive at the intended close-out deliverables, namely their academic supervision/mentoring, and the "emotional contract" (to do with student commitment) of the project. The intended final deliverables by the Student Team should be shared with the Academic Supervisor who can dedicate their remaining supervision time to encouraging/ supporting the team to deliver on the subject matter aspects of that agreement. The Student Team and the Professional Practice Team agree upon the appropriate handover (both content and method) with the Client. This *may require* a Client Close-Out meeting—this is not automatic and subject to Student Team performance and engagement (Arthur et al. 2022).

Figure 10 below illustrates the many months long planning and inception process for each AP semester of 12 weeks, plus 2 weeks of academic assessment. This is challenging at times as the Annual Leave seasons of academics (typically in August, as at other times there is intensive teaching, research and consultancy, administration, and outreach etc. demands on them) does not align with that of the Professional Practice team.

The following sections will be providing an overview about the following four modules and three programmes of study:

- International Year in Multi-Disciplinary Innovation (IYiMDI, undergraduate Level 5)—DE5013 Creative Cities
- MA/MSc Multi-Disciplinary Innovation (MDI, Level 7)—DE7024 Design-Led Innovation Practice
- MSc Disaster Management and Sustainable Development (MSc DMSD, Level 7)—KE7003 Subject Exploration

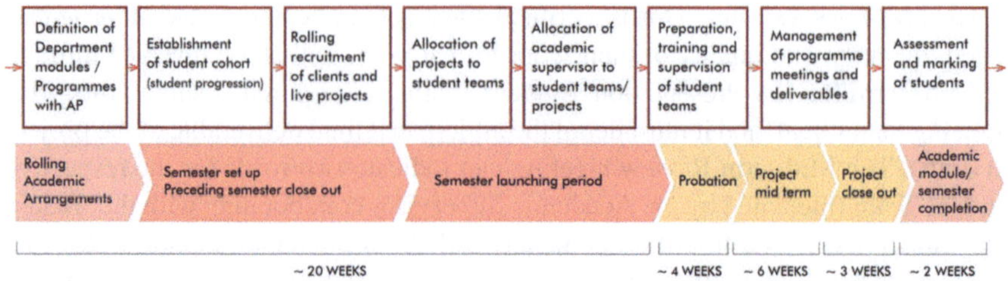

Fig. 10 The advanced practice 'live project' process in the faculty of engineering and environment at Northumbria University (adapted from: Arthur et al. 2022)

These will be critically evaluated to highlight how the PFNA informs the module structure, and how sustainability and is embedded within the delivery of contents. This, in turn, will inform the final reflection about how design projects and approaches to teaching can be further aligned to the institutional commitment to climate action. One point to note also is the argument by Kornelakis and Petrakaki (2020:290) that there are arguably *'different types of employability skills, as identified by employers' surveys, and… [when matched] with specific small-group teaching activities, drawing on pedagogic theory and practice [their study] concludes that, beyond work-integrated learning, traditional small-group teaching activities can go a long way towards bridging the gap between graduates' skills and labour market needs.'*

International Year in Multidisciplinary Innovation (IYiMDI)

The programme represents an opportunity for students from both and non-Design background to complement their curriculum with an optional year of study abroad, which can be reflected in their final degree award title upon request. Delivered for two academic years to date it aligns to some extent to an optional work placement, as it is a one-year award for NU students who have successfully completed their undergraduate (UG) Level 5 studies and intend to embark on a year of study that provides them with a valuable experience abroad (one semester), as well as a range of skills and design thinking approaches that enhance their employability prospects (Northumbria University 2022d).

Different modules provide students with key insights about fundamental design thinking approaches to research with focus on city-based studies. The first semester is delivered in Amsterdam and the second one is based in Newcastle. Conscious of the fact that the international education may be considered economically viable but not sustainable from a mobility and energy consumption-related perspective, and that actions to mitigate the environmental impact should be taken (Rumbley 2020), a range of measures have been considered when planning the international activities for this programme. For this reason, taught delivery aligns to Northumbria's extended studio approach to teaching, and activities in the first semester—when students are based on the Amsterdam campus—are hybrid (blended). This means that one Northumbria member of staff from Newcastle is based in Amsterdam for the entire semester and attends all studio sessions alongside staff from the Amsterdam campus, and other members of staff from the Newcastle campus delivering online.

Creative Cities Module of Study

This is one of the modules of study which are offered and reflect the inter-disciplinary ethos of the International Year programme that is delivered in Newcastle. As a city-focussed project-based module of studies, it explores different urban environments—intended as multi-faceted contexts that constitute the background for comparative evaluations, research and analysis. In this respect, students are required to bring onboard the experience gained in the previous semester in Amsterdam. However, it runs alongside the Design-inspired Research Methods module which is designed to deliver key research skills and design thinking practices that are then applied to the project-based module. This 'tandem' structure of the semester allows students to strengthen their research abilities and apply critical thinking to approach problems and define solutions (Fig. 11).

Activities include lectures, seminars, and tutorials in which proactive engagement is crucial for the progression of projects, as the practice-informed Creative Cities module asks students to:

– Work within multi-disciplinary teams.

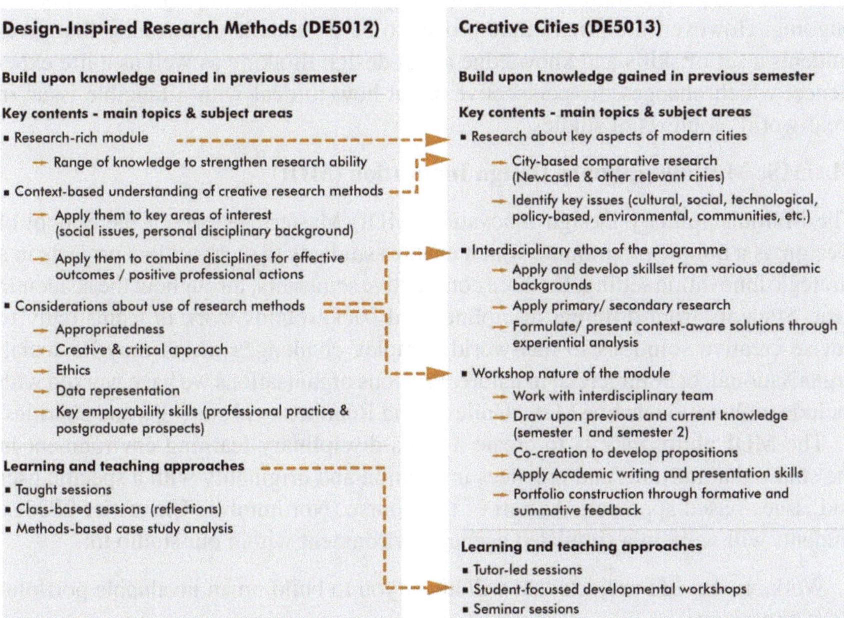

Fig. 11 Diagram showing the research-informed relationship between the DE5012 and DE5013 modules of study (*source* authors)

- Apply design thinking methods to key urban-related issues (cultural, social, technological, policy-based, or environmental).
- Utilise comparative primary and secondary research, applying personal reflection and design thinking to urban contexts to identify solutions (actions, practices, etc.).

This is applied to two assignments (formal components) for this module:

- Assignment 1—City-Based Research proposal (worth 30% of the final grade). Here students explore, understand, and analyse urban contexts to outline opportunities for innovation.
- Assignment 2—Comparative City Report (worth 70% of the final grade). Here students expand on the opportunities identified in the Assignment 1 through comparative research, critical analysis and research-informed evaluation of how proposed solutions could be successfully applied to chosen urban context/s.

The combination of written, visual, digital and creative material is key to support the communication and presentation of ideas throughout the design process. Figure 12 identifies key insights from staff and students about the DE5013 module, as well as key issues and proposed solutions that inform contents and approaches to teaching as a response to previous years' experience.

It must be appreciated that this module has been implemented for only two years and further experimentation with the delivery of contents and project briefs is still ongoing. However, the programme proves to be a valuable opportunity that gives students a set of skills and knowledge about design thinking as well as a life experience, which changes the perspective about how to deal with a tangible issue in 'real-world' contexts of studies.

MA/MSc Multidisciplinary Design Innovation (MDI)

The Multidisciplinary Design Innovation (MDI) Masters, as part of the School of Design, is a unique programme which enables students to work on live projects in a strategic innovation setting, for three consecutive semesters, throughout the academic year. Students from different disciplines and backgrounds work in teams daily to devise creative solutions to real-world complex challenges which may be social, organisational, or commercial in nature. Previous organisations we have worked with include multinationals like Mars, Unilever and Realme, local start-ups and charities.

The MDI philosophy is to create a cross-disciplinary learning environment in the studio that nurtures and practices innovation and originality with a specific user and issues-based approach. As part of this course (Northumbria University 2023e), students will work in a simulated agency environment within our studio to:

- Work on real-life projects that will allow you to build up an invaluable portfolio of experience.
- Blend design, technology, and business and develop your entrepreneurial mindset.
- Learn to manage change and innovation in a strategic way.
- Build up an invaluable portfolio of experience and network of contacts that will span multiple industries.

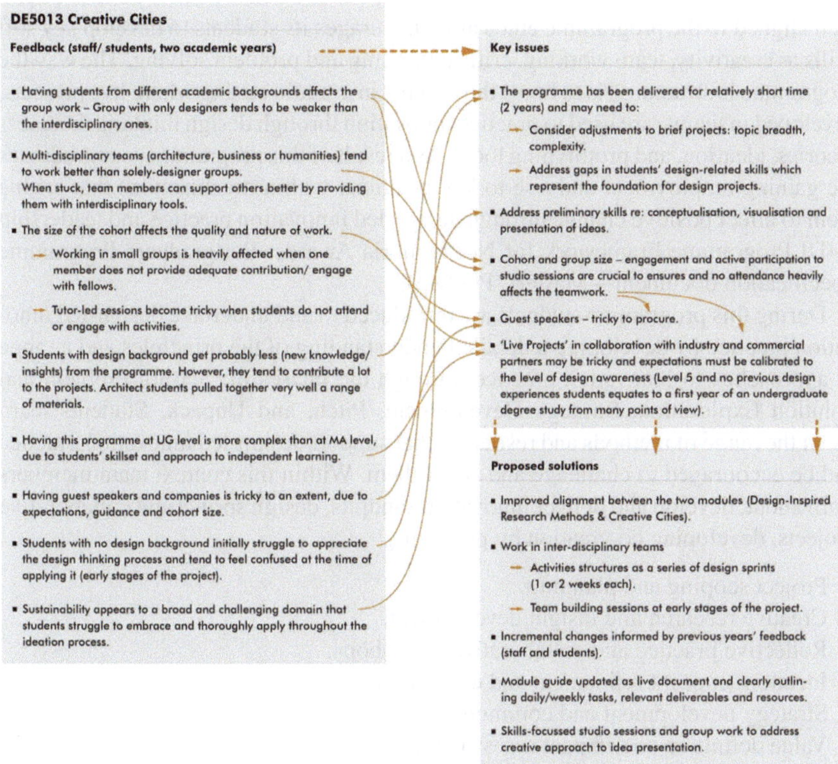

Fig. 12 Creative Cities module feedback from students and staff (*source* authors)

– Create strategic value through adopting a range of innovative approaches and resources—highly valued by future employers.

Students study one module per semester to allow flexibility around live projects and related timelines to experience practical involvement in multiple aspects of innovation, design-led co-creation, and strategic entrepreneurial leadership. These include project scoping and planning, creative research and insight development, ideation and concept refinement, strategy development and communication, value definition, and pitching and reporting to clients. The assessment strategy is designed to consolidate theory-rich project learning. Assessment methods include portfolios and project reports which through reflective writing enable students to consolidate their learning for the end of semester submission. This arrangement allows students to concentrate on a 'live' project developed in teams daily to maximise learning, team collaboration, co-creation, project development and final work outputs.

Referring to relevant pedagogical theories discussed earlier, MDI embodies its long-established principle of 'learning by doing' which shapes the students' learning experience daily in a more autonomous, dynamic, and experiential manner. It follows that the theory behind Kolb's Experiential Learning Theory (ELT) model (1984) is well aligned to the programme ethos as it encourages its students to develop key soft skills as creativity, team working, critical thinking and problem solving. The way the programme is structured is to boost those skills in a studio setting where live projects, developed in teams, are used to practice innovation through design thinking, business theories, ideation, and prototyping tools. As a result of this programme, our graduates are gaining employment and rise to key positions within organisations that enable them to affect positive change through design-led innovation practice and leadership (MDI Programme Framework for Northumbria Awards: Postgraduate Programme Specification document—Masters, PG Dip).

During this programme students are introduced to and undertake design-led innovation practice by developing a detailed understanding of the principles and a range of approaches to innovation practice, through the phases of: Set-up, Problem and Solution Exploration, Strategy Development, Pitch, and Unpack. Students learn about the range of methods and resources that support design-led innovation practice and be encouraged to challenge and adapt them. Within this context team members learn about, develop and deliver innovation sandpits, design sprints and collaborative projects, developing co-creation by practicing:

- Project scoping and planning.
- Creative research and insight development.
- Reflective practice and collaborative workshops.
- Ideation, concept refinement and evaluation.
- Strategy development and communication.
- Value definition and impact assessment.
- Client pitches and reporting.

In this first semester module students are normally required to work on 4–6 projects with various partners and focussing on social, organisational or product/service innovation based on the agreed brief. In most cases, sustainability is an underlying theme and requirement embedded into the students' work as it guides their project development and outputs accordingly.

As already mentioned in the IYiMDI section, PFNA informs the module specifications and the expected module outcomes (MLOs) aligned to three key areas:

1. Knowledge and understanding.
2. Intellectual and professional skills and abilities.
3. Personal values and attributes.

In more practical terms this means students are expected to achieve an understanding and application of design-led innovation approaches and employ research methods to produce insights through co-creation to develop innovative solutions. In addition, an application of critical and reflexive thinking is required to enable students to challenge and develop established ways of behaving and working in a multidisciplinary innovation context.

Those learning objectives are embedded into two sustainability projects, used here as a sample, which focus on UN SDGs and demonstrate how PFNA pillars are also integrated in the module learning and delivery strategy to boost research activities for instance, and through teamwork and live projects enhance employability prospects for students involved:

Project 1—Northumbria Carbon Reduction Project.

- Explore, understand, analyse how the University is tackling its carbon reduction programme.
- Outline opportunities and propose ideas to the university become more sustainable.
- Develop narratives and ideas following the 4 proposed themes of enquiry.
- Opinions on the University's carbon performance and activity.
- Sustainable solutions that should be assigned to carbon-reduction at Northumbria.
- What can we do at Northumbria to further reduce scope 1,2 and 3 carbon emissions.
- Structures and methods for delivery and engagement.
- Define strategies and support activities to stretch the boundaries of the carbon reduction solutions and countermeasures to climate change.
- Required use of visual, digital, and creative material to support the communication and presentation of ideas throughout the development process.

Project 2—Northumbria University Sustainability Challenge

- Explore, understand, analyse how the University is tackling its sustainability commitments.
- Outline opportunities and propose ideas around staff and student engagement and recognition.
- Define strategies and support activities to stretch the boundaries of the sustainability agenda.
- Required use of visual, digital, and creative material to support the communication and presentation of ideas throughout the development process.

The qualitative student feedback below confirms the aim of the programme and how research and innovation practice are conducted in a studio setting which simulates an agency/consultancy environment (Fig. 13).

The summarised academic staff feedback shown in Fig. 14 clearly indicates the importance of teamwork to maintain a satisfactory learning experience for students, who in turn are better enabled to produce pertinent and well-developed project outputs which satisfy the project brief requirements and the project partner.

MSc Disaster Management and Sustainable Development

This Master's programme, which has been running for about two decades, has different strands (disaster risk reduction, sustainable development, integrated emergency planning, humanitarian development, health and wellbeing etc.) which are developed through core and then some limited number of optional modules over a 12-month period, including a master's dissertation at the end. In the case of the "with Advanced Practice", an additional 3 months (60 credits) module is inserted [see

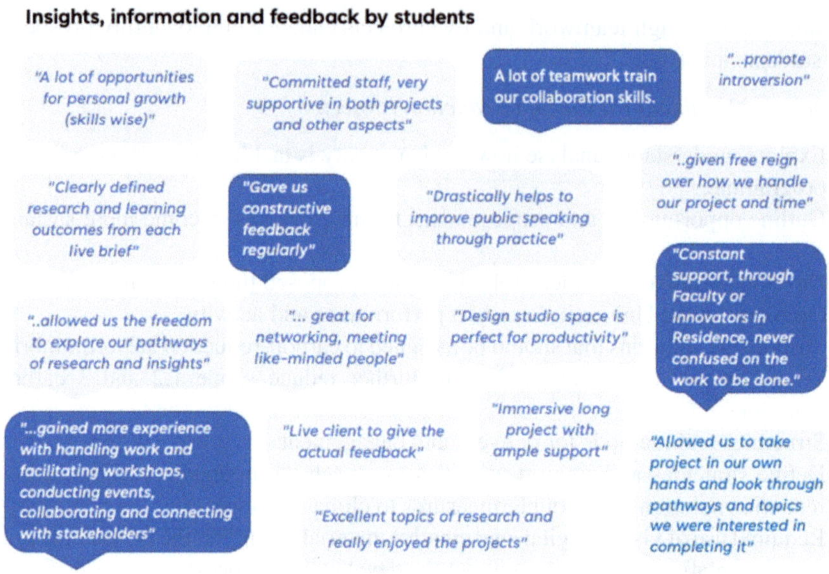

Fig. 13 Feedback from MDI students (*source* authors)

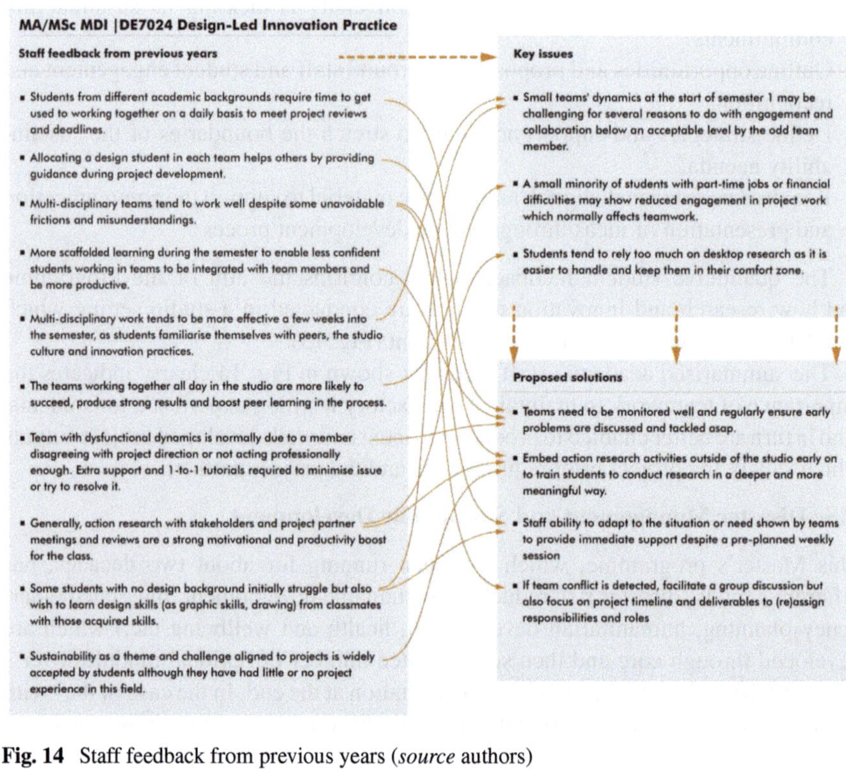

Fig. 14 Staff feedback from previous years (*source* authors)

above in an earlier section] before the students undertake their Masters Dissertation module.

One of the core modules is on Project Development/Management, where the LogFrame (LF matrix, problem and solutions trees, and stakeholder analysis) analytical approach widely used in the international development sector and governments is central to it, with simulated and real clients in projects where students are asked to develop solutions though not necessarily implement or 'manage' operationally. The Subject Exploration module (KE7003) follows on from core modules on Themes in Sustainably Development and Disaster Risk Reduction and Response. The focus here is to encourage and enable individual students to develop a focussed and specialised investigation based purely on secondary data into something contested, controversial, cutting edge, philosophically/conceptually and/or technologically or societal or economically uncertain, dynamic, and innovation rich.

The assessment is in the form of an individual 3000-word critical report or essay (which in principle could also take the form of a business plan or application for funding). The idea is that students get moved out of their (and the disciplines') "comfort zone" as the topics covered in any given year are astonishingly diverse (approved by the Module Tutor after the students submits a 1–2-page formative sketch or at least by e-mail or verbally in 1–1 meetings).

The Teaching and Learning Strategy is that the Module Tutor presents (in lectures) up-to-date 'quirky' research discussed with the students in class seminars, though in practice lectures are also quite interactive and seminars often require significant academic staff input rather than being student-led. For the past years, the Module Tutor has brought in both Alumni in the Disaster Management and Sustainable Development broad field to present their professional practice or research as well as Early Career Researchers globally (and in those cases in a blended format) through networks (such as the Disaster and Development Network we host at Northumbria University). The role of the Module Tutor is to help 'translate' areas of the external speakers' presentations which may not be fully clear to students or assume prior knowledge which they do not possess or may have "forgotten about" from previous core modules. The tutor also acts as being a sounding board, moderator, and instigator (to at times being educational 'agent provocateur') for the student-centred discussions—which are crucial for peer learning and typically involving both international and UK students (Fig. 15).

Alongside the other MSc DMSD core modules, KE7003 also sets up the Geography and Environmental Sciences Department's specific version of the Advanced Practice semester (KE7037) for those students that choose the prolonged version of the degree. Students have developed projects as teams and—as exceptions for sound reasons of employability and career destination, for instance—individually, which have often tackled sustainability issues. These range from educational to technical ones of (embodied) carbon reduction in a hospital trust or feasibility studies for energy conservation and integration of on-site renewable energy sources in historic buildings. Both the assessments submitted, and the mostly qualitative feedback received point to steep but extremely rewarding personal skills development process and outcomes perceived by both students and tutors. However, significant group/ team

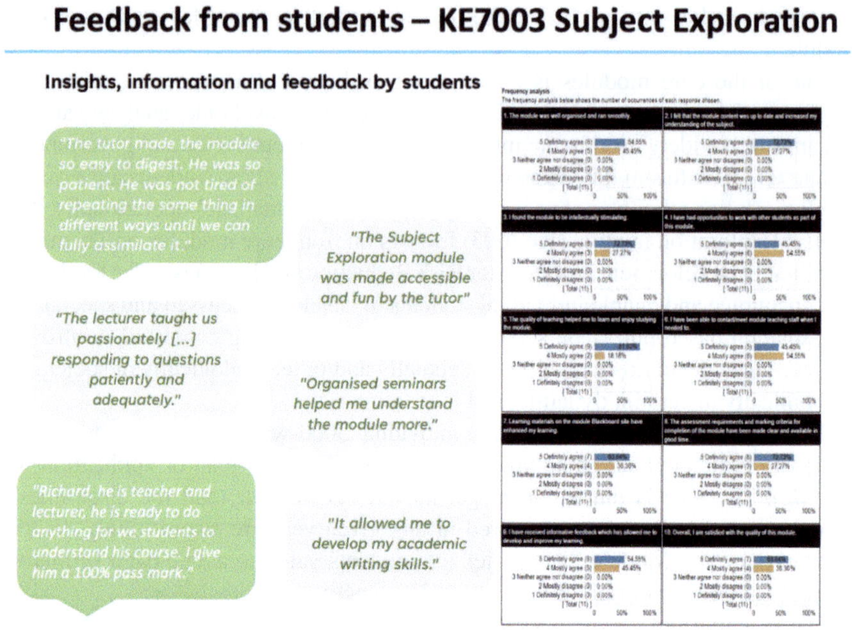

Fig. 15 Feedback from KE7003 students (*source* authors)

working challenges and dynamics with temporary or even full-on breakdowns can happen at times. This normally requires significant interventions up to slitting groups or turning specific 'live' projects into simulated or even to desk-top ones. In those circumstances staff step into the breech if an external delivery is critical for reputational damage mitigation to safeguard the university's/department's/programme's reputation with external and professional partners. On the other hand, there are also plenty of successful examples of direct employment with 'live project' partners into the targeted professional sector, towards Doctoral programmes, or research and consultancy outputs co-authored by some of those students with staff.

We can also demonstrate some of the value of the Advanced Practice (AP) semester and the (managed and supervised) 12 weeks long Live Project fir this MSc programme by excerpts from the individual reflective accounts they students have to individually submit at the end of the AP semester, take below from students on the two (amongst others from other academic Departments) Faculty-award winning (all international, from different countries and different genders) students teams in academic year 2022–2023:

A female migrant refugee Taiwanese student, who experienced significant issues as a student team leader in this context, which necessitated the mitigating involvement of the module tutor to address and manage some of issues with the [original, and then subsequently split] group interactions, wrote: "*In this reflective report, by practicing self-reflection in line with the 5R framework of reflection, I presented my experience of multicultural teamwork challenges during this Live Project Advance Practice semester, and how I responded to it. Later, the reasons I expressed my viewpoints in certain ways would be addressed. After self-analysis, my re-constructing plans going forward in the final stretch of my Masters degree and beyond are presented in the end. I argued that reflective practice should be promoted as an ongoing development of inter-personal awareness; however, without a safe environment, self-reflection would just bring out an expected but fake successful self-awareness development.*"

A female student from Malaysia wrote: "*From my understanding through research and discussions with the project client, Hartlepool Borough Council as the hosting authority for the Cleveland Emergency Planning Unit for four local authorities, namely Hartlepool Borough Council, Middlesbrough Council, Stockton on Tees Borough Council, and Redcar* and *Cleveland Borough Council, as part of the Cleveland Local Resilience Forum, I was able to apply my skills and knowledge gained from my past work experience, as well as concepts taught in the Project Management and Integrated Emergency Management (IEM) modules into my Advanced Practice semester as part of my MSc Disaster management with Sustainable Development with Advanced Practice Programme. This had been an added advantage for me to link the theories with a real-life project.*"

A male student on the same student team from Nigeria wrote: "*I have had the opportunity to enhance and apply the knowledge and abilities I gained during the taught portion of my programme and also through this Advanced Practice semester experience, to assist the Cleveland Local Resilience Forum in considering further how to tackle selected risks, as well as to gain new management and professional "industry" knowledge and abilities to increase my employability. Even though I first did not initially appreciate the project, it included a number of stages that together resulted in something which I was really pleased with and impressed by, and the client was by their own statement happy with. The many processes of the project were thoughtfully and skilfully designed to ensure that I left with a life-long of learning experience. My team and I met with our client, the Engineering* and *Environment Advanced Practice Live Projects cohort industrial supervisor as well as programme manager and coordinators, as well as our [two] academic supervisors on a number of occasions. The discussions were useful as I acquired additional knowledge and understanding about the best way to carry out the project. A technical report for the Cleveland local resilience forum (LRF) was produced over the course of the project and has since been submitted. The report was quite detailed and included suggested preventive measures for selected risks listed in the Cleveland Community Risk Register. In arriving at the desired results of this project I actually used and learnt more about numerous project management tools. The technical report provided a thorough explanation of the fundamental preventive measures that my team members and I proposed. This was summarized into a hierarch of controls and a flowchart, as advised by the client.*"

6 Analysis and Conclusions

This reflective commentary seeks to critically dissect how difficult but also gratifying and necessary it can be to embed multi-disciplinary activities into (UK-based) higher education teaching. In addition, an account on how much was accomplished is illustrated through some case studies, but also what can and should be done to improve the current situation towards a truly student-centric and industry-aware approach to teaching with increased emphasis on sustainability.

First, an important premise should be made to capture the current situation educators find themselves. In a heavily digitalised world where most consumers (including young users and students) find themselves overwhelmed with information through different social and media platforms, there is the widespread tendency to hide behind the screen and become too familiar with the daily habit of consuming individually large volumes of information and media content. This phenomenon can be regarded as a typical user-in-reactive mode which does not normally allow for much critical thinking, or a prolonged attention span due to the fast rate at which information is digested. A further behavioural side-effect can be witnessed in the reduced amount of direct and meaningful face-to-face interpersonal communication as digital communication is often used as an appealing and easy-to use-alternative. Thus, students entering higher education are less prone to be in a proactive and creative mode showing a tangible spirit of curiosity and initiative. An individual's willingness to develop their skills and knowledge in a committed, outward, and collaborative manner is not necessarily the norm any longer as it challenges the acquired reactive student's posture of being absorbed and often distracted by digital content and entertainment throughout the day.

Arguably, the biggest challenge that higher education institutions are currently facing is how to devise methods to counteract the above-mentioned trend by training students to be once again in proactive mode through active learning which is also student-centred and dynamic in nature (Holdsworth et al. 2022). If this happens in a setting where open collaboration, creativity (Cropley 2001), and teamwork take place then it is more likely that students' experience and learning will be enhanced. It is worth adding that involving mature students with a remarkably diverse employment, academic or vocational study backgrounds (e.g., on taught Master's degrees or at post-graduate research level) and life experiences, can often inspire undergraduate students and help foster this culture of peer-to-peer interactions and exchange of views and good practices.

The case studies presented here strongly suggest that collaborative patterns and peer support become key factors in the learning and taught processes, as they provide students with key skills and insights towards the creation of solutions to real-world problems (Alam 2022). In this respect, the application of a Design Thinking approach and methods is instrumental in informing the way students explore given challenges (design briefs) through primary and secondary research, outline viable solutions (ideation), test them (through prototypes), and suggest concepts that can then be updated through feedback (from peers, staff, and partner organisations). This creates a virtuous circle that promotes innovation while reinforcing students' skills and awareness about professional practices. It is also a way of conducting research and project development as a series of practical investigation tasks and exercises, which make the whole experience interesting and rewarding for committed students involved. Based on students' feedback during the projects described it is often found that motivational and engagement levels were significantly high. This can also be confirmed by students' body language and behavioural patterns that tutors, as authors in this instance, noticed around the studio (or study hub) on a daily or weekly basis.

Working in teams is of primary importance for the activities and students' work developed in design and innovation projects, as peer learning is a massive boost to students' development on top of what is being taught and practiced during the weekly module sessions. Teamwork is particularly relevant when dealing with multidisciplinary teams as it is a tedious and challenging job to align team members to a common vision considering their different mindsets, skills, and habits. This in turn leads to considerations about studio practice and culture, alongside how studio and facilities are better utilised to support and facilitate the group activities.

Based on the teaching experiences gained by the authors it can be concluded that human factors and sustainability guidelines need to be placed at the centre of this design approach for projects that focus on tackling SDG challenges, but also for those where sustainability is not necessarily the prime objective. In both cases the ability to maintain a mindset that induces students to act and develop projects responsibly through sustainability and circular economy principles is a significant objective that universities should aspire to. Northumbria University is committed to this vision, as clearly outlined in previous sections.

Based on the authors' knowledge and 'live' project lead and management experience, a useful example of an integrated design process framework focussed on sustainability is proposed by considering key parameters as staff involvement alongside project partners and stakeholders' input. The diagram below illustrates a typical project which could be run between 4 to 8 weeks in length. There are obviously many permutations of the project key milestones (as reviews, workshops, project discussions) which would be planned before the start of the project. In addition to that, as the project unfolds it may be necessary to review the project delivery and timings of pre-planned events due to unforeseen circumstances or simply based on the productivity of the teams involved (Fig. 16).

What emerges from the diagram is the approach required to set up and manage projects which have a significant sustainability focus (SDG-focussed) or simply require a sustainable mindset (SDG-inspired). This is an underlying theme that provides guidelines to facilitate research, ideation, and final proposal development accordingly.

The role of lecturers and tutors as facilitators in the learning process is critical as sustainability work and related project outputs are complex in nature, very multifaceted and often associated with systemic and wicked problems which are exceedingly difficult to address in any industrial, organisational, and societal domain. It follows that divergent and convergent DT is normally needed throughout different project development phases to handle complexity in a focussed and less chaotic

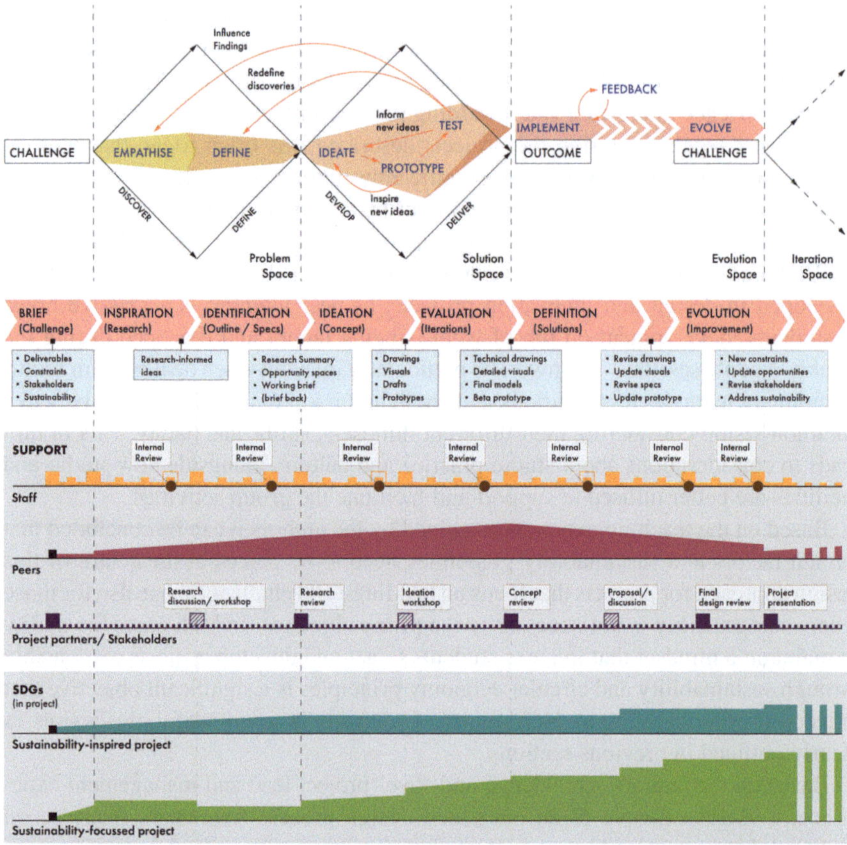

Fig. 16 Integrated diagram of design processes in relation to project support and SDG relevance (*source* authors)

manner, and to enable teams to arrive at conclusions in a more complete/holistic, and structured manner. Having said that, the proposed diagram does not consider the inherent fluidity of the design process in real life which requires teams not only to faithfully follow the project planner but also to move between phases and tasks back and forth as they deem to be appropriate, logical, and beneficial to the actual evolution of the project itself.

For the above-mentioned reasons, the role of peer-learning/ support throughout the process is essential to nurture the dynamics between students, tutors, and project partners but also to combine capabilities and valid approaches towards the successful realisation of project aims. The role of language and social interaction (e.g., studio sessions / studio culture) is a crucial aspect of applied teaching which cannot be neglected at any stage of a project. If handled well it also allows for independent study/learning to be embedded into project work where team members have the opportunity for individual self-reflection and self-expression of their capabilities in freer but time constrained settings. It is also worth pointing out that interactions with MKOs (peers and staff) are important enablers which support the ideation process, enhancing problem solving skills, if learners are provided with relevant instructions and timely project guidance.

There are some limitations to be considered in the planning and application of guidance provided in this study, in relation to learning or project activities around sustainability and multi-disciplinary approaches. These can be summarised as:

- The limited amount of primary pedagogic research with university students that the authors relied on involving their respective cohorts of students. In addition, a more extensive and less detailed central administrative survey.
- The suitable nature of educational theories which was drawn upon without conducting a full thematic analysis.
- Whilst thematic research on pedagogic approaches within other European institutions has been undertaken, a full comprehensive mapping of HE approaches and strategies would be required.
- Only a partial understanding and application of how UN SDGs' potential could be unlocked through the mentioned live projects has been presented in this contribution.
- The time allocated for live projects, normally spanning from 1 to 4 weeks, does not allow students and staff to fully develop and detail the proposed solutions answering the set brief.
- A better-defined university wide approach to illustrate what experiential learning stands for and how live projects should be developed and managed internally is currently being developed and so far, only available as a draft.

Final Recommendations

The proposed recommendations should be regarded as a starting point for academics willing to enhance multi-disciplinary innovation and soft skills practice in higher education programmes with a focus on sustainability.

Sustainability and SDGs—how to embed them within programmes and modules:

- Acknowledge the critical role of education in delivering the SDGs, and their value in governments, businesses, and society at large.
- Devise winning strategies to create sustainability-focussed projects and integrate SDGs into student learning and curriculum.
- Use SDGs as strategic leverage for innovation across the programmes.
- Use 'real-world' challenges and scenarios to inform project briefs that leads to human-centred and environment-aware solutions to challenges but also to grasp opportunities. This prepares students for challenges and demands from the professional sector and ensures impact, both at academic and 'industry'/ societal level.

- Review project learning, outputs, and overall effectiveness of module content in relation to student and staff feedback. Evaluation should consider students' performance, experience, and overall academic attainment, not just their 'satisfaction' as 'customers' of HE.
- Adopt an incremental/ scaffolding approach to module contents and project briefs to ensure that learning is aligned to progressive stages of complexity (challenging tasks).
- Provide students with insights about research methods at the initial stages of projects, modules, and programmes.
- Analysis to a requisite academic level (standard, depth, detail, and thoroughness) is key to processing research-informed insights to derive context-specific proposals. In this respect, both taught and supervised sessions are critical in shaping this.
- Practicing with representation techniques such as drawing, visuals, storyboards etc. is central for the communication and presentation of ideas. Multi-disciplinary student cohorts may have different levels of awareness and skills. Staff have the key role of supporting students in the process of building their creative confidence through practice.
- Prototyping (both digital and physical models) is a central stage across the Design disciplines which often requires more time than expected, especially if inexperienced students are involved.
- Proactive engagement and 'can do' attitude must be nurtured through project briefs, partnerships with companies and live projects. This is particularly relevant at master's level.
- Digital tools (e.g., digital canvasses, info-graphics apps, drawing and rendering apps) should be used in line with Technology Enhanced Learning and to respond to the needs of the professional sectors.
- Co-creation educational design strategies are critical aspects of the staff-supervised studio activities. In this respect, cross-disciplinary insights and collaborative work within inclusive interdisciplinary teams contribute to students' curriculum and reinforce their employability prospects.
- Bring students' feedback onboard to constantly evolve academic programmes and modules. In this regard, central University (digital) surveys at the end of the semester (as well as additional local ones where and when permitted), alongside Student-Staff Programme Committees scheduled in each semester, and teaching staff review meetings should inform continuous and cyclical programme performance reviews with requisite quality and enhancement actions at the end of the semester or academic year in order to proactively improve the modules and programmes the next time they run.

Primary educational research (workshops and/or conversation with current and past students to discuss their experience) is part of the next steps in the critical reflection process to identify areas for improvement moving forward. However, the line of argument presented in this self-reflective contribution has been undertaken through the central university feedback mechanism strategy and has allowed the authors to identify critical areas for embedding students' active engagement into co-creation. In this regard, surveys and formal meetings scheduled along the academic year are crucial to acquire quantitative and qualitative insights from students.

To put it more simply, Education for Sustainable Development around experiential and authentic learning should be devised through the learning-by-doing approach to provide a friendly yet professional environment where students interact, share knowledge, develop skills, and gradually with a sustainable mindset to evolve as the next-generation junior 'green practitioners' our world requires.

References

Acklin C (2010) Design-driven innovation process model. Design Manag J 5(1):50–60

Alam A (2022) Mapping a sustainable future through conceptualization of transformative learning framework, education for sustainable development, critical reflection, and responsible citizenship: an exploration of pedagogies for twenty-first century learning. ECS Trans 107:9827. https://doi.org/10.1149/10701.9827ecst

Andersson N (2009) Designed Umeå. Industrial design education at the Umeå Institute of design. Infotain & Inforbooks Sweden AB and Umeå Institute of Design: Umeå

Arrowsmith C, Cartwright W (2019) Embedding employability skills in the curriculum and extending into postgraduate programs. In: Walkington H, Hill J, Dyer S (eds) Handbook for Teaching and Learning in Geography. Edward Elgar, Cheltenham, pp 414–429

Arthur J et al. (2022) Draft AP handbook for engineering & environment (E&E) advanced practice provision. J Arthur Consulting Ltd & Northumbria University: Newcastle upon Tyne

Asante G (2018) Effective design methodologies. dmi:review—design management and innovation 29:2:10–15

Ball J (2019) The double diamond: a universally-accepted depiction of the design process. https://www.designcouncil.org.uk/our-work/news-opinion/double-diamond-universally-accepted-depiction-design-process/

Bereton M, Cannon D, Mabogunje A, Leifer L (1996) Collaboration in design teams: how social interaction shapes the product. In: Cross N, Christiaan H, Dorst K (eds) Analysing design activity. Wiley, Chichester

Biggs J, Tang C (2011) Teaching for quality learning at university, 4th edn. Maidenhead, The Society for Research into Higher Education / Mc Graw Hill

Blair B, Cummings A, Dunbar T, Hayward D, Woodman J (2008) BAU-WOW! A model for creative practice, thinking, learning, research and innovation in the 21st century. In: Drew L (ed) The student experience in art and design higher education: drivers for change. Group for Leaning in Art and Design / Jill Rogers Associates Ltd., Cambridge, pp 65–95

Bond TG (2008) Cognitive development and school readiness: In: Salking NJ (ed), Encyclopedia of educational psychology. SAGE. https://sk.sagepub.com/reference/educationalpsychology/n43.xml

Brown T (2008) Design thinking, in harvard business review, June 2008. https://readings.design/PDF/Tim%20Brown,%20Design%20Thinking.pdf

Budge K (2012) A question of values: why we need art and design in higher education. Art Design Commun Higher Educ 11(1). https://doi.org/10.1386/adch.11.1.5_1

Bueno M, Ramckers L (2003) Understanding people in new ways. Personas in context: forging a stronger link between research and its application in design. Eindhoven, Philips, Design Research

Chaiklin S (2003) The zone of proximal development in Vygotsky's analysis of learning and instruction. In: Kozulin A, Gindis VS, Ageyev VS, Miller SM (eds) Vygotsky's educational theory in context. Cambridge University Press, Cambridge, pp 39–64

Crisp AR (2010) Reflections on design education. In: Boks C, McMahon C, Ion W, Parkinson B (eds), When design education and design research meet… proceedings of the 12th international conference on engineering and product design education. Norwegian University of Science and Technology: Trondheim
Cropley AJ (2001) Creativity in education & learning. Kogan Page, London
Cross N (2011) Design thinking. understanding how designers think and work. Oxford, Berg Publishers
Datar S and Bowler CN (2015) Design thinking and innovative problem solving. In: Canals J (eds) Shaping entrepreneurial mindsets. Palgrave Macmillan, London. https://doi.org/10.1057/9781137516671_7
Davidson G, Glaves P, Kotter R, Wilkinson J (2016) Student work placements as a focus for building partnerships between universities and sustainable development stakeholders. In: Leal Filho W, Brandli L (Eds), Engaging stakeholders in education for sustainable development at university Level (93–108). Springer: Cham. https://doi.org/10.1007/978-3-319-26734-0_7
Dawson P, Andriopoulus C (2021) Managing change, creativity and innovation, 4th edn. London, SAGE
Design Council (2019) Framework for innovation: design council's evolved double diamond. https://www.designcouncil.org.uk/our-work/skills-learning/tools-frameworks/framework-for-innovation-design-councils-evolved-double-diamond/
DMI Review Staff (2018) Design and the SDGs. Dmi:review—design management and innovation 29:1:8–11.
Donald WE, Ford N (2022) Fostering social mobility and employability: the case for peer learning. Teaching Higher Educ J.https://doi.org/10.1080/13562517.2022.2145467
Dong A (2005) The latent semantic approach to studying design team communication. Des Stud 26:445–461
Dorst K (2011) The core of "design thinking" and its application. Des Stud 32:521–532. https://doi.org/10.1016/j.destud.2011.07.006
Droga D, Shah B (2022) Keeping up with customers' increasingly dynamic needs. https://hbr.org/2022/09/keeping-up-with-customers-increasingly-dynamic-needs
Fang J, O'Toole J (2023) Embedding sustainable development goals (SDGs) in an undergraduate business capstone subject using an experiential learning approach: a qualitative analysis. Int J Manag Educ 21(1):100749
Gedenryd H (1983) Frames of mind: the theory of multiple intelligences. Heinemann, London
Goldschmidt G (1995) The designer as a team of one. In: Cross N, Christiaan H, Dorst K (eds), Analysing design activity, Chichester: Wiley
Harper R, Rodden T, Rogers Y, Sellen A (2008) Being human. Human-Computer Interaction in the Year 2020. Microsoft Research Ltd.: Cambridge
Holdsworth S, Hegarty K, Sandri O (2022) Assessment design for interdisciplinary, professional and sustainability education. In: Leal Filho W, Portela de Vasconcelos CR (eds), Handbook of best practices in sustainable development at University Level https://doi.org/10.1007/978-3-031-04764-0_8
Holgate P (2016) Developing a curriculum for engagement: architectural education at Northumbria University. https://northumbria.academia.edu/PeterHolgate
Hummels C, Vinke D (2009) Developing the competence of designing intelligent systems. Eindhoven University of Technology Department of Industrial Design, Eindhoven
IDEO LLC (2012) Design thinking for educators, 2nd edition. Riverdale
IDEO LLC (2015) The field guide to human-centered design. https://www.ideo.com/post/design-kit
IDEO LLC (2023) Design thinking defined, https://designthinking.ideo.com/
Jenkins A, Healy M, Zetter R (2007) Linking teaching and research in disciplines and departments. The Higher Education Academy, York
Julier G (2017) Economies of design. Sage, London
Kane M, Trochim WMK (2007) Concept mapping for planning and evaluation. Sage, London

Kelley D, Kelley T (2013) Creative confidence: unleashing the creative potential within us all. Crown Business.

Kornelakis A, Petrakaki D (2020) Embedding employability skills in UK higher education: Between digitalization and marketization. Industry Higher Educ 34(5):290–297. https://doi.org/10.1177/0950422220902978#tab-contributors

Lake D, Flannery K, Kearns MA (2021) Cross-disciplines and cross-sector mixed-methods examination of design thinking practices and outcome. Innov High Educ 46:337–356. https://doi.org/10.1007/s10755-020-09539-1

Landry L (2020) What is human-centered design? Harvard business school online, 15th December 2020. https://online.hbs.edu/blog/post/what-is-human-centered-design

Luka I (2019) Design thinking in pedagogy: frameworks and uses. Euro J Educ 54(4):499–512. https://doi.org/10.1111/ejed.12367

McLaughlin JE, Chen E, Lake D, Guo W, Skywark ER, Chernik A et al (2022) Design thinking teaching and learning in higher education: experiences across four universities. PLoS ONE 17(3):e0265902. https://doi.org/10.1371/journal.pone.0265902

Mcleod S (2023a) Kolb's learning styles and experiential learning cycle. https://www.simplypsychology.org/learning-kolb.html

Mcleod S (2023b) Jean Piaget's theory and stages of cognitive development. https://simplypsychology.org/piaget.html

Mcleod S (2023c) Lev Vygotsky's Sociocultural theory of cognitive development. https://simplypsychology.org/vygotsky.html

Morgan T, Jaspersen LJ (2022) Design thinking for student projects. Sage, London

Northumbria University (2015) Action learning. https://www.northumbria.ac.uk/about-us/academic-departments/newcastle-business-school/newsletter/action-learning-1/

Northumbria University (2020a) X studio. https://www.northumbriax-studio.com/

Northumbria University (2020b) Press Release—Among the Best in World for Sustainability. https://newsroom.northumbria.ac.uk/pressreleases/among-the-best-in-world-for-sustainability-2993418

Northumbria University (2021a) Sustainability living lab. https://one.northumbria.ac.uk/service/cs/es/Pages/SustainableLab.aspx

Northumbria University (2021b) Education for sustainable development. https://one.northumbria.ac.uk/service/cs/es/Pages/EducationSusDev.aspx)

Northumbria University (2022a) Why deliver ESD? https://one.northumbria.ac.uk/service/cs/es/Pages/WhyDeliverESD.aspx

Northumbria University (2022b) Environmental sustainability. Supporting the UN Sustainable Development Goals. https://www.northumbria.ac.uk/about-us/environmental-sustainability/

Northumbria University (2022c) Environmental sustainability policy 2022/023. https://northumbria-cdn.azureedge.net/-/media/services/campus-services/documents/pdf/sustainability-documents/environmental-sustainability-policy.pdf?modified=20221111132656%C2%A0%C2%A0%20%E2%80%8B

Northumbria University (2022d) Year in international multidisciplinary innovation. https://www.northumbria.ac.uk/international/amsterdam-campus/undergraduate-study-abroad/multidisciplinary-innovation/

Northumbria University (2023a) Blackboard: digital learning space. https://www.northumbria.ac.uk/about-us/university-services/it-services/student-digital-induction/blackboard/

Northumbria University (2023b) Programme framework for Northumbria awards. https://www.northumbria.ac.uk/study-at-northumbria/programme-framework-for-northumbria-awards/

Northumbria University (2023c) Programme Framework for Northumbria Awards. https://one.northumbria.ac.uk/sites/pfna/Pages/home.aspx

Northumbria University (2023d) Partnerships for the goals. https://www.northumbria.ac.uk/about-us/environmental-sustainability/partnerships-for-the-goals/

Northumbria University (2023e) Multidisciplinary innovation MA/MSc. https://www.northumbria.ac.uk/study-at-northumbria/courses/multidisciplinary-innovation-dtfmdy6/

Nous Group (2023) How human-centred design can transform government service delivery during COVID and beyond. https://nousgroup.com/ca/insights/hcd-government-service-delivery/
Orr S, Shreeve A (2019) Art and design pedagogy in higher education: knowledge. Values and Ambiguity in the Creative Curriculum, Routledge, London
People & Planet (2022) https://peopleandplanet.org/university-league
Piaget J, Cook MT (1952) The origins of intelligence in children. International University Press, N.Y
Programme Framework for Northumbria Awards (PFNA), https://one.northumbria.ac.uk/sites/pfna/Pages/home.aspx
Quality Assurance Agency (2014) Education for sustainable development: guidance for UK education providers. https://www.qaa.ac.uk/the-quality-code/education-for-sustainable-development
Read JM (2005) Teaching user-focused innovation. Design. Manage Rev 16(4):80–84
Rumbley L (2020), Internationalization of higher education and the future of the planet. Int Higher Educ 100:2–34 https://ejournals.bc.edu/index.php/ihe/article/view/14237
Smith L (1996) Critical readings on Piaget. Routledge, London
Smith PK, Cowie H, Blades M (1998) Understanding children's development, 6th edn. John Wiley & Sons, Chichester
Tamarack Institute (2022) Tool | Experimenting with human-centered design: workbook. https://www.tamarackcommunity.ca/library/experimenting-with-human-centered-design
Thompson P. (2019) Foundations of educational technology. Oklahoma State University Libraries, https://doi.org/10.22488/okstate.19.000002
Townson D (2017) The seven tenets of human-centred design. https://www.designcouncil.org.uk/our-work/news-opinion/seven-tenets-human-centred-design/
United Nations (2022) The 17 goals. https://sdgs.un.org/goals
Val E, Gonzalez I, Iriarte I, Beitia A, Lasa G, Elkoro M (2017) A design thinking approach to introduce entrepreneurship education in European school curricula. Des J 20:754–766. https://doi.org/10.1080/14606925.2017.1353022
Verenikina I (2003) Understanding scaffolding and the ZPD in educational research. https://ro.uow.edu.au/edupapers/381
Veryzer RW (2000) (20000) Design and consumer research. Design Manag J Acad Rev 1:64–73
Vygotsky LS (1978) Mind in society: the development of higher psychological processes. Harvard University Press, Cambridge, MA
Wilson E (2022) Learning from the past: what can we learn from Polytechnics? https://www.hepi.ac.uk/2022/01/25/learning-from-the-past-what-can-we-learn-from-polytechnics/#:~:text=The%20Further%20and%20Higher%20Education,universities%20in%20a%20unified%20system

Putting Sustainable Development at the Heart of Transnational Education: Principles and Practice from a UK-UAE Partnership

Andrew Reeves, Manjeet Ridon, and Stuart Lawson

Abstract Transnational Education (TNE) is a significant and growing area of activity for higher education institutions, through which universities establish overseas partnerships or campuses to enable international students to study close to home to achieve internationally recognised qualifications. Whilst TNE is frequently understood as being income-driven and has been critiqued from the stance of post-colonialism, there is arguably significant potential for its adoption to positively contribute to the sustainable development agenda embodied in the UN Sustainable Development Goals (SDGs). This potential contribution is addressed here by discussing the experience of De Montfort University (DMU) in the UK, in seeking to put sustainable development at the heart of its TNE approach when establishing its DMU Dubai campus in the United Arab Emirates (UAE), launched in 2021. A range of innovative curriculum-based, co-curricular and operational practices at DMU Dubai are highlighted, that extended the Education for Sustainable Development initiatives already underway at DMU's UK campus and contextualised them for the UAE environment. These practices are used to illustrate a proposed framework of principles and indicators that can shed light on how TNE activity can address the SDGs. In so doing, the paper introduces 'Transformative TNE' as a concept which highlights an aspirational approach to TNE practice which delivers educational, civic and socio-economic benefits aligned to the sustainable development agenda.

Keywords Sustainable development goals · Transnational education · International higher education · Education for sustainable development

A. Reeves (✉) · M. Ridon · S. Lawson
De Montfort University, The Gateway, Leicester LE1 9BH, UK
e-mail: areeves@dmu.ac.uk

W. Leal Filho et al. (eds.), *The Contribution of Universities Towards Education for Sustainable Development*, World Sustainability Series,
https://doi.org/10.1007/978-3-031-49853-4_22

1 Introduction

Since the adoption in 2015 of the United Nations Sustainable Development Goals (SDGs) for 2030, the profile of the sustainable development agenda has risen significantly for Higher Education Institutions (HEIs) across the globe. Universities can play a key role in addressing the SDGs, through their research, teaching and learning, operational practices and through engagement with society at large through thought-leadership, innovation and civic activities (SDSN Australia/Pacific 2017). Education for Sustainable Development (ESD) is a long-standing agenda through which HEIs can play a transformative role in enabling strong action on the SDGs through taught course provision and wider community engagement (UNESCO 2020). This paper explores how an aspect of the Internationalisation agenda for higher education, namely 'Transnational Education', could potentially contribute towards the SDGs and ESD.

Internationalisation is an area of strategic importance for many universities, encompassing activities ranging from international student recruitment, overseas student mobility opportunities, live projects involving engagement with multinational companies, overseas staff and student exchange, and collaboration international research projects. Knight (2008, p. xi) defined internationalisation as "a process of integrating an international, intercultural and global dimension into the purpose, functions (teaching, research and service), and delivery of higher education at the institutional and national levels". There is frequently strong governmental support for this agenda, as demonstrated by the UK Government's International Education Strategy, which highlights two key benefits of UK universities engaging in this activity: a direct economic benefit through the export of education provision and enhanced international relationships supporting other UK policy goals (UK Government 2021).

Transnational Education (TNE) is "the delivery of an educational award in a country other than that in which the awarding body is based" (UUKi 2018). Historically, HEIs from English-speaking countries such as the UK, USA and Australia have dominated TNE provision, responding to increasing demand from students in emerging economies across the globe, particularly in Asia. For UK HEIs, the latest available data suggests that over 500,000 students are studying annually via TNE, with student numbers increasing and projected to continue doing so (UUKi 2023). The forms of TNE offered by UK HEIs were recently documented by Bennell (2019): the dominant form in terms of student enrolments is "Franchised, validation and articulation" (FVA) provision, with degrees awarded either by the provider institution or host partner; this is followed by Overseas Distance Learning (ODL) courses, most commonly offered by high-status 'redbrick' HEIs, for which international students study remotely; finally, International Branch Campuses (IBCs, also called 'Overseas Branch Campuses') make up less than 10% of provision.

As a form of education provision with a built-in international focus, TNE in all its forms offers potential to address challenges highlighted by the SDGs via delivery of

ESD. This potential contribution is addressed in this paper by discussing the experience of De Montfort University (DMU), a UK HEI which established its 'DMU Dubai' IBC in 2021. This new branch campus in the United Arab Emirates (UAE) is representative of an emerging trend within international student mobility of an increasing preference to travel regionally for overseas study rather than only to traditional host countries such as the US and UK, with the UAE specifically rising to the world's 7th ranking host country by student numbers (Tsiligkiris and Ilieva 2022). This paper aims to shed light on how TNE can address ESD and the SDGs through discussion of curriculum-based, co-curricular and operational practices employed at DMU Dubai, aligned to DMU's sustainable development commitments. We align this to the introduction of principles and performance indicators for 'Transformative TNE', a proposed aspirational approach to the practice of TNE which delivers educational, civic and socio-economic benefits aligned to the sustainable development agenda.

2 'Transactional' TNE—Supply and Demand Factors

TNE has emerged through a global marketplace for international higher education provision shaped by supply and demand-side forces.

From a demand perspective, TNE has been characterised as "the most consumer-driven form of education delivery in the world today" (McBurnie and Ziguras 2006, p. 1). Provision is responsive to growing aspirations for higher education from students in countries which often do not have well-established HE systems offering a range of courses across all disciplines. The prestige and perceived quality associated with degrees awarded by provider institutions and countries such as the UK and USA motivates students to select those providers, alongside 'soft power' considerations such as the chance to engage with (for example) US culture (Ghosh et al. 2021). Students can incur lower costs by remaining in their home country or region, avoiding higher tuition fees associated with being based overseas at a provider institution and minimising relocation and accommodation costs (Papageorgiou et al. 2021). In contrast, from a supply perspective, there is a strong financial incentive that attracts many universities to compete to recruit these students. Often, this is building upon already large-scale recruitment of international students to a provider's home campus, enabling universities to grow their income through offshore activities, and diversify their income streams, reducing risks to their future financial viability. The experience of adopting online and remote teaching and learning approaches during the period of restrictions on movement due to the Covid-19 pandemic further highlighted the potential to generate income from students across international borders for many HEIs.

Literature on TNE has tended to focus on operational challenges and business models, in line with its development to date as an opportunity for income generation for HEIs, through the large-scale delivery of predominantly business school programmes. Whether or not TNE is viable in a host country depends upon not just educational need, but a viable marketplace, including sufficient student interest for the courses offered (Papageorgiou et al. 2021). For example, in the case of TNE provision in Greece, key success factors have included delivery of a wide range of programmes using the local language, making higher education readily accessible and more affordable for learners (ibid).

For IBCs specifically, the literature highlights a number of distinctive challenges as they start up, mature and develop (e.g. Hill and Thabet 2018), making their operation differ from 'traditional' HEIs based in one location. These challenges include staff recruitment, retention and development (Salt and Wood 2014), ensuring quality assurance and quality enhancement in their programmes (Bennett et al. 2010), managing the relationship between host and provider institutions (Hill and Thabet 2018) and meeting the requirements of regulatory bodies and government in host countries (Hickey and Davies 2022). Strong performance in external rankings affecting student recruitment can also be a key factor at regional level—for example, the Knowledge and Human Development Agency (KHDA) is a regulatory body that operates a Higher Education Classification (HEC) rating of all IBCs in Dubai (KHDA 2022). For long-term success, Hickey and Davies advocate a holistic approach (with the host institution engaging in research and service alongside teaching), offering a wide range of courses at undergraduate and postgraduate levels and co-curricular activities to offer a rounded student experience beyond the formal curriculum.

3 TNE and Sustainable Development

Beyond the transactional rationale for TNE for students and institutions involved in delivery, what are the wider benefits that might relate to the sustainable development agenda?

One key benefit highlighted in TNE literature is that the approach offers enhanced access to high quality higher education for students through an increase in capacity within host countries (Bennell 2019; UUKi 2018; Papageorgiou et al. 2021; Hickey and Davies 2022). This apparent benefit comes with two key caveats—firstly, that this is typically an expansion of private higher education provision (Hill et al. 2014), bringing associated risks of restricting access to HE to those that cannot afford it (Papageorgiou et al. 2021); and secondly that the programmes provided on the basis of supply and demand (with business-related courses often dominating) might align weakly with the wider sustainable development needs for knowledge and skills in the host country. This contrasts with the potential for a more supply-driven and strategic approach to capacity building through TNE, such as a UK-Philippines partnership to develop HE provision related to government priority areas for national development (Tsiligkiris and Ilieva 2022).

A second key benefit of TNE is often articulated by provider institutions and their governments as 'global engagement' (Bennell 2019), enabling knowledge exchange, research partnerships and more to develop between provider and host institutions and their countries (e.g. Shams 2016). This could be viewed as an alignment with

SDG17 (Partnerships for the Goals) as an enabler of collaborative research, service and knowledge exchange towards other SDGs and the public good. The key constraint here is overcoming the challenges around running TNE partnerships as holistic HEIs that move beyond teaching provision to the wider functions of universities, in particular given that they operate in a competitive marketplace where wider activities might increase operational costs and therefore student fees.

The above benefits are however very rarely expressed in the terminology of sustainable development. Literature that explicitly explores this linkage is scarce, with 'sustainability' or a sustainable approach most frequently referring to the longevity of a business model (e.g. Shams 2016; Hickey and Davies 2022). Amongst the rare linkages found in literature are the claim that TNE provision can offer a reduced carbon footprint compared to study overseas (Tsiligkiris and Ilieva 2022) and that the capacity building function of TNE is of benefit to SDG4 on enabling access to quality education (Papageorgiou et al. 2021).

A point of leverage that might lead to ESD being explicitly addressed in TNE practice is the growing pressure by HE sector bodies and professional standards for engagement with sustainable development in taught courses. In the UK context, this includes guidance for addressing ESD jointly developed by AdvanceHE and QAA (2021), bodies concerned with professional development of academics and quality of taught course provision respectively. If ESD is addressed by design through learning outcomes, assessments and learning activities in taught courses in a TNE provider country, then this should also apply to the same courses when delivered in a host country. The challenge for provision is then about pedagogy and contextualisation—adapting provision to the local context as much as is viable within the confines of the validated course being delivered (Hickey and Davies 2022). The same logic applies to the Equality Diversity and Inclusion (EDI) agenda as applied to internationalisation (Killick 2017) and initiatives to Decolonise higher education (see Bhambra et al. 2018). Given the high priority afforded to these issues at TNE provider institutions, this could provide impetus to address them within practice through TNE partnerships. Putting together the ESD and EDI agendas is likely to lead to a more equitable stance of engaging with stakeholders prior to launching programmes, ensuring strong relevance and contextualisation by design. This latter point also highlights that leverage to address ESD may also come at the TNE host country level, through courses that respond to Government or industry priorities towards social responsibility, global citizenship or otherwise.

TNE can also be analysed in terms of the purposes of education. In his influential book "Sustainable Education", Sterling (2001) highlights four functions of education: Socialisation (replicating society and culture); Vocational (preparing people for employment); Liberal Humanist (developing the potential of individuals); Transformative (facilitating change towards a better and livable world). All are seen by Sterling as having a role to play, but he argues that ESD puts a much stronger focus on the Liberal Humanist and Transformative functions of education. Viewed from this perspective, TNE risks being overly vocational in its respond to demand-side forces for provision to enable career advancement. An analogous risk emerges when looking at the process of TNE provision. SDSN Australia/Pacific (2017) highlights that universities can address sustainable development via operations, learning and teaching, research and external leadership—the latter two areas of activity may not be prioritised where teaching is the overwhelming focus, meaning that the integrated approach to ESD (EUA 2021) drawing on civic contributions, research-informed teaching and co-curricular experiences may be lost. Literature on ESD highlights the value of the whole institutional experience (e.g. AdvanceHE/QAA 2021) as part of the Liberal Humanist function of HE, placing strong value on co-curricular and extra-curricular learning which might be absent or weakened in some TNE settings.

Overall, literature published to date reveals that while TNE's importance has been considered from pedagogic and management perspectives, the potential transformative impact of TNE as a driver for ESD and action on the SDGs is under-explored.

4 Transformative TNE

Thus far, TNE has been introduced as a market-driven approach to education provision, whose intention is more transactional than transformational. Here we put forward an alternative paradigm, 'Transformative TNE', which can be understood as TNE delivered with the purpose of addressing the Sustainable Development agenda and the SDGs. Our purpose here is to introduce an aspirational framework that can serve as a counterpoint to traditional transactional framings of TNE.

Given the diversity of contexts in which TNE takes place and various formats of delivery, our approach is to set forward six principles, which are flexible enough to be applied in any setting, but which align with the agendas of ESD and the SDGs.

Our starting point for this is the recent UNESCO (2020) Roadmap for ESD, which sets out five priority areas for the global implementation of Education for Sustainable Development: 1: Advancing Policy; 2: Transforming Learning Environments; 3: Building Capacities of Educators; 4: Empowering and Mobilizing Youth; 5: Accelerating Local Level Actions. These priorities were articulated towards a policymaker and policy-oriented audience, so we have drawn upon these to put forward related principles more oriented towards the role of HEIs.

In addition, we have added one further principle, an explicit focus on EDI issues of (and related concepts such as Equity, Justice and Mutuality), which risk not being made explicit if only the five UNESCO principles are used as a basis. This principle reflects the strong commitment to these agendas within HEI operations and that addressing inequalities are a key aspect of the UN SDGs. Furthermore, TNE partnerships are frequently based in societies where social inequalities (e.g. via income, gender or ethnicity) are prevalent, meaning that conscious consideration is required in terms of how partnership activities are delivered. This principle makes explicit an over-arching principle of the UN SDGs to "leave no one behind" (UN 2023). Our list of six proposed principles is summarised in Table 1, alongside illustrative example SDGs to which they link.

Table 1 Principles for transformative TNE

Principle	Example SDG linkages	UNESCO/UN link
1. Policy engagement	#SDG16 (Good Governance) #SDG5 (Gender Equality)	ESD Priority 1
2. Sustainability in operations	#SDG7 (Clean Energy) #SDG12 (Consumption/ Production)	ESD Priority 2
3. Capacity Building	#SDG4 (Education) #SDG8 (Economy)	ESD Priority 3
4. Empower students as change agents	#SDG4 (Education) #SDG2 (Zero Hunger)	ESD Priority 4
5. Integrated civic action	#SDG11 (Cities and Communities) #SDG13 (Climate Action)	ESD Priority 5
6. Equity and mutual learning	#SDG10 (Reduced Inequalities) #SDG17 (Partnerships)	"Leave no one behind"

- **Principle 1 on Policy Engagement** highlights the role of HEIs in engaging with government policy locally, regionally and (inter)nationally, providing evidence for development, spaces for discussion and seeking to influence policies that further sustainable development.
- **Principle 2 on Sustainability in Operations** emphasises internal policies and practices to minimise environmental impacts (e.g. working towards net zero carbon emissions), alongside people-focussed aspects of sustainability such as appropriate polices on EDI.
- **Principle 3 Capacity Building** relates to developing the capacity of the higher education system to serve the sustainable development needs of the locality or region, whether through provision of courses, development of staff or the wider educational system infrastructure.
- **Principle 4 to Empower Students as Change Agents** can be related in part to the employability-related purpose of higher education, but goes further by adopting a holistic ESD approach, empowering students to challenge unsustainable practices and develop sustainable approaches in their professional and personal lives.
- **Principle 5 Integrated Civic Action** emphasises the role of the institution in its place and how through links to its stakeholder communities it can harness teaching, learning and research for social good.
- **Principle 6 Equity and Mutual Learning** reflects an ethos that recognises and challenges inequality and injustice, and works towards a mutually beneficial learning approach between institutions (Hickey and Davies 2022)

Taken together the principles offer terminology to both articulate the benefits of TNE practice and to provide a challenge to enhance or expand practice in certain areas, such as by developing local partnerships with businesses or government for mutual benefit. We next illustrate these principles through an illustrative case study.

5 Case Study—DMU Dubai International Branch Campus

5.1 Introduction to DMU Dubai

De Montfort University (DMU) is a former polytechnic university based in Leicester, UK, offering undergraduate and postgraduate courses across four faculties (Arts Design and Humanities; Business and Law; Health and Life Sciences; Computing, Engineering and Media). DMU has over 25,000 enrolled students and close to 3000 staff. Both "International" activity and "Sustainability and the SDGs" are strategic priorities at the university, each overseen by a pro-vice chancellor role on the university's leadership team. Prior to the launch of DMU Dubai, DMU had over ten TNE partnerships with more than 2500 students enrolled. These other partnerships used some type of FVA model, with Dubai being DMU's first IBC arrangement.

The launch of DMU Dubai in Autumn 2021 was the outcome of DMU's strategy to grow international reputation and impact by establishing global TNE hubs of teaching, research and enterprise. The commitment to the UN SDGs was embedded in this vision along with the understanding that each global TNE hub would respond to the sustainability agenda in a way that is relevant and contextualised to its location. From the outset, DMU Dubai would offer programmes from all four of DMU's Faculties, offering an unusually broad-based curriculum at launch for a new IBC. The strategic decision to locate its first IBC in Dubai, was based on market intelligence and aligned with other opportunities and major institutional initiatives such as DMU's engagement and sponsorship of the Dubai EXPO2020, which was originally scheduled for 2020, but took place a year later owing to the global pandemic.

IBCs are prevalent in the UAE which hosts the highest number of IBCs worldwide, with the country's 37 IBCs making up 19% of the world share (Lawton and Katsomitros 2012). They can be the most expensive and time-consuming form of TNE, especially in the early years following launch, when universities are establishing student administration and academic management systems, setting-up campus facilities, establishing their reputation in the host country, and deepening their knowledge of local and regional regulatory and assurance bodies. For DMU Dubai, this involved the setup of a leadership team comprising of Leicester and Dubai-based staff and active engagement from DMU's senior leadership through visits to the Dubai campus in its formative years.

Along with the set-up of teaching and academic governance structures, IBCs also must contextualise their curriculum so that it is sensitive to the local culture and background of its student population. This is especially relevant for Western IBCs such as DMU Dubai that acknowledge the importance of cultural diversity,

Table 2 Case study practices and aligned principles

Case study	#1 Policy Engagement	#2 Sustainability in Operations	#3 Capacity Building	#4 Empower Students as Change Agents	#5 Integrated Civic Action	#6 Equity and Mutual Learning
1: MENA UN PRME conference	X	X	X	X	X	
2: Internal policies and principles		X	X	X		X
3: DMU Leicester and Dubai links	X	X	X	X		X

internationalisation and the current trend towards decolonisation, in which the predominance of Western-centric and European knowledge and pedagogy is contested and challenged (Bhambra et al. 2018).

Below three case study practices are introduced, highlighting how this commitment to sustainable development was put into practice, with linkages to the Transformative TNE principles highlighted as relevant (e.g. "**#1: Policy Engagement**"). A visual summary of the practices and linkages to principles is provided in Table 2.

5.2 *Practice 1: MENA UN PRME Conference*

In March 2022, DMU Dubai hosted the Middle East and North Africa (MENA) Regional Chapter of the UN PRME (Principles of Responsible Management Education) Conference—a 3-day event linked to the delayed Dubai World Expo 2020. This was instigated as a catalyst for the development of projects and partnerships between attendees and to strengthen the connection and collaboration of the regional PRME MENA Chapter Network, UN Global Compact Networks and other UAE-based partner universities. The in-person event comprised keynote talks, panel discussions, a gala reception and networking opportunities, with over 70 external regional stakeholders contributing. It was attended by DMU Dubai staff and was open to students. Some events were focussed on students, such as a panel discussion on opportunities for civic participation with guests from a range of regional UN agencies and NGOs (**#4 Empower Students as Change Agents**).

Since the event, contacts from the conference have visited students on campus as part of teaching delivery, have taken part in follow-up panel discussion events and have hosted students on placements, adding regional contextualisation and a sustainability focussed to DMU Dubai's teaching (**#4 Empower Students as Change Agents**). It has thus served to forge links between DMU Dubai and policymakers, NGOs and regional UN agencies, enhancing DMU Dubai's reputation (**#1 Policy Engagement; #5 Integrated Civic Action**). This level of policymaker engagement so early after an institution's launch was unprecedented for a Dubai-based international university.

One of the most significant reasons for organising the conference was to encourage Dubai academic staff to engage in collaborative interdisciplinary research related to the SDGs and to grow their networks. Nine published conference papers were the result of 13 DMU Dubai staff (approx. 50% of all staff) from all programmes collaborating with each other or with colleagues from other institutions. For two staff it was their first conference paper publication. Significant benefits of this include: encouraging a collaborative research culture; kick-starting early-stage research careers of academic staff; and demonstrating to students, staff and external stakeholders that the TNE staff are of the calibre to present at an international, UN-affiliated conference **(#3 Capacity Building*).***

The conference cemented DMU Dubai's commitment to the UNPRME framework and principles (UNPRME 2023), which include a commitment to addressing sustainable development through operational matters. Given that programmes from all four DMU Faculties in Dubai have signed up to PRME, this has established a lever to strengthen operational sustainability across campus-based and other issues (**#2 Sustainability in Operations**).

5.3 Practice 2: Internal Policies and Principles

Analysis of the UAE TNE market and student and employer expectations highlighted the importance of employability, empowering women for professional roles and reflecting this in strong external ratings of DMU Dubai (e.g. KHDA 2022). This inspired development of a bespoke strategy for employability, through which four principles with the acronym SEED (Sustainability, Employability, EDI and Decolonisation) were embedded. This included the DMU Dubai internship preparation programme (for students from levels 4 to 7) which alongside traditional career-focussed activities (e.g. on CV development, communication) included sessions on Sustainability Literacy and EDI. Addressing gender inequalities is a key aspect of SEED, in particular as most DMU Dubai students are female—this was reflected via a week-long event of teaching-related activities on addressing social inequalities and a dedicated 'Women's Entrepreneurship Day' event, with several inspirational speakers. Partnerships with local and regional businesses were set up to contextualise the curriculum for a range of programmes (e.g. interior design, accountancy and finance) employing the SEED principles, enhancing the relevance of teaching for future careers and infusing this with sustainable development challenges (**#4 Empower Students as Change Agents; #6 Equity and Mutual Learning).**

The experience of addressing ESD and sustainability through a range of activities in Dubai is being translated into policies to be applied in future at Dubai and for all DMU's other international partnerships, including TNE delivered via IBCs and other modalities. These include referring to sustainability as a strategic commitment in partnership agreements, requiring ESD to be addressed in programme (re)validation processes and devising context-specific plans for student and staff engagement on ESD (**#2 Sustainability in Operations; 3 Capacity Building; #4 Empower Students as Change Agents**).

5.4 *Practice 3: DMU Leicester and Dubai Links*

In January 2022, all DMU Dubai staff including the leadership team took part in one-day Carbon Literacy Training, which supports participants to develop understanding of climate change issues and develop personal and group-based plans to take action. (**#3 Capacity Building; #4 Empower Students as Change Agents**). This initiative seeded a culture of addressing sustainability across the taught curriculum, provided motivation for operational enhancements to practice (**#2 Sustainability in Operations**) and gave DMU Dubai a platform for enhanced policy engagement through accredited bronze status as a Carbon Literate Organisation (**#1 Policy Engagement**).

A group of Dubai students established a Sustainability focussed student society. They were supported to develop projects on campus by staff and initiated a 'SMART Green Wall' project, using a small-scale solar panel installation and rainwater harvesting to supply a mini-garden outside the main campus library. This project applied students' subject-based knowledge from several disciplines (e.g. engineering, interior design), provided a visual exemplar of the university's sustainability commitments and was funded by DMU Leicester's small grant fund for SDG projects—an illustration of an inclusive and equitable approach that permitted students in Dubai to also apply. The Dubai students were invited to present their project by joining the DMU Student Sustainability Showcase online (**#4 Empower Students as Change Agents; #6 Equity and Mutual Learning**).

The 'Decolonising DMU' initiative has been received well by staff and students at the Dubai campus and much was learned in terms of effective and inclusive contextualisation of curriculums. This knowledge has been shared back to inform programme design at DMU Leicester through a workshop delivered in February 2022 (**#6 Equity and Mutual Learning**).

5.5 *Contextual Influences on Practice*

The above examples illustrate aspects of what can be achieved where there is a commitment to sustainable development as part of the establishment of a new TNE IBC. However, internal and contextual factors play a key factor on the implementation of ESD in any institution (e.g. see EUA 2021), so it is important to highlight some significant factors at play in this case, as understood by the authors through their roles.

Firstly, DMU has a strong strategic commitment to "Sustainability and the SDGs", which is highlighted as a cross-cutting theme in its institutional strategy. Although this did not create a direct push for action, it did give permission and strong backing once pursued by staff leading initiatives. Support capacity was also available at DMU Leicester via dedicated ESD staffing (a 50% part-time academic lead role) and a Sustainability Team (2 full time roles, plus student internships) with an associated operational budget, which enabled interventions such as Carbon Literacy Training provision and grant funding for student societies. An opportunity offered by DMU's link to the Dubai World EXPO created a significant stakeholder engagement opportunity and DMU's status as a Global Hub for SDG16 enabled UN dignitaries and their contacts to be enlisted to take part in events. The personal commitment to this agenda by individuals concerned, including many of the DMU Dubai academic staff, a highly engaged student society and the Provost leading the IBC also played a key part.

6 Discussion

6.1 Challenges to Implementing Transformative TNE

The above discussion has introduced the practice and rationale behind TNE as-it-is, which whilst bringing some significant benefits, is delivered through a predominantly transactional ethos. We have contrasted this to a Transformative TNE approach, oriented towards international collaboration to address the UN SDGs and highlighted some examples of practice through one IBC-based TNE partnership of how this can be put this into practice.

The prospects for widespread adoption of Transformative TNE face several limitations. A key issue is whether institutions (whether in host or provider countries) wish to develop TNE partnerships at all, and if so, what their driving motivations are. In the current global HE environment, the profit motive, although it can be critiqued on a number of grounds for its impact on how TNE is delivered, is a key catalyst for international education partnerships being developed, rather than HEIs continuing to deliver courses in their local area. However, if TNE provision is to play a transformative role in capacity building, then in line with principle #6, this should be a mutually supportive relationship, not an extractive one (Hickey and Davies 2022), which develops towards a HE system offering a wide range of courses and access to HE being widened across societies. This raises questions about the longevity of partnerships or a transitional structure towards the kind of equal partnership arrangement that Hickey and Davies (2022) highlight as being employed in some cases.

A second key issue is whether TNE operating models enable the holistic role of HEIs, incorporating mutually beneficial teaching, research and civic activity. Transformative TNE, as put forward here, is linked to this rounded vision of the role of HEIs in their wider communities and more-limited approaches to ESD may be achievable in settings where, for example, Online Distance Learning dominates.

A final issue, which we address next is the pragmatic question of how existing or planned TNE partnerships can be influenced to go further to address sustainable development in their design and implementation.

6.2 Indicators for Transformative TNE

Decision making within HEIs is frequently guided by quantitative evidence, and for TNE, student numbers and income levels are the key metrics used to judge success. Further metrics are needed to provide a well-rounded picture of the value of TNE, in the same way that a 'triple bottom line' approach of social, environmental and economic value can better reflect the holistic value of business activity.

This could occur in part via consideration in already existing frameworks with significant reach, such as the Times Higher SDG Impact Rankings (THE 2022) or SDG Accord (EAUC 2023) and this would undoubtedly achieve some leverage. For DMU Dubai specifically, the KHDA framework for Dubai HEIs offers rankings out of 1000 points and up to 50 points are available for each of environmental sustainability and inclusion (aligned to **#2 Sustainability in Operations** here). At this time where practice to address Sustainable Development via TNE is at an early stage of development, our aim is to offer examples of indicators which might be adapted and adopted by institutions seeking to lead in this area, or which might have value if used within wider frameworks or accreditation schemes (see proposed Indicators in Table 3).

These have been put forward based upon the practices already done or under consideration for DMU's TNE partnerships. Such process and outcome-based indicators can offer a basis for qualitative and/or quantitative reporting of impacts and activities related to the SDGs and ESD within TNE partnerships, and consideration of their adoption has potential as a driver of enhanced activity.

7 Conclusions

TNE provision has been developed as a transactional model to meet emerging demand for higher education provision. It holds transformational potential to address societal challenges linked to sustainable development, but this requires a distinctive focus and significant strategic effort, with all partners concerned seeing this as part of the value and purpose of the partnership.

Table 3 Indicators to support implementation of transformative TNE

Principle	Indicators (P = process-based; O = outcome-based)
1. Policy engagement	TNE partners can demonstrate engagement with policymakers, businesses and governance processes in region on sustainable development issues. (P)
2. Sustainability in operations	Commitment to operational sustainability and carbon management of host campus in partnership agreement (P) All TNE partners address sustainable development in published strategy (P)
3. Capacity building	Staff at TNE host partner are trained on ESD pedagogic approaches (P)
4. Empower students as change agents	Academic Quality processes ensure ESD is addressed in TNE courses (P) Students take part in co-curricular projects addressing the SDGs (O)
5. Integrated civic action	Students take part in projects that support local action for sustainable development in the TNE host locality (O)
6. Equity and mutual learning	Strategy to widen participation addressing barriers to HE participation is part of TNE strategy (P) Joint academic work beyond taught course delivery or student–student interaction between TNE partners (O)

This paper sets out some concepts to work towards this vision—**Principles** to help frame TNE partnerships; **Indicators** that can be used to gauge if and how a transformative approach is being adopted; examples of **Practices** that demonstrates some of the actions associated with a commitment to addressing sustainable development challenges within TNE.

A key limitation of this work is that our proposed principles and indicators are at an early stage of formulation and that our examples of implementation are drawn from a single case study developed in a specific context. Further research is therefore needed on a range of issues raised here:

- clarifying the range of practices to address sustainable development via TNE that are effective in differing contexts;
- identifying the outcomes of ESD engagement for students and staff within TNE partnerships;
- exploring principles and indicators which can be effectively adopted through TNE partnerships to leverage action on ESD;
- mapping out the extent to which TNE institutions are able to operate holistically as HEIs, achieving research-related and civic impacts;
- identifying if and how TNE is a pathway to accessible and inclusive development of HE provision, in line with #SDG4 on Quality Education.

Overall, although TNE has not to date been formulated as a solution strategy for sustainable development, we argue that it can play a transformative role in addressing the SDGs, if implemented with an ethos focussed on the wider societal benefits of higher education.

References

AdvanceHE/QAA (2021) Education for sustainable development guidance. https://www.advance-he.ac.uk/teaching-and-learning/education-sustainable-development-higher-education#guidance

Bennell P (2019) Transnational higher education in the United Kingdom: an up-date. Int J Educ Dev 67:29–40. https://doi.org/10.1016/j.ijedudev.2019.03.003

Bennett P, Bergan S, Cassar D, Hamilton M, Soinila M, Sursock A, Uvalic-Trumbic S, Williams P (2010) Quality assurance in transnational higher education. ENQA workshop report 11. ENQA (European Association for Quality Assurance in Higher Education)

Bhambra GK, Gebrial D, Nişancıoğlu K (2018) Decolonising the university. Pluto Press

EAUC (2023) SDG Accord. [Online] Available from: https://sdgaccord.org

EUA (2021) Environmental sustainability of learning and teaching. European Universities Association. https://eua.eu/resources/news/653-sustainable-learning-and-teaching-eua-learning-teaching-thematic-peer-groups-focus-on-grande-challenges.html

Ghosh S, Lee J, Haupt J (2021) Understanding the attraction of the microcampus: a quantitative investigation of students' motivations to enroll in transnational education. J Comparat Int Higher Educ 13(5):53–71. https://doi.org/10.32674/jcihe.v13i5.3679

Hickey R, Davies D (2022) The common factors underlying successful international branch campuses: towards a conceptual decision-making framework. Globalisation Societies Educ.https://doi.org/10.1080/14767724.2022.2037072

Hill C, Cheong KC, Leong YC, Fernandez-Chung R (2014) TNE–Trans-national education or tensions between national and external? A case study of Malaysia. Stud High Educ 39(6):952–966. https://doi.org/10.1080/03075079.2012.754862

Hill C, Thabet RA (2018) Managing international branch campuses: lessons learnt from eight years on a branch campus in Malaysia. Int J Educ Manag 32(2):310–322. https://doi.org/10.1108/IJEM-04-2017-0079

KHDA (2022) Rating of higher education institutions in Dubai—summary report 2022. Knowledge and Human Development Agency. https://web.khda.gov.ae/en/Resources/Publications/Higher-Education-and-UQAIB/Rating-of-Higher-Education-Institutions-in-Dubai-%E2%80%93?ext

Killick D (2017) Internationalization and diversity in higher education: Implications for teaching learning and assessment. Bloomsbury Publishing

Knight J (2008) Higher education in turmoil: the changing world of internationalization. Brill

Lawton W, Katsomitros A (2012) International branch campuses. Data Dev. Lond. Obs. Borderless Higher Education

McBurnie G, Ziguras C (2006) Transnational education: issues and trends in offshore higher education. Routledge

Papageorgiou F, Makellaraki V, Tsiligiris V (2021) Transnational education as enabler of access to quality higher education and its link to UN SDG4: the case of Greece. In: Importing transnational education 65–81. Palgrave Macmillan Cham. https://doi.org/10.1007/978-3-030-43647-6_5

Salt J, Wood P (2014) Staffing UK university campuses overseas: lessons from MNE practice. J Stud Int Educ 18(1):84–97. https://doi.org/10.1177/1028315313483773

SDSN Australia/Pacific (2017) Getting started with the SDGs in universities: a guide for universities higher education institutions and the academic sector. Sustainable development solutions network—Australia/Pacific Melbourne

Shams SR (2016) Sustainability issues in transnational education service: a conceptual framework and empirical insights. J Global Market 29(3):139–155.https://doi.org/10.1080/08911762.2016.1161098

Sterling S (2001) Sustainable Education—Re-visioning learning and change. Schumacher Briefing no6. Schumacher Society/Green Books Dartington, UK

THE (2022) Times higher impact rankings 2022. https://www.timeshighereducation.com/impactrankings

Tsiligkiris V, Ilieva J (2022) Global engagement in the post-pandemic world: challenges and responses. Perspective from the UK. Higher Educ Quart *76*(2):343–366. https://doi.org/10.1111/hequ.12390

UK Government (2021) International education strategy (2021 update). https://www.gov.uk/government/publications/international-education-strategy-2021-update

UN (2023) Leave no one behind. Available from: https://unsdg.un.org/2030-agenda/universal-values/leave-no-one-behind

UNESCO (2020) Education for sustainable development: a roadmap. https://unesdoc.unesco.org/ark:/48223/pf0000374802

UNPRME (2023) Six principles. PRME chapter UK and Ireland. https://www.unprme.org.uk/six-principles

UUKi (2018) Transnational education: global location local innovation. Universities UK International. https://www.universitiesuk.ac.uk/topics/international/transnational-education-global-location

UUKi (2023) The scale of UK higher education transnational education 2020–21. Universities UK International. https://www.universitiesuk.ac.uk/universities-uk-international/insights-and-publications/uuki-publications/scale-uk-higher-education-transnational-3

Latin American Knowledge Production on Education for Sustainable Development: An Analysis of Policy Documents Using the Overton Software

Thais Dibbern, Denis dos Santos Alves, Evandro Coggo Cristofoletti, Priscila Berbert Rampazzo, Ludmila Pioli, Milena Eugênio da Silva, and Milena Pavan Serafim

Abstract Higher Education Institutions (HEIs) play an important role in achieving Sustainable Development, especially due to their efforts in producing and disseminating knowledge oriented toward solving problems involving social, economic, and environmental dimensions. Therefore, this study aims to understand and analyze to what extent Education for Sustainable Development (ESD) has been addressed in Latin American policy documents, considering the regional knowledge production on the subject. Methodologically, in addition to the literature review, the software Overton.io was used to identify policy documents that address ESD and content analysis, being linked to a qualitative-quantitative approach. As for the preliminary results, among the 253,389 Latin American policy documents in the platform, only 279 (0.11%) explicitly mention the term "Education for Sustainable Development". From the data collected and the analysis developed, it was possible to observe that this low percentage of policy documents dealing with ESD may be connected to the

T. Dibbern (✉) · D. dos Santos Alves · E. C. Cristofoletti · L. Pioli · M. E. da Silva
Department of Science and Technology Policy, University of Campinas, Campinas, Brazil
e-mail: tdibbern@unicamp.br

D. dos Santos Alves
e-mail: d233514@dac.unicamp.br

E. C. Cristofoletti
e-mail: evcoggo@unicamp.br

L. Pioli
e-mail: lpioli@unicamp.br

M. E. da Silva
e-mail: m184650@dac.unicamp.br

P. B. Rampazzo · M. P. Serafim
School of Applied Sciences, University of Campinas, Limeira, Brazil
e-mail: pberbert@unicamp.br

M. P. Serafim
e-mail: milenaps@unicamp.br

W. Leal Filho et al. (eds.), *The Contribution of Universities Towards Education for Sustainable Development*, World Sustainability Series,
https://doi.org/10.1007/978-3-031-49853-4_23

absence of a consensus in the literature on the term ESD, that is, of a clear definition that could work as a starting point for the formulation of public policies.

Keyword Higher education institutions · Education for sustainable development · Policy documents · Latin America

1 Introduction

The achievement of the Sustainable Development Goals (SDGs) requires an organized and joint effort between Higher Education Institutions (HEIs), governments, and other sectors of civil society. This effort must be generated from the definition of cross-cutting strategies of knowledge production and innovation, which leverage the fulfillment of these goals, which are transdisciplinary and at the frontier of knowledge so that we can, in a short time, redesign practices and measures that drive development under sustainable bases.

Since HEIs are one of the important actors in achieving Sustainable Development, mainly due to their role in the education of people and the generation and dissemination of knowledge oriented towards the solution of problems involving social, economic, and environmental dimensions, this study aims to understand and analyze, still in an exploratory way, how Education for Sustainable Development (ESD) has been incorporated in Latin American policy documents, given the regional academic expertise on the subject. The research questions are: Is this expertise used in public policy formulation? If so, what is the contribution of Latin American academia?

There is the understanding that Education for Sustainable Development refers to an educational approach that aims to promote awareness and understanding of environmental issues to develop skills and values that lead to adopting sustainable practices. In other words, it is an approach aligned with the SDGs, intending to empower individuals regarding the different dimensions to which the concept of Sustainable Development is linked.

Based on these questions, the chapter is organized as follows: The first section discusses the Education for Sustainable Development to contextualize and ground the data analysis. Then, the study's methodology is presented, focusing on collecting and analyzing (quantitative and qualitative) data on policy documents that address ESD. To this end, was used the Overton.io software considered to be the largest and most comprehensive database for the use of research on policy documents. On this basis, the third section presents and analyses the results, highlighting how ESD has appeared in Latin American policy documents and the contribution of universities in the region. The chapter ends with a conclusion that points to future reflections and research questions, given the exploratory nature of this study.

2 Education for Sustainable Development and Universities

This section addresses the institutionalization of ESD, its concepts, and the role of universities in the international and regional framework. Despite its relevance, the process of institutionalizing ESD has occurred very slowly and is connected with environmental discussions, particularly with the concept of Environmental Education. This term emerged in 1965 at the Conference on Education at Keele University, Great Britain (Bastos and Souza 2013). At the Stockholm Conference in 1972, recommendation n. 92 of the conference's final report spoke of the need to build a program of environmental education (UN 1973); it was only in 1974, in the Environmental Education Seminar held in Finland, that Environmental Education was effectively recognized. In the same year, at the International Congress on Environmental Education and Training, the International Strategy for Environmental Training for the 1990s was defined as the embryo of ESD (Bastos and Lemes 2017).

In 2002, the United Nations Johannesburg Conference was of utmost relevance to the institutionalization of ESD; due at that time, education ceased to be just a complement/aid (as demonstrated by the absence of action plans in deliberations and previous conferences of the United Nations) and gains prominence in the process, either as an instrument or objective. At the same meeting, the Decade of Education for Sustainable Development (DESD) was proposed and approved a few months later by the General Assembly of the United Nations, lasting from 2005–2014 (UN 2005).

The International Implementation Plan of the United Nations Decade of Education for Sustainable Development points out that the main objective of the Decade was the search for a global vision of a reality in which "everyone has the opportunity to benefit from quality education and to learn the values, behavior, and lifestyles required for Sustainable Development and positive social transformation" (UN 2005, p. 65). The Resolution that approved the DESD considered education "the key—*sine qua non*—for sustainable development" (idem, p. 35). Therefore, ESD becomes a major pillar for achieving sustainable development. The following are key requirements for ESD (UNESCO 2021): (i) is interdisciplinary and holistic; (ii) is value-driven, i.e., has clear values and principles; (iii) encourages critical thinking and problem-solving; (iv) uses multiple methods; (v) participates in the decision-making process, i.e., the learner participates in deciding how they will learn; (vi) is applicable; (vii) is locally relevant.

In the last year of the DESD, the United Nations Educational, Scientific and Cultural Organization (UNESCO) held the World Conference on Education for Sustainable Development in Aichi-Nagoya (Japan). It launched the Global Action Programme for ESD. The program aimed to "generate and scale up ESD and accelerate progress towards sustainable development" in 4 years (2015–2019). In New York, in 2015, the Sustainable Development Summit took place. At this meeting, the 193 countries that make up the United Nations defined the 17 Sustainable Development Goals (SDGs) with a deadline of 2030 (therefore, the commitment also became known as Agenda 2030).

Two years later, with the Global Action Programme for ESD, UNESCO launched another guidance document entitled Education for Sustainable Development Goals. This document brings clarification to some issues regarding ESD, for instance, by clearly presenting that ESD aims to "develop competencies that enable people to reflect on their actions, taking into account their current and future social, cultural, economic and environmental impacts, from a local and global perspective" (UNESCO 2017, p. 7). In this presentation, it is argued that ESD has the following characteristics: (a) "ESD is a holistic and transformative education that addresses learning content and outcomes, pedagogy and learning environment" (UNESCO 2017, p. 7); (b) Therefore, beyond the inclusion of isolated subject matters, such education should create "learner-centered, interactive teaching and learning contexts" (idem, ibid.); (c) "It requires an action-oriented transformative pedagogy that supports self-directed learning, participation and collaboration" and "problem-solving" (ibid.); (d) being, "inter and transdisciplinary"; (e) realizing "the connection between formal and informal learning" (ibid.).

It is essential to highlight the evolution of the characteristics pointed out in the Plan for this guidance document, such as the inclusion of the terms "transformative" and "transdisciplinary", as well as reiterating the holistic aspects, favoring critical thinking and problem-solving, the participation and learning of the learner in the decision-making process, applicability in the sense of making the connection between the various environments and a group of key competences that seek to be formed in the students.

In 2018, UNESCO initiated discussions on the continuity of the work to promote ESD, intending to build a framework that would result in its institutionalization. It is a topic of substantial international scope. Its success and continuity depend on a solid articulation between the most diverse levels: global, regional, national, and subnational. In 2019, the 40th General Meeting of UNESCO took place, a space in which the next steps for ESD at the international level were deliberated.

In parallel, on Latin American soil, most South American states' institutionalization of Environmental Education is unquestionable (Loureiro 2008; Carvalho 2008; Berchin et al. 2021). Nevertheless, the same cannot be pointed out about the ESD. This educational practice presents a certain resistance since its recognition at the regional level, despite its convalidation by the heads of South American states (Berchin et al. 2021; Bastos 2016). These conclusions may be the result, to a greater or lesser extent, of the need for more conceptual consensus in the academic and public policy literature. Moreover, the concept determined for ESD influences its applications in studies, university extension projects, and everyday educational practices, reinforcing the need for conceptual consensus.

Based on a preliminary bibliometric study linked to the content analysis technique (Alves 2022), it was identified that research on ESD does not present a consensus about the concept. The five main lines present in this debate are presented in Table 1.

Despite the different lines and perspectives, it has been observed that common to all understandings is the use of the dimensions of sustainable development (SD): social, economic, environmental, and institutional (Bastos and Souza 2013; Dyer et al. 2006; Gallopín 2003; Nascimento 2012).

Table 1 Theoretical and conceptual lines of ESD

Lines	Description
Synonym of Environmental Education	Studies understand ESD as synonymous with Environmental Education (EE), given the high level of similarity between the two educational practices, or even as mere "official" use and propagated by the United Nations (Berchin et al. 2021; Bonnet 2006)
Part of the Environmental Education	Studies that state the ESD and EE would have significant overlaps. Therefore the ESD would be under the EE, or the reverse, as well as both, have great proximity but without making them synonymous (Barreto and Vilaça 2018; Branco et al. 2011; Setti and Azeiteiro 2016)
Evolution of Environmental Education	Studies understand that the origin of ESD is Environmental Education. Therefore, ESD would be an improved form of EE—some even point out an overcoming of EE, being then ESD the education of a broad and balanced character between the environmental, social, and economic pillars (Ruiz et al. 2019; Campello and Silveira 2016; Cañellas 2006; Correa and Ashley 2018)
"Umbrella" education	Studies that claim that ESD is an education that brings together all educations that contribute to sustainable development or that overlaps other educations—other studies will claim that there is an attempt to make ESD one ample educational practice that addresses all issues necessary for Sustainable Development, thus becoming a big "umbrella" of educational approaches, such as Peace Education, Education in/for Human Rights, Environmental Education, Education for the Future, Education for Sustainable Consumption, among others (Bajaj and Chiu 2009; Engjellushe 2013)
Given concept	In the fifth group, we can add all the studies that present ESD as a given concept. In other words, without entering into conceptual discussions or even merging etymological constructions, forming (or perpetuating) constructions of the type: "Education for Sustainability", "Environmental Education for Sustainable Development" (Olmo-Fernandez 2021; Silva and Teixeira 2019; Burmeister et al. 2012; Esteban 2020)

Source Adapted from Alves (2022)

UNESCO characterizes ESD as an educational practice that contributes to transforming humanity's thinking towards a 'sustainable future' (2021a), meaning the inclusion of 'key issues of sustainable development in teaching and learning' (2021b). ESD is also understood as an element part of the 2030 Agenda and a key enabler of the whole Agenda (*idem, ibid.*). The concept of ESD generates reflections about its definitions, as does Uchôa (2016); the scholar points out that the concept emerged in an environment politically guided by a US vision and thus disregards a diversity of approaches, including naturalizing the environmental issue. In counterpoint, based on the conceptual debate exposed, we take the liberty of pointing out that criticism is important for the evolution of the educational practice itself and, therefore, should not be taken as "perpetual" characteristics nor ignored. Since there is no consensus, on the one hand, among the disadvantages is the risk of the various lines of understanding following very different paths of action practice; on the other hand, among the advantages is the possibility, through uncertainty and, therefore, doubts, to improve and welcome different perspectives of knowledge in this wide debate. Therefore, based on the objectives that guide us in this chapter, it is necessary to specify the role that higher education institutions play in the Latin American scenario.

In the political arena, many Latin American countries have incorporated the 17 SDGs and 169 goals into their policies, mainly through monitoring indicators adapted from the global ones in the agenda. In the scope of university policies and science and technology policies, reference to the SDGs can be found in institutional planning, teaching programs, studies, funding guidelines, laboratories, and research centers, among other spaces intended to contribute to different dimensions of the 2030 Agenda. Among these dimensions, as highlighted above, is the ESD.

In this context, universities play an important role through research and teaching (UNESCO 2016), especially in terms of the discursive contribution related to Sustainable Development in public policies, as presented in Priority Action Area 1: Advancing Policy on Education for Sustainable Development: a roadmap (UN 2022). Thus, understanding and analyzing the extent to which ESD has been addressed in Latin American policy documents enables us to understand, from one perspective, the forms and level of incorporation of this educational practice in public policies; on the other, to observe how discursive contribution of universities to these policies, particularly universities in the region, is made. In order to facilitate the understanding of the incorporation of ESD in legal, educational documents, UNESCO has made a methodological suggestion defining eight issues or themes that should be easily identified in these documents to define that the practice is present—being: (a) cultural diversity and tolerance, (b) gender equality; (c) human rights (d) peace and non-violence, (e) climate challenge, (f) environmental sustainability, (g) survival and human well-being, (h) consumption and sustainable production. These themes are important as categories to be mobilized in analyzing the policy documents extracted via Overton.io.

3 Methodology

In order to achieve its aims, the present study was based on exploratory research with a qualitative-quantitative approach. In addition to the literature review, the software Overton.io was used to identify Latin American policy documents that address the ESD. The software, created in 2019, maps and identifies the use of research (identifies the mention of papers) in policy documents, constituting a base that allows collecting information and data related both to policy documents (such as source, country of origin, type of document, topic, and SDGs) and the research located in them (Journals, authors, universities, funding, among others). The Overton platform is a comprehensive database covering policy documents from 182 countries (Overton,

S/D). By policy documents, the software considers reports, working papers, legislation, clinical trials, and other related publications. Such documents come mostly from governments and parliaments but also from International Governmental Organizations (IGOs) such as the United Nations, UNESCO, and think tanks. The software has a natural coverage limitation (i.e., it covers a certain number of policy documents), it is easier to capture English language documents, and, depending on the research objectives, it may imply the need for disambiguation and data cleaning. However, in addition to constituting the largest and most comprehensive base of this type, it allows producing quantitative and qualitative analyses on the use of policy research, opening up possibilities for research in this regard (Maleki and Holmberg 2022; Szomszor and Adie 2022; Pinheiro et al. 2021).

By applying initial filters on Overton.io, such as South and Central America and ESD, 279 policy documents mentioning ESD were found. Next, a third inclusion criterion—Latin American universities—was applied to capture policy documents referencing academics or expertise from these universities. Twenty policy documents that reference Latin American universities were found. After selecting these policy documents, they were extracted and worked on the visualization of the keywords contained in the 279 and then in the 20 policy documents.

Visualization (Van Eck and Waltman 2014) is an important approach to analyzing bibliometric networks, such as networks of citation relations between publications, networks of co-authorship relations between researchers, or networks of co-occurrence relations between keywords. In this paper, the analysis of keyword clusters was divided into two parts: (i) analysis of the 265 keywords contained in the 279 policy documents from South and Central America that mention ESD; (ii) analysis of the 55 keywords contained in the 20 policy documents that reference articles or expertise from Latin American universities. Keywords are ways of summarizing the main topics addressed by a text. Through these keywords, we can identify ideas under which research is oriented. We chose the graph structure to analyze the keywords of each set of policy documents. Graphs are structures for representing Complex Networks composed of sets of vertices and edges. This theme is relevant to the Data Science area and can be approached through different methods.

Each policy document has a set with three associated keywords. Keywords are automatically associated with documents by Overton.io. When a policy document references articles or expertise from more than one Latin American university, this document appears more than once as a result of the search. In this case, we count the document only once (resulting in 20 documents), and we consider a single set of Top Topics related to the document, with all associated words in all occurrences of the search. We create a list of keywords and evaluate the number of times each word appeared in some policy document; we call this list vertices. We create a list called edges: each element of this second list is formed by a link and counts the number of times two words appear together in the same document. The following examples illustrate part of the vertices and edges lists for the 265 words of the first analysis (Table 2) and the 55 words of the second analysis (Table 3). Considering the 279 political documents, the keyword that appears more often is "human activities",

Table 2 Top 10 vertices and edges for the 265 words of the first analysis

Vertices	
Human activities	142
Culture	42
Politics	37
Education	32
Economy	23
Natural environment	21
Nature	17
Cognition	17
Sustainability	15
Psychological concepts	14
Edges	
Culture—human activities	25
Economy—human activities	15
Human activities—politics	13
Branches of science—human activities	10
Human activities—natural environment	9
Culture—politics	8
Cognition—human activities	7
Brazil—education	7
Cognition—culture	6
Earth sciences—nature	6

Source Authors

accounted for in 142 documents; the words "culture" and "human activities" appear together in 25 documents.

A graph structure was created where each vertice has a size proportional to the number of occurrences of the keyword in the set of documents. The width of an edge is associated with the frequency with which the two associated keywords are cited together. A widespread network property is the community structure (groups or clusters), which divides vertices into groups according to the density connection. Identifying communities in networks is very useful since vertices belonging to the same community tend to have similar properties.

The structures were implemented in Python (Table 4) through the *igraph* package. We used the function *DictList* to construct the graph with this package. In the grouping stage, for the constitution of the communities, we used the *leidenalg* package, with functioning dependent on the *igraph* package. The main function of the *leidenalg* package is *find_partition*, which is responsible for finding the optimal partition using Leiden's algorithm (Traag et al. 2018). We use the modularity-based partitioning method. The flexibility of adaptation and scalability justifies the choice of implementation in Python programming language and using these packages.

Table 3 Top 10 vertices and edges for the 55 words of the second analysis

Vertices	
Human activities	9
Education	8
Economy	4
Sustainable development goals	3
Culture	3
Interdisciplinarity	3
Preschool	3
Sustainability	2
Agriculture	2
Human nature	2
Edges	
Climate resilience—human activities	1
Education—human activities	1
Education—sustainable development goals	1
Culture—education	1
Culture—human activities	1
Culture—interdisciplinarity	1
Economics—interdisciplinarity	1
Agriculture—economic development	1
Economy—investment	1
Human activities—social institutions	1

Source Authors

Table 4 Importing packages and using the main functions

```
from igraph import Graph as IGraph
import leidenalg as la
ig = IGraph.DictList(vertices, edges)
communities = la.find_partition(ig, la.ModularityVertexPartition)
```

Source Authors

Leiden's algorithm (Traag et al. 2018) is a graph partitioning algorithm that detects communities in weighted graphs. In the first step, the algorithm applies a modularity-based optimization technique to group vertices of the graph in such a way that the modularity of the network is maximized. The concept of modularity was proposed to measure the network structure by dividing it into modules called groups, clusters, or communities. This measure is calculated through the relationship of each vertice with its neighboring vertices. Networks with high modularity have connections with high density between vertices of the same cluster and connections with low density between vertices of different clusters; that is, modularity is based on the idea that good partitioning should maximize the number of connections within communities and minimize the number of connections between communities. In the second step, the algorithm uses a clustering method to combine the communities identified in the first step into more extensive and stable groups. This process is repeated several times until there are no more significant gains in network modularity.

Based on this methodological design, extracting and clustering the keywords from the policy documents was possible. The results will be presented and discussed below.

4 Education for Sustainable Development and the Contribution of the Latin America Research

This section aims to identify and analyze the academic community's contribution (and knowledge flows) to the topic of Education for Sustainable Development through policy documents in Latin America.

Therefore, from the methodological design, it was possible to find more than 252 thousand policy documents produced by Latin American institutions from almost 7 million documents. Of the 252 thousand policy documents in Latin America, only 279 mention ESD (0.11%), of which 93.2% (260) come from the Government and 6.8% (19) from Think Tanks (Fig. 2). As shown in Fig. 1, these documents were produced between 2006 and 2022.

Through analysis of these documents, it was found that only 20 include the contribution, in some way, of Latin American academics or groups of researchers. In this sense, we verified the direct or indirect contribution of 15 Latin American universities in these documents. Similarly to the general Latin American documents, there is also a predominance of documents from government (17) and think tanks (03) (Fig. 3).

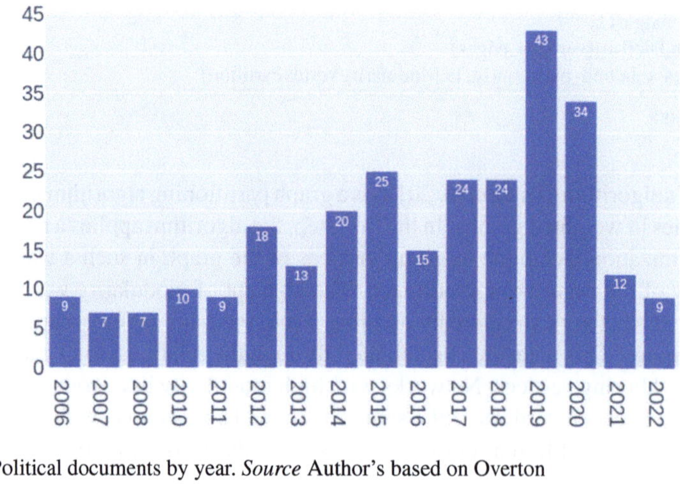

Fig. 1 Political documents by year. *Source* Author's based on Overton

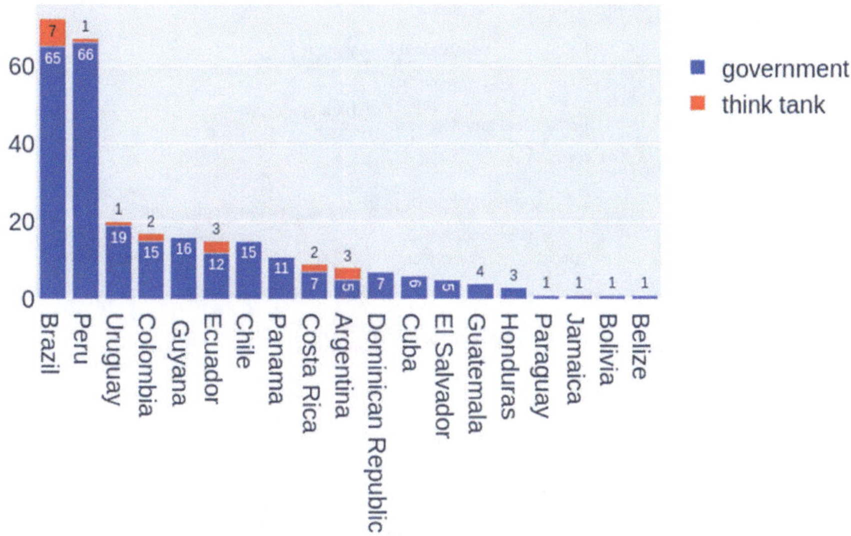

Fig. 2 Political documents by source and country. *Source* Author's based on Overton

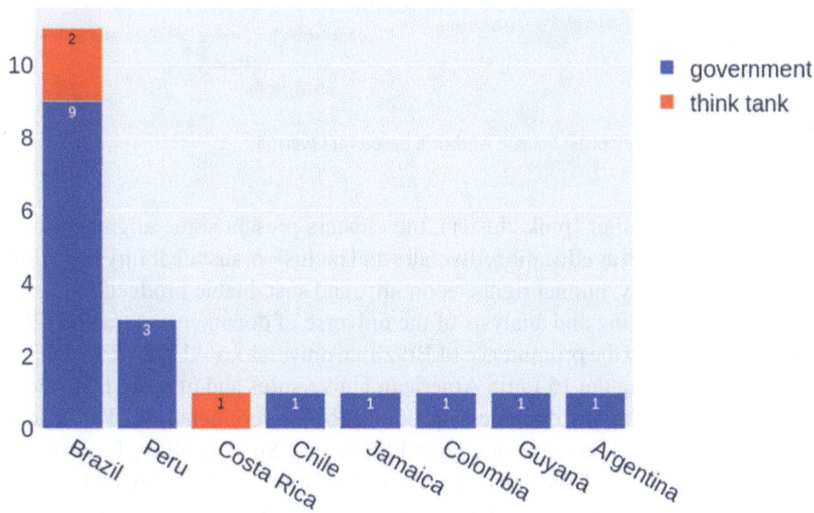

Fig. 3 Political documents that reference the Latin American academic community, by country and source. *Source* Author's based on Overton

From these 20 policy documents, 55 different keywords connected through 68 links (edges) are extracted. As described in the methodology, the Leiden Algorithm has been applied to group these keywords, resulting in 07 clusters (Fig. 4). The Figure below shows the main clusters, one being an outlier.

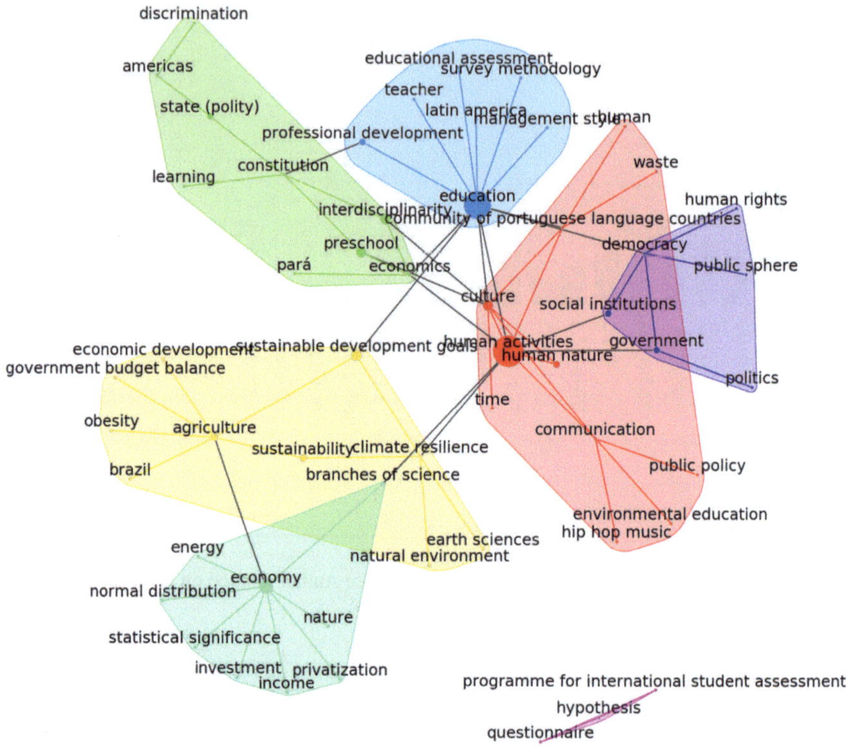

Fig. 4 Clustering of keywords. *Source* Author's based on Overton

Except for the outlier (pink cluster), the clusters present some alignment of the main ESD topics such as education, diversity and inclusion, sustainability and climate resilience, democracy, human rights, economy, and sustainable production.

Based on the reading and analysis of the universe of documents extracted (20), it was possible to verify the prominence of Brazilian universities—USP, UFRJ, UFMA, UFBA, UFC—among the 15 Latin American Universities and of some ESD themes most highlighted by policy documents produced by Governments and Think Tanks. These themes can be grouped into four blocks: (1) Sustainability, Environment, and Oceans; (2) Energy Efficiency; (3) Basic Education, Curriculum, and Capacity Building; (4) Education in Human Rights, focusing on citizenship training of children and youth.

Regarding the first block, the debate on *sustainability, environment, and oceans*, this can be exemplified by the report produced within the South-South Framework for Scientific and Technical Cooperation in the South and Tropical Atlantic and Southern Oceans. This cooperation included the participation of academics from the Federal University of São Paulo and the Federal University of Bahia, among others. There was a meeting of stakeholders from academia, the Ministry of Science, Technology, and Innovation (MCTI), and international institutions to discuss the topic. The report mentioned above was generated and became a subsidy for the MCTI to propose a set of actions and scientific and technological programs that seek to stop the contamination of the oceans. The Ministry used the expected results proposed for the next decade to define short, medium, and long-term actions. The concern with ESD is a significant issue in guaranteeing the concreteness of the expected results. Thus, the report discussed the education and science model needed to preserve a clean, healthy, predictable, safe, sustainable, and productive ocean for future generations.

In the *energy efficiency* block, there is a study that includes several researchers and universities, in particular researchers from the University of Buenos Aires, who contributed to the preparation of the report "Energy efficiency in Latin America and the Caribbean: situation and outlook", coordinated by the Natural Resources and Infrastructure Division of the Economic Commission for Latin America and the Caribbean (ECLAC/UN), in partnership with the German Federal Ministry for Economic Development and the German Technical Cooperation Agency. This study analyzed the experiences of national energy efficiency policies of 26 countries in Latin America. It served as knowledge for the Government of Jamaica, mainly through its Ministry of Science, Energy, and Technology, to structure its energy efficiency guidelines.

The third block refers to *basic education prepared for diversity and inclusion.* From the direct participation of Brazilian academics in the rapporteur of the National Education Council (CNE) and contributions of referenced academic research, the Ministry of Education of Brazil structured the National Curriculum Guidelines for Basic Education (2013). In addition to the curricular guidelines for basic, infant, primary, secondary, and vocational education, they are also composed of 7 chapters that establish principles and complementary and operational standards for each of the themes: (i) Field Basic Education, (ii) Education of Ethnic-Racial Relations and Afro-Brazilian and African History and Culture Teaching; (iii) Youth and Adult Education, in general, and for those in the situation of freedom deprivation; (iv) Indigenous School Education; (v) Human Rights Education; (vi) Environmental Education and (vii) Quilombola School Education.

The chapters are rich in detailing contents that should be incorporated and stimulated and skills and competencies that should be fostered in the students' formative process. An example can be seen in the chapter on environmental education (BRASIL 2013, p. 383), which points out that the curricular planning of the Educational Institutions must "stimulate: (a) integrated, multidimensional view of the environmental area, considering the study of biogeographic diversity and its vital ecological processes, the political, social, economic, psychological influences, among others, in the relationship between society, environment, nature, culture, science and technology". To achieve this vision, it is proposed among a set of actions "to promote the implementation of: (b) observation and study of nature and its operating systems … (c) projects and activities, including artistic and recreational ones, which value the sense of belonging of human beings to nature, the diversity of living beings, the different local cultures, the oral tradition, among others". These guidelines and actions are derived from international frameworks, national legislation, studies, and theoretical constructs underpinning each theme. Academics and rapporteurs prepare each chapter for the National Education Council who, in addition to their own expertise, add their expertise and technical-scientific knowledge to the process of preparing their opinion.

Finally, the last block is *human rights education and training on environmental issues and the civic knowledge and engagement of students* (children and young people). The study in question, produced by an international network of academics and experts in Educational Assessment (International Association for the Evaluation of Educational Achievement-IEA), including Latin Americans (researchers from Chile, Colombia, Peru, and the Dominican Republic), presents and monitors (every 7-year cycle) the understanding of countries on issues such as the role of students for environmental sustainability, social interactions at school and the use of new social media for civic engagement. As with the other blocks, this report was the basis for guidelines set by the Ministry of Education of the Government of Peru.

The approximations between academic and scientific production and policy documents prepared by governments or think tanks bring quality to the formulation of guidelines and decision-making. At the same time, it undoubtedly re-signifies the academic view of social problems and qualifies scientific and academic production as a whole. In this aspect, Weible (2008) highlights that three basic functions support the use of scientific knowledge in public policy-making: firstly, the cognitive function, which is related to the process of learning and transforming participants' beliefs about the causes and possible solutions of the problem; secondly, the political function, which refers to the strategic role of scientific knowledge in justifying or legitimating decisions and courses of action; thirdly, the instrumental function of scientific expertise, which considers the participants' need to understand the outcomes that conflict with their beliefs, following the data and policy implications, regardless of participants' normative and strategic preferences. These functions are intertwined and may co-occur, depending on contextual circumstances. Our analysis does not permeate under these elements. However, they can serve as a basis for the development of future studies.

As noted above, the policy documents are predominantly from the Ministries of Education, Science, Technology and Innovation, Energy, and Environment from several Latin American countries; however, two ESD issues—Culture of Peace and Gender Equality—did not appear. This point deserves further study, particularly as the gender issue is so widely debated and discussed by higher education institutions but has not appeared as firmly in policy documents as other themes.

5 Final Remarks

This chapter aims to conduct an exploratory analysis of how Education for Sustainable Development (ESD) has been included in Latin American policy documents in view of the regional academic expertise on the subject. To this end, we have taken up the debate on the social commitment of universities, considering the theoretical framework of ESD. In this sense, it was possible to understand the relationship between the production of knowledge flows and the formation of increasingly more relevant expertise to propose solutions to the problems that involve the social, economic, environmental, and institutional dimensions of Sustainable Development. Universities stand out as one of the main actors in achieving this agenda, becoming fundamental in knowledge production and the relationship between its application in the scope of public policies.

From the theoretical framework used, it was also possible to verify difficulties in the institutionalization of ESD in Latin America, given the absence of a conceptual consensus in the literature. Therefore, five theoretical and conceptual lines were identified to define and contextualize this practice in the regional context studied: i. Synonym of Environmental Education; ii. Part of Environmental Education; iii. Evolution of Environmental Education; iv. "Umbrella" Education; and, v. Given concept. Based on the above, the questions "Is this (scientific) expertise being used in the public policy-making process? If so, what is the contribution of Latin American academia?" were countered by the methodological procedures used, given the use and analysis of information collected via software Overton.io.

Specifically on the data collected, it was possible to observe a low percentage of policy documents dealing with ESD (0.11%), which may be connected with this absence of a consensus in the literature, i.e., of a clear definition that could function as a starting point for the formulation of public policies. Such analysis also allowed us to observe, from the verification of 20 specific documents, that the evident contribution of universities in policy documents on ESD is still very incipient. However, such a contribution was highly relevant (when contemplated) for advancing the themes that constitute ESD. Four thematic blocks have been identified concerning this analysis: (1) Sustainability, Environment, and Oceans; (2) Energy Efficiency; (3) Basic Education, Curriculum, and Capacity Building; (4) Human Rights Education with a focus on citizenship training of children and youth; among these, gender has not been contemplated as strongly as the others.

Despite the limitations of the exploratory research carried out (inherent limitations to the use of the software, as well as methodological), it sheds light on the different ways of apprehending the use and impact of scientific knowledge in policy documents, considering the use of alternative metrics that, in our case, were collected via Overton. Therefore, the use of this tool is visualized in the scope of other themes beyond the ESD, observing its limits and restrictions. Furthermore, based on the results obtained, a series of questions can be formulated in order to deepen how this knowledge has been used in the field of public policy-making, such as how governments/think tanks/intergovernmental organizations view such interaction, how the ESD has been incorporated in other contexts (Europe, North America, Asia, and Africa) and the process of constituting a definition of the term.

References

Alves D (2022) The presence of target 4.7 of 2030 Agenda in Brazil: data on the national indicator

Bajaj M, Chiu B (2009) Education for sustainable development as peace education (September 8, 2009). Peace Change 34(4):441–455

Barreto LM, Vilaça, MTM (2018) Controversies and consensus on environmental education and education for sustainable development. Res Soc Dev 7(5):e975167. https://doi.org/10.17648/rsd-v7i5.223

Bastos AM (2016) Dossiê DEDS (2005 a 2014): a década da educação para o desenvolvimento sustentável no Brasil: uma análise pela perspectiva estratégica [Dossier DEDS (2005 to 2014): the decade of education for sustainable development in Brazil: an analysis from a strategic perspective]. Thesis (Doutorado em Educação Escolar). Faculdade de Ciências e Letras, Universidade Estadual Paulista "Júlia de Mesquita Filho". Araraquara

Bastos AM, Lemes S (2017) A educação para o desenvolvimento sustentável no contexto curricular da rede pública de ensino do governo do estado de São Paulo: uma breve reflexão pela perspectiva da década da educação para o desenvolvimento sustentável da UNESCO (2005–2014) [Education for sustainable development in the context of the São Paulo state government's public school curriculum: a brief reflection from the perspective of UNESCO's decade of education for sustainable development (2005–2014)]. Revista on Line De Política E Gestão Educacional (19). https://doi.org/10.22633/rpge.v0i19.9382

Bastos AM, Souza CBG (2013) A educação e a sustentabilidade: o desafio de um paradigma e a década da educação para o desenvolvimento sustentável da UNESCO (2005–2014) [Education and sustainability: the challenge of a paradigm and UNESCO's decade of education for sustainable development (2005–2014)]. Revista Ibero-Americana De Estudos Em Educação 8(1):208–240. https://doi.org/10.21723/riaee.v8i1.6484

Berchin II, de Aguiar Dutra AR, Guerra J (2021) How do higher education institutions promote sustainable development? A literature review. Sustain Dev, 1–19.https://doi.org/10.1002/sd.2219

Bonnet M (2006) Education for Sustainable Development: a coherent philosophy for environmental education? Camb J Educ 29(3):313–324. https://doi.org/10.1080/0305764990290302

Branco AFV , Linard Z, Sousa AC (2011) Educação para o Desenvolvimento Sustentável e Educação Ambiental [Education for Sustainable Development and Environmental Education]. Conexões—Ciência e Tecnologia, [S.l.], pp 25–31, maio 2011. https://doi.org/10.21439/conexoes.v5i1.439

Burmeister M, Rauch F, Eilks I (2012) Education for sustainable development (ESD) and chemistry education. Chem Educ Res Pract 13:59–68

Campello L, Silveira V (2016) Educação para o desenvolvimento sustentável (eds) e o greening das universidades [Education for sustainable development (eds) and the greening of universities]. RTJ, São Paulo, V. 5, N.2, p. 549–572, Mai.- Ago. https://doi.org/10.5585/rtj.v5i2.464

Cañellas AJC (2006) El desarrollo sostenible y la educación para el desarrollo. Universidad de las Islas Baleares [Sustainable development and education for development. University of the Balearic Islands]. Pedagogia Social, pp 31–49

Carvalho I (2008) A Educação Ambiental no Brasil. Educação Ambiental. Salto para o Futuro. Ano XVIII Boletim 01—Março de 2008 [Environmental Education in Brazil. Environmental Education. Leap into the Future. Year XVIII Bulletin 01—March 2008]

Correa MM, Ashley PA (2018) Sustainable development, sustainability, environmental education and education for sustainable development: reflections for undergraduate education. REMEA—Revista Eletrônica Do Mestrado Em Educação Ambiental 35(1):92–111. https://doi.org/10.14295/remea.v35i1.7417

Dyer A, Selby D, Chalkley, B (2006) A centre for excellence in education for sustainable development. J Geogr Higher Educ 30:307–312. https://doi.org/10.1080/03098260600717406

Engjellushe E (2013) Education for sustainable development. European. J Sustain Dev 2(4):227

Esteban A (2020) Educación sustentable no formal para conservar los manglares en zonas costeras con estudiantes de Sociología, UAGro [Non-formal sustainable education to conserve mangroves in coastal areas with Sociology students, UAGro]. RIDE. Rev. Iberoam. Investig. Desarro. Educ, Guadalajara , v. 10, n. 20, e018, jun

Gallopín GC (2003) Sostenibilidad y desarrollo sostenible: un enfoque sistémico [Sustainability and sustainable development: a systemic approach]. Cepal/Gobierno de los Países Bajos

Loureiro C (2008) Proposta Pedagógica. Educação Ambiental. Salto para o Futuro. Ano XVIII Boletim 01—Março de 2008 [Pedagogical proposal. Environmental Education. Leap into the Future. Year XVIII Bulletin 01—March 2008]

Maleki A, Holmberg K (2022) Comparing coverage of policy citations to scientific publications in Overton and Altmetric.com. Informaatiotutkimus 41(2–3):92–96. https://doi.org/10.23978/inf.122592

Nascimento EP (2012) Trajetória da sustentabilidade: do ambiental ao social, do social ao econômico [The trajectory of sustainability: from environmental to social, from social to economic]. Estudos Avançados 26(74):51–64. https://doi.org/10.1590/S0103-40142012000100005

Olmo-Fernández M, Rodríguez-Jiménez C, Santos-Villalba M, Gómez-García G (2021) Educar para el desarrollo sostenible en el contexto universitario: un análisis bibliométrico [Educating for sustainable development in the university context: a bibliometric analysis]. Formación Universitaria 14(3):85–94. https://doi.org/10.4067/S0718-50062021000300085

Overton (S/D) Is your work influencing policy?. Available at https://www.overton.io/

Pinheiro H, Vignola-Gagné E, Campbell D (2021) A large-scale validation of the relationship between cross-disciplinary research and its uptake in policy-related documents, using the novel overton altmetrics database. Quant Sci Studies 2(2):616–642. https://doi.org/10.1162/qss_a_00137

Ruiz et al (2019) Educación para el desarrollo sostenible: el papel de la universidad en la Agenda 2030 [Education for sustainable development: the role of the university in Agenda 2030]. Revista Prisma Social

Setti AFF, Azeiteiro UM (2016) Education for Sustainable Development in Brazil: Challenges for Inclusive, Differentiated and Multicultural Education. In: Castro P, Azeiteiro U, Bacelar-Nicolau P, Leal Filho W, Azul A (eds) Biodiversity and Education for Sustainable Development. World Sustainability Series. Springer, Cham. https://doi.org/10.1007/978-3-319-32318-3_15

Silva CA, Teixeira SF (2019) Educação Ambiental no Brasil: reflexões a partir da Década da Educação para o Desenvolvimento Sustentável das Nações Unidas (2005–2014) [Environmental Education in Brazil: reflections from the United Nations Decade of Education for Sustainable Development (2005–2014)]. Educação (UFSM)

Szomszor M, Adie E (2022) Overton: a bibliometric database of policy document citations. Quant Sci Studies 3(3):624–650. https://doi.org/10.1162/qss_a_00204

Traag VA, Waltman L, Van Eck N-J (2018) From Louvain to Leiden: guaranteeing well-connected communities. Sci Rep 9(1):5233. https://doi.org/10.1038/s41598-019-41695-z10.1038/s41598-019-41695-z

Uchôa RS (2016) A Década da Educação para o Desenvolvimento Sustentável da UNESCO: uma análise sob a ótica da Educação Ambiental Crítica [UNESCO's Decade of Education for Sustainable Development: an analysis from the perspective of Critical Environmental Education]. Thesis (Master in Education). Universidade Federal do Estado do Rio de Janeiro

UNESCO—United Nations Educational, Scientific and Cultural Organization (2021) Educação para o Desenvolvimento Sustentável. Disponível em https://pt.unesco.org/fieldoffice/brasilia/expe Acessado em 26/08/2021

UNESCO (2016) Education 2030 agenda framework for action. Paris: 2015

UNESCO (2017) Education for sustainable development goals

UN—United Nations (2022) Education for Sustainable Development: a roadmap

UN—United Nations (1973) Report of the united nations conference on the human environment, Stockholm, 5–16 June 1972

UN—United Nations (2005) United nations decade of education for sustainable development, 2005–2014: draft international implementation scheme

Van Eck NJ, Waltman L (2014) Visualizing bibliometric networks. In: Ding Y, Rousseau R, Wolfram D (Eds), Measuring scholarly impact: methods and practice, pp 285–320. Springer. https://doi.org/10.1007/978-3-319-10377-8_13

Weible CM (2008) Expert-based information and policy subsystems: a review and synthesis. Policy Stud J 36(4):615–635. https://doi.org/10.1111/j.1541-0072.2008.00287.x

Dimensions of Sustainability Consciousness Among Business Students: Exploring the Contributions of Business Schools in Nigeria

Agu Godswill Agu, Tade Daniel Omotosho, and Onwubiko Ngozi Dike

Abstract The desire for a sustainable planet has triggered intensive efforts towards achieving sustainable development goals. Higher education institutions have continued to play a prominent role in this direction. Previous research has examined the roles of specific branches of business schools (e.g., Marketing education, Accounting Education, Entrepreneurship education etc.) in imparting sustainability-oriented knowledge that is necessary for sustainable behaviour. However, the sustainability literature is currently lacking a comprehensive presentation of the dimensions of contributions of business schools in stimulating sustainability consciousness among university students, especially in the context of developing countries. To close this gap, this study leverages the knowledge-attitude-behaviour model and the self-control theory in understanding, through a qualitative survey, the dimensions of sustainability consciousness in business students. By focusing on graduate students of a foremost state-owned business school in Nigeria, the study contributes to the scarce literature on the contributions of universities towards sustainable development.

Keywords Sustainable development · Sustainability consciousness · Dimensions · Students · Business schools · Nigeria

A. G. Agu (✉)
Department of Marketing, Abia State University, Uturu, Nigeria
e-mail: godswill.agu@abiastateuniversity.edu.ng

T. D. Omotosho
Faculty of Business, Management and Leadership, Coventry University, Wrocław, Poland
e-mail: tade.omotosho@coventry.ac.uk

O. N. Dike
Department of Management Sciences (Marketing Program), Rhema University, Aba, Nigeria
e-mail: onwubiko_dike@rhemauniversity.edu.ng

W. Leal Filho et al. (eds.), *The Contribution of Universities Towards Education for Sustainable Development*, World Sustainability Series,
https://doi.org/10.1007/978-3-031-49853-4_24

1 Introduction

The global journey towards sustainable development has come a long way. Beginning with the historic 1972 Stockholm Conference on the Human Environment, the Brundtland Commission report of 1987, the Millennium Development Goals (MDGs) of 2000, and the 2030 Agenda for Sustainable Development which led to the launching of the 17 Sustainable Development Goals (SDGs) in 2015, the United Nations has consistently promoted the idea of sustainable development. This consistency is in recognition of the fact that countries of the world have pursued largely unsustainable development goals, and there is an urgent need to change the narrative (Tulder et al. 2021) by streamlining actions towards achieving sustainable development. To achieve the 17 SDGs with the 169 integrated targets and 231 indicators, the support of various sectors of the global economy remains inevitable. Prominent among the sectors is the education sector, which has been linked to virtually all the SDGs (Vladimiravo and LeBlanc 2015). Thus, beyond its standalone recognition as a sustainable development goal (SDG 4), education introduces knowledge that is relevant to influencing the lifestyle of citizens towards sustainable practices (Leal Filho 2022; Agu et al. 2022a, b).

Within the education sector, higher education institutions (HEIs) have continued to exhibit encouraging signs of commitment towards sustainable development since they occupy a strategic position in transforming the lifestyles of the global active population (including undergraduate and graduate students) towards a sustainable daily living. For instance, by embedding sustainability into the curriculum of disciplines across HEIs, students' consciousness towards sustainability has been tremendously influenced (Agu et al. 2022a, b; Ovais 2023). This consciousness affects various aspects of life, including consumption decisions (Agu et al. 2022b; Longoria et al. 2021), entrepreneurial pursuit (Edokpolor 2020; Romero-Colmenares and Reyes-Rodriguez 2022), and environmental consciousness (Lin et al. 2022) among others. Again, universities across the globe encourage research activities on sustainable development, promote the idea of sustainable campus through various approaches, and engage in community activities that support sustainability (Leal Filho 2022; Purcell et al. 2019; Homer and Khor 2022). These efforts have the ultimate goal of actualizing the societal, environmental and economic sustainability, which represent the pillars of sustainable development (United Nations 2023).

Of interest to this study is the role of business schools in promoting sustainable development across the globe. For instance, business schools are committed to the 2007 United Nations' Principles for Responsible Management Education (PRME) which aims at transforming business and management education to actualize the SDGs working through six principles—purpose, values, methods, research, partnership, and dialogue (PRME 2020). Although many business schools in Nigeria are not yet signatories to PRME, their teachings, research, outreach programmes and community engagements align with the PRME mandate of arousing sustainability consciousness in students (Agu et al. 2022a, b). In fact, for about a decade now, most conferences organized by business schools in Nigeria have themes that center around the SDGs. The ultimate goal is to grow a population of sustainability ambassadors from among relevant stakeholders such as educators, government officials, politicians, business people, consumers, and students who will pursue the SDGs in their various life endeavours.

The aim of this study is to explore, from the perspective of students, how business schools have influenced their sustainability consciousness and to report the various dimensions of the consciousness. Weybrecht (2022) expresses the need for more research that discloses the effort of business schools in achieving a sustainable society. Previous research has examined the roles of specific branches of the business school (e.g., Marketing education, Accounting Education, Entrepreneurship education etc.) in imparting sustainability-oriented knowledge and behaviour (Agu et al. 2022b). However, the sustainability literature is currently lacking a comprehensive presentation of the various dimensions of contributions of business schools in stimulating sustainability consciousness among university students, especially in the context of developing countries. Besides, the few studies that examined how business schools are pursing the SDGs utilized reports published by business schools and focused on their strategies of engagement (see Weybrecht 2022; Miotto et al. 2020). Research that explores students' sustainability consciousness is currently scarce (Ovais 2023). Little is known currently about students' accounts of the extent business schools have influenced their consciousness towards sustainability and the dimensions of the consciousness. This knowledge is relevant since it gives a more reliable and authentic account of the achievement of business schools in transforming students' lives towards sustainability, and given the important role of students in achieving sustainable development (Murray 2018). Besides, Ferrer-Estévez and Chalmeta (2021) expressed the need for more qualitative findings on the learning of SDGs.

Therefore, this qualitative study contributes to empirical literature on the role of business schools in achieving the SDGs by looking at how various strategies of business schools' engagement in the SDGs have influenced students' sustainability consciousness. The focus is on the dimensions of consciousness created by the SDGs-related activities of business schools through the study period of university students. In the opinion of Ovais (2023), sustainability consciousness can be examined in terms of sustainability knowledge, attitude, and behaviour. This opinion guides this study, since in trying to modify human behaviour, the expected behavioural changes could be divided into three continuous processes—knowledge, attitude, and behaviour (Bettinghaus 1986) cited in Liu et al. (2016). So, in trying to develop students' sustainability consciousness through their various programmes, business schools target to influence these three processes positively. Furthermore, this paper argues that the knowledge created by business education is capable of helping students to develop self-control over unsustainable emotions, actions, and desires, thus improving their sustainability behaviour. Generally, those who exercise self-control are able to regulate their behaviour, especially their emotions, actions and desires (Blankstein and Polivy 1982). Therefore, by integrating the knowledge-attitude-behaviour model and the self-control theory in exploring the contributions of business schools in stimulating sustainability consciousness of students in a developing country, this study makes theoretical and practical contributions.

Specifically, the following research questions agitate the study:

1. Through what activities of business schools do students acquire sustainability knowledge?
2. Does the knowledge acquired influence attitude formation and exercise of self-control which are relevant for sustainability-conscious behaviour?
3. What sustainability-conscious behaviour of students can be attributed to the activities of business schools?

To address these questions, the rest of the chapter proceeds as follows: the next section is the literature review on education for sustainable development, business schools and sustainability consciousness. This is followed by theoretical development and a proposed conceptual framework. In section three, the method of data collection is presented, while section four includes analysis of data and discussion of results. The last section is for conclusion and recommendations, including theoretical/practical implications and areas for further studies.

2 Literature Review: Education for Sustainable Development, Business Schools, and Sustainability Consciousness

Education for sustainable development (ESD) refers to a special education programme of the United Nations designed to impart the tenets of sustainable development in people, whether formally, non-formally or informally (Council of the European Union 2010). It ensures that every human being acquires the knowledge, skills, attitudes and values necessary to shape a sustainable future by including development issues such as sustainable consumption, poverty reduction, biodiversity, climate change, and disaster risk reduction into teaching and learning (UNESCO 2014). ESD covers pedagogical processes, validation of knowledge, and the functioning of educational institutions (da la Riva et al. 2015), with the aim of providing a coherent interaction between education, public awareness, and training that creates a more sustainable future (UNESCO 2012). The four thrusts of ESD as outlined by UNESCO (2014) are:

i. Improving access and retention in quality basic education
ii. Reorienting existing educational programmes to address sustainability
iii. Increasing public understanding and awareness of sustainability
iv. Providing training to all sectors of the workforce

Generally, ESD integrates the five pillars of education—learning to know, learning to do, learning to live together, learning to be, and learning to transform oneself and society (UNESCO 2014), the principles and values that underlie sustainable development, and the three spheres of sustainability—environment, society, and economy (UNESCO 2012). This integration empowers people to change the way they think and work towards a sustainable future, promotes sustainability literacy, and ignites sustainability consciousness (Agu et al. 2022a, b; Caeron and Decamps 2014; UNESCO 2020).

Different schools pursue ESD in various ways, which are largely related. The business school stands out, given its role in building future business executives and entrepreneurs, and thus receives increasing global attention about how it pursues sustainability. Therefore, sustainability business education aims to transform society by re-orienting business education and helping students to develop knowledge, skills, values, and behaviours that drive sustainable development (UNESCO 2020; Agu et al. 2022a, b). A recent study by Godemann et al. (2023) identified significant developments in the strategies business schools use to implement sustainability, noting that business schools have improved their sustainability engagement in many areas such as education offerings and teaching methods, campus practices and engaging stakeholders. Also, Fang and O'Toole (2023) found that business students achieved an enhanced awareness and knowledge of SDGs, acquired key employability skills and developed new global perspectives during an experiential learning programme in which SDGs were integrated. In an attempt to confirm which SDGs are the most cited in the sustainability reports of the best top 50 business schools in the world, Miotto et al. (2020) found SDGs 16 (peace, justice and strong institutions), 10 (reduce inequality), 17 (partnership for the goals) and 4 (quality education) as the most reported goals. They conclude that "business schools focus on social responsibility strategies mainly to define the professional standards to train future leaders qualified to manage organizations with a social, economic and environmental positive impact for all the stakeholders and capable of shaping a better world" page 1. This was corroborated by Jun and Moon (2021) who found that business schools in South Korea increased in the absolute number and proportion of sustainability-linked courses between 2013 and 2019 with emphasis on "ethics" and "corporate social responsibility". Rusinko (2015) proposed a matrix of how business schools integrate sustainability, and suggests that business schools deliver sustainability higher education by using their existing structures and or by creating new structures. The author further notes that business schools' sustainability higher education focus is either broad (cross-disciplinary) or narrow (discipline-specific). This study corroborates Rusinko (2015)'s suggestion by identifying the core dimensions of students' sustainability consciousness delivered by business schools (see Table 2).

Currently, there is a growing scholarly attention to sustainability in the context of developing countries (Ulrich et al. 2018). However, little research attention has been given to some specific areas of sustainability (Nwaigwe et al. 2022). Whether Nigerian business schools' effort to integrate sustainability in their programmes is yielding any positive results in terms of building sustainability consciousness among business students deserves scholarly attention. According to UNESCO (2020), sustainability implies concern for the environment, economic and social elements of development. Thus, sustainability consciousness is seen as an individual's experience and awareness of sustainable development, including thoughts, feelings, and behaviours (Gericke et al. 2019). It refers to the combination of knowledge, attitude, and behaviour about sustainability (Legault and Pelletier 2000; Olsson et al. 2016) in Ovais (2023:3). While describing it as awareness of sustainability issues, Raymundo et al. (2019) describe it as understanding the fragility of the environment and the importance of its protection, thinking in terms of an ecological consciousness. Here, sustainability consciousness describes a mindset that is sensitive to the environmental, economic and social concerns of society with a strong intent to behave in ways that are helpful in addressing these concerns.

A number of empirical studies have examined how various disciplines within business schools encourage students to imbibe the sustainability culture. In a study involving undergraduate and graduate students, Agu et al. (2022a, b) report that sustainability-oriented marketing education influences students' attitude, subjective norms, perceived behavioural control, and consciousness towards sustainable consumption. Similarly, studies affirm the positive effect of education for sustainable entrepreneurship on students' intention to engage in sustainable entrepreneurship (Romero-Colmenares and Reyes-Rodríguez 2022; Agu 2021; Gupta and Matharu 2022). Besides, Ertas and Ozkan (2017) found evidence to support the efficacy of environmental accounting education in inducing environmental consciousness among students. Furthermore, Greenland et al. (2022) disclose that sustainable management education influences students' sustainability perception. The present study takes a holistic perspective of the business school with a view to disclosing its role in promoting sustainability among graduate students.

2.1 Theoretical Development and Conceptual Framework

The knowledge-attitude-behaviour (KAB) model measures the effectiveness of information gained about a particular phenomenon in modifying peoples' lifestyles and specifically by helping them to develop a predisposition which leads to a desired behaviour change. Scholars in the medical profession have extensively utilized the KAB model (Liu et al. 2016). Within the sustainability and education literature, the KAB model has been applied (Ovais 2023; Schrader and Lawless 2004). Like most concepts, scholars are yet to reach a consensus on the definition of knowledge, thus its definition remains elusive (Bolisani and Bratianu 2018). However, Nonaka and Takeuchi's (1995; p. 87) definition of knowledge as "justified true belief" stands out. In the context of this study, sustainability knowledge refers to awareness, skills and information which one acquires through the learning process (education) about sustainable development. Studies affirm the positive effect of sustainability knowledge on attitudes towards sustainability (Saleh et al. 2022; Agu 2021; Ovais 2023).

While several definitions of attitude exist in the literature, the definition by Ajzen and Fishbein (2000:3) that attitude "refers to the evaluation of an object, concept, or behavior along a dimension of favor or disfavor, good or bad, like or dislike," is adopted here. Attitude is assumed to have three components comprising the affective, behavioural, and cognitive (Ostrom 1969), and is affirmed by previous studies as a driver of sustainable behaviour (Agu et al. 2022a, b; Saleh et al. 2022). Behaviour refers to the actions and decisions that people take in response to a particular situation. Therefore, sustainable behaviour is a deliberate and effective action that leads to the conservation of natural and social resources (Tapia-Fonllem et al. 2017). A person who is sustainability conscious, engages in actions and decisions that are geared towards the societal, environmental and economic wellbeing of the present and future generations. This paper, therefore, argues that business schools, through the various programmes such as Marketing, Accounting, Business Administration/ Management, Banking and Finance etc., inculcates sustainability knowledge (education) that shapes the attitude of students positively towards sustainability, and which, in turn, leads to the desired goal of sustainability conscious behaviour.

The Self-control theory was propounded by Gottfredson and Hirschi (1990). Writing from the perspective of criminology, the authors postulate that the essential element of criminality is the absence of self-control. They argue that people with high self-control are conscious of the long-term consequences of their behaviour, while those with low self-control do not (Burt 2020). Again, self-control is learned, and once learned, is highly resistant to change (Gottfredson and Hirschi 1990). This highlights the role of knowledge (education) in building self-control. Therefore, knowledge gained through business education can influence the degree of self-control, which is necessary for ensuring a sustainability-conscious behaviour (Nguyen et al. 2019). Self-control has been defined by Jiang et al. (2023) as an individual's ability to control thoughts, feelings, desires, and actions in the face of momentary more tempting alternatives.

Self-control helps one to discontinue bad habits, resist temptations and overcome first impulses. It is the capacity for altering one's own behavior, especially to bring them into line with standards such as ideals, values, morals and social expectations, and to support the pursuit of long-term goals (Baumeister and Tierney 2012) in Whelan et al. (2022). Thus, self-control helps people to overcome the temptation to engage in unsustainable actions and decisions. It regulates behaviour positively (Blankstein and Polivy 1982). Given the nature of sustainable behaviour, which requires a reasonable degree of restraint in order to protect the environment and promote the economic and social wellbeing of the present and future generation, a high degree of self-control becomes imperative in actualizing it.

Figure 1 depicts a conceptual framework that emanated from the reviewed literature and analyzed data.

3 Methods

This study adopted the qualitative research design approach based on a sustainability consciousness paradigm. It leveraged the interpretivist research approach to explore, from the standpoint of target informants (Borg et al. 2023), the role of business schools

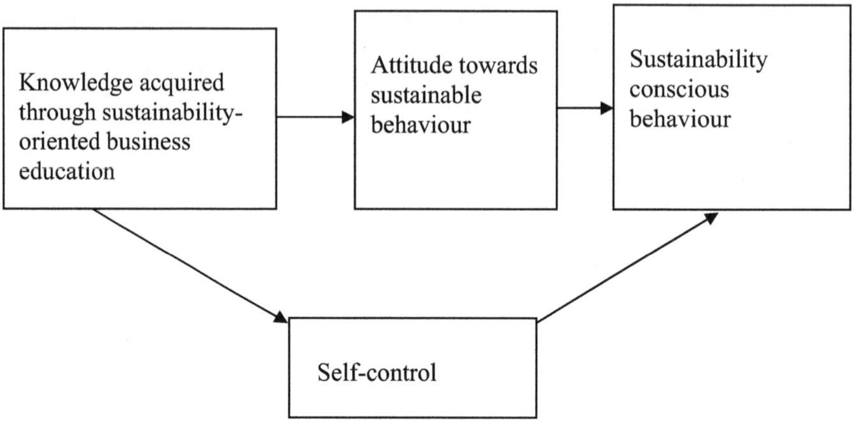

Fig. 1 Conceptual framework of knowledge, attitude, self-control, and sustainability conscious behaviour. *Source* Authors' conceptualization, 2023

in building sustainability consciousness in students. Given the context-specific nature of the study and the need to gain a holistic overview of Nigerian business students' sustainability-oriented behaviour, the qualitative research design becomes suitable (Corbin and Strauss 2008). Qualitative research gathers participants' experiences, perceptions, and behaviour (Tenny et al. 2022), and uses open-ended and conversational communication to generate relevant data on how business education influences students' attitude, self-control, and sustainability consciousness (Eriksson and Kovalainen 2016).

The research context is business school, which has been proven to be one of the avenues for imparting relevant knowledge for sustainable behaviour. The business school, through its various programmes such as Marketing, Management, Business Administration, Accountancy, Finance & Banking, Entrepreneurship, Insurance, Tourism etc., provides sustainability-centered teachings that are expected to re-orientate the students towards sustainable behaviour. The undergraduate curriculum in various business schools across the globe still lacks, to a reasonable extent, courses devoted to sustainable development (Agu et al. 2022a, b; Edwards et al. 2020). However, the graduate curricula in most disciplines have moved beyond mere having sustainability-oriented topics to creating courses that are focused on sustainable development. Thus, this study explores how these teachings have affected the students' level of sustainability consciousness by impacting their attitude and level of self-control, by asking them to draw on their personal experiences.

3.1 Sample and Data Collection

To examine the dimensions of sustainability consciousness among business students in Nigeria, the study focused on graduate students in a foremost business school in the South-east of Nigeria. The business school harbours five progarmmes, including Marketing, Management, Accountancy, Banking & Finance, and Economics. However, only students from the first four were covered since there is an ongoing debate in the university about the suitability of the Department of Economics in the business school. In neighbouring universities within the zone, Economics is housed in the faculty of social sciences. Therefore, to qualify for inclusion in the study, the subject must be a graduate student, specifically, masters or doctoral, in any of the four departments. By focusing on this group of students, it enabled information on the various dimensions of sustainability consciousness resulting from the knowledge acquired through business education to be obtained.

As a result of the small size of the target population, the study adopted the census approach to arrive at the number of participants for the study. Thus, every member of the population was contacted for the purpose of collecting data (Harding 2006). Masters and doctoral students of the four departments who were admitted in the 2019/ 2020 academic session, and who had completed their course work were involved. Being experienced in the phenomenon of sustainable development and wanting to advance overall knowledge on the topic, the participants were contacted through their emails and phone numbers. Information from the various heads of department indicates that a total of 32 graduate students qualified to participate in the study.

Bearing in mind the need to reach the objectives of the study and achieve theoretical saturation, an interpretative approach was used to collect, analyze then provide insights into the data (Corley and Gioia 2004) in Ratten (2023). Theoretical saturation—the point where collection of additional data adds no new insight– was reached after the 18th interview. The participants were assigned numbers 1 to 18 for privacy and confidentiality reasons. As shown (Table 1), 6 or 33.33% of participants were female, while 12 or 66.67% were males. Again, 5 or 27.78% of the participants aged between 25–35 years, 11 or 61.11% aged 36–45 years and 2 or 11.11% aged 46–60 years. A greater number of the participants, 16 or 88.89%, were married, while 2 or 11.11% were single. 3 or 16.67% of participants were drawn from Banking & Finance, while 5 or 27.78% were drawn from each of the other three disciplines—Marketing, Management, and Accountancy. While 12 or 66.67% of participants were doctoral (Ph.D.) candidates, 6 or 33.33% were master's degree (MSc) students.

In-depth interviews were conducted via telephone or fact-to-face to explore the participants' personal experiences of how the business education has influenced their overall lifestyle towards sustainable development. The entire data collection period lasted for two months with each interview taking an average of 39 min. The participants approved that the interviews be audio recorded. Thereafter, the interviews were transcribed into text. The same interview protocol was administered on the participants after it was approved by the Ethics, Research and Publications Committee of the lead author's university. Thereafter, the researchers returned the transcript

Table 1 Demographic analysis of the respondents

Interviewee number	Gender	Age (years)	Marital status	Discipline (department)	Programme
1 Paul	Male	36–45	Married	Management	Doctoral
2 Ubani	Male	25–36	Married	Marketing	Doctoral
3 Kama	Male	36–45	Married	Accountancy	Doctoral
4 Fayol	Male	36–45	Married	Accountancy	Doctoral
5 Chima	Male	25–35	Single	Marketing	Masters
6 Chosen	Female	25–35	Married	Marketing	Masters
7 Okocha	Female	36–45	Married	Management	Doctoral
8 John	Male	46–60	Married	Banking	Doctoral
9 Eberechi	Female	25–35	Married	Management	Masters
10 Rev	Male	36–45	Married	Marketing	Masters
11 Steve	Male	36–45	Married	Management	Doctoral
12 Uche	Male	36–45	Married	Management	Doctoral
13 Onwuka	Male	36–45	Married	Accountancy	Doctoral
14 Blessing	Female	36–45	Married	Banking	Masters
15 Ejem	Male	46–60	Married	Banking	Doctoral
16 EMMA	Male	36–45	Married	Accounting	Doctoral
17 Hon	Female	36–45	Married	Marketing	Doctoral
18 Ngwo	Female	25–35	Single	Accounting	Masters

Source Authors' Compilation, 2023

to the participants to review for clarity and validation (Ratten 2023; Madichie and Agu 2022). Minor corrections were effected based on the observations made by few participants.

3.2 Data Analysis

To analyze the generated data, a widely used interpretative approach- deductive thematic analysis- was adopted. Thus, the researchers immersed themselves in the data in order to identify common ideas or themes that emerged based on the phenomenon under investigation and that resonate with the research question(s) posed in the study (Peterson 2017). While a number of suggestions exist in the literature on the steps in thematic analysis, the Braun and Clarke's (2006) method, which is an iterative process consisting of six steps: becoming familiar with the data, generating codes, generating themes, reviewing themes, defining and naming themes, and locating exemplars (producing the reports), was adopted given its simplicity and

Table 2 Coded thematic frames of dimensions of business students' sustainability consciousness

Themes	Emerging subthemes	Code/Description
Sustainability knowledge acquisition channels	Academic activities Community engagements Administrative policies	Seminars, lectures, research, workshops/trainings, conferences, and public/inaugural lectures Host community-oriented corporate social responsibility, and peace-building engagements Cost reduction policies, students' welfare schemes, and environmental protection policies
Sustainability knowledge acquired by students	Environmental sustainability Sustainable resources allocation Sustainable consumption Sustainable entrepreneurship Climate change Corporate social responsibility Sustainability reporting Ethical decision/consumption Social equality/justice	Various areas of sustainability knowledge acquired by business students
Dimensions of sustainability consciousness	Sustainability-oriented career behaviour Environmental consciousness Sustainable consumption Social responsibility/economic consciousness Self-control for sustainable outcomes	Actual sustainability actions engaged by students as a result of knowledge gained through sustainability business education

Source Authors' own compilation (2023)

popularity. Therefore, to understand the common themes emerging from the generated data on students' experiences of the role of business education in building their sustainability consciousness, first, transcripts were read over a number of times, thereby enabling an initial identification of ideas, familiarization process of the data, and generation of relevant codes (Madichie and Agu 2022; Ratten 2023).

In line with the research questions and the theories grounding the study, the open coding approach and manual labeling were deployed to identify the common phrases in the transcripts. The phrases were classified based on the students' course of study (discipline) and type of graduate programme (Master of Science/MSc or Doctor of Philosophy/PhD). Multilayered expressions of the business students' experiences were then examined that outlined the interdependence across constructs (Ratten 2023). Emerging themes were written down and cross-matched before arriving at key themes (Madichie and Agu 2022). The emergent themes are presented in Table 2, while some direct quotes from the participants are also captured which support the emergence of the themes. Two independent qualitative research experts were presented with the data and the emerging themes in order to establish the reliability of the findings (Ilac 2018). Their suggestions were taken into consideration before the final report.

4 Findings and Discussion

4.1 Activities of the University Through Which Sustainability Knowledge Was Acquired

On the responses of participants regarding the activities of the university through which they gained the knowledge of sustainability, three key themes emerged: Academic activities, community engagements, and university administration/ policies.

First, all the participants mentioned academic activities such as teaching/ lectures, research, seminars, workshops/training programmes, conferences, and public/inaugural lectures that draw the attention of the university stakeholders to issues of sustainability.

> I remain being exposed to ethical and sustainable marketing practices during my course work on 'Contemporary Issues in Marketing'. P2

Secondly, community engagements of the university such as host community-oriented social responsibilities, and periodic university-community interactions which aim at building and sustaining peaceful co-existence of the parties.

> This University has a long history of employing indigenous first class graduates from the host community, and has also created direct and indirect economic empowerment of indigenes of community. P8

Thirdly is the administrative style of the university management which is cost-reduction oriented, and which prioritizes policies on students' welfare and environmental protection.

> My University has in the last couple of months drastically reduced its power supply to different faculties of the University as they grapple with the challenge of poor funding. P1

> The university management came up with a policy to prevent exploitation of students, especially in the area of textbook purchase. P5

> I was a member of the joint university taskforce on environmental sanitation. Several policies that aim at protecting the environment within and around the university were implemented. P11

The findings corroborate previous studies which affirmed the engagement of business schools in the listed activities as means of actualizing sustainability development (Godemann et al. 2023; Fang and O'Toole 2023).

4.2 Sustainability Knowledge Acquired

The business school through the explored programmes imparts sustainability knowledge in the graduate students. A perusal of the curricula of the four disciplines indicates an absence of a stand-alone course termed "sustainability marketing/accounting/management/finance/banking. However, across the disciplines are sustainability-focused courses. For instance, there are "Contemporary Issues in Marketing", "Contemporary Issues in Banking & Finance", "Contemporary Issues in Accounting", "Ethical accounting, marketing, management, banking and finance", as well as "Corporate Reporting". Besides, some relevant aspects of sustainability appear in the course outlines of most other courses offered by the graduate students of the disciplines. For instance, the course "Organizational behaviour" has the topic "organizations and environment, while "Advanced Marketing Communication," a PhD course in Marketing has "ethical and legal environment for marketing communications". Furthermore, the interviews provide the following core areas of sustainability knowledge gained by the participants: environmental sustainability, sustainable resource allocation, sustainable consumption, sustainable entrepreneurship, climate change, corporate reporting on sustainability, corporate social responsibility, ethical decision/consumption, and social equity and justice. Few of the participants specialize in these areas.

> So far as a researcher and an accountant, my interest has been in the area of reporting by corporate entities on their sustainability performance and impact. Participant 3 (P3)

P1 and P2 who are doctoral students of Management and Marketing, respectively, stated as follows:

> I have been exposed to several sustainability issues including climate change, renewable energy, waste management, and sustainable business/entrepreneurship practices. P1

> I am currently running a PhD program in Marketing and the main area of sustainability I'm exposed to is the effect of marketing activities on the environment as it relates to packaging, promotion, features and components of the products. P2

The findings support previous research that business schools have the tendency to integrate sustainability into graduate programmes (Godemann et al. 2023; Jun and Moon 2021; Miotto et al. 2020; Rusinko 2010).

The interviews further validate the assertion that the sustainability knowledge of participants has positive influence on their attitude towards sustainable development. All the participants rated the influence of the knowledge on attitude and sustainability consciousness very high.

> I would rate it highly, because I think I am more sustainability conscious now than ever. P1

> Being able to live an environmentally and socially conscious life style, I can rate the efficacy of the knowledge in inducing sustainability consciousness very high. P7

The findings provide support to previous research on the positive impact of sustainability knowledge on attitude and sustainability consciousness (Saleh et al. 2022; Agu 2021; Ovais 2023).

4.3 Dimensions of Sustainability Consciousness

The study explored the ways sustainability knowledge acquired via business education has affected the students. Sustainability-oriented career behaviour, environmental consciousness, sustainable consumption, social responsibility/economic consciousness, and self-control for a sustainable outcome were identified.

4.3.1 Sustainability-Oriented Career Behaviour

The graduate students explained how the knowledge they acquired about sustainable development helps them to act sustainably in their chosen career. Being able to influence the behaviour of people in their work place or business is an important aspect of the sustainable development pursuit. The respondents acknowledged that the knowledge of sustainability they possess has improved their career behaviour towards sustainable development. The following quotes from doctoral students who also are lecturers provide examples of such sustainability-oriented career behaviour:

> The knowledge so far garnered on issues of sustainability has no doubt impacted my personal and career life, particularly with regards to decisions on consumption of materials such as paper. I make efforts to reduce paper use by encouraging my supervisees to send soft copies (instead of hard copies) for review and corrections. P3

> I go ahead to teach what I've learnt and pass it on to the undergraduate students I teach as well as talk about it in informal meetings with friends and colleagues. P2

> It has been a comprehensive transformation in teaching and research, thus, an improvement on my quality of service delivery. P18

> The knowledge of sustainability influences my decision-making in my place of work as an educator. I educate and enlighten the younger generation about sustainable behaviour. This is to provoke the mindset towards environmental friendliness and preserving natural resources. P9

Other respondents disclosed various career-related influences of the acquired knowledge.

> I now incorporate sustainability principles which promote more responsible and ethical practices, positive relationship with colleagues, and positive work culture in my place of work. P4

> The knowledge has made me to be more aware of the need to integrate sustainability into my business practices. As a researcher I have devoted more attention to sustainable issues as a way of contributing to the global quest for a sustainable society. P2

> It has shaped my behaviour in the area of lean (waste) management for more profitability. P7

The finding supports previous research (Romero-Colmenares and Reyes-Rodríguez 2022; Agu et al. 2021; Gupta and Matharu 2022).

4.3.2 Environmental Consciousness

Environmental consciousness was a common expression among the 18 participants. They noted that the sustainability-focused business education reawakened their consciousness towards a safe environment.

> My understanding of the principles of ecosystems and the impacts of human activities on the environment has helped me to contribute towards natural resource management, reducing greenhouse gas emissions and adherence to regulations, and policies related to environmental protection in my area. P12

> It has made me to be cautious and conscious of the work environment and to make it more habitable and conducive. P7

> My attitude towards waste disposal has been affected. We take the pains to ensure that the manner we dispose our wastes does not negatively impact the environment. P1

> I am conscious of the effect of the products I use on my health and the environment and I engage in proper disposal of waste. P3

> My knowledge of sustainability has helped me to embrace green marketing and environmental consciousness. P13

The findings are supported by Ertas and Ozkan (2017), Lin et al. (2022) and Greenland et al. (2022).

4.3.3 Sustainable Consumption

Being consumers of products (goods, services, events, ideas, places etc.), the interviews establish that participants' consciousness towards sustainable consumption has enhanced as a result of the sustainability knowledge acquired via business education.

> In my place of work, we implement strategies that reduce the consumption of fossil fuel thereby ensuring less emission of greenhouse gas (GHG) to the atmosphere. Electronic means of communication are now in use instead of the hitherto, bulky paper files dispatched to staff members during monthly meetings. P1

> I am now very careful of the type of products I buy. Whether they are produced by firms that adhere to sustainability rules, including recycling and environmental safety influences my purchase decision. P15

> My buying habit has been greatly influenced. The knowledge I gained helps me to form the necessary judgment about firms to patronize and the products to consume. P6

Previous research supports these findings (Agu et al. 2022a, b; Longoria et al. 2021).

4.3.4 Social Responsibility and Economic Consciousness

The political officeholders and business owner-managers among the participants stressed their consciousness towards giving back to society as a means of improving the lives of people around them. A master's degree student who owns a small enterprise made the following expression:

> My knowledge of sustainability has helped me to embrace the practice of being socially responsible to the host community of my organization. I now know that I'm ethically bound to perform some social responsibilities to my host community. Besides ensuring proper preservation and protection of the host environment, my firm provides periodic economic empowerment to indigenous people through scholarships, employment, and patronage of the local suppliers etc. P10

Other respondents, who are political officeholders said:

> It helps me to ensure inclusive and equitable distribution of national resources to my constituents, thereby ensuring hope for the future. P15

Besides, the knowledge of sustainability is confirmed to induce conscious economic behaviour.

> We prioritize social equity and justice in making policies that will have positive effect on the economic wellbeing of people. Recently, we enacted a policy that seeks to empower entrepreneurial people with disabilities. The intention is to support their means of livelihood and promote their wellbeing. My knowledge of the SDGs encouraged me to initiate this idea. P17

Previous studies (Leal Filho 2022; Purcell et al. 2019; Homer and Khor 2022; Miotto et al. 2020) provide support for the finding.

4.3.5 Self-Control

The interviews confirmed the positive role of sustainability knowledge via business education in helping the participants develop self-control. For example, consuming less, sacrificing for the economy, and engaging in good environmental behaviour are conscious acts that have long-term positive effect on the society. All the participants, while responding to the question "Has the sustainability knowledge you acquired so far helped you to form self-control that has a positive impact on achieving sustainable development," replied in the affirmative.

> Yes. It has made me become more conscious and control certain ways of doing things. P2

> Surely, with every sense of modesty, my knowledge of sustainability today has instilled some form of self-control in me with regards to consumption of certain resources and I believe this contributes positively towards the much needed global sustainable development. P1

> Yes. To a very large extent, I exercise a higher level of self-control in my waste disposal behaviour, and strive to consume only environmentally friendly brands, no matter how enticing and affordable the alternative brands are. P8

These findings are in alignment with Nguyen et al. (2019), Blankstein and Polivy (1982) and Whelan et al. (2022).

A summary of how business schools influence the sustainability consciousness of students is presented in Table 2.

5 Conclusions

This research sought to explore the dimensions of sustainability consciousness in business schools, thereby responding to the call for more empirical studies on the role of universities in promoting sustainable development (Leal Filho 2022). The study provides timely insights into how a business school in a developing country, specifically Nigeria, integrates and inculcates sustainability and the areas of sustainability consciousness emanating from these actions. The results support the notion that business schools are reliable vehicles for actualizing the SDGs. Thereby making relevant contributions to theory and practice about the role of business schools and sustainability consciousness. The finding suggests that business schools play a role in inculcating sustainability knowledge that stimulates a positive attitude and high self-control towards sustainability consciousness. This finding will help policymakers and business educators enrich their curricula and step up sustainability-focused teachings and actions. The findings further emphasize how business schools deploy academic programs, community engagements, and administrative policies to create environmental, social, and economic sustainability knowledge that induces sustainability-oriented career behavior, environmental consciousness, sustainable consumption, social responsibility/economic consciousness, and self-control among graduate students. Practically, therefore, there is a need to encourage and emphasize sustainability-oriented research, cases, experiential learning, inaugural lectures, and events. These will increase the number of sustainability ambassadors at universities.

Theoretically, this research extends the literature on the KAB model and self-control theory in the context of sustainable development. It validates the applicability of KAB (Ovais 2023) and self-control theory (Nguyen et al. 2019) in sustainability-conscious research, thereby making significant contributions to theory. So far, no study has integrated these two theories in the context of business schools and sustainable development in Nigeria.

To take this study further, future and ongoing studies are required that can reduce the limitations of the present study. Firstly, the small sample size and scope will limit generalizability. But this can be addressed by enlarging the scope of future studies. Secondly, the methodology, which is qualitative, might affect the application of the findings. Further studies are required to validate the findings of this study by employing other methods, such as quantitative and mixed methods. Again, the study focused only on the experiences of graduate students of the faculty of management sciences (business school) of the case study university. Studying undergraduates, which have a larger population, and extending to other faculties will provide more insights in this area of study. Finally, future research should explore the level of research interest of university students in specializing in and conducting their theses, dissertations, or project research in the area of sustainable development. This will further indicate the students' consciousness towards sustainability.

Acknowledgements The lead author is grateful to Professor Walter Leal Filho and the editorial team for the invitation to contribute to this book project as part of the 100 papers to accelerate the UN's SDGs.

References

Agu GA (2021) A survey of business and science students' intentions to engage in sustainable entrepreneurship. Small Enterp Res 28(2):206–227

Agu GA, Etuk SG, Madichie NO (2022a) Exploring the role of sustainability-oriented marketing education in promoting consciousness for sustainable consumption. Sustainability 14(13):8077. https://doi.org/10.3390/su14138077

Agu GA, Etuk SG, Egele AE, Ibekwe FC, Nnamocha AO (2022b) Determinants of intentions towards sustainable consumption among university students in Nigeria. The role of sustainability-oriented marketing education and morality. Afr J Bus Econ Res 17(4):29–49

Ajzen I, Fishbein M (2000) Attitudes and the attitude-behavior relation: reasoned and automatic processes. Eur Rev Soc Psychol 11(1):1–33. https://doi.org/10.1080/14792779943000116

Blankstein KR, Polivy J (1982) Emotions, self-control, and self-modification. In: Blankstein KR, Polivy J (eds) Self-control and self-modification of emotional behavior: advances in the study of communication and affect, vol 7. Springer, Boston, MA. https://doi.org/10.1007/978-1-4615-9260-0_1

Bolisani E, Bratianu C (2018) The elusive definition of knowledge. In: Bolisani E, Bratianu C (2018) Emergent knowledge strategies: strategic thinking in knowledge management. Springer, Cham, pp 1–22

Borg J, Scott-Young CM, Borg N (2023) What generation Z needs: the role of project-based organizations in creating career sustainability. Int J Manag Proj Bus 16(3):571–591. https://doi.org/10.1108/IJMPB-12-2022-0273

Braun V, Clarke V (2006) Using thematic analysis in psychology. Qual Res Psychol 3:77–101

Burt CH (2020) Self-control and crime: beyond Gottfredson & Hirschi's theory. Annu Rev Criminol 3(1):43–73

Caeron JC, Decamps A (2014) The sustainability literacy test: can universities be sure that they are producing sustain-ability literate graduates? In: Proceedings of the world conference on education for sustainable development, Aichi-Nagoya, Japan,10–12 November 2014

Council of the European Union (2010) Council conclusions on education for sustainable development. 3046th Education, Youth, Culture and Sport Council meeting Brussels, 18 and 19 November 2010

Corbin J, Strauss A (2008) Basics of qualitative research: techniques and procedures for developing grounded theory, 3rd edn. Sage Publications, Inc. https://doi.org/10.4135/9781452230153

de la Riva GA, Espinosa Fajardo CC, Juárez Nájera M (2015) Sustainability in engineering education: An approach to reach significant learning and character skills. In: Sustainability in higher education. Elsevier Ltd., New York City, NY, USA, pp 97–125

Edokpolor JE (2020) Entrepreneurship education and sustainable development: mediating role of entrepreneurial skills. Asia Pac J Innov Entrep 14(3):329–339. https://doi.org/10.1108/APJIE-03-2020-0036

Edwards DB, Sustarsic M, Chiba M, McCormick M, Goo M, Perriton S (2020) Achieving and monitoring education for sustainable development and global citizenship: a systematic review of the literature. Sustainability 12(4):1383. https://doi.org/10.3390/su12041383
Eriksson P, Kovalainen A (2016) Qualitative methods in business research, 2nd edn. Sage
Ertas FC, Ozkan O (2017) The effects of environmental accounting education on students' environmental consciousness. In: Conference: IV. International Symposium on Accounting and Finance, At: Ohrid-Macedonia
Fang J, O'Toole J (2023) Embedding sustainable development goals (SDGs) in an undergraduate business capstone subject using an experiential learning approach: a qualitative analysis. Int J Manag Educ 21(1):100749
Ferrer-Estévez M, Chalmeta R (2021) Integrating sustainable development goals in educational institutions. Int J Manag Educ 19:100496
Gericke N, Boeve-de Pauw J, Berglund T, Olsson D (2019) The sustainability consciousness questionnaire: the theoretical development and empirical validation of an evaluation instrument for stakeholders working with sustainable development. Sustain Dev 27(1):35–49
Godemann J, Nguyen B-N, Herzig C (2023) Business schools' transformation towards sustainability: empirical insights from UN PRME signatories. Int J Sustain High Educ. https://doi.org/10.1108/IJSHE-07-2022-0223. (ahead-of-print)
Gottfredson MR, Hirschi T (1990) A general theory of crime. Stanford University Press
Greenland S, Saleem M, Misra R, Mason J (2022) Sustainable management education and an empirical five-pillar model of sustainability. Int J Manag Educ 20(3):100658. https://doi.org/10.1016/j.ijme.2022.100658
Gupta N, Matharu M (2022) Examining the enablers of sustainable entrepreneurship – an interpretive structural modelling technique. J Entrep Emerg Econ 14(5):701–726. https://doi.org/10.1108/JEEE-10-2021-0416
Harding J (ed) (2006) vols. 1–0. SAGE Publications, Ltd. https://doi.org/10.4135/9780857020116
Homer ST, Khor KS (2022) Sustainable campus using concept mapping: a bottom-up approach engaging both staff and students. Int J Sustain High Educ 23(3):645–665. https://doi.org/10.1108/IJSHE-02-2021-0059
Ilac EJD (2018) Exploring social enterprise leadership development through phenomenological analysis. Soc Enterp J 14(3):268–288
Jiang X, Zhang H, Wang T, Zhang C (2023) The association of self-control, self-efficacy, and demographic characteristics with home-based E-learning behavior in nursing and midwifery undergraduates: a cross-sectional study under the COVID-19 epidemic. Nurse Educ Today 120:105628. https://doi.org/10.1016/j.nedt.2022.105628
Jun H, Moon S (2021) An analysis of sustainability integration in business school curricula: evidence from Korea. Sustainability 13(5):2779. https://doi.org/10.3390/su13052779
Leal Filho W (2022) Pursuing sustainable development across disciplines. In: Leal Filho W, Portela de Vasconcelos CR (eds) Handbook of best practices in sustainable development at university level. World Sustainability Series. Springer. https://doi.org/10.1007/978-3-031-04764-0_27
Lin W, Nguyen L, Tran N, Huang C (2022) Impact of action-oriented education course on students' environmental awareness. Appl Environ Educ Commun 21(4):331–347
Liu L, Liu YP, Wang J, An LW, Jiao JM (2016) Use of a knowledge-attitude-behaviour education programme for Chinese adults undergoing maintenance haemodialysis: Randomized controlled trial. J Int Med Res 44(3):557–568. https://doi.org/10.1177/0300060515604980
Longoria LC, Formes IL, Sovens DC, Perez JJ (2021) Promoting sustainable consumption in higher education institutions through integrative co-creation process involving stakeholders. Sustain Prod Consum 28:445–458
Madichie N, Agu GA (2022) The role of universities in scaling up informal entrepreneurship. Ind High Educ. https://doi.org/10.1177/09504222221101548
Miotto G, Blanco-Gonzales A, Diez-Martin F (2020) Top business schools legitimacy quest through the SDGs. Heliyon 6(11):e05395

Murray J (2018) Student-led action for sustainability in higher education: a literature review. Int J Sustain High Educ 19(3). https://doi.org/10.1108/IJSHE-09-2017-0164

Nguyen TD, Dadzie CA, Chaudhuri HR, Tanner T (2019) Self-control and sustainability consumption: findings from a cross cultural study. J Int Consum Mark 31(5):1–15. https://doi.org/10.1080/08961530.2019.1576152

Nonaka I, Takeuchi H (1995) The knowledge-creating company: How Japanese companies create the dynamics of innovation. Oxford University Press, New York

Nwaigwe NG, Ofoegbu GN, Dibia NO, Nwaogwugwu CV (2022) Sustainability disclosure: impact of its extent and quality on value of listed firms in Nigeria. Cogent Bus & Manag 9(1). https://doi.org/10.1080/23311975.2022.2079393

Ostrom TM (1969) The relationship between the affective, behavioral, and cognitive components of attitude. J Exp Soc Psychol 5(1):12–30. https://doi.org/10.1016/0022-1031(69)90003-1

Ovais O (2023) Students' sustainability consciousness with the three dimensions of sustainability: does the locus of control play a role? Reg Sustain 4(1):13–27

Peterson BL (2017) Thematic analysis/interpretive thematic analysis. https://doi.org/10.1002/9781118901731.iecrm0249

PRME (2020). What We Do. PRME. https://www.unprme.org/what-we-do

Purcell WM, Henriksen H, Spengler JD (2019) Universities as the engine of transformational sustainability toward delivering the sustainable development goals: "Living labs" for sustainability. Int J Sustain High Educ 20(8):1343–1357. https://doi.org/10.1108/IJSHE-02-2019-0103

Ratten V (2023) Gender and climate change: a sustainable tourism perspective. Thunderbird Int Bus Rev. https://doi.org/10.1002/tie.22351

Raymundo MG, Matheus C, de Moraes Hoefel JL (2019) Awareness of sustainability issues. In: Leal Filho, W. (eds) Encyclopedia of sustainability in higher education. Springer, Cham. https://doi.org/10.1007/978-3-030-11352-0_2

Romero-Colmenares LM, Reyes-Rodríguez JF (2022) Sustainable entrepreneurial intentions: Exploration of a model based on the theory of planned behaviour among university students in north-east Colombia. Int J Manag Educ 20(2). https://doi.org/10.1016/j.ijme.2022.100627

Rusinko CA (2010) Integrating sustainability in management and business education: a matrix approach. Acad Manag Learn & Educ 9(3):507–519

Saleh MS, Mehellou A, Huang M, Briandana R (2022) The influence of sustainability knowledge and attitude on sustainable intention and behaviour of Malaysian and Indonesian undergraduate students. Res Comp Int Educ 17(4):677–693. https://doi.org/10.1177/17454999221126712

Schrader PG, Lawless KA (2004) The knowledge, attitudes, & behaviors approach how to evaluate performance and learning in complex environments. Perform Improv 43(9):8–15

Tapia-Fonllem C, Fraijo-Sing B, Corral-Verdugo V, Ortiz Valdez A (2017) Education for sustainable development in higher education institutions: its influence on the pro-sustainability orientation of Mexican students. SAGE Open 7(1). https://doi.org/10.1177/2158244016676295

Tenny S, Brannan JM, Brannan GD (2022) Qualitative study. In: StatPearls Treasure Island. StatPearls Publishing. https://www.ncbi.nlm.nih.gov/books/NBK470395

Tulder RV, Rodrigues SR, Mirza A, Sexsmith K (2021) The United Nation's SDGs: can multinational enterprises lead the decade of actions? J Int Bus Policy 4(1):1–21

Ulrich EH, Ivan N, Henny R, Anna W, Linda MK, Laurens K (2018) Sustainability transitions in developing countries: stocktaking, new contributions and a research agenda. Environ Sci Policy 84:198–203. https://doi.org/10.1016/j.envsci.2017.11.009

UNESCO (2012) Education for sustainable development: building a better, fairer world for the 21st century. www.unesco.org

UNESCO (2014) Education for sustainable development: sourcebook. www.sustainabledevelopment.un.org

UNESCO (2020) Education for sustainable development: empowerment through knowledge, skills and values to protect the planet. www.unesco.org. Accessed 5 June 2023

Vladimirova K, LeBlanc D (2015) How well are the links between education and other SDGs covered in the United Nation flagship reports? A contribution to the study of science-policy interface on education in the United Nation systems. DESA Working Paper Number 146. www.sustainabledevelopment/un.org
Weybrecht G (2022) Business schools are embracing the SDGs – but is it enough?- How business schools are reporting on their engagement in the SDGs. Int J Manag Educ 20(1):100589
Whelan E, Golden W, Tarafdar M (2022) How technostress and self-control of social networking sites affect academic achievement and wellbeing. Internet Res 32(7):280–306. https://doi.org/10.1108/INTR-0

Sustainable Health and Functional Foods

Vijaya Deshmukh and Rajashri Kanad Gutte

Abstract Given the challenges posed by climate change and the growing global population, it is crucial to prioritize the sustainability of agricultural food production. This necessitates the development of efficient manufacturing and processing systems to meet the ongoing demand for high-quality food that fulfils essential human needs. These imperative holds significance in both social and scientific spheres. The pronouncer market pressure supported Functional foods to become a lucrative segment that is increasing drastically with huge market potential and providing value-added nutrition beyond basic food. Also used for commercial purposes around the world. This research aims to provide detailed information about some innovations, the latest trends, and strategies followed by functional food manufacturers to ensure sustainable food production while sticking to the sustainable framework in development of functional foods emphasizing environmental impact. The focus of this study is to emphasize the significance of including sustainable food systems in the curricula of universities and agriculture institutes as a strategy to attain the Sustainable Development Goals. By integrating principles of sustainability into education, we can raise awareness, enhance knowledge, and foster practical skills essential for the establishment of a resilient and sustainable food system.

Keywords Functional foods · Innovations · Sustainable food production · Healthy food · Etc.

1 Introduction

The European Food Safety Authority (EFSA) defines functional foods that have been enriched or fortified with specific nutrients or additives that provide health benefits beyond the typical nutrients found in foods. These foods can also be natural foods that already contain beneficial compounds that promote health and prevent disease.

V. Deshmukh (✉) · R. K. Gutte
Institute of Management & Research, MGM University, Aurangabad, Maharashtra 431003, India
e-mail: veedesh@gmail.com

W. Leal Filho et al. (eds.), *The Contribution of Universities Towards Education for Sustainable Development*, World Sustainability Series,
https://doi.org/10.1007/978-3-031-49853-4_25

Some common examples of functional foods include probiotics, omega-3 fatty acids, and fibre-rich foods (The European Parliament, & the Council of the European Union 2006).

Functional nutrition aims to promote optimal health and well-being by utilizing wholesome nutrient-dense ingredients in one's diet. By emphasizing natural, minimally processed foods, functional nutrition can help you get the nutrients your body needs to thrive. Plus, sustainable food production practices can help reduce the environmental impact of modern agriculture and promote a healthier planet for generations to come (Gallotti and Lavelli 2020; Ogori et al. 2021; Šojić et al. 2020).

Functional nutrition is becoming increasingly popular among consumers who are looking for ways to improve their health and well-being, while also supporting sustainable agriculture and economic growth. Many food manufacturers are also recognizing this trend and are investing in the development of functional foods that provide specific health benefits, such as improved digestion, better immune function, and increased energy levels. By promoting the use of natural, whole-food ingredients and sustainable production practices, functional nutrition has the potential to create a healthier and more sustainable food system for everyone (Putnik et al. 2020).

Not surprisingly, new socio-demographic trends (e.g., longer life expectancy, promotion of healthy lifestyles, better healthcare, etc.) supported the functional sector to become an increasingly lucrative segment of the food industry with a rapidly growing market.

Functional beverages made with indigenous fruits are a great example of how functional nutrition can support sustainable economic growth. These fruits are often rich in nutrients and antioxidants but may be economically underutilized due to a lack of market demand or limited accessibility. By creating functional juices and other beverages made with these fruits, food manufacturers can introduce new and unique Flavors to the market, while also supporting local farmers and promoting sustainable agriculture. In addition to their appeal to health-conscious consumers, these products are also sought after by individuals looking for beverages that offer specific health benefits, such as enhanced digestion or immune function. As a result, the market for functional beverages has experienced significant growth, positioning it as a profitable segment within the food industry (Duralija et al. 2021; Roselló-Soto et al. 2018).

In recent years, functional drinks have become increasingly popular in the juice and beverage industry, as they offer a diverse array of nutrients and biologically active compounds that can promote various health benefits. These benefits may include strengthening the immune system, aiding digestion, and supporting overall health and wellness. Some common examples of these ingredients include vitamins, minerals, fibre, unsaturated fatty acids, polyphenols, carotenoids, and antioxidants.

Probiotics and prebiotics are also popular functional ingredients that can help support digestive health and improve gut microbiome balance. By incorporating these functional ingredients into juices and other beverages, manufacturers can create products that not only taste great but also provide important health benefits for consumers (Hinestroza-Córdoba et al. 2020; Kandylis 2021).

As food science and nutrition continue to advance, there is an increasing recognition of the potential health benefits that food can offer beyond fulfilling basic nutritional needs. A growing number of individuals are actively seeking out foods that can help prevent or manage nutrition-related diseases, including heart disease, diabetes, and obesity. Furthermore, there is an increasing demand for food products that have the potential to enhance both physical and mental well-being (Martirosyan and Singh 2015). These products may encompass those that boost energy levels, uplift mood, or improve cognitive function. As a result, there has been a notable increase in the availability of functional foods and specialized diets that cater to specific health concerns (Beardsworth and Keil 1992).

Creating functional food is a multifaceted and expensive undertaking that demands meticulous planning and research. To gain useful insights into the target audience's needs and preferences, as well as the competitive environment, quantitative and qualitative marketing studies can be conducted. By gathering data on consumer behaviour, market trends, and product performance, one can make informed decisions that increase the likelihood of success (Jousse 2008).

The low success rates of new international functional foods are frequently linked to inadequate customer knowledge management and a lack of collaboration among different disciplines involved in the product development process. To address this, effective knowledge management is crucial in providing stakeholders with timely and relevant information, enabling the identification and resolution of potential issues before they escalate. Additionally, collaboration across functional disciplines ensures that all aspects of product development are harmonized and aligned with the overarching strategy. By leveraging the diverse expertise and perspectives of various teams, the likelihood of developing successful functional foods that cater to customer needs and preferences is significantly increased (Khan et al. 2013).

The commercial success of functional products ultimately depends on taste, appearance, price, and health claim appeal to consumers.

In this paper we have highlighted some techniques that are employed in the development of functional foods, what is environmental sustainability, what is Sustainable Health and how can it be achieved, latest trends and strategies being followed, Latest trends and strategies for sustainable food production, Need of Education for Sustainable Food and Nutrition etc., from this study we have categorised paper in two parts. One is for how sustainability can be achieved at industrial level and second will be for university level by education.

Some techniques employed in the development of functional foods are—

1. **Formulation and blending**

Formulation and blending are indeed cost-effective methods for developing new functional foods with added vitamins and minerals. This approach has been used for many years to address deficiencies and improve the nutritional value of foods. With the right combination of ingredients, it is possible to create functional foods that promote health, prevent disease, and support overall well-being. For example, some foods fortified with vitamin D. (Lavelli et al. 2021).

2. **Microencapsulation**

Microencapsulation is a fascinating technology that has many applications in diverse fields, such as food, pharmaceuticals, and cosmetics. By encapsulating small particles, droplets, or gases in a protective coating, we can control their release, protect them from degradation, and enhance their stability. This can lead to longer shelf life, improved bioavailability, and better sensory properties of the final product (Mehta et al. 2022).

The use of microencapsulated food ingredients for controlled-release applications is a promising alternative to solve the major problem of food ingredient delivery faced by food industries, and in solving micronutrient deficiencies, especially in countries like India where they are widely prevalent. (Zarn 1995).

3. **Nutrigenomics**

Nutrigenomics is an exciting field that explores the relationship between our genes, nutrition, and health. It recognizes that each person's genetic makeup is unique and that our bodies respond differently to the foods we eat and the nutrients we consume. By studying how our genes influence our metabolism, researchers can identify dietary patterns and personalized nutrition strategies that can optimize health, prevent disease, and improve quality of life. Nutrigenomics is a promising area of research that has the potential to revolutionize the way we approach nutrition and wellness. (German et al. 2011).

4. **Micronization and Micro fluidization**

Products, such as poor solubility and low nutrient bioavailability, while also improving their texture. Microstructure modification through micronization and micro fluidization also enhances the digestibility and sensory attributes of cereal-based products, making them more appealing to consumers. By implementing these advanced techniques, we can improve the nutritional quality and overall value of cereal-based products, which can positively impact public health. (Mert 2020).

4. **Environmental Sustainability**

To ensure food security for all living beings, it is crucial to commence preparations for forthcoming challenges and address the consequences of climate change. To this end, the journal "Environmental Sustainability" endeavours to disseminate eco-friendly techniques that can mitigate climate change and enhance agricultural production through sustainable means. (Arora 2019).

The implementation of policies and frameworks, at global, national, and regional levels, is crucial for tackling climate change. To achieve success, it is essential to work together and create unified policies instead of solely focusing on state or national-level policies. While international policies on climate change and agriculture are vital, it is equally important to implement these policies at the ground level to see real changes (Betoret et al. 2022).

2 What is Sustainable Health and How Can Be Achieved?

Sustainable health encompasses the idea of individuals and communities sustaining their well-being and good health in the long run while minimizing negative impacts on the environment and conserving limited natural resources (Kruk et al. 2018).

Attaining sustainable health requires a comprehensive approach that considers the social, economic, and environmental aspects affecting health outcomes. This includes ensuring access to nutritious food, safe drinking water, unpolluted air, and affordable healthcare. By addressing these factors, we can work towards building a healthier society that also preserves the planet for future generations. To achieve this, it is crucial to adopt a holistic and integrated approach that recognizes the interdependencies between human health, environmental health, and social equity. This means considering the impacts of our actions on all these aspects and striving for sustainable solutions that promote the well-being of individuals and the planet (Mortimer et al. 2018).

This can involve promoting healthy lifestyles, reducing exposure to environmental toxins, promoting sustainable agriculture and food systems, and advocating for policies that support universal access to healthcare and education. Additionally, we must also work to reduce our carbon footprint, conserve natural resources, and protect biodiversity, as these are essential components of sustainable health (Takachi et al. 2008).

To achieve sustainable health, it is essential for individuals, communities, and governments worldwide to collaborate. Through this collective effort, we can create a healthier world that is equitable and environmentally sustainable, benefiting future generations.

To promote sustainability in food production, functional food manufacturers are embracing innovative strategies and technologies. These advancements aim to ensure that food production practices are environmentally friendly and capable of meeting the growing demand for functional foods.

Some of the latest trends and strategies being followed are—

- Plant-based proteins: Many functional food manufacturers are switching to plant-based proteins, such as pea, soy, and hemp, as they are more sustainable and require less water, land, and resources than animal-based proteins (Tachie et al. 2023).
- Sustainable packaging: In response to the growing awareness of the importance of waste reduction and environmental impact, companies are now prioritizing sustainable packaging materials. This includes the adoption of biodegradable plastics and recycled materials, which provide more eco-friendly alternatives. By actively integrating these sustainable packaging options, companies are taking significant strides towards building a greener and more sustainable future for our planet (Berg, et al. 2020; Versino et al. 2023).

- Regenerative agriculture: This is a holistic approach to farming that aims to restore soil health, promote biodiversity, and sequester carbon in the soil. Many functional food manufacturers are partnering with farmers who follow regenerative agriculture practices to source sustainable ingredients (Peter et al. 2020).
- Precision agriculture: This involves using technology, such as drones, sensors, and machine learning, to optimize crop yields, reduce waste, and minimize the use of pesticides and fertilizers (Zaman 2023).
- Circular economy: Functional food manufacturers are increasingly recognizing the importance of the circular economy system, which aims to minimize waste and maximize resource efficiency. This system focuses on designing products and processes that can be reused, repaired, or recycled. By adopting circular economy principles, these manufacturers are actively working towards reducing their environmental impact and improving the sustainability of their supply chains. Through these proactive measures, they are contributing to the creation of a more sustainable future (Yang et al. 2023).
- Sustainable health: To promote both personal and planetary health, it is important to make sustainable choices. One way to do this is by consuming functional foods, which offer additional health benefits beyond basic nutrition. By incorporating more plant-based proteins into our diets, we can reduce our dependence on animal products, which can have a significant environmental impact (Lonnie and Johnstone 2020). Moreover, opting for a variety of fruits and vegetables is a sustainable decision since they generally require fewer resources to cultivate than animal-based products. Furthermore, selecting minimally processed foods that come in eco-friendly packaging can also contribute to reducing waste and promoting sustainable practices. By implementing these minor adjustments, we can prioritize our health and simultaneously demonstrate our concern for the environment.

3 Latest Trends and Strategies for Sustainable Food Production

Phenolic compounds not only offer natural health benefits as compounds found in food but also serve as great additives and preservatives. The metabolic engineering of microbes offers a sustainable way for producing phenolic compounds on a larger scale, with high commercial availability and minimal environmental impact. This approach offers a practical alternative to traditional plant extraction and chemical biosynthesis techniques (Mark et al. 2019).

Integrating plant-based sources of protein, such as beans and lentils, into our meals can offer a multitude of health advantages. Furthermore, superfoods like berries, leafy greens, and nuts are brimming with nutrients and antioxidants. When it comes to beverages, herbal teas and hydrating options like coconut water can also be excellent choices. However, it is crucial to seek advice from a healthcare professional before making any significant modifications to your diet. By integrating Nutritional

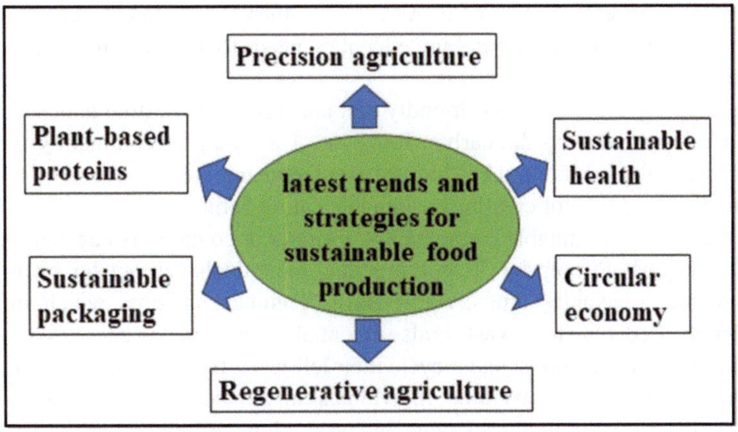

Fig. 1 Latest trends and strategies for sustainable food production

Psychology into education, students gain valuable insights into the significance of a balanced diet, the influence of specific nutrients on mental health, and techniques for fostering optimal nutrition and psychological well-being. This knowledge empowers individuals to make informed dietary decisions and embrace healthy eating habits that promote their mental well-being (Sloan 2022).

As the field of Nutritional Psychology continues to grow, it is likely to become a trending topic in educational settings. The recognition of the connection between mental health and nutraceutical functional ingredients offers exciting opportunities for research, education, and practical applications to improve overall well-being (Gutte and Deshmukh 2023) (Fig. 1).

4 Need of Education for Sustainable Food and Nutrition

Becoming food system literate is a process that takes time and commitment to developing a suite of knowledge and skills that will allow effective working with others for a better understanding of food systems and how to intervene in them. Food systems literacy is not only concerned with gaining knowledge of the food system (i.e., the acquisition of facts), it is the acquisition of knowledge that will help to organize, investigate, and critique food system facts and values (Pope et al. 2021).

Food is a necessity for survival, but the food industry is significantly harming the environment. To tackle this problem, it is essential to include sustainable food and nutrition education in school curriculum, as this is the best way to educate younger generations.

Although there is a lot of research on sustainable development education, there is not enough information about sustainable food and nutrition education. Therefore, practical knowledge of how to incorporate sustainable food and nutrition habits in schools is currently inadequate. As a result, there is an urgent need for more information on this topic. Agriculture education needs to focus on sustainable foods as a priority sector.

Encouraging a more climate-friendly diet and sustainable eating habits can play a major role in reducing the carbon footprint of a school or community promote sustainable living habits for a brighter future (Tippmann 2020).

Here are some tips for creating sustainable eating habits:

By promoting sustainable eating habits, a school or community can significantly decrease its carbon footprint. To achieve this, individuals can opt for plant-based meals such as vegetables, fruits, whole grains, legumes, and nuts, which are more eco-friendly. Reducing food waste is also essential by encouraging people to take only what they need and compost and recycle their leftovers. By supporting local farmers and purchasing locally-grown food, individuals can reduce transportation emissions. Additionally, using minimal packaging and promoting the use of reusable containers can reduce packaging waste. Lastly, educating students about sustainable eating habits and their positive impact on the environment is crucial for encouraging long-term change. Choosing regionally produced and seasonal food is beneficial because it results in shorter transportation routes and less storage time, which ultimately reduces greenhouse gas emissions (Westmecklenburg and Oekom e. V. 2019) (Fig. 2).

The education system can integrate healthy nutritional planning by introducing the MyPlate program. This program emphasizes the importance of consuming a balanced diet that includes fruits, grains, vegetables, and proteins. Initially, students will acquire knowledge about the classification of foods into distinct food groups. Subsequently, they will engage in an activity that involves arranging food items on a plate based on their respective categories. Further, they will participate in exercises that facilitate the recognition of wholesome foods and determining appropriate portions. Additionally, they will have the opportunity to monitor their food

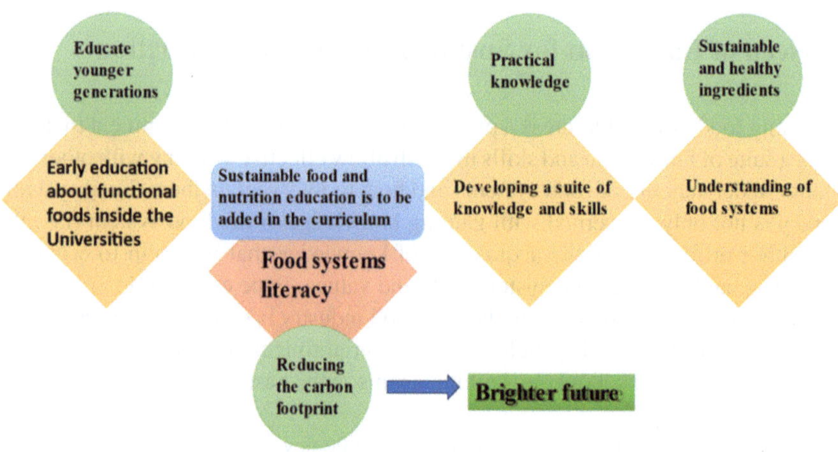

Fig. 2 Sustainable food and nutrition education

consumption and devise strategies for maintaining a healthy diet (U.S. Government's Website).

Varela-Candamio et al. (2018), This study delves into the generation, production, and consumption process of functional foods, which involves not just transferring knowledge about the health benefits of these products but also the multiple roles of rural women as producers, educators/advisors, and buyers of these foods. The involvement of rural women in the entire process of functional foods is crucial in building their competitive advantage as local entrepreneurs. The strategic formulation of green business models was strengthened using a dynamic perspective through the road mapping methodology, which included forming alliances with knowledge suppliers (researchers), identifying with reference groups (investors, clients), and managing intellectual capital (structural, human, relational). Ultimately, the study found that rural women are better positioned than others to establish businesses based on functional foods from a niche formulation (Varela-Candamio et al. 2018).

In the sustainable development sector, the utilization of nutraceuticals necessitates higher education to thrive. By placing students at the core of the learning process, this approach effectively reinforces the principles of education quality in an innovative manner, while also promoting perspectives aligned with Higher Education for Sustainable Development (HESD).

The impact of participation in co-creation has been shown in the context of environmental engineering programs at universities in Ecuador. This specific case has provided evidence that implementing strategies based on co-creation models is crucial for enhancing learning and teaching methodologies, especially within the framework of Higher Education for Sustainable Development (HESD) (Perello-Marín et al. 2018).

In 1992, approximately 31 years before, during the Earth Summit held in Brazil, the concept of Education for Sustainable Development (ESD) was acknowledged as a means to achieve a harmonious equilibrium between "human and economic well-being, cultural traditions, and the preservation of the Earth's natural resources" through educational initiatives (Barth and Michelsen 2013).

Agusdinata (2022) presented a framework and case study that illustrate a design and engagement process where the university assumes a leadership role in co-creating solutions for Sustainable Development Goals. The framework supports university-led initiatives by harnessing three key elements:

(i) recognizing the inherent synergies among SDGs,
(ii) employing various approaches for identifying, designing, and implementing solutions, and
(iii) involving stakeholders through different modes of engagement and interaction.

By integrating human-centered design (HCD) and shared-action learning (SAL), they documented a case study led by a large, public, research-oriented university, highlighting the active participation of diverse stakeholders in the co-creation process to discover innovative solutions (Agusdinata 2022).

5 Conclusion

To sum up, sustainable health is an essential aspect of a healthy lifestyle, and functional foods can aid in achieving this objective. By integrating plant-based proteins and a diverse range of fruits and vegetables into our diets, can decrease our dependence on animal-based products and reduce our environmental footprint. Food science and technology can help address problems for food production and waste. Additionally, selecting minimally processed foods in eco-friendly packaging can lead to a more sustainable lifestyle.

By implementing these minor adjustments, we can prioritize our health while also considering the planet, guaranteeing a healthier future for ourselves and future generations.

Green technologies, artificial intelligence, biotech, and agricultural advances help tame emissions, and waste while safeguarding limited resources. One must adopt technology for sustainable growth and an improved version of the food.

Effective collaboration in the food system requires individuals to possess certain knowledge and skills, which can be obtained through education and training. The best delivery methods for this education would need to be studied and prioritized for future implementation.

At the community level, we have an exciting opportunity for universities to collaborate and tap into the potential of Sustainable Development Goals (SDGs) by co-creating solutions and implementing them together. It's crucial for us to keep driving the agenda forward by designing and implementing even more effective solutions at the local community level. While focusing on SDG synergies, we also need to address the trade-offs that might arise between different solutions. One key trade-off is the balance between economic opportunities and environmental considerations. It is important to find a model that not only offers greater economic opportunities but also considers the impact on the environment. By carefully managing these trade-offs, we can create a holistic and sustainable approach to development (Agusdinata 2022).

6 Limitations

There are several limitations that may be associated with a study on sustainable health using functional foods. Some of these limitations include:

1. The results may not be representative of the larger population.
2. Longer-term studies are needed to determine the effectiveness and sustainability of incorporating functional foods into a person's diet.
3. Confounding factors: There may be other factors that can influence sustainable health outcomes, such as lifestyle habits, genetics, and environmental factors. Controlling for these confounding factors can be challenging, and their influence on the study outcomes may not be fully accounted for.

4. The results may not be applicable to other universities or educational institutions, or to individuals outside of the academic setting as there is very less research available on university point of view.
5. compliance and adherence to guidelines given in the research can be challenging, particularly in a university setting where students have diverse schedules and lifestyles. Non-compliance or incomplete adherence can impact the validity of the results.

Further research with larger sample sizes, longer durations, and rigorous study designs is needed to overcome these limitations and provide more robust evidence on the role of functional foods in promoting sustainable health.

References

Agusdinata DB (2022) The role of universities in SDGs solution co-creation and implementation: a human-centered design and shared-action learning process. Sustain Sci 17(4):1589–1604. https://doi.org/10.1007/s11625-022-01128-9

Arora NK (2019) Impact of climate change on agriculture production and its sustainable solutions. Environ Sustain 2(2):95–96. https://doi.org/10.1007/s42398-019-00078-w

Barth M, Michelsen G (2013) Learning for change: an educational contribution to sustainability science. Sustain Sci 8(1):103–119. https://doi.org/10.1007/s11625-012-0181-5

Beardsworth A, Keil T (1992) The vegetarian option: varieties, conversions, motives and careers. Sociol Rev 40(2):253–293. https://doi.org/10.1111/j.1467-954X.1992.tb00889.x

Berg P, Feber D et al (2020) The drive toward sustainability in packaging-beyond the quick wins. Mc Kinsey & Company

Betoret E, Barrera C, Seguí LN et al (2022) Sustainable strategies in the development of functional foods, Innovation Strategies in the Food Industry (Chap. 9), 2nd edn. Academic Press, Cambridge, MA, pp 145–156. https://doi.org/10.1016/B978-0-323-85203-6.00006-2. ISBN 9780323852036

Duralija B, Putnik P, Brdar D, Bebek Markovinović A, Zavadlav S, Pateiro M, Bursać Kovačević D (2021) The perspective of croatian old apple cultivars in extensive farming for the production of functional foods. Foods 10(4):708. https://doi.org/10.3390/foods10040708

De Zarn TJ (1995) Food ingredient encapsulation. In: Encapsulation and controlled release of food ingredients, CS symposium series, vol 590. American Chemical Society, Washington, DC, pp 74–86

Gallotti F, Lavelli V (2020) The effect of UV irradiation on vitamin d2 content and antioxidant and antiglycation activities of mushrooms. Foods 9(8):1087. https://doi.org/10.3390/foods9081087

German JB, Zivkovic AM, Dallas DC, Smilowitz JT (2011) Nutrigenomics and personalized diets: what will they mean for food? Annu Rev Food Sci Technol 2:97–123. https://doi.org/10.1146/annurev.food.102308.124147

Gutte RK, Deshmukh V (2023) Sectional study of nutritional psychology to identify the significance of the connection between mental health and nutraceutical functional ingredients. Funct Food Ingred Mental Health 1(5):1–13. https://doi.org/10.31989/ffimh.v1i5.1100

Hinestroza-Córdoba LI, Duarte Serna S, Seguí L, Barrera C, Betoret N (2020) Characterization of powdered lulo (*Solanum quitoense*) bagasse as a functional food ingredient. Foods 9(6):723. https://doi.org/10.3390/foods9060723

Jousse F (2008) Modeling to improve the efficiency of product and process development. Compr Rev Food Sci Food Saf 7(1):175–181. https://doi.org/10.1111/j.1541-4337.2007.00033.x

Kandylis P (2021) Grapes and their derivatives in functional foods. Foods 10(3):672. https://doi.org/10.3390/foods10030672

Khan RS, Grigor J, Winger R, Win A (2013) Functional food product development – opportunities and challenges for food manufacturers. Trends Food Sci Technol 30(1):27–37. https://doi.org/10.1016/j.tifs.2012.11.004, ISSN 0924-2244

Kruk ME, Gage AD, Arsenault C, Jordan K, Leslie HH, Roder-DeWan S, Pate M (2018) High-quality health systems in the sustainable development goals era: time for a revolution. Lancet Glob Health 6(11):e1196–e1252. https://doi.org/10.1016/S2214-109X(18)30386-3

Lavelli V, D'Incecco P, Pellegrino L (2021) Vitamin D incorporation in foods: formulation strategies, stability, and bio accessibility as affected by the food matrix. Foods 10(9):1989. https://doi.org/10.3390/foods10091989

Lonnie M, Johnstone AM (2020) The public health rationale for promoting plant protein as an important part of a sustainable and healthy diet. Nutr Bull 45(3):281–293. https://doi.org/10.1111/nbu.12453

Mark R, Lyu X, Lee JJL, Parra-Saldívar R, Chen WN (2019) Sustainable production of natural phenolics for functional food applications. J Funct Foods 57:233–254. https://doi.org/10.1016/j.jff.2019.04.008, ISSN 1756-4646

Martirosyan DM, Singh J (2015) A new definition for functional food by FFC: creating functional food products using new definition. Funct Foods Health Dis 5(6):209–223. https://doi.org/10.31989/ffhd.v5i6.183

Mehta N, Kumar P, Verma AK, Umaraw P, Kumar Y, Malav OP, Lorenzo JM (2022) Microencapsulation as a noble technique for the application of bioactive compounds in the food industry: a comprehensive review. Appl Sci 12(3):1424. https://doi.org/10.3390/app12031424

Mert ID (2020) The applications of microfluidization in cereals and cereal-based products: an overview. Crit Rev Food Sci Nutr 60(6):1007–1024. https://doi.org/10.1080/10408398.2018.1555134

Mortimer F, Isherwood J, Wilkinson A, Vaux E (2018) Sustainability in quality improvement: redefining value. Future Healthc J 5(2):88–93. https://doi.org/10.7861/futurehosp.5-2-88

Ogori AF, Amove JM, Aduloju P, Sardo G, Okpala COR, Bono G, Korzeniowska M (2021) Functional and quality characteristics of ginger, pineapple, and turmeric juice mix as influenced by blend variations. Foods 10(3):525. https://doi.org/10.3390/foods10030525

Perello-Marín MR, Ribes-Giner G, Pantoja Díaz O (2018) Enhancing education for sustainable development in environmental university programmes: a co-creation approach. Sustainability 10(2):158. https://doi.org/10.3390/su10010158

Peter N, Nicole C et al (2020) Front Sustain Food Syst Sec Agroecol Ecosyst Serv 4–2020. https://doi.org/10.3389/fsufs.2020.577723

Pope H, de Frece A, Wells R, Borrelli R, Ajates R, Arnall A, Häsler B (2021) Developing a functional food systems literacy for interdisciplinary dynamic learning networks. Front Sustain Food Syst 5:747627. https://doi.org/10.3389/fsufs.2021.747627

Putnik P, Bezuk I, Barba FJ et al (2020) Sugar reduction: Stevia rebaudiana Bertoni as a natural sweetener. In: Agri-food industry strategies for healthy diets and sustainability. Academic Press, London, UK, pp 123–152

Regionaler Planungsverband Westmecklenburg and Oekom e. V. (2019) Klima Sparbuch Westmecklenburg, 2019/20. Oekom Verlag, München

Roselló-Soto E, Barba FJ, Putnik P, Bursać Kovačević D, Lorenzo JM, Cantavella-Ferrero Y (2018) Enhancing bioactive antioxidants' extraction from "horchata de chufa" by-products. Foods 7(10):161. https://doi.org/10.3390/foods7100161

Sloan E (2022) Food technology magazine, Article magazine/issues/2022/April/features/top-10-functional-food-trends, 1 April 2022. https://www.ift.org/news-and-publications/food-technology

Šojić B, Tomović V, Kocić-Tanackov S, Kovačević DB, Putnik P, Mrkonjić Ž, Đurović S, Jokanović M, Ivić M, Škaljac S, Pavlić B (2020) Supercritical extracts of wild thyme (*Thymus serpyllum* L.)

by-product as natural antioxidants in ground pork patties. LWT 130:109661. https://doi.org/10.1016/j.lwt.2020.109661
Tachie C, Nwachukwu ID, Aryee ANA (2023) Trends, and innovations in the formulation of plant-based foods. Food Prod Process Nutr 5(1):16. https://doi.org/10.1186/s43014-023-00129-0
Takachi R, Inoue M, Ishihara J, Kurahashi N, Iwasaki M, Sasazuki S (JPHC Study Group) (2008) Fruit and vegetable intake and risk of total cancer and cardiovascular disease: Japan public health center-based prospective study. Am J Epidemiol 167(1):59–70. https://doi.org/10.1093/aje/kwm263
The European Parliament, & the Council of the European Union (2006) Regulation (EC) no 1924/2006 of the European Parliament and of the council on nutrition and health claims made on foods. Off J Eur Union 404:9–25
Tippmann M (2020) Education for sustainable food and nutrition – towards criteria for German secondary schools 3(1). Page/Article: 5. https://doi.org/10.5334/glo.28
U.S. Government's Website. http://www.choosemyplate.gov
Varela-Candamio L, Calvo N, Novo-Corti I (2018) The role of public subsidies for efficiency and environmental adaptation of farming: a multi-layered business model based on functional foods and rural women. J Clean Prod 183:555–565. https://doi.org/10.1016/j.jclepro.2018.02.109, ISSN 0959-6526
Versino F, Ortega F, Monroy Y, Rivero S, López OV, García MA (2023) Sustainable and bio-based food packaging: a review on past and current design innovations. Foods 12(5):1057. https://doi.org/10.3390/foods12051057
Yang M, Chen L, Wang J, Msigwa G, Osman AI, Fawzy S, Yap P (2023) Circular economy strategies for combating climate change and other environmental issues. Environ Chem Lett 21(1):55–80. https://doi.org/10.1007/s10311-022-01499-6
Zaman QU (2023) Precision agriculture technology: a pathway toward sustainable agriculture (Chap. 1). In: Precision agriculture. Academic Press, Cambridge, MA, pp 1–17. https://doi.org/10.1016/B978-0-443-18953-1.00013-1, ISBN 9780443189531

Using Fashion as a Medium to Engage Business Students with Sustainability

Ann Hindley, Konstantina Skritsovali, and Katherine Neary

Abstract This chapter will provide examples of sustainability projects, practices and initiatives in a North-West England Business School that is a Principles for Responsible Management Education (PRME) signatory. Examples include such as fast-fashion, production processes, pollution, waste, the circular economy, technological advancements, and changes to human behaviour.

Keywords Fashion · SDGs · PRME · ESD · Engagement

1 Introduction

Although Education for Sustainable Development (ESD) has been criticised as vague, it has supported the Sustainable Development Goals (SDGs) as a key enabler in education policy and the curriculum (UNESCO 2019). Knowledge and understanding of sustainability are critical to graduate futures; to agency (Fang and O'Toole 2023), work-readiness and career-resilience (Borg et al. 2021). Essential to teaching and learning is the delivery of sustainability content via a vehicle that facilitates the scaffolding of learning and student engagement.

As an everyday essential, students have knowledge and experience from browsing or purchasing clothing and accessories, making fashion a highly suitable vehicle to carry sustainability messages. Over the past few decades, the fashion industry has witnessed a substantial increase in clothing production, accompanied by a decline in profit margins and a heightened environmental impact. From 2000 to 2015, the production of garments doubled, but the utilization rate, which refers to the number of times clothing is worn before being discarded, decreased by 36% during the same period (Ellen MacArthur Foundation 2023).

The question of how fashion can be used as a medium to engage business students with sustainability, is the focus of this chapter. It begins with a review

A. Hindley (✉) · K. Skritsovali · K. Neary
Liverpool Business School, Liverpool John Moores University, Liverpool, UK
e-mail: a.m.hindley@ljmu.ac.uk

W. Leal Filho et al. (eds.), *The Contribution of Universities Towards Education for Sustainable Development*, World Sustainability Series,
https://doi.org/10.1007/978-3-031-49853-4_26

of literature on sustainability agendas, pedagogical approaches, engagement, and behaviour change. This is followed by a qualitative descriptive approach that explores delivery of and student reaction to sustainability and fashion, both in the formal and informal curriculum. This includes vignettes that illustrate content delivery aims and the opportunities for transformational learning and action; balancing negative impacts with positive solutions or actions that assist in reducing cognitive dissonance (Festinger 1962a, b). The chapter finishes with limitations and conclusions with recommendations for future research.

2 Sustainability Agendas

As part of developing a sustainability curriculum it is important to outline and consider key sustainability terminology and agendas. Explaining the role and objectives of current agendas to students forms part of the framework necessary to understand sustainability challenges and change seeing activities. In the area of business and management studies the following five initiatives are of relevance.

Education for Sustainable Development (ESD)

As part of the ongoing learning process, it is useful to incorporate the ESD pedagogy as a foundation of sustainability teaching. The focus on action, enablement and innovation helps when planning a progressive and purposeful curriculum (UNESCO 2020 p14). As part of this process the learning environment, capacity of educators are significant (UNESCO 2020, pp. 28–30). The ESD website provides a diverse range of resources to support evidence-based teaching, linked to the broader UN sustainability agenda (UNESCO 2023).

Sustainable Development Goals (SDGs)

For a period of 15 years, from 2015 to 2030, the 193 member states of the UN signed up to the UN SDG agenda, which aims to achieve progress in addressing global sustainability challenges (UN 2023). The 17 SDGs provide a comprehensive framework to introduce students to the complex and interconnected nature of global sustainability challenges. The SDG framework and connected resources are well suited to support independent student-led study.

PRME

The Principles of Responsible Management Education (PRME) aim to support Business and Management Schools in promoting sustainability as part of their practice. PRME is a UN Global Compact initiative and thus has existing synergy with the SDG framework. An established range of networks, resources and events are available for Business Schools who join this global community (PRME 2023). Participating in and engaging with PRME is useful to align teaching and student activity with the 6 PRME principles, which link into the wider SDG agenda in terms of research, teaching and business school organisational culture.

Circular Carbon Economy (CCE)

Distinct from Climate Change Education (CCE) is Circular Carbon Economy (CCE), which is the focus here. Circular Carbon Economy (CCE) emphasises the importance of committing to a new process, in pursuit of global sustainability. The principles focus on reaching climate goals by encouraging stakeholders to change their behaviour to minimise carbon emissions (KAPSARC 2020). Within sustainability teaching, its inclusion helps students to connect the effect of stakeholder decisions and the compounding impact to the planet and those who live here. The resource base provides interactive visualisation tools that are particularly powerful teaching tools embedding an appreciation for the importance of reducing carbon (CCE 2023a, b).

Systems Thinking

When situating these concepts and initiatives within sustainability teaching it is important to provide students with a relatable academic structure. One useful approach is to applying systems thinking as a method of showing the interconnected relationships and impacts between stakeholder groups (Williams et al. 2017). This enables students to begin to understand the impact of different types of production and method of trade within fashion global value chains. Within the scope of business teaching, it is then interesting to introduce alternatives, and then consider the impact on the broader, often complex stakeholder map as part of a systematic approach to these complex challenges (Gregory and Miller 2014). This approach helps the student to understand the power and influence they have as consumers. It is this placement of self, and the visual connectivity to other stakeholders which can then be used begin a process of critical self-reflection on purchasing choices and their environmental and social impacts.

3 Pedagogical Approaches

Educational bodies have published frameworks and quality standards to assist individuals in acquiring the knowledge, skills, and attitudes necessary to live, work, and act sustainably. ESD highlights the responsibility of educational institutions to tackle social, economic, and environmental problems by incorporating sustainable practices into their curricula (Rieckmann 2018). The purpose of ESD in higher education is to help students develop general sustainability skills. This includes fostering systems thinking and future thinking, cultivating problem-solving; preparing them to deal with uncertainty and working collaboratively; appreciating different cultures, and being empathetic (Bianchi et al. 2022).

In the Education for Sustainable Development in Higher Education guide, the QAA and Advance HE provide a set of guidelines that can be used in curriculum design, teaching, learning and assessment approaches (QAA 2021). The Green Competences Framework (GreenComp) developed by the European Commission introduces 12 competences organised into four areas that can help learners become empathetic, responsible and caring world leaders (Bianchi et al. 2022). These frameworks share a common belief that tutors hold responsibility in arranging and executing their teaching approach to equip students with the necessary knowledge, skills, and competencies to address sustainability concerns.

Acknowledging diversity within teaching cohorts it is important that tutors are open to different pedagogical approaches which should be based on the teaching and learning objectives and the specific educational context. For instance, Cotton and Winter (2010) suggested several pedagogical approaches, such as role-plays and simulations, group discussions, stimulus activities, debates, critical incidents, case studies, reflexive accounts, personal development planning, critical reading and writing, problem-based learning, fieldwork, and modelling good practice. Lozano et al. (2017) synthesized 12 pedagogical approaches grouped into three categories: Universal approaches that have been used across various disciplines and contexts; Community and social justice approaches specifically designed to address social justice and foster community-building and, environmental education approaches rooted in environmental sciences and practices.

4 Engagement and Behaviour Change

There is general understanding that motivation is an antecedent to student engagement, with the more empowered and engaged students channelling energy into learning (Bond et al. 2020). Student engagement is a multidimensional phenomenon (Fredricks et al. 2004) with four perspectives dominating: behavioural, psychological, socio-cultural and holistic (Kahu 2013; Kahu and Nelson 2018). An individualised experience, engagement is influenced by context (Aparicio-Flores et al. 2020). By example, traditional lectures as a method of delivery can be perceived to be less engaging, perhaps even leading to academic boredom (Mann and Robinson 2009; Sharp et al. 2018).

If indicators of engagement are involvement, participation and interaction (Bond et al. 2020), then drowsiness, yawning, slouching and vacant stares are clear indicators of academic boredom (Sharp et al. 2021). This boredom leads to a surface approach to learning which can result in lower student achievement (Trigwell et al. 2012). Therefore, it is no surprise that the most frequently identified research domain is educational technology (Bond et al. 2020). Edutainment multimedia is used to attract attention, create emotional involvement and engage audiences with real-life situations that aims to achieve self-efficacy and personal growth (De Fossard 2008).

Multimedia entertainment includes passive and active experiences such as reading books, watching videos, listening to music, using hypertext and playing games (Cacciaguerra et al. 2006). Guest lectures, field trips and videos, for example, can provide the real-life examples that De Fossard (2008) suggests builds empathy for others and shows how behaviour can be accomplished. This usefully engages attention and allows theory to be introduced at a later point (De Fossard 2008). Within a sustainability strategy game Emblen-Perry (2018) embeds the expectations that students will challenge their own values, but this does depend on the exposure to issues within the game.

Even where engagement is achieved, this does not necessarily lead to behaviour change as this can be difficult and unpleasant (Festinger 1962a, b). Any feelings of incompetence or immorality (Rokeach 1973), misconceptions or ideas inconsistent with personal beliefs and values result in a state of dissonance (Festinger 1962a, b). This negative arousal is more likely to arise when linked to self-concept (Aronson 1969) and ego-involvement, where an attitude or identity is tied to self-concept (Carpenter 2019). As "consistency with self-esteem is probably a more compelling consideration than consistency with logic or reality" (Rokeach 1968, p. 20), the challenge is to harness dissonance to nudge students towards behaviour change (McGrath 2020).

Vignette 1—Sustainability and the Circular Economy

A Sustainability and Circular Economy module is delivered to Level 5 business management undergraduates. Key theory from Carroll (2016)[1], Elkington (2018)[2] and Webster[3] underpins the module. Models are examined in the context of fashion with news, documentaries, reports and campaigns providing currency and vibrancy. The SDGs are linked to content across the module. The aim is to harness cognitive dissonance to change behaviour of individuals and for these (as consumers or employees) to recognise how they could influence businesses to change their behaviour.

Lessons Learnt

The fashion theme is used as everyone wears clothes and therefore, has an opinion. The module content includes real-life fashion industry examples. It builds empathy with others, such as industry employees or local community. A wide mix of multimedia is included to create engagement and participation. Negative industry impacts are balanced with positive solutions to show how behaviour can be accomplished.

SDG1 No Poverty

Negative: Visuals from Clean Clothes Campaign show a breakdown of costs for a t-shirt made in Bangladesh illustrating how low wages keep workers in poverty.[4]

---

[1] Carroll, A. B. (2016). Carroll's pyramid of CSR: taking another look. *International journal of corporate social responsibility*, *1*(1), 1–8.

[2] Elkington, J. (2018). 25 years ago I coined the phrase "triple bottom line." Here's why it's time to rethink it. *Harvard business review*, *25*, 2–5.

[3] Webster, K. (2021). A circular economy is about the economy. *Circular Economy and Sustainability*, *1*(1), 115–126.

[4] https://cleanclothes.org/poverty-wages.

Positive: Fair Trade fashion producer People Tree aims to "support producer partners' efforts towards economic independence and control over their environment and to challenge the power structures that undermine their rights to a livelihood".[5]

SDG2 Zero Hunger

Negative: In a survey of workers during the Covid pandemic, 60% had worked their usual hours but had nonetheless received pay cuts, leading to 80% having missed meals to feed their children.[6]

Positive: Agraloop™ and BioFibre™use left-overs from the food and medicine industries, with the stems and leaves of crops such as hemp, flax, banana and pineapple being spun into yarn.[7]

SDG3 Good Health and Wellbeing

Negative: The case of Rana Plaza in Bangladesh and the death of 1,134 people is examined, with recent reports in The Guardian suggesting there are continuing issues with worker safety.[8]

Positive: A European Environmental Bureau report seeks a wellbeing economy approach that benefits workers, consumers, and those dealing with fashion waste.[9]

SDG4 Quality Education

Negative: Children who are in the supply chain and how this affects their schooling is examined via UNICEF[10] and consideration of Modern-Day Slavery via the International Labour Organisation.[11]

Positive: Mango and the charity Save the Children have joined together to promote education in Bangladesh that will benefit around 4000 women and children.[12]

---

[5] https://www.peopletree.co.uk/about-us.

[6] https://www.workersrights.org/wp-content/uploads/2020/11/Hunger-in-the-Apparel-Supply-Chain.pdf.

[7] https://circularsystems.com/agraloop.

[8] https://www.theguardian.com/world/2023/apr/24/10-years-on-bangladesh-rana-plaza-disaster-safety-garment-workers-rights-pay.

[9] https://eeb.org/wp-content/uploads/2022/03/Summary-Wellbeing-Wardrobe-A-wellbeing-economy-for-the-fashion-and-textile-sector-March-2022.pdf.

[10] https://www.unicef.org/protection/child-labour.

[11] https://www.ilo.org/global/about-the-ilo/newsroom/news/WCMS_855019/lang--en/index.htm.

[12] https://fashionunited.com/news/culture/mango-and-save-the-children-join-forces-to-promote-education-and-sport-in-bangladesh/2023022852536.

SDG5 Gender Equality

Negative: Fashion Revolution[13] references the gender wage gap and sticky floor, while The Guardian investigates women forced to have sex with male supervisors to keep their jobs.[14]

Positive: The Clean Clothes Campaign is committed to challenge gender inequality as women have little voice and are denied the right to join a union.[15]

SDG6 Clean Water and Sanitation

Negative: The film RiverBlue investigates toxic chemicals released into waterways resulting in chemical foam and rivers coloured by dye that children swim in and people water their crops with.[16]

Positive: In 2022 Fashion for Good launched the D(r)ye Factory of the Future Project to accelerate a shift from wet to dry dying, aiming for a 95% decrease in water consumption.[17]

SDG7 Affordable and Clean Energy

Negative: According to Changing Markets, as demand for fossil-fuelled transport reduces the oil companies are increasingly seeing synthetic textiles as a future revenue stream.[18]

Positive: The International Wool Textile Organisation promotes the use of the natural fibre wool over fossil-fuel based textiles,[19] although the Ethical Consumer has wool on its best to avoid list.[20]

SDG8 Decent Work and Economic Growth

Negative: The exploitative nature of fast fashion retailer Boohoo is now usefully available as a case study by Goddard and Freeman (2023).[21]

---

[13] https://www.fashionrevolution.org/the-fashion-industry-needs-to-break-with-its-gender-and-womens-rights-problems/.

[14] https://www.theguardian.com/news/2020/aug/20/fashion-industry-jeans-lesotho-garment-factory-workers-sexual-violence.

[15] https://cleanclothes.org/about/gender-policy.

[16] https://riverbluethemovie.eco/.

[17] https://fashionforgood.com/our_news/fashion-for-good-launches-drye-factory-of-the-future-project-%E2%80%8E/#:~:text=Fashion%20for%20Good%20launches%20the,wet%20to%20mostly%20dry%20processing.

[18] https://www.changingmarkets.org/wp-content/uploads/2021/01/FOSSIL-FASHION_Web-compressed.pdf.

[19] https://iwto.org/wear-wool-not-fossil-fuel/#:~:text=Synthetic%20fibres%2C%20which%20include%20polyester,the%20total%20global%20fibre%20market.

[20] https://www.changingmarkets.org/wp-content/uploads/2021/01/FOSSIL-FASHION_Web-compressed.pdf.

[21] Nassar, M., Goddard, T., & Freeman, R. (2023). Walk the Talk: The Boohoo Case Study. *Journal of Macromarketing*, 02761467231164960.

Positive: B-Corp Certification is holistic and meets high standards of verified accountability, transparency and performance on employee benefits.[22]

SDG9 Industry, Innovation and Infrastructure

Negative: The United Nations Economic Commission for Europe identifies cotton farming as responsible for 24% of global insecticides and 11% of global pesticides.[23]

Positive: Outdoor adventure brand Patagonia offers do-it-yourself guides on the repair of clothes, spare parts or gives the opportunity to request a repair free of charge.[24]

SDG10 Reduced Inequalities

Negative: Forced labour produced cotton and cotton-based goods from the Uyghur region are entering international supply chains.[25]

Positive: Community Clothing is a social enterprise that creates employment with partners in some of the most deprived former textile areas of the UK.[26]

SDG11 Sustainable Cities and Communities

Negative: The Ghanian city of Accra has traded in "Obroni Wawu" or 'dead white man's clothes' for two decades[27] leading to toxic landfill.[28]

Positive: Australia's textile, clothing and footwear (TCF) industry explores the role between textiles, clothing, waste management and sustainable cities.[29]

SDG12 Responsible Consumption and Production

Negative: According to Ethical Consumer, 75% of Gen Zs believe influencers are somewhat accountable for disposable fashion rising and ultra-cheap garments such as Missguided's £1 bikini.[30]

Positive: The interactive Fashion Footprint Calculator from thredUP[31] measures the carbon, energy and water footprint for each wear of a garment.[32]

---

[22] https://www.bcorporation.net/en-us/certification.

[23] https://unece.org/forests/events/sdg-9-fashion-industry-innovation-and-infrastructure.

[24] https://eu.patagonia.com/gb/en/repairs/.

[25] https://www.shu.ac.uk/helena-kennedy-centre-international-justice/research-and-projects/all-projects/laundered-cotton.

[26] https://communityclothing.co.uk/pages/about-us.

[27] https://www.youtube.com/watch?v=bB3kuuBPVys.

[28] https://www.abc.net.au/news/2021-08-12/fast-fashion-turning-parts-ghana-into-toxic-landfill/100358702.

[29] https://ausfashioncouncil.com/sdg-11-cities-and-fashion-part-1/.

[30] https://www.ethicalconsumer.org/fashion-clothing/what-fast-fashion-why-it-problem.

[31] https://www.thredup.com/fashionfootprint/.

[32] https://www.thredup.com/bg/p/fashion-footprint-sources.

SDG13 Climate Action

Negative: The Synthetics Anonymous Report (Changing Markets Foundation)[33] tackles greenwashing, suggesting green claims are flouted at a rate of 96% for H&M, 89% for ASOS and 88% for M&S.[34]

Positive: The Fashion Industry Charter for Climate Action[35] envisions achieving net zero emissions by 2050.[36]

SDG14 Life Below Water

Negative: The European Environment Agency estimates up to 500,000 tonnes of synthetic microfibres enter the global marine environment annually, usually from washing processes.[37]

Positive: Fish skin, a by-product of the food industry, is an alternative sustainable raw material for fashion and enhanced innovation and sustainable design of fish-leather products is discussed.[38]

SDG15 Life on Land

Negative: While most of Shein's returns end up in landfill,[39] Burberry received a backlash from incinerating £28.6 m of accessories, perfume and unsold clothes in 2017.[40]

Positive: The Italian town of Prato has been recycling clothing for over 100-years that can be turned into thread for new garments.[41]

SDG16 Peace and Justice

Negative: For suppressing negative online reviews, the US fast fashion retailer Fashion Nova was fined $4.2 m by the Federal Trade Commission.[42]

---

[33] https://changingmarkets.org/.

[34] https://changingmarkets.org/wp-content/uploads/2021/07/SyntheticsAnonymous_FinalWeb.pdf.

[35] https://unfccc.int/sites/default/files/resource/Fashion%20Industry%20Carter%20for%20Climate%20Action_2021.pdf.

[36] https://unfccc.int/climate-action/sectoral-engagement/global-climate-action-in-fashion/about-the-fashion-industry-charter-for-climate-action.

[37] https://www.eea.europa.eu/publications/microplastics-from-textiles-towards-a/microplastics-from-textiles-towards-a.

[38] https://www.springerprofessional.de/en/sdg-14-life-below-water/17384708.

[39] https://www.theguardian.com/fashion/2022/apr/10/shein-the-unacceptable-face-of-throwaway-fast-fashion.

[40] https://www.bbc.co.uk/news/business-44885983.

[41] https://www.bbc.co.uk/news/av/world-europe-55267992.

[42] https://www.ftc.gov/news-events/news/press-releases/2022/01/fashion-nova-will-pay-42-million-part-settlement-ftc-allegations-it-blocked-negative-reviews.

Positive: The International Labour Organisation Convention No. 190 is the first international treaty to recognise the right of all to work free violence and harassment.[43]

SDG17 Partnerships for the Goals

Negative: With poor oversight by other governments there are limited ways to police factories that farm out work or understate problems during audits because workers are coached.[44]

Positive: Organisations and initiatives working towards change Textiles 2030 (WRAP)][45]; Ellen MacArthur Foundation[46]; UN Alliance for Sustainable Fashion.[47]

Vignette 2—Hair Extensions and Dead White Man's Clothes

Teaching sustainability to students who have varying levels of interest in the subject can be challenging. Capturing interest and attention is necessary as part of a process which seeks to inform, explore and promote critical reflection. The choice of content is an important consideration when thinking about the learning outcomes you are seeking to achieve. The fashion context provides a valuable landscape to connect students with a number of social and environmental sustainability issues. There are a number of thought-provoking options, which are likely to be new, yet relatable for students.

The notion of stakeholder responsibility is an important part of the ESD agenda. The use of case studies, like those outlined here support the teaching process by helping students situate themselves and the brands they value within global sustainability challenges. Igniting an awareness of, and passion for increased accountability, fairness and responsible processes is a significant learning outcome which supports academic skills development and future employability.

Lessons Learnt

Case study examples, especially those where a person or group of people share their story were welcomed by students. The opportunity to learn about, and form a connection to another person, and then begin to understand their role in relation to a product they, as a consumer, bought was powerful. Importantly, their age and consumer preferences were significant. Students were more interested and responsive to the examples which follow, than content on coffee production, for example, as many of them weren't coffee drinkers.

---

[43] https://www.ilo.org/global/topics/violence-harassment/lang--en/index.htm#:~:text=ILO%20Convention%20No.,gender%2Dbased%20violence%20and%20harassment.

[44] https://www.nytimes.com/2013/12/23/world/americas/buying-overseas-clothing-us-flouts-its-own-advice.html.

[45] https://wrap.org.uk/taking-action/textiles/initiatives/textiles-2030.

[46] https://ellenmacarthurfoundation.org/.

[47] https://unfashionalliance.org/.

Hair Extensions

A case study example of the sourcing patterns for hair extensions proved to be relatable and somewhat uncomfortable for students. Students either directly bought or knew someone who used human hair extensions. This previously overlooked purchasing decision provided opportunities to explore historic injustices linked to the sourcing of human hair. This example from Refinery 29 (2019) illustrated issues linked to poverty and power, producing a strong objection from students to the eventual distribution of profits. However, it effectively conveyed the human interconnectedness of our purchasing decisions.

The use of video facilitated student understanding of the impact on women selling their hair and space for reflection on the tangible impact of students' own fashion-based purchasing patterns. The story telling approach within the documentary enabled students to see reasons behind stakeholder behaviour. It introduced the importance of choice, fairness and financial payment for goods traded. Following the hair sourcing process and financial transactions enabled students to consider why the price consumers pay is important to others in the supply chain. The inclusion of human emotion as part of this transaction was powerful. The value of what is being sold, at each stage of the journey from the source to the consumer took on a deeper level of meaning. In a society where students access and buy products easily, seeing how money received by those at the beginning of the global value chain was important. Specifically, how the financial renumeration impacted life choices and opportunities for necessities like food, education, and healthcare.

Focusing student's attention on the importance of the trading process, builds on their interest in the quality, price and aesthetics of what they buy and ultimately how it makes them feel. Thus, the value students place on products they relate to enables the teaching process to become transformational and reflective. Students see the product they have felt on their own body or others as something that holds value in different ways. The importance of the product components is extended, and the price is no longer just about how much it personally costs them. This builds interconnectedness, from the micro level through to macro and then back to individual purchasing choices.

The importance of how business is conducted, how stakeholders make decisions and how these decisions are informed leads into the role of global sustainability campaigns like the SDG's. Once students understand the role and importance of systems thinking, discussions in seminars become dynamic, enthusiastic, and critically reflective. Enthusiasm to know more is a powerful motivator and students can apply their initial understanding and extend their critical reflection through connections with scholarly work and thought-provoking articles during independent study. This in turn exposes students to examples of academic writing and areas of debate.

Lessons Learnt

As part of an undergraduate education, understanding how to use both contextual and academic sources is important for the successful completion of assessments, and achieving learning outcomes. It can be challenging to do this, when the scope and complexity of global issues is either not sufficiently understood, or if the student feels unable to compartmentalise this complexity prior to discussing certain elements within their assessments.

Dead White Man's Clothes

Often, when encouraging students to think and write critically, it can be difficult to foster the necessary feelings which precede original assessment content and engagement with appropriate sources. The area of textile waste generated by the fast fashion industry provides an additional example for students to understand the interconnectedness of people involved in the fashion industry. When teaching, maintaining engagement with subject matter through resources which create a level of personal connectivity between people impacted by the lifecycle of a fashion product is an important stage in the learning process.

Made for television documentaries provide a relatable resource to educate students about the impact of their purchasing patterns and appetite for new garments (e.g. ABC News In-Depth 2022). These show the reality of textile waste exports to sub-Saharan Africa from America, Australia and Europe, with students witnessing the local community emotion associated with the consistent and unsustainable volumes of fast fashion waste. The small proportion of textile waste that is suitable for re-sale is disturbing and this unease holds student interest as the identifiable brands, issues of over production, and de-compostable plastic-based fibres are connected to multiple SDG's in the areas of water, land and air pollution and negative social, cultural economic impact.

Using fashion as a sector to understand global sustainability issues is an effective way to show how a product changes in terms of value and impact throughout its life cycle. From the way in which raw materials are sourced, the product is manufactured and the scale at which it is produced can be seen to create positive and negative social and environmental impacts. Furthermore, the visual impact of the scale of the issue shown through the film captures student's attention, promoting critical thinking.

When this is appreciated, the mission behind current sustainability campaigns including the SDG's and CCE become valued as part of the interconnected systems and the impact on stakeholders within the fashion industry.

ABC News In-Depth (2022). *Dead White Man's Clothes: The Environmental Disaster that is Fuelled by Used Clothes and Fast Fashion.* YouTube (Online Video). Available from https://youtu.be/bB3kuuBPVys (Accessed 1/6/23).

Refinery 29 (2019). *The truth about where hair extensions come from.* YouTube (Online Video). Available from https://youtu.be/VlZ1SWLBfPE (Accessed 1/6/23).

Vignette 3—Experiential Learning

In February 2023, 16 final-year undergraduate Business students participated in a four-day field trip to Valencia, Spain. The students were taking a module in Corporate Social Responsibility and were introduced to notions such as social innovation, frugal innovation, responsible marketing, responsible leadership, UN SDGs and many more. Whilst students had a theoretical understanding of the topic, they had never experienced how contemporary businesses and institutions engage with social innovation in day-to-day business. Interestingly, students were among the cohorts whose university life was affected by the lockdown imposed by the Covid-19 pandemic. As a result, they had very limited opportunities for face-to-face internships and placements or socializing with their peers. Therefore, the trip aimed to allow students to explore the application of social innovation practices in the Spanish business context.

Valencia was selected due to the multiple collaborations of the local university with local businesses and communities on areas such as social innovation and, inclusivity and diversity in the workplace. As a result, participating students would not only be immersed in a cultural, international experience but also experience the application of the topics in real life. The trip and the agenda were co-organised and agreed upon between the host and the visiting universities, ensuring consistency between the teaching syllabus, the student background and the organised visits and talks. The agreed agenda consisted of an introductory lecture designed to welcome students and introduce them to the culture and history of the city. That was later followed by visits to local social enterprises, private and public organisations, all of which are pioneers in social innovation and social responsibility (UN SDG 9-Industry, Innovation and Infrastructure; UN SDG 12 Responsible Consumption and Production). In addition, visits were arranged to give students the opportunity to take part in culinary and cultural experiences that were organised to complement the internationalisation of the student experience.

Lessons Learnt

Pre-Trip

Since participating students were selected from three different business programmes, some had never met with others throughout their studies. Students were therefore invited to three meetings prior to the trip, with the core aim to create a sense of community. During the meetings, students were familiarised with the trip agenda and expectations were set. As a matter of health and safety, tutors and students joined an online channel of communication that was used to share information about the trip, ask questions and support each other before, during and after the trip. A pre-trip survey was shared with students, and they were encouraged to anonymously share their expectations from the trip, highlight activities that stood out when reviewing the trip agenda and share feedback to allow organizing tutors to make any last-minute adaptations to the itinerary. For instance, questions such as ‘What are your expectations from this trip?’; ‘What were the main interests/incentives for you to join the trip?’; ‘Are there any activities that stood out for you?’ were some of the areas that students were invited to reflect on.

During-Trip

On the first day of the trip, students attended a welcome lecture at the local university. This was an introduction to the social, historical, and cultural underpinnings of the city. Later, they visited on-site projects on sustainable sports design and, innovation in sustainable transportation. During the second day of the trip, they visited a local national park where an electricity supplier introduced them to initiatives adopted to preserve wildlife (UN SDG 7 Affordable and Clean Energy, UN SDG 15 Life on Land). The day ended with a visit to a food cooperative that has recently introduced artificial intelligence to achieve sustainability and efficiency in their supply chain (UN SDG 9 Industry, Innovation and Infrastructure). On day three, the group visited the local port to attend a presentation on the challenges of international trade and its implication on life below water (UN SDG 14 Life Below Water). In the afternoon, they attended a talk at the Stock Exchange Building on responsible finance and accountability. During the final day, the group visited a pioneering fashion supplier that designs and produces jeans sustainably and responsibly, before visiting a local social enterprise that helps disadvantaged people and social groups to reintegrate into society (UN SDG 1 No Poverty, UN SDG 2 No Hunger, UN SDG 8 Decent Work and Economic Growth).

During the visit, representatives of the host company delivered a presentation of the business-as-usual practices that often take place in conventional fashion supply chains and provided examples of the implication on people and the natural environment. They then showcased how they have been applying technology and artificial intelligence to design and produce net zero jeans. As a result, the field trip complemented the existing knowledge and understanding on the impact of the fashion industry supply chain on people and the planet. Students agreed that the visit to the fashion manufacturing plant was the most thought provoking and this was due to their ability to relate to the product.

Post-Trip

After the end of the field trip, students were invited to a debriefing session that encouraged reflection on their learning and overall experience. The session included but was not limited to sharing the experiences and insights gained at the trip. During the session students were invited to consider their original motivations to join the trip. To support discussions, a survey was conducted to gain student perceptions about the field trip. Similarly, organising tutors joined a debriefing meeting to evaluate the trip outcomes and consider areas of improvement for future trips.

Student reflections on the trip highlighted the impact that experiential learning has on learners. Students reported that the experience, knowledge and understanding of the applications of social innovations and sustainability on business they gained during the trip were much more powerful than many modules that they attended in the classroom during the last year of their studies. Their justification was that the actual visits to businesses and the opportunities to witness first-hand how sustainability and social innovation is applied in business, led in memorable experiences. Many also claimed that the way they perceived of impact and business success changed. For instance, some of the students claimed that they had always perceived success based on financials. However, witnessing success, innovation and impact from a triple bottom line perspective, made them revisit and reconsider their views.

5 Limitations

Supporting academic knowledge and experience, anecdotal evidence from students is often used to help inform module planning and experiential activities. By example, the qualitative descriptive approach allows the vignettes to outline features and benefits of sustainability using fashion examples that have engaged students. However, although such anecdotal evidence can provide stories and testimonies, it is biased here by its small sample size in comparison to the overall cohort size. It is further noted that evidence-based decision-making or prior beliefs on sustainability or fashion could be influencing students and their level of engagement or behaviour. These factors have not been explored in this chapter but are worthy of future study.

6 Conclusions

For ESD and other sustainability agendas to be achieved, pedagogical approaches in higher education need to encourage student engagement with sustainability, which aims for transformational change at the personal (at home) and employee (at work) level. As future leaders and managers with decision-making roles or as peer-level influencers, for example, our students need to be aware of actions needed to enhance positive impacts and reduce negative impacts on planet and people. As all students wear clothing, they have knowledge of fashion from the consumer and retail perspective, making the fashion industry a highly suitable topic for providing engaging opportunities to illustrate environmental, social and governance impacts arising from fashion global value chains. A rich vein of resources can be combined with the SDGs to connect broader contextual and academic resources, while balancing negative impacts with positive action. The vignettes highlight opportunities to illustrate sustainability through the medium of fashion and the flexibility offered as one part of a module or course; across an entire module or course; or as part of an international experiential fieldtrip.

Future empirical research is recommended to further explore student learning and behaviour change relating to sustainability and the use of fashion as a medium to engage business students. Notwithstanding the biases in using anecdotal evidence, rigorous scientific analysis that samples the entire cohort about sustainability learning and behaviour within and outside of the university is needed. Evidence-based decision-making and prior beliefs on sustainability or fashion need to be considered as factors influencing students and their level of engagement or behaviour. Furthermore, the long-term impact of learning and behaviour change needs to be studied beyond the undergraduate degree. By example, considering the relationship or influence they have with sustainability in their employment or their personal lives.

References

Aparicio-Flores MP, Esteve-Faubel JM, Esteve-Faubel RP & Alvarez-Teruel JD (2020) High academic self-efficacy and dispositional empathy in future teachers. Sustainability 12(17). https://doi.org/10.3390/su12176728

Aronson E (1969) The theory of cognitive dissonance: a current perspective. In: Advances in experimental social psychology, vol 4, pp 1–34. https://doi.org/10.1016/S0065-2601(08)60075-1

Bianchi G, Pisiotis U, Cabrera Giraldez M (2022) GreenComp the European sustainability competence framework. Publications Office of the European Union: Luxembourg. https://doi.org/10.2760/13286

Bond M, Buntins K, Bedenlier S, Zawacki-Richter O, Kerres M (2020) Mapping research in student engagement and educational technology in higher education: a systematic evidence map. Int J Educ Technol High Educ 17(1):1–30. https://doi.org/10.1186/s41239-019-0176-8

Borg J, Borg N, Scott-Young CM, Naderpajouh N (2021) The work readiness–career resilience linkage: implications for project talent management. Int J Manag Proj Bus 14(4):917–935. https://doi.org/10.1108/IJMPB-04-2020-0129

Cacciaguerra S, Roccetti M, Salomoni P (2006) Multimedia entertainment applications. In: Furht B (ed) Encyclopedia of multimedia. Springer US, Boston, MA, pp 510–518

Carpenter CJ (2019) Cognitive dissonance, ego-involvement, and motivated reasoning. Ann Int Commun Assoc 43(1):1–23. https://doi.org/10.1080/23808985.2018.1564881

CCE (2023a) Technology Mindmap Wheel. CCE. https://www.cceguide.org/technologies/. Accessed 31 May 2023

CCE (2023b) CCE Visualiser: Scenarios. CCE. https://www.cceguide.org/cce-visualiser/. Accessed 31 May 2023

Cotton D, Winter J (2010) 'It's not just bits of paper and light bulbs': a review of sustainability pedagogies and their potential for use in higher education. In: Jones P, Selby D, Sterling (eds) Sustainability education: perspectives and practice across higher education. Routledge, London. https://doi.org/10.4324/9781849776516

De Fossard E (2008) Using edutainment for distance education in community work, vol 3. SAGE Publications, India. https://doi.org/10.4135/9788132108467

Ellen MacArthur Foundation (2023) Fashion and the Circular Economy. Deep Dive. https://ellenmacarthurfoundation.org/fashion-and-the-circular-economy-deep-dive. Accessed 1 July 2023

Emblen-Perry K (2018) Enhancing student engagement in business sustainability through games. Int J Sustain High Educ 19(5):858–876. https://doi.org/10.1108/IJSHE-05-2017-0075

Fang J, O'Toole J (2023) Embedding sustainable development goals (SDGs) in an undergraduate business capstone subject using an experiential learning approach: a qualitative analysis. Int J Manag Educ 21(1):100749. https://doi.org/10.1016/j.ijme.2022.100749

Festinger L (1962) Cognitive dissonance. Sci Am 207(4):93–106. https://www.jstor.org/stable/24936719

Festinger L (1962b) A theory of cognitive dissonance. Stanford University Press, Stanford, CA

Fredricks JA, Blumenfeld PC, Paris AH (2004) School engagement: potential of the concept, state of the evidence. Rev Educ Res 74(1):59–109. https://doi.org/10.3102/00346543074001059

Gregory A, Miller S (2014) Using Systems Thinking to Educate for Sustainability in a Business School. Systems 2(3):313–327. https://doi.org/10.3390/systems2030313

Kahu ER (2013) Framing student engagement in higher education. Stud High Educ 38(5):758–773. https://doi.org/10.1080/03075079.2011.598505

Kahu ER, Nelson K (2018) Student engagement in the educational interface: understanding the mechanisms of student success. High Educ Res & Dev 37(1):58–71. https://doi.org/10.1080/07294360.2017.1344197

KAPSARC (2020) CCE Guide Overview. KAPSARC. https://www.cceguide.org/guide/. Accessed 31 May 2023

Lozano R, Merrill MY, Sammalisto K, Ceulemans K, Lozano FJ (2017) Connecting competences and pedagogical approaches for sustainable development in higher education: a literature review and framework proposal. Sustainability 9(11):1889. https://doi.org/10.3390/su9101889
Mann S, Robinson A (2009) Boredom in the lecture theatre: an investigation into the contributors, moderators and outcomes of boredom amongst university students. Br Edu Res J 35(2):243–258. https://doi.org/10.1080/01411920802042911
McGrath A (2020) Bringing cognitive dissonance theory into the scholarship of teaching and learning: topics and questions in need of investigation. Scholarsh Teach Learn Psychol 6(1):84. https://doi.org/10.1037/stl0000168
PRME (2023) PRME Secretariat. https://www.unprme.org. Accessed 31 May 2023
QAA (2021) QAA and Advance HE launch landmark new guidance on Education for Sustainable Development. https://www.qaa.ac.uk/news-events/news/qaa-and-advance-he-launch-landmark-new-guidance-on-education-for-sustainable-development. Accessed 29 June 2023
Rieckmann M (2018) Learning to transform the world: key competencies in Education for Sustainable Development. In: Leicht A, Heiss J, Byun WJ (eds) Issues and trends in education for sustainable development. UNESCO, France, pp 39–59
Rokeach M (1968) A theory of organization and change within value-attitude systems. J Soc Issues 24:13–33. https://doi.org/10.1111/j.1540-4560.1968.tb01466.x
Rokeach M (1973) The nature of human values. Free Press, New York
Sharp JG, Hemmings B, Kay R, Atkin C. (2018) Academic boredom, approaches to learning and the final-year degree outcomes of undergraduate students. J Further High Educ 42(8):1055–1077. https://doi.org/10.1080/0309877X.2017.1349883
Sharp JG, Zhu XT, Matos M, Sharp JC (2021) The Academic Boredom Survey Instrument (ABSI): a measure of trait, state and other characteristic attributes for the exploratory study of student engagement. J Further High Educ 45(9):1253–1280. https://doi.org/10.1080/0309877X.2021.1947998
Trigwell K, Ellis RA, Han F (2012) Relations between students' approaches to learning, experienced emotions and outcomes of learning. Stud High Educ 37(7):811–824. https://doi.org/10.1080/03075079.2010.549220
UN (2023) The Sustainable Development Goals. United Nations. https://www.un.org/sustainabledevelopment/sustainable-development-goals/. Accessed 31 May 2023
UNESCO (2019) Framework for the Implementation of Education for Sustainable Development (ESD) beyond 2019. https://unesdoc.unesco.org/ark:/48223/pf0000370215. Accessed 7 Apr 2023
UNESCO (2020) Education for Sustainable Development: A roadmap. UNESCO. https://unesdoc.unesco.org/ark:/48223/pf0000374802.locale=en. Accessed 31 May 2023
UNESCO (2023) Education for Sustainable Development. UNESCO. https://www.unesco.org/en/education-sustainable-development. Accessed 31 May 2023
Williams A, Kennedy S, Philipp F, Whiteman G (2017) Systems thinking: a review of sustainability management research. J Clean Prod 148(1):866–881. https://doi.org/10.1016/j.jclepro.2017.02.002

Teaching Sustainability in Agriculture to Students of Development

M. Manjula and Divya Sharma

Abstract The Post Graduate Programme in Development, at Azim Premji University, Bangalore, India aims to nurture reflective development practitioners. The programme curriculum is designed to impart students the skills to employ multidisciplinary and transdisciplinary approaches in understanding, analyzing, and formulating strategies to overcome development challenges. The University offers a course on Sustainable Agriculture and Development to equip students of development with ideas of sustainability in agriculture. The 2030 Agenda for sustainable development has duly acknowledged the pivotal role of agriculture in addressing hunger and poverty. Concerns of sustainability of agriculture arise out of the need to nurture and promote agricultural systems, technologies, and practices that are environmentally benign, socially inclusive, and economically viable. Education for sustainable development identifies five priority action areas and various competencies that universities must address through their academic programmes. The Sustainable Agriculture and Development course at Azim Premji University is aligned with these priority areas and competencies. The pedagogy in the course recognizes the knowledge-practice-policy interconnectedness. Competencies for sustainable development are nested within these broad dimensions. This chapter aims to capture the pedagogical experience of teaching sustainability in agriculture to students of development.

Keywords Education for sustainable development · Competencies · Sustainable agriculture

M. Manjula (✉) · D. Sharma
School of Development, Azim Premji University, Bengaluru, India
e-mail: manjula.m@apu.edu.in

D. Sharma
e-mail: divya.sharma@apu.edu.in

W. Leal Filho et al. (eds.), *The Contribution of Universities Towards Education for Sustainable Development*, World Sustainability Series,
https://doi.org/10.1007/978-3-031-49853-4_27

1 Introduction

Education for sustainable development (ESD) is vital for achieving the Sustainable Development Goals (SDGs) and global development agenda 2030. It has been recognised that education will play a key function in creating the transformation required for achieving the 17 SDGs (United Nations 2022). Governments, civil society, and other agencies of development require young professionals that are equipped with competencies to become change-agents for development. ESD talks about five priority areas—policy, learning environment, educator capacity building, youth, and local level action. These five priority action areas provide a comprehensive framework for advancing ESD's role in achieving (SDGs). ESD opens immense opportunities for universities to capture the space for SDG-related education through reshaping existing curriculum and introducing new educational programmes.

Universities/Higher Education Institutions are uniquely positioned to take the lead in the multi-stakeholder, cross sectoral design, and implementation of policies and programmes on SDGs. Universities, especially those of the transdisciplinary nature, engage in teaching, cutting edge research, practice, and advocacy across various domains and sectors. Universities as centres of expertise and as neutral stakeholders, can play the role of an influential change agent in advancing the SDG agenda. They create social footprints through their student, teachers, alumni, neighbourhood, and campus networks (SDSN Australia/Pacific 2017). SDGs help universities reshape and reimagine their role in society and emerge as hubs facilitating societal transformation (SDSN Australia/Pacific 2017).

The past couple of decades, universities across the globe have evolved from mere centres of knowledge production to being active agents in the development process.[1] Case stories of universities' transformation to centres of education for sustainable development can serve as an insightful repository for emerging institutions. However, there is a dearth of literature capturing these innovations and transformations in developing nations. Hallinger and Chatpinyakoop (2019) highlight that there is a geographical imbalance in the contribution towards literature related to higher education for sustainable development (HESD). For the decade 1998–2018, the total number of articles in this field is estimated at 1459. Of this, 16% of the literature has been authored in the developing countries. The country wise distribution of literatures from developing nations, stand at India (25), China (47), Malaysia (40), South Africa (40), Brazil (37), Mexico (18), and Turkey (15). There is a need to bring in more representation from developing countries to have a range of context specific cases of HESD.

[1] The United Nations—Higher Education Sustainability (HEI)—partnership between several UN entities and higher education institutions—chaired by UN Department of Economic and Social Affairs (UN DESA) and a non-profit organization, Sulitest Association was a game changer in this transformation. The UN launched HEI in the run up to the 2012, Rio+ 20 conference.

The current chapter is an attempt to document the experience, strategies, and process of ESD adopted in one of the courses on sustainable approaches for agriculture in an Indian university. The pedagogical approach, units, modules, and assessments that are designed to build critical competencies for the postgraduate students are discussed in this paper. These competencies are essential for bringing in transformational change in India's agricultural sector. Higher education in India is undergoing massive transformation. Several universities in India, are redesigning and modifying their curriculum to cater to diverse needs of development. The focus of this chapter is the Azim Premji University, a private transdisciplinary university at Bengaluru, Karnataka, India.

Azim Premji University and the Sustainability Space

Azim Premji University, nested within the Azim Premji Foundation was started in 2011. The university operates with the mandate of education for social change with the aim to contribute to a more just, equitable, sustainable, and humane society. The degree, postgraduate, and short-term teaching programmes are focused on domains of human development viz., education, livelihoods, development, governance and policy, public health, and sustainability. The University's work cuts across research, teaching, and community engagement.

Master's in Development

The Master's in Development at Azim Premji University aims to nurture empathetic and reflective development practitioners. The programme follows multidisciplinary and transdisciplinary approaches to help students understand different theories of development, analyze important developmental challenges of today, and examine possible strategies to resolve them (Program Brochure, Azim Premji University). The two-year programme prepares students to work towards social change which promotes equity, justice, and sustainability.

The chapter details the rationale, pedagogy and learning outcomes of a course titled 'Sustainable Agriculture and Development', which is offered as an elective for students of MA Development. This course falls at the interface of sustainability and livelihood. The course attempts to situate agriculture in the larger development context. It deals with the subject of sustainability in agricultural production, with a perspective of transforming agriculture into a sustainable livelihood option for resource poor smallholder farmers.

2 Teaching Sustainability in Agriculture: Rationale from a Development Perspective

The role of agriculture in development has undergone a paradigm shift over the years. The structural transformation theories of economies view agriculture as a precursor to industrial growth. Early development theorists restricted the role of agriculture to ensuring food supply and maintaining food prices and nominal wage costs. However, a shift in development thinking in the 1960s sought a more central role for agriculture in development, with a focus on equity, employment, and pro-poor productivity growth.

The 2030 Agenda for sustainable development recognizes the importance of agriculture in addressing hunger and poverty. With the world population expected to reach 9.1 billion and a projected global economic growth of 2.9%, food production needs to grow by almost 70% to keep pace with this demand (FAO 2009). Green revolution-led agricultural intensification was the answer to global food scarcity during the 1960s. This tripled cereal production globally and had a significant impact on reducing hunger and poverty.

However, this was not without trade-offs on the environment and long-term agricultural productivity. Agriculture is potentially the most prominent sector where the tension between development and environment is more apparent, as it uses about 70% of global fresh water and 42% of land (FAO 2017). Traditional strategies focused on agricultural intensification and expansion impose environmental costs on the sector and society at large, resulting in negative externalities such as depletion of ground water, decline in soil fertility, soil erosion, loss of biodiversity, agrochemical pollution, and desertification.

The Sustainable Agriculture and Development course at Azim Premji University is designed to help students of development (i) gain a deeper understanding of the theories of agricultural development; (ii) get exposed to different agricultural production systems and their sustainability implications; (iii) critically analyze the role of the existing institutional and policy environment for promoting 'sustainable agriculture for development' and (iv) understand the role played by the 'third sector'—*civil society organizations, community-based organizations (FPOs), private sector and donor agencies-* in promoting sustainable agriculture for development. The course is designed to impart multiple competencies and equip students who aspire to contribute to facilitating transformation in the sector at various levels.

There are 63 State Agricultural universities in India, which aim to provide agricultural education, agricultural research, and agricultural extension (Reddy 2023). Though it is important to facilitate specialization in the field, it also brings in the risk of creating knowledge silos. The course at APU helps in creating a wider knowledge base which is essential for breaking siloed thinking, joining the dots, and understanding the bigger picture. Being a generalist (as opposite to specialist) helps in lateral thinking (Epstein 2021), which is important for innovative solutions and transformation. It adopts an action oriented transformative pedagogy that encourages collaborations across diverse disciplines (Rieckmann 2018).

Pedagogical Aspects:

The course has a balanced mix of theory and practice. Students engage in theorising and conceptualisation through perspective building, discourse analysis and hands on operationalisation of meta concept of sustainability. The course exposes them to varied systems of agriculture, the opportunity to assess production systems on the sustainability framework, and hands on experience at farming. It also exposes them to the role of government, civil society, grass root agencies and markets in the sustainable agriculture space. This is imparted through expert talks, sharing of experiences of programmes, and discussion of field based real case studies. The case studies help students to critically reflect on the efforts by the different agents of development in making agriculture sustainable.

On-field exposure and hands on practice in farming are integral elements in the pedagogy. Students are required to apply the learning from classes to assess sustainability of different farming systems through observation and interaction with the farmers. The course overall equips the students to reflect on the relevance of agriculture for development, issues of agricultural sustainability, the existing constraints, and opportunities for promoting sustainable agriculture. The role of different agents of development—government, civil society, markets—are also dealt with. At the end of the course, the students apply the learnings from the course and develop a blueprint for transformation to sustainable agricultural systems in select geographies/agro-ecologies. It has been widely accepted that the competency required for sustainable development are systems thinking, anticipatory, normative, strategic, collaboration, critical thinking, self-awareness, and integrated problem solving (Rieckmann 2018).

The pedagogy of the course can be broadly classified into domains of knowledge, practice, and policy (Fig. 1). The subsequent section elaborates some of the pedagogical approaches in the course.

Knowledge:

This domain emphasizes building basic knowledge related to sustainability and agricultural systems. This is transacted over four modules:

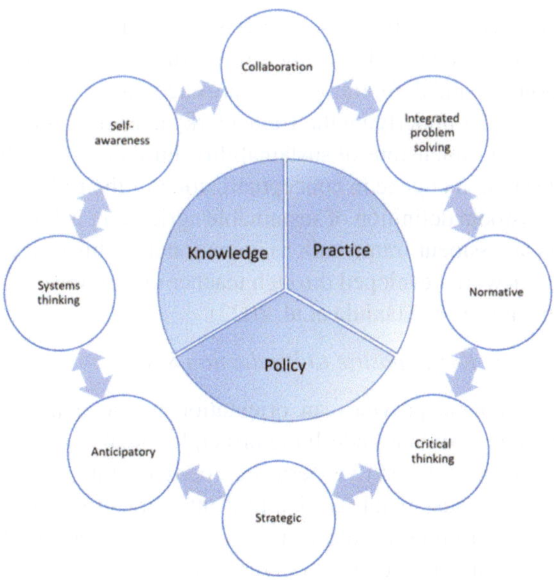

Fig. 1 Framework for pedagogy: knowledge, policy, and practice (*Source* Author's' Reflection)

Module K1: Perspective Building: Sustainable Agriculture-What it means for Development?

The module provides an orientation to the important theories of agricultural development and the evolution of thoughts on the role of agriculture for development. It orients the students to thc different pathways in which agriculture influences development in purely agrarian economies, transition economies, and in urbanized economies. The students learn about the role of agriculture as an economic activity, as a livelihood option, and as a provider of food security and environmental services. It also deals with the challenges in enlisting 'agriculture for development' in the larger development agenda.

Module K2: Discourse Analysis: 'Sustainability' in Agriculture:

The transformative role of agriculture in achieving Agenda 2030 will be dictated by the understanding of what constitutes sustainable agriculture. There is no commonly accepted definition of sustainable agriculture. The meaning and concept of sustainable agriculture has been influenced and shaped by political and scientific discourses. Programmes and strategies for achieving sustainability in agriculture at global, national and sub nation levels are based on this understanding. Discourse analysis exposes the students to the political and scientific discourses on sustainability in agriculture. Students are exposed to the foundational positions driving the discourses, the evolution of these discourses over time, and the major narrative changes and rationale underlying them. The similarities and contestations in the different discourses are also dealt with as part of the discourse analysis.

Module K3: Operationalizing Sustainability:

Following the discourse analysis, the course exposes the students to some of the existing frameworks for measuring sustainability in agriculture. Classroom discussions around these frameworks are encouraged to gain a critical understanding of the principles underlying the framework, areas of emphasis, and the balance among the various dimensions of sustainability. Further, to this, the students through classroom discussion engage in conceptualization of their idea of sustainable agriculture. This classroom definition of sustainable agriculture is further critically reflected upon and an assessment framework to operationalize this is developed in class. Sustainability framework developed through teacher moderated classroom discussion is presented in Annexure (Manjula et al. 2022).

Module K4: Exposure to Production Systems:

This course provides an orientation to the major systems of agriculture in the country and their underlying principles, goals, and processes. Agroecology, organic farming, regenerative farming, permaculture, conservation agriculture, natural farming, sustainable agriculture intensification, climate smart agriculture, biodynamic farming, agroforestry, are to mention a few. Concerns of sustainability of agricultural production systems arise out of the need to nurture and promote agricultural technologies and practices that are environmentally benign, socially inclusive, and economically viable while ensuring food and nutritional security to increasing population.

Practice:

This domain emphasizes engaging with the concepts learnt in the previous module and applying them in real-life situations. This is transacted through five modules:

Module P1: In-Class Application of Sustainability Framework:

The framework developed in-class is tested and applied to assess the performance of the different agricultural systems. The existing systems of agriculture that are promoted as 'sustainable' are critically analyzed by applying sustainability principles. Traditional agriculture systems, agroecological systems, conservation farming, organic agriculture, conventional agricultural systems, and others are taken up for detailed analysis. The analysis helps students to develop a critical understanding of these production systems. It offers valuable insights into the performance and tradeoffs of the different production systems across various dimensions of sustainability.

Module P2: On-Field Assessment of Farming Systems:

The students apply the sustainability framework developed in-class to real world farming systems. This requires visits and interaction with farmers in the field. This is a one-day exercise and involves field level observation, surveys and interviews with the farming community (Fig. 2), and on-field measurements for sustainability. The analysis and outcomes of the assessment are discussed during class. This helps students to develop a firsthand experience of the challenges in achieving sustainability in farming practices. This provides an opportunity to gain understanding of the ground level success/failure of government policies and programmes. On ground exposure to the role of markets and civil society in enabling and/or constraining the transition to sustainability in agriculture in the region is another takeaway from this exercise.

Module P3: Hands on Farming Experience:

Students are assigned plots within the campus where they engage in farming. Land preparation, crop choices, crop variety choices, choice of farming practices are all decided and executed by the students. The students maintain field records of the farm and the practices (Fig. 3). Details of the various process, stages of crop growth, pest and disease sighting, management practices adopted, watering, harvesting, post harvesting, and sales are noted. The effort, processes and results are rewarded through continuous assessment of farming activities and the plots.

Module P4: Exposure to Role of Third Sector: Sustainable Agriculture 'In-Practice':

The role of the 'third sector' in agriculture in India is substantive, especially following the structural adjustment policies. This has resulted in a clear shift in the role of the state from that of provider of services to that of an enabler and coordinator across sectors. This provides opportunities for the emergence of a 'third sector'

Fig. 2 Students doing on-field assessment of the framework created in the class. (Place: Thanachegare village, Ramnagara District, Karnataka, India) (*Source* Audio-Visual Division, Azim Premji University)

Fig. 3 *Students doing farming activity on allotted plots within Azim Premji University* campus, Bengaluru, India (*Source* Audio-Visual Division, Azim Premji University)

comprising of civil society organisations (CSOs) including nongovernmental organisation (NGOs) and community-based organisations (CBOs). They play a key role in enhancing the political economy of agriculture for development and are important governance structures for effective implementation of programmes. In India, CSOs have played a key role in promoting sustainable agriculture initiatives since the 1980s. They receive funds from government or donor organisations and implement specific projects. They are either professional development organisations or grassroot member organisations. The course exposes the students to the ground level initiatives of these organisations in promoting and nurturing sustainable agriculture. It also studies factors that enable/constrain the role of these organisations in promoting the agenda of 'agriculture for development'.

Module P5: Developing a Blueprint for Sustainable Agriculture:

This is designed as a test of the overall learnings and relevance for development practice of the course. Students select a specific geography/agroecology for this purpose. Further, they engage in problematisation around issues of sustainability in agriculture in the selected site using system thinking approaches. Based on this, they develop a blueprint of programme or an intervention that would address the challenges they have identified at the problematisation stage. The programme/intervention should be locally suitable, context specific pathways for transitioning the existing farming systems in the intervention sites to sustainable production systems. The students develop this as a complete project document, with rationale, objectives, theory of change, log frames, and the budgetary requirement. The proposals are evaluated on criteria of relevance, suitability, and technical cum financial viability. Engaging in the process builds the students capabilities to emerge as programme managers in the field of sustainable agricultural practice. This also builds their capacities as change agents in the domain of agriculture for sustainable development.

Policy:

This domain exposes the students to various agriculture related policies, the political economy of policy formulation, and aspects of implementation. It also orients the students to issues of intersectionality and inclusion challenges in sustainable agriculture.

Module PO1: Exposure to Institutional and Policy Environment for Agriculture:

An enabling policy environment and strong governance systems are essential to realise the agenda of 'agriculture for development'. Policies, institutions, and governance systems should not only respond to the changing macro-economic context, but also be inclusive for equitable growth and development. The course takes stock of the agricultural policies in India and critically examines the political economy context of these policies. It further highlights the opportunities for having market-based policy instruments to incentivize environmentally sustainable production practices. Students critically reflect on these policies and programmes and identify the challenge areas and suggest pathways for improvement.

Module PO2: Exposure to Inclusion Challenges in Agriculture:

The course also deals with issues of class, caste, gender, geographies, and differential development outcomes dictated by these social determinants. It exposes students to the differential impact globalisation and trade liberalisation has had on the resource poor small holder farmers. The unit seeks to address issues of social inclusion, the role and position of women farmers, increasing feminisation of agriculture, existing gender inequalities in agriculture and the need for gender mainstreaming in the agricultural sector. It also delves into aspects of inclusion, changing socio-economic context of the rural landscapes, and their impact on agricultural sustainability.

Competencies and Modules

The modules covered in the course target all the competencies that are essential for sustainable development. These competencies equips a student to understand the bigger picture, the interconnectedness of various systems and to contextualize solutions for long term. Though the modules are designed to address three broad domains of knowledge, policy and practice, they cater to multiple competencies simultaneously. These competencies are mapped against the relevant modules in Table 1. For example, in the module of *perspective building,* students are exposed to different narratives around development and agriculture internationally and in India. This helps them understand the regional and sectoral interconnectedness, nudge them to juxtapose these within current Indian agricultural context. Engagement with these modules help them anticipate problems as well as solution trajectories.

Multiple viewpoints on the same subject are presented which creates a foreground for thinking critically. Given vast geographical and socio-economic differences, students are encouraged to understand the diversity of perceptions that various stakeholders might hold in terms of what sustainable agriculture should look like in practice. In *discourse analysis* the students are exposed deeper into the debates around sustainable agriculture. They delve with the origin and evolution of definition of sustainability in agriculture. They deal with questions like: why diverse groups—*scientists, international political actors, government, among others*—emphasize on different, and at times competing aspects of sustainability? This helps them recognize that normative thinking is essential to understand varying and at times competing claims, for problematisation as well as for creating integrated solutions.

Though these competencies are subjective in nature, the modules are informed by diverse pedagogical approaches. Multiplicity of approaches helps in reemphasising the criticality of these competencies for sustainable development. Assessments are graded around these competencies at both individual and group level.

Table 1 Learning modules mapped to competencies for sustainable development

Learning modules	Competencies							
	Systems thinking	Anticipatory	Normative	Strategic	Collaboration	Critical thinking	Self-awareness	Integrated problem solving
Perspective building	●	●	●	●		●		
Discourse analysis	●		●			●		
Operationalizing sustainability	●	●	●	●	●	●	●	●
Hands on farming experience	●	●		●		●		●
Exposure to production systems	●		●			●	●	
In-Class application of sustainability framework on production systems	●				●	●	●	●
Policy exposure to institutional and policy environment for agriculture	●		●	●		●	●	
Exposure to inclusion challenges in agriculture	●		●	●		●	●	
On-Field assessment of real time farming systems			●		●	●	●	●

(continued)

Table 1 (continued)

Learning modules	Competencies							
	Systems thinking	Anticipatory	Normative	Strategic	Collaboration	Critical thinking	Self-awareness	Integrated problem solving
Region specific problematization	●	●	●			●	●	●
Developing a blueprint for transitioning to sustainable agriculture	●	●	●	●	●	●	●	●
Exposure to role of 'third sector': sustainable agriculture 'In-Practice'	●			●		●		●

Source Authours' reflection

3 Challenges

Teaching sustainability is a challenging task. It requires an open mind and engagement with multiple and competing viewpoints in a neutral way. Diversity of the classroom adds to this challenge. Engaging with the idea of sustainability requires a comprehensive understanding of phenomenon that may not have visibility in the immediate future. In developing countries, where agricultural issues revolve primarily around food and livelihood security, talking about sustainability feels far removed from issues that are now and here. The equity question, the responsibility and accountability to ensure sustainability is a major bone of contention. Students debate why this burden needs to be passed on to marginal farmers (less than two hectares of land), whose contribution to the problem is negligible.

In a country like India, where small and marginal farmers contribute to 89.4% of the total agricultural households (Agricultural Census 2015–16), justice and livelihood become prominent debates, making it difficult to appreciate the kind of path that is required for long term solution. Sustainability debates are challenging to handle in a classroom, where students have first-hand experience of agricultural distress and farmer's suicide in their own villages. Students often ask which is the right path in agriculture? Can we think in terms of natural farming and upscale it for India? Can a farmer in Vidarbha who is going through debt crisis think about sustainability in agriculture?

These are philosophical questions, which has no tailormade right answers. During the course, students are nudged to learn and unlearn a lot of concepts and viewpoints. These competencies require questioning one's own learning pathways and mental models accumulated over the past, which at times, leaves the students confused. The sustainability paradigm makes it difficult to reach and offer a quick solution, as the competencies require facilitation of openness and inclusion, where consensus building is challenging but is fruitful in longer run. The course on sustainable agriculture and development aims to build the competencies to challenge the old thought patterns and create room for openness and inclusivity.

4 Limitations

Sustaiability in agriculture being a vast and critical subject tend to attarct students with vaired expectations. The course not only strives to cater to the broader understanding of the sustaiability paradigm, but also aims to impart skill sets that can be readly used and enhance empolyability of the students. The pedagogical approches discussed here evolves with every new student cohort. Since students prioritise employability, the skill sets tends to be viewed as more tangible learning outcomes. Thus students fall in the self limiting trap of focussing on skill building and lose sight of competencies.

Futher, the diversity of the cohort adds to the dynamism of expectations, compounding the challenges of making the course relevent and inetersing year after year. Current cohort of twenty three students (July 2023–Dec 2023) come from eleven states of India. These states widely vary in terms of their agroecologies, socio-economic development and priorities which influences the perception of sustainabilty. The divesity also makes it challanging to trace the impact and effectiveness of a course which is determined by each student's expectations and engagement with the learning outcomes.

5 Conclusions

Sustainable development is critical to India's growth pathway and HESD will be key to the nations progress. India requires context-specific development interventions. The pedagogy framework and competencies for sustainable development need to be priorities for all higher education institutions. The effectiveness of various fameworks that captures those linkages also needs further investigation. The course on Sustainable Agriculture and Development highlights that it is possible to remap existing courses in universities to cater to HESD. Linking competencies and pedogogical approaches is critical to ESD. Emphasizing the relevance of these competencies in curriculum design will build the human resources required for effecting larger societal transformation. Documented evidence of Indian univeristy experiences in bringing this linkage is limited.

The pedagogical approach in the Sustainable Agriculture and Development course, emphasizes the open mindedness required for any course that aims to offer these competencies. The journey from philosophically articulating a concept and embedding that in a context specific case throws up lots of challenges. At the end of the course, the learner is cognizant of and is open to the fact that there is no ready and perfect solution for any problem.. There is continued effort in improving the relevance of the course as a means of perspective and skill building in the field of sustainability in agriculture in India.

Majority of the students attending the Masters's in Development program come from agrarian backgrounds. The course relies on continued student feedback which extends outside of their university life and into their professional engagements. Student feedbacks and the career pathways they choose becomes testing ground for the relevence of content, pedagogy and learning outcomes in the course. Hence, its improtatnt to trace the cohorts to build evidence for the effectiviness of the pedagogical approaches in building competencies and skills that helps bridge the gap between thoery and practice. Understanding these connections are critical for Indian context as we have the largest youth population in the world.

A competent, skilled and empathetic youth population are integral and key for realising India's dream of sustainable development for all. The Universites serves as their professional training ground. Hence there is a need for continued effort at designing and refining programs aimed at building competenices that cater to education for sustainable development. A logical next step in this direction would be designing a dynamic feedback systems that enables tracking of student cohorts over a period of time. These feedbacks will provide valuable insights to redesign and improve the relevance and effectiveness of courses aimed at achieving education for sustaianble development.

Acknowledgements The authors would like to thank all the students who had registered for Sustainable Agriculture and Development Course, for their feedback and engagement, which contributed to mproving the relevance and padagogical approaches in the course. Authors acknowledge the support extended by various individuals and functions at the Azim Premji University for their support. Ms. Anuradha Nagaraj, Faculty, for copy editing the chapter, Victor and IMF team for making experiential farming possible on the campus, V. Manikantan and Rajesh Joseph, Faculty, for facilitating the on-site farm assessment in diverse agroecologies and D.C. Shashwat and team for audio-visual documentation of student's farming experience.

Disclaimer The content and opinions expressed are that of the author(s), and are not necessarily endorsed by/do not necessarily reflect the views of Azim Premji University.

Annexure-sustainability framework used in field assessments

	Dimensions/ Indicators	Ratings scales		
		1	3	5
Environment	On-farm external input use	Chemical	Both	Natural
	Presence of earthworms	No		Yes
	Presence of honeybees	No		Yes
	Type of soil	Sandy	Clay	Laterite/Loamy
	Change in groundwater table—past decade	Deep	Remains same	Closer to Surface
	Changes in surface water availability	Lesser months than before	Remains same	More months than before
Economic	Average yield of the major crops?	<than potential yield	Same as potential yield	>than potential yield
	Share of agriculture income in household income	<25%	25–50%	=/>50%
	Number of crops grown in a year	One or two	Three	More than three
	Source of credit	No source	Informal	Formal

(continued)

(continued)

	Dimensions/ Indicators	Ratings scales		
		1	3	5
	Alternate livelihoods opportunities	None	Two in addition to crop-based agriculture	More than two in addition to crop-based agriculture
	Size of the land holding	<2.5 Acre	2.6–5 Acre	>5 Acre
Social	Ownership of land holding	Lease	Joint ownership with family	Sole/name of husband and wife/own
	In whose name is the land registered		Male	Female
	Who usually makes decisions related to agriculture?	Only men	Men leading decision in consultation with women	Joint decisions through consensus
	Wage difference between male and female labour	Two times	<Two times	No difference
	Membership in collectives	No		Yes
	Dominance of caste in access to productive resources	Always	Sometimes	Never
Governance & perception of climate variability	Number of schemes availed for agriculture in the last two farming seasons?	None	One to two	More than two
	Number of interactions with the extension officer in the last two farming seasons	None	Once or twice	More than twice
	Distance to the PHC	>5 km	Within 5 km	In the village
	Distance to the nearest veterinary dispensary	>5 km	1–5 km	<1 km
	Perception of rainfall pattern—compared to parent's time?	Different		Same

(continued)

(continued)

	Dimensions/ Indicators	Ratings scales		
		1	3	5
	Perception of the impact of weather on farm-based livelihoods	Negative	None	Positive

References

Agricultural Census (2015–16) Table 1. Number and Area of Holding by Size Class. https://agcensus.dacnet.nic.in/NL/nationalt1sizetable1.aspx. Accessed 18 July 2023

Epstein D (2021) Range: why generalists triumph in a specialized world. Penguin

Food and Agricultural Organization (FAO) (2009) Global agriculture towards 2050. In: How to feed the world 2050, October 2009. https://www.fao.org/fileadmin/templates/wsfs/docs/Issues_papers/HLEF2050_Global_Agriculture.pdf. Accessed 27 July 2023

Food and Agricultural Organization (FAO) (2017) Water for sustainable food and agriculture: a report produced for the G20 Presidency of Germany, October 2017. https://www.fao.org/3/i7959e/i7959e.pdf. Accessed 29 July 2023

Hallinger P, Chatpinyakoop C (2019) A bibliometric review of research on higher education for sustainable development, 1998–2018. Sustainability 11(8):2401

Manjula M, Manikandan V, Sharma D (2022) Building farm resilience through natural farming – reflections from field. Leisa India, 21 September 2022. https://leisaindia.org/building-farm-resilience-through-natural-farming-reflections-from-field/. Accessed 27 July 2023

Programme Brochure. Masters in Development, Azim Premji University. https://azimpremjiuniversity.edu.in/programmes/ma-in-development

Reddy AA (2023) Can NEP recommendations transform state agricultural universities? – The wire science. Can NEP Recommendations Transform State Agricultural Universities? – the Wire Science, 3 January 2023. https://science.thewire.in/education/nep-agricultural-university-transform/. Accessed 27 July 2023

Rieckmann M (2018) Learning to transform the world: key competencies in education for sustainable development. Issues Trends Educ Sustain Dev 39:39–59

SDSN Australia/Pacific (2017) Getting started with the SDGs in Universities: a guide for Universities, Higher Education Institutions, and the Academic Sector. Australia, New Zealand and Pacific Edition. Sustainable Development Solutions Network – Australia/Pacific, Melbourne. https://ap-unsdsn.org/wp-content/uploads/University-SDG-Guide_web.pdf

United Nations (2022) The sustainable development goals report 2022. Department of Economic and Social Affairs. https://unstats.un.org/sdgs/report/2022/The-Sustainable-Development-Goals-Report-2022.pdf

Behavioral Foundations of Sustainable Development at Universities

Julia M. Puaschunder

Abstract This paper presents three behavioral insights for researching and teaching sustainability at universities: In a behavioral field experiment at a North American university campus, individuals were found to engage in sustainability actions if facing social status loss prospects. In line with (Kahneman and Tversky 1979) prospect theory holding losses looming larger than gains, social status loss prospects appear as powerful motivating factors to boost sustainability. In a university collaboration, behavioral online and paper-pen campus survey studies showed that when individuals judge alternative choices, presenting alternatives concurrently improves decision-making outcomes for future orientation and intergenerational balance. The joint alternative presentation mode therefore can boost sustainability. The third study used academic laboratory experiments, in which students played an economic trust game followed by a common goods game. The repeated interaction in two standard economic games provided evidence for trust and reciprocity leading to individuals contributing more to common goals. Educating future leaders at universities in a climate of trust and reciprocity may thus help elicit future contributions to sustainable development. Combined these studies prove behavioral economics to offer powerful ways to foster a sustainable society and pursue the Sustainable Development Goals.

Keywords Behavioral economics · Behavioral insights · Common goods · Economics · Energy · Leadership · Management · Organizational behavior · Policy · Reciprocity · Recycling · Social status · Sustainability · Tax · Trust

J. M. Puaschunder (✉)
Graduate School of Arts and Sciences, Columbia University, 535 W 116Th St #109, New York, NY 10027, USA
e-mail: Julia.Puaschunder@columbia.edu
URL: http://www.juliampuaschunder.com; http://blogs.cuit.columbia.edu/jmp2265

W. Leal Filho et al. (eds.), *The Contribution of Universities Towards Education for Sustainable Development*, World Sustainability Series,
https://doi.org/10.1007/978-3-031-49853-4_28

1 Introduction

Ever since their existence, universities have been focal points to address the most pressing contemporary problems of our times. As a gateway to international information exchange and a laboratory of innovative ideas with unknown outcomes in basic research and a practical focus on applied sciences, academics at universities are predestined to find groundbreaking solutions for global problems. For the nature of scientists' independent thinking and integrity in reporting hard facts, academics have also become change agents in science diplomacy acknowledging that there are global problems that are too large, complex and important to be concerned by national politics and country denominations (Puaschunder forthcoming b).

One of the most pressing contemporary concerns around the globe is climate change. Since the 1950s onwards, scientists have noticed a trend in rising world surface temperatures. Since the 1990s, there has been wide acceptance of the human-made impact of CO_2 emissions on climate change. Leading scientists around the world incepted the concept of Sustainable Development at the 1992 United Nations "Conference on Environment and Development," also known as Earth Summit. The Rio Declaration consists of 27 principles to foster sustainable development for all humankind and was signed by 175 countries around the world. Environmental sustainability was also a key focal point of the *United Nations Millennium Declaration* of 2000, followed by the *United Nations 2030 Agenda for Sustainable Development* of 2015 (United Nations Department of Economic and Social Affairs 2015; United Nations Department of Economic and Social Affairs Sustainable Development; United Nations Sustainable Development Goals). The United Nations Conferences of the Parties (COP) on climate change are at the forefront to address sustainable development from a global warming perspective with particular attention to climate justice (Puaschunder 2020b).

Universities have become major change agencies and driving factors in addressing Sustainable Development. Providing hard facts about the natural environment, anticipating future living conditions, finding legally sound implementation methods for redistribution mechanisms to compensate those who lose the most and the fastest from global warming are only some positive ways in which universities have changed the world for the better (Puaschunder 2020b). Fostering the Sustainable Development Goals and with an international outreach as well as the luxury of being able to focus on non-monetary gains, are just some privileges of universities to steer betterment. One of the most powerful forces that academia offers in catalyzing change is through teaching a cadre of upcoming scholars and students from all disciplines, which makes a compelling case that universities offer mind-changing experiences and transformative roles in society.

One of the hallmarks of innovative sciences in the field of economics was the behavioral economics revolution. Starting in the 1950s as a theoretical critique of rational choice models, which assumed all human beings strive for efficiency and rationality all the time, behavioral economics first drew attention to deviations from rationality and the non-applicability of rational choice models for depicting the actual behavior of humans (Simon 1956; Simon and Bartel 1986). During the 1970s, Amartya Sen formalized the rational choice critique and published powerful examples of how economics needs a reality check and backtesting of its core axioms of rationality, efficiency and time consistency for actual real-world relevancy and external validity of the standard rational choice claims (Sen 1971, 1977). In subsequent work over decades, Amartya Sen debunked some of the hallmarks of economic science assumptions (Sen 1993, 2002a, b).

By 1979, two psychologists, Daniel Kahneman and Amos Tversky presented a line of laboratory experiments at universities that proved the rational choice theory to be inaccurate to explain the decision-making patterns of individuals on a constant basis (Kahneman and Tversky 1979). The following 1980s and 1990s saw a revolution of behavioral aspects for economic sciences (Kahneman and Tversky 1983, 1992; Kahneman et al. 1991). First laboratory experiments at universities but later also field experiments became the norm to cartograph a new order by which humans are 'economic'—often striving for efficiency and economic ideals, but in reality, oftentimes muddling through a complex world and thereby only taking the quick best alternative of choices given the complex world human face and in light of their natural cognitive limitations (Kahneman and Tversky 2000). The change in the field stemmed from academic institutions that rewrote economics for accuracy and predictability for actual behavior (Kahneman and Tversky 2000). Several Nobel Prizes crowned the accomplishment to change a field as dominant as economics from within but also from different disciplines adding to bundle efficiency with accuracy and predictability. First sociologist George Akerlof (2001), Nobel Prize Committee (2002), political scientist Ostrom (2008), and Thaler (2017) all gained Nobel prizes to rewrite economics and pursuing a goal to find ways how to 'nudge' individuals, communities and—most recently—leaders to help others to make better choices by the guidance of behavioral insights (Akerlof 2001; Kahneman 2008; Nobel Prize Committee 2002; Ostrom 2009; Puaschunder uaschunder 2020a; Thaler 2017; Thaler and Sunstein 2008).

This paper captures three behavioral economics studies that were conducted at universities that found powerful 'nudges' for sustainable choices (Puaschunder 2015a, 2017a; Puaschunder and Schwarz 2012). A field experiment outlined the role of social status prospects for pro-social choices and provides evidence that in particular social status drops are associated with sustainable decisions (Puaschunder 2017a). The field experiment was recently independently replicated in a large-scale field experiment in a different population and setting, validating the core findings beyond the university context (Gangl et al. 2022c). This serves as a powerful example of how universities can serve as breeding grounds for large-scale change in the general population and around the world.

The second survey study was conducted at three different universities in paper-and-pen format as well as online. Behavioral international online and paper-pen campus survey studies that were prepared in Europe and held in Asia and international online showed that when individuals visually bundle different choices and review them together next to each other, they become more intergenerationally balanced (Puaschunder 2019a; Puaschunder and Schwarz 2012). The joint decision-making advantage can thus improve sustainability. The study conducted in different settings on three continents underlines the impetus of universities for science diplomacy in mutual understanding of scientific facts, leaving aside national interests or cultural boundaries.

The third laboratory experiment at a Behavioral Decision Science Laboratory at a North American campus, in which students played a standard economic trust game followed by a common goods game revealed that trust and reciprocity lead to individuals contributing more to common goals. Educating future leaders at universities in a climate of trust and reciprocity may thus help trickle down in the economy and nurture a broad-based consensus on sustainable development in the 21st century.

This paper first addresses the role of universities in sustainable development. The methods section then presents three previous studies at universities that offered behavioral 'nudges' to change behavior towards sustainability. All three studies were conducted about a decade ago and will be reflected upon most novel developments, such as a powerful replication of a field experiment and the current shift from 'nudges' to self-determined leadership in behavioral economics. The discussion ends with a future prospect of how behavioral sciences can reinvent itself towards a profession of oversight and accountability as well as self-determined leadership to foster an even stronger a sustainable society.

2 Universities for Sustainable Development

Universities are innovation catalysts. Academic research is aimed at studying the most recent trends around the world by exchange in international networks. Globalization and digitalization have perpetuated the international character of innovative research in the 21st century. Universities are predestined to find creative solutions for the most pressing contemporary problems of our times in their role as independent intermediaries between public and private sector interests. The integrity of science and the hard-fact-driven scientific competition also enables to add science diplomacy for the alleviation of complex challenges to conventional global politics and international diplomacy. The product of universities is the education of students, who bring in critical reflection of the state-of-the-art academic conduct, highest productivity levels and the prospect of becoming future leaders in all sectors of life.

Universities and academic circles have a history of addressing sustainable development. Since the "Our Common Future Report" of the Brundtland Commission (1987) advocating for sustainability in equity on natural resources with future generations, the urge to conserve and protect the earth has steadily risen. Natural resources and animal livestock were soon found to be endangered by economic production and industry development. In the predicament between economic productivity and environmental conservation yet also acknowledging the importance of development around the world for humanity, the Rio Declaration of 1992 coined the term "sustainable development." Leading scientists around the world incepted the concept of Sustainable Development at the 1992 United Nations "Conference on Environment and Development," also known as Earth Summit in an attempt to harmonize international development with respect for sustainability of the earth and natural resources as well as animal habitats. Further advancements in the integration of sustainable development in economic productivity for international development came from universities and academic scholars in the so-called 'triple bottom line' movement of sustainable development that advocates for businesses to integrate economic growth, environmental preservation and societal well-being in whole-rounded sustainable development (Leal Filho et al. 2023). Concepts like Corporate Social Responsibility further advanced the notion of imbuing social care into all industries and layers of economic productivity (Puaschunder 2019b; forthcoming c). Socially Responsible Investment was meant to harness the power of finance for sustainability (Puaschunder 2017b). Universities have a long history of breaking the most novel trends in political divestiture—advocating for the act of removal of finance from politically-incorrect markets or harmful market behaviors (Puaschunder 2015b, 2016, 2019b). Most recent developments include Financial Resilience and digitalization for fostering sustainable development (Leal Filho et al. 2023; Puaschunder 2023).

Academic settings predestine innovation and heterodox leadership in breaking the most novel trends and contesting the status quo of modi operandi. As such, universities were vital in the contemporary critique of the neoclassical rational choice model in economics. From the 1950s on a theoretical critique started with revising neo-classical concepts—such as efficiency and rationality—for their real-world relevant applicability (Simon 1956; Simon and Bartel 1986). Individual decision-makers were depicted as satisficing their choices and being constrained by the overall complexity of life (Simon 1956). What followed was a systematic revision of neo-classic economics hallmarks further propelling a behavioral backtesting of actual decision makers' choices (Sen 1971, 1977, 1993). The theoretical formalizations led to a line of powerful university experiments that proved humans deviate from rational choice in a constant and systemic way (Kahneman and Tversky 1979, 1983, 1992). Laboratory experiments found that people care more about the present than any other prospect (Laibson 1997). They evaluate choices based on future reference points and become risk-seeking in the eye of previous losses (Kahneman and Tversky 1979, 1983, 1992). People have a mental accounting for financial assets and tend to value things they possess higher for the mere act of possession (Kahneman et al. 1991; Thaler 1999). In the long line of subsequent laboratory and field experiments, many of them conducted in university settings, many human deviations from what rational choice theory would predict were found (Puaschunder 2020a). Behavioral economists started to make plans to 'nudge' and 'wink' people in better choices, meaning to employ little changes to their choice architectures and mental concepts that would help them—oftentimes implicitly—to tend towards a choice that is more sustainable and simply better for themselves, often over time but also in relation to their communities and the greater societal compound (Thaler and Sunstein 2008; Puaschunder 2020a). With the rising popularity of behavioral economics over the past few decades, in which four Nobel Prizes were granted to behavioral sciences, the academic community not only changed the landscape of modern economics. Behavioral economists also became powerful change agents that crafted policies and governance around the world. Most recent developments feature the widespread popularity of behavioral economics in the general populace and attempts to move from 'nudges' to self-determined leadership (Puaschunder forthcoming a).

The application of behavioral economics in the sustainable development domain on university campuses is widespread. To name a few direct behavioral intervention studies that tested sustainability parameters, laboratory experiments at university campuses are quite common to stringently investigate behavioral effects and the environment (e.g., Fox and Dayan 2004; Meng and Trudel 2017). In field interventions, nudges were installed in visual prompts to increase recycling behavior at a large university (e.g., Austin et al. 1993). Watchful eyes were tested to increase payment honesty in a university coffee room setting (e.g., Bateson et al. 2006) and reduce littering at a university campus (e.g., Bateson et al. 2015; Durdan et al. 1985; Ernest-Jones et al. 2011). Field experiments were also used to study behavioral interventions and nudges to reduce recycling at a university campus (e.g., Cosic et al. 2018; McCoy et al. 2018; Sussman et al. 2013).

The international, multi-disciplinary and wide-reaching character of universities allows for studying large-scale effects and international trends in behavioral aspects of sustainable development with public and private sector applications. For example, an overview of behavioral interventions is provided in the following papers: Abrahamse and Matthies (2012) review strategies to promote pro-environmental behavior including information (knowledge, tailored, normative, social learning, modeling), goal setting, commitment, prompting and feedback. Dear et al. (2019) conduct a meta-analysis of 15 experiments from 13 research papers on the reduction of anti-social behavior in the presence of visual environmental cues in order to provide 'nudges' as a low-cost behavioral intervention strategy. Dwyer et al. (1993) conduct a meta-analysis of 54 studies on environmental preservation during the 1980s with particular attention to behavioral cues and design intervention to promote sustainability finding that commitment, demonstration, and goal-setting strategies are generally most effective in encouraging environmentally-responsible behavior, while consequence conditions were effective during experimental settings. Kolodko et al. (2016) provide an extensive overview of applied behavioral insights that improve environmentalism including situational and social contexts, promoting cooperation, design interventions—ranging from information, institutions, incentives, identity—as well as the actual design in packaging and disposal, accessibility, attractiveness, monetary incentives, penalties, recycling, personalization, transparency, community identity and consequences, among other. Lehman and Geller (2004) provide an extensive review of behavioral intervention literature and directives to expand intervention targets and increase longer-term maintenance of pro-environmental behavior. Schwartz et al. (2019) discuss intrinsic (pro-environmental and prosocial) and extrinsic (financial incentives and reputational motivation) drivers of sustainable consumption by focusing on a wide range of behavioral interventions employed in public policies by private organizations and governments. Staddon et al. (2016) synthesize empirical evidence from 22 studies of behavioral change intervention's impact on energy saving and find that social and physical opportunities can help save energy.

As for concrete behavioral economics hallmarks' applicability, the following papers provide evidence of the external validity of prospect theory's core findings: Levy (1997) reviews social psychology and experimental economics to capture changes to the external validity of prospect theory-based hypotheses as well as large-scale applications of prospect theory's core assumptions on international politics, foreign policy, interactive decision making and political and strategic foreign policy contexts. McDermott et al. (2008) discuss prospect theory's origins in evolutionary psychology in order to model context-dependent risk aversion and risk-seeking behavior to apply prospect theory's core findings to problems relating to sustainable development ranging from politics, voting, legislator behavior, international bargaining and conflict as well as economic development and reform.

As for international outlooks, the following papers provide an academic overview of the global validity of behavioral insights in the environmental domain. Halvorsen (2012) analyzed household recycling activities across 10 OECD countries revealing that the belief in recycling being good for the environment and viewing recycling as a civic duty are the most powerful motivating nudges as well as recycling service supply being effective to change households' behavioral patterns. Silvi and Padilla (2021) use a dataset measuring individual adoption of eight pro-environmental behaviors in the European Union and find intrinsic motivation to be a dominant factor underlying environmentalism. Soyez (2012) studies national cultural values' influence on personal pro-environmental value orientations empirically by including over 1000 consumers in five nations finding eco-centric value orientation to be important in the US, Canadian, German, and Australian samples, while an anthropocentric value orientation is salient in the Russian sample, characterized by collectivistic values.

The following part analyzes three concrete behavioral interventions that were conducted at university campuses employing field studies, survey studies as well as laboratory experiments. The findings highlight the importance of social status, trust and reciprocity for communities to foster pro-social and environmental norms. Bundling two time prospects to evaluate choices for now and later concurrently can raise intergenerational harmony in the sustainable development domain. The findings are reflected upon the newest developments in behavioral sciences advocating for self-determined leadership directives and individualism in behavioral choices.

3 Behavioral Sustainable Development

3.1 Ethical Decision-Making Under Social Uncertainty

Two field experiments at a North American university campus investigated the role of social status in sustainable development. Puaschunder (2017a, b) studied prospect theory for social status gains and losses. Social status loss prospects were hypothesized to drive social responsibility in the environmental domain (Puaschunder 2017a, b). The individual reference point and drops in social status opportunities by removing social norms and social status cues from the environment were hypothesized to 'nudge' people into sustainability compliance (Puaschunder 2017a, b).

The two field experiments operationalized social status drops by signages that were placed next to recycling bins in university dormitories as well as signages that were placed in a library next to self-controllable light switches. The signs in the dormitory garbage collection areas either displayed the university logo and the sustainability initiative logo of the university plus the university logo or only the sustainability initiative logo next to the recycling bins. The signage in the libraries either held the university logo and the sustainability initiative logo of the university plus the university logo plus the instruction 'Please avoid losing energy by switching off lights after use' or only the sustainability initiative logo plus the instruction 'Please avoid losing energy by switching off lights after use' next to the self-controllable light switches. After a period of several weeks of exposure to the signages, which showed a positive impact, the removal of the signages even furthered the positive effect of social status insignia removal on pro-social behavior in the environmental domain. Overall, the Puaschunder (2017a, b) study provided evidence for sustainability being influenced by environmental cues, especially after the removal of social status prospects from the environment.

The finding of social status drops in the removal of social cues from the environment was recently replicated in an independent large-scale study performed by a research team at the Institute for Advanced Studies (IHS) in Vienna, Austria, European Union. Starting in 2019, field experiments were planned (Walter et al. 2019) and executed in 2021 in a field experiment including 400 garbage areas in Austrian community housing to test the impact of signage in garbage areas littering of the inhabitants (Gangl et al. 2021). The design included 440 waste disposal areas in 89 social community housing projects in 15 different districts of Vienna, Austria, covering almost 30,000 apartments and over 70,000 inhabitants (Gangl et al. 2022c). Gangl et al. (2022c) show that displaying a nature picture is most effective in reducing litter following a picture displaying watchful eyes and these effects persisted and even became stronger in the after-exposure period in the seven weeks following the intervention (Gangl et al. 2022c). The replication (Gangl et al. 2022c) of the initial results at a North American university campus (Puaschunder 2017a, b) underlines the robustness of the findings and serves as a powerful example how ideas can inspire academics from one continent to another via personal contacts.

An additional field experiment by a research team at the Institute for Advanced Studies (IHS) (Gangl et al. 2022b) tested the effect of visual cues on recycling behavior in public locations.

In a different project, researchers at the Institute for Advanced Studies (IHS) in Vienna Austria replicated the second part of the Puaschunder (2017a, b) study in using cards as an intervention to elicit energy light consumption conscientiousness (Abstiens et al. 2022; Gangl et al. 2022a). In a field experiment in workplace settings, reminders of energy consumption (push-adds and informational reminders) were placed and tested as reminders to switch off the lights and other energy-consuming devices (e.g., heating and computer screen) when leaving work (Abstiens et al. 2022).

Over all studies, one can sum up that social norms and environmental cues appear to elicit pro-social sustainability choice patterns.

In order to find explanations for the impact of the environment on human decision-making, Puaschunder (2020a) offers four studies that prove that people discount differently when being in an 'environmental' mindset. An online survey finds that individuals tend to feel they make the most out of their time when spent in the natural environment in comparison to economic time use such as being productive at work and social time spent with others (Puaschunder 2020a). Valuing time spent outdoors in the open environment the longest may also elicit preciousness feelings that one wants to conserve—raising also the wish to keep the environment clean by being sustainable.

In combination, these results add to mental accounting theories' context dependency and prove social status drops outlined by mere signage to heighten pro-social environmentalism (Thaler 1999; Puaschunder 2017a, b, 2020a). Innovative basic research conducted at a North American university stood the test of replication in a different time, national setting and more diverse population (Puaschunder 2017a, b; Gangl et al. 2022c). Together the findings hold invaluable insights on incentives to nudge individuals into positive societal change and bestow decision-makers with sustainability.

3.2 *The Future is Now: How Joint Decision-Making Curbs Hyperbolic Discounting*

In an international collaboration, one pre-study and two survey studies were conducted to test the influence of presentation modes on intergenerational equitable decision-making in the policy domain (Puaschunder and Schwarz 2012). Surveys were first handed out in a paper-and-pen format to generate a scale of public policies with different time foci and social responsibility levels (Puaschunder and Schwarz 2012). In total 36 fictitious public policies were generated from a United Nations catalog of public policies that included the areas of 'Culture, 'Economics,' 'Education,' 'Foreign Aid,' 'Infrastructure,' and 'Environment' (Puaschunder and Schwarz 2012). The fictitious policies were then rated by 120 public policy experts and students from a North American university and a European university. The policies were then presented at a mainland China university campus paper-and-pen and online in an Amazon Mechanical Turk (MTurk) online survey format (Puaschunder and Schwarz 2012). Respondents in mainland China and online were given the choice to allocate fictitious tax allocations towards these policies that ranged from very short-sighted to very long-term oriented (Puaschunder and Schwarz 2012).

The presentation mode of the policies varied between individual evaluation, combined evaluation and bundled evaluation (Puaschunder and Schwarz 2012). This behavioral intervention only varied the display of the options (Puaschunder and Schwarz 2012). The policies were scrambled in between and within subjects in order to avoid any other influence factors and specific preferences (Puaschunder and Schwarz 2012).

The study found the same effect in the online and the paper-and-pen setting as well as held across cultures (Puaschunder and Schwarz 2012). Evaluating two policies with different outcome times—one that would benefit current and one that would favor future generations—together elicited more intergenerationally equitable choices when surveying 223 Chinese individuals and 374 international respondents online (Puaschunder and Schwarz 2012). The joint alternative presentation mode therefore can boost sustainability (Puaschunder and Schwarz 2012). At the same time, this bundling effect increases rationality and therefore crowds out social responsibility (Puaschunder and Schwarz 2012). The joint alternative thus allows for implementing sustainable development but also implies a trade-off between future-orientation and social responsibility in intergenerational decisions (Puaschunder and Schwarz 2012).

Policy makers are advised to consider multi-faceted decision schemata going from single evaluations to combined and bundled ones in order to avert social responsibility drops (Puaschunder and Schwarz 2012). Tax allocations for socially-responsible policies are an additional proposal for using the joint decision-making advantage without trading off from social responsibility (Puaschunder and Schwarz 2012). In the real-world application, age-differentiated decision consortia and in particular inviting all ages to the decision table can help overcome social responsibility drawbacks. Overall, this study proves that international collaboration in academia provides fruitful global validity of an effect that can aid in fostering intergenerational equity.

3.3 Trust and Reciprocity Drive Social Common Goods Allocation Norms

The third university-based study included a behavioral economics university laboratory experiment in the realm of tax psychology (Puaschunder 2015a). Tax psychology has become a vibrant field within behavioral economics to study human drivers of tax compliance in order to curb tax evasion in positive settings (Kirchler 2007).

In behavioral economics laboratory experiments at a North American campus, social cues that elicit common goals compliance were tested in order to derive behavioral insights on how to improve tax honesty (Puaschunder 2015a). In a behavioral decision-making laboratory experiment, 256 participants played a standard economic trust game followed by a common goods game (Puaschunder 2015a). In the trust game, one person, player 1, could trust by giving an amount of money to a player 2 anonymously enabled via computers. Player 2 can then decide how much to return to Player 1 or even defect by completely keeping the received payment. Money transfers were incentivized in the experiment by doubled payouts for each of the recipients, first player 2 in the initial transfer, and then player 1 in the returned money from player 2 (Puaschunder 2015a). This standard trust game captures the degree of trust in humans (Puaschunder 2015a). As the typical set-up of these economic studies does not shed light on the future consequences of the trusting, rewarding or defecting experience of the players, Puaschunder (2015a) tested the subsequent consequences of the trusting (of player 1 on player 2) and the reciprocal return willingness (of player 2) (Puaschunder 2015a). The public common goods game studies collective action problems. In this classic economics laboratory experiment, each participant chooses how much to contribute to a common pool that equally returns benefits to all participants (Puaschunder 2015a). The ideal outcome would occur if everyone contributes the maximum amount, but self-interests interfere with people saving for themselves without any public contributions, which lowers the outcomes for everyone.

Puaschunder (2015a) used an in-between-subjects design, in which players persisted from game 1 (trust game) to game 2 (common good game) in order to capture the impact of differing trust and reciprocity experiences—of rewarding and defecting outcomes—on subsequent common goods contributions. In a non-deception laboratory experiment, all participants were informed that the other players also received the same amount from which they had to decide individually how much to keep for themselves or contribute to the same common pool (Puaschunder 2015a). All contributions of both players of this game were added up in a common pool, which was factored by 1.6 and then split evenly among all players (Puaschunder 2015a).

In 256 participants playing a standard economic trust game followed by a common goods game, evidence was found for trust and reciprocity from the trust game leading to individuals contributing more to the common pool in the common goods game (Puaschunder 2015a). The more trust and reciprocity were practiced in the trust game and experienced via reciprocity, the more common goals compliance was exhibited (Puaschunder 2015a).

Eliciting trust and reciprocity in societies can thus create a socially-favorable environment that supports social tax ethics norms (Puaschunder 2015a). The findings provide behavioral insights derived at a university laboratory that helps leverage trust and reciprocity as interesting tax compliance antecedents (Puaschunder 2015a). As first shown in a university setting, the results could now be tested for their external and global validity. Future studies could see if the impact of trust on common goods compliance now also holds for other contexts, settings and nationalities.

In their wider application, the results speak for trustful environments and reciprocity norms at universities as a vital breeding ground for setting society on a path to sustainable development. Future laboratory experiments and field studies could derive further implications for governmental-citizen relations. Policy makers and public servants are advised to establish a service-oriented customer atmosphere with citizens breeding trust and reciprocity in order to reach common societal goals, such as the Sustainable Development Goals.

4 Conclusions

This paper presented the outcomes of three different studies in the realm of sustainability conducted at universities on three continents.

Study 1 comprised two field experiments staged at a North American university campus that finds that social status drops elicit pro-social and sustainable behavior. The study was replicated and found convincing evidence that the field studies at a North American university are generalizable on a different continent and varying populations in the field.

Study 2 included university settings in Asia, Europe and North America. A pre-test generated a scale for fictitious public policies that ranged from short-term to long-term orientation as well as from egoistic profit-maximizing to more socially responsible choices. In subsequent paper-and-pen as well as online survey studies, tax allocation preferences for the policies were measured in mainland China and online with an international sample. The results show that by presenting two time-diverse policies closely together in a bundled representation that allows to evaluate both at the same time physically next to each other, people tend to prefer more internationally-balanced choices. The bundled presentation mode also makes decision-makers more rational. It is therefore advisable to make decisions in age-varying consortia and evaluate them together in order to elicit intergenerational equity. At the same time, social responsibility should be factored into decisions per se in order to contribute to common goals. The results serve as powerful nudges to present the Sustainable Development Goals and in particular, the specific targets that demand for an intergenerationally-conscientious consensus.

Study 3 reports the results of a university behavioral economics decision sciences laboratory that found trust and reciprocity to help strengthen social common goods allocation norms in the tax compliance domain.

In their entirety, the three behavioral economics studies highlight the important role universities can play in fostering sustainable development. International collaborations and inspiring results but also the integrity and intermediary role of academics are all credentials that predestine universities as vital breeding grounds of societal change. In particular in their function to combine research with the teaching of a new cadre of academics, public and private sector leaders and executives, universities can become catalysts of social transformation. In the non-immediate results-driven basic research but also in the applied science pledge, academics can foster sustainable development most innovatively as well as with a widespread impact for this generation and the following.

As for limitations of the presented effects, the standard behavioral economics limitations apply. Field experiments may elucidate an effect with high external validity but comparatively lower internal validity. The results presented in study 1 may therefore hold credible behavioral patterns that are exhibited by a multitude of people but if the effect is directly related to the intervention could be investigated more precisely in a stringent laboratory setting. Online surveys have the general problem of controlling who the subjects are and overseeing their behavior. Study 2 may therefore be limited in its sampling. Vast intercultural differences exist in serving as respondents in studies and the international angle of the study may therefore weaken the actual effect, which held overall. The internal validity of controlled laboratory experiments is relatively high compared to other modes of measurement, yet the global validity and generalizability of effects found in laboratory settings have to stand the test of time and further exploration for applicability to real-world settings. Study 3 is therefore limited in its potential to describe actual real-world relevant behavior and warrants for further exploration of the effect of trust for tax compliance, also in the international arena.

Over all studies and papers in general one has to mention limitations of the field of behavioral sciences as well. Most recent replication crises have warned about overgeneralizing results but call for more self-determined decision making. Replicated results should be communicated to the general public in a palpable way in order to help the general populace understand their choice patterns so that they can make decisions informed and self-determined. In this democratization of behavioral insights lies the future grand change that an as important and global endeavor as sustainable development deserves.

In the future, behavioral economics is prospected to become a mainstream solution to capture the economic calculus of human beings. Since its inception in the 1950s, behavioral economics has leveraged into a prominent way to actually hold how humans tend to react to a certain environment. Many powerful laboratory experiments, survey studies and field experiments have elucidated a wide catalog of decision-making predictions. The subsequent behavioral insights revolution made use of these choice patterns by crafting well-curated design architectures that should get people to behave in a certain pro-social and sustainable way.

With the advent of behavioral insights and the prominence of behavioral nudges and winks, however, came a widespread knowledge of certain manipulations. Most recent calls now advocate to rather focus away from implicit guidance that curbs certain tendencies or heightens the likelihood of favorable choices to a more educated and self-determined behavioral leadership solution (Liscow and Markovits 2022; Puaschunder 2020a). The rationale is that individuals should be given the opportunity to be educated about behavioral tendencies and then make an informed decision about what suits them and their lifestyles in harmony with their environment. Internet help tools, like AI-enhanced search engines, could be empowered with behavioral choice patterns that aid people find the right choice based on their needs, history and community. Deriving information on the baseline and putting one's own choice in relation to the norm, will help create the intrinsic circumstances under which decision-makers exhibit wise individually-accustomed yet hopefully also social responsibility choices when nurturing an awareness for the long-term benefits of pro-social norms. Self-determined behavioral leadership and mastery of one's own choice power can thereby be seen as a new generation of behavioral economics and a revolution in self-determined decision leadership. Reclaiming one's decision-making autonomy and endowment-driven choices can thereby be seen as tomorrow's cost-effective, easily implementable human-innate power conditions that help promote sustainable consumption of the broad populace on a daily basis.

Overall, university research grants recommendations on how to breed positive social forces that work toward sustainable development. Behavioral economics results have served as an innovative way to better understand the socio-dynamics around environmental ethicality in the sustainability domain in many university-led activities in research and teaching. While main parts of the behavioral economics community are currently employed to find the generalizability and validity of the found effects, a concurrent revolutionary wave is starting to incept the idea of self-determined, self-endowed choices based on the broad understanding of behavioral insights of the general populace. This democratization of behavioral economics accounts for the newest revolution in decision science and warrants integrating behavioral insights into leadership theory and practice. All these endeavors will hopefully foster sustainable development on a global level with benefits for the earth of this generation and the following.

References

Abrahamse W, Matthies E (2012) Informational strategies to promote pro-environmental behaviour: Changing knowledge, awareness and attitudes. In: Steg L, Berg AvD, Groot JD (eds) Environmental psychology: an introduction. British Psychological Society, Blackwell, NJ

Abstiens K, Gangl K, Keinprecht M, Laa E, Plank K (2022) Energiesparen am Büroarbeitsplatz. Verhaltenswissenschaftliche Maßnahmenentwicklung anhand eines praktischen Beispiels. November Research Report, Institute for Advanced Studies (IHS), Vienna, Austria

Akerlof GA (2001) Nobel Prize Lecture. Behavioral Macroeconomics and Macroeconomic Behavior. December 8, 2001, Aula Magna, Stockholm University. https://www.nobelprize.org/prizes/economic-sciences/2001/akerlof/lecture/

Austin J, Hatfield DB, Grindle AC, Bailey JS (1993) Increasing recycling in office environments: the effect of specific, informative cues. J Appl Behav Anal 26(2):247–253

Bateson M, Nettle D, Roberts G (2006) Cues of being watched enhance cooperation in a real-world setting. Biol Let 2(3):412–414

Bateson M, Robinson R, Abayomi-Cole T, Greenlees J, O'Connor A, Nettle D (2015) Watching eyes on potential litter can reduce littering: evidence from two field experiments. PeerJ: Life & Environ 3:1443

Brundtland Commission (1987) Report of the world commission on environment and development: our common future. https://sustainabledevelopment.un.org/content/documents/5987our-common-future.pdf

Cosic A, Cosic H, Ille S (2018) Can nudges affect students' green behaviour? A field experiment. J Behav Econ Policy 2(1):107–111

Dear K, Dutton K, Fox E (2019) Do 'watching eyes' influence antisocial behavior? A systematic review & meta-analysis. Evol Hum Behav 40(3):269–280

Durdan CA, Reeder GD, Hecht PR (1985) Litter in a University Cafeteria: demographic data and the use of prompts as an intervention strategy. Environ Behav 17(3):387–404

Dwyer WO, Leeming FC, Cobern MK, Porter BE, Jackson JM (1993) Critical review of behavioral interventions to preserve the environment: research since 1980. Environ Behav 25(5):275–321

Ernest-Jones M, Nettle D, Bateson M (2011) Effects of eye images on everyday cooperative behavior: a field experiment. Evol Hum Behav 32(3):172–178

Fox S, Dayan K (2004) Framing and risky choice as influenced by comparison of one's achievements with others: the case of investment in the stock exchange. J Bus Psychol 18(3):301–321

Gangl K, Abstiens K, Gsottbauer E, Kirchler E, Riener G, Seifert M, Walter A (2022a) Energiekrise – was tun? Verhaltenswissenschaftliche Empfehlungen. IHS Policy Brief 10, July 2022

Gangl K, Grosch K, Walter A (2021) Mehr Sauberkeit im Gemeindebau: Ergebnisse eines verhaltensökonomischen Feldexperiments in den Müllbereichen, Research Report, Institute for Advanced Studies (IHS), Vienna, Austria

Gangl K, Spitzer F, Walter A (2022b) Abfalltrennung und Littering im öffentlichen Raum. Ein verhaltensökonomisches Feldexperiment. February Research Report, Institute for Advanced Studies (IHS), Vienna, Austria

Gangl K, Walter A, Van Lange PAM (2022c) Implicit reminders of reputation and nature reduce littering more than explicit information on injunctive norms and monetary costs. J Environ Psychol 84(101914):1–9

Halvorsen B (2012) Effects of norms and policy incentives on household recycling: an international comparison. Resour Conserv Recycl 67:18–26

Kahneman D (2008) Thinking, fast and slow. Farrar, Straus & Giroux, New York

Kahneman D, Knetsch JL, Thaler RH (1991) The endowment effect, loss aversion, and status quo bias. J Econ Perspect 5(12):1325–1347

Kahneman D, Tversky A (1979) Prospect theory: an analysis of decision under risk. Econometrica 47(2):263–291

Kahneman D, Tversky A (1983) Choices, values, and frames. Am Psychol 39(4):341–350

Kahneman D, Tversky A (1992) Advances in prospect theory: cumulative representation of uncertainty. J Risk Uncertain 5:297–324

Kahneman D, Tversky A (2000) Choices, values, and frames. Cambridge University Press, Cambridge

Kirchler E (2007) The economic psychology of tax behavior. Cambridge University Press, Cambridge

Kolodko J, Read D, Taj U (2016) Using behavioural insights to reduce littering in the UK. Cleanup Britain. http://cleanupbritain.org/WBS-Report-for-CLUB.pdf

Laibson D (1997) Golden eggs and hyperbolic discounting. Quart J Econ 11:443–477

Leal Filho W, Lange Salvia A, Beynaghi A, Fritzen B, Ulisses A, Veiga Avila L, Shulla K, Vasconcelos CRP, Moggi S, Mifsud M, Anholon R, Rampasso IS, Kozlova V, Iliško D, Skouloudis A, Nikolaou I (2023) Digital transformation and sustainable development in higher education in a post-pandemic world. Int J Sust Dev World. https://doi.org/10.1080/13504509.2023.2237933

Lehman PK, Geller ES (2004) Behavior analysis and environmental protection: accomplishments and potential for more. Behav Soc Issues 13(1):13–33

Levy JS (1997) Prospect theory, rational choice, and international relations. Int Stud Quart 4(1):87–112

Liscow Z, Markovits D (2022) Democratizing Behavioral Economics. Yale J Regul 39(3):1274–1342. https://www.yalejreg.com/print/democratizing-behavioral-economics/

McCoy K, Oliver JJ, Borden DS, Cohn SI (2018) Nudging waste diversion at Western State Colorado University: application of behavioral insights. Int J Sustain High Educ 19(3):608–621

McDermott R, Fowler JH, Smirnov O (2008) On the evolutionary origin of prospect theory preferences. J Polit 70:335–350

Meng MD, Trudel R (2017) Using emoticons to encourage students to recycle. J Environ Educ 48(3):196–204

Nobel Prize (2002) The Sveriges Riksbank prize in economic sciences in memory of Alfred Nobel 2002. https://www.nobelprize.org/prizes/economic-sciences/2002/summary/

Nobel Prize Committee (2002) Foundations of behavioral and experimental economics: Daniel Kahneman and Vernon Smith, Nobel Prize in Economics documents 2002, Nobel Prize Committee. https://www.nobelprize.org/prizes/economic-sciences/2002/summary/

Ostrom E (2008) Nobel Prize lecture: beyond markets and states: polycentric governance of complex economic systems, 8 December 2009 at Aula Magna, Stockholm University. https://www.nobelprize.org/prizes/economic-sciences/2009/ostrom/lecture/

Puaschunder JM (2015a) Trust and reciprocity drive common goods allocation norms. In: Proceedings of the Cambridge business & economics conference, UK: Cambridge University; Proceedings of the 2015 6th international conference of the association of global management studies at Alfred Lerner Hall of Columbia University, New York: The Association of Global Management Studies; Proceedings of the Academy of Management

Puaschunder JM (2015b) When investors care about politics: a meta-synthesis of political divestiture studies on the capital flight from South Africa during Apartheid. Bus Peace Sustain Dev 5(24):29–52

Puaschunder JM (2016) The role of political divestiture for sustainable development. J Manag Sustain 6(1):76–91

Puaschunder JM (2017a) Ethical decision making under social uncertainty: an introduction to Überethicality. Sustain Prod Consum 12:78–89

Puaschunder JM (2017b) Socio-psychological motives of socially responsible investors. Glob Corp Gov: Adv Financ Econ 19:209–247

Puaschunder JM (2019a) Intergenerational equity: corporate and financial leadership. Edward Elgar, Cheltenham, UK & Northampton, MA

Puaschunder JM (2019b) When investors care about causes: learning from political divestiture during Apartheid and the Sudan crisis for the transition to renewable energy via green bonds. In: Lukhele F (ed) Apartheid in South Africa: reflections and lessons, pp 1–28. Nova, Hauppauge, NY

Puaschunder JM (2020a) Behavioral economics and finance leadership: nudging and winking to make better choices. Springer Nature, Cham

Puaschunder JM (2020b) Governance and climate justice: global south and developing nations. Palgrave Macmillan, New York; Springer Nature, Cham

Puaschunder JM (2023) Human rights online: towards a new generation of human rights in the virtual world. 75 years of the Universal Declaration of Human Rights (UDHR) historic edition. United Nations, New York

Puaschunder JM (forthcoming a) Behavioral economics for all: from nudging to leadership

Puaschunder JM (forthcoming b) Science diplomacy online: implementing sustainable development in the virtual world. In: Leal W, Kautisch S, Wall T, Williams K, Rewhorn S, Paul S (eds) Digital technologies to implement the UN sustainable development goals. Springer Nature, Cham

Puaschunder JM (forthcoming c) The future of resilient finance: finance politics in the age of sustainable development. Palgrave Macmillan

Puaschunder JM, Schwarz, G (2012) The future is now: how joint decision-making curbs hyperbolic discounting but blurs social responsibility in the intergenerational equity public policy domain. The Situationist Law and Mind Sciences. https://thesituationist.wordpress.com/2012/12/17/the-situation-of-intergenerational-equity/

Schwartz D, Milfont TL, Hilton D (2019) The interplay between intrinsic motivation, financial incentives and nudges in sustainable consumption. In: A research agenda for economic psychology. Edward Elgar Publishing

Sen AK (1971) Choice functions and revealed preference. Rev Econ Stud 38(3):307–317

Sen AK (1977) Rational fools: a critique of the behavioral foundations of economic theory. Philos Public Aff 6(4):317–344

Sen AK (1993) Internal consistency of choice. Econometrica 61(3):495–521

Sen AK (2002a) Consistency of choice. In: Sen AK (ed) Rationality and freedom. Harvard University Press, Cambridge, MA, pp 207–224

Sen AK (2002b) Goals, commitment, and identity. In: Sen AK (ed) Rationality and freedom. Harvard University Press, Cambridge, MA, pp 35–56

Silvi M, Padilla E (2021) Pro-environmental behavior: social norms, intrinsic motivation and external conditions. Environ Policy Gov 31(6):619–632

Simon HA (1956) Rational choice and the structure of the environment. Psychol Rev 63(2):129–138

Simon HA, Bartel RD (1986) The failure of armchair economics. Challenge 29(5):18–25

Soyez K (2012) How national cultural values affect pro-environmental consumer behavior. Int Mark Rev 29(6):623–646

Staddon SC, Cycil C, Goulden M, Leygue C, Spence A (2016) Intervening to change behaviour and save energy in the workplace: a systematic review of available evidence. Energy Res Soc Sci 17:30–51

Sussman R, Greeno M, Gifford R, Scannell L (2013) The effectiveness of models and prompts on waste diversion: a field experiment on composting by cafeteria patrons: prompts, models, and cafeteria composting. J Appl Soc Psychol 43(1):24–34

Thaler RH (1999) Mental accounting matters. J Behav Decis Mak 12(3):183–206

Thaler RH (2017) Nobel Prize lecture: from cashews to nudges: the evolution of behavioral economics, 8 December 2017 at the Aula Magna, Stockholm University. https://www.nobelprize.org/prizes/economic-sciences/2017/thaler/lecture/

Thaler RH, Sunstein CR (2008) Nudge: Improving decisions about health, wealth, and happiness. Yale University Press, New Haven

United Nations Department of Economic and Social Affairs (2015) Sustainable development: the 17 goals. https://sdgs.un.org/goals

United Nations Department of Economic and Social Affairs Sustainable Development. Transforming our world: The 2030 Agenda for Sustainable Development. https://sdgs.un.org/2030agenda

United Nations Sustainable Development Goals. https://www.undp.org/sustainable-development-goals?gclid=EAIaIQobChMIiKOl4syJgAMVq5paBR3eXASHEAAYAyAAEgLqz_D_BwE

Walter A, Gangl K, Sonntag A, Kocher MG (2019) Verhaltensökonomische Maßnahmen für mehr Sauberkeit im Gemeindebau, November Research Report, Institute for Advanced Studies (IHS), Vienna, Austria

Unified Sustainability Consciousness Through DISEMBARC Framework—Reverberations of India's Mission LiFE

Parul Rishi and Ruchi Mudaliar

Abstract Unified sustainability consciousness (USCon) refers to a shared cognition and awareness among individuals and society about the importance of sustainability and the need to protect the environment and preserve natural resources for future generations. This consciousness encompasses a holistic view of sustainability that considers the interconnections between psychological, philosophical, economic, social, and environmental factors and recognizes the impact of human actions on the planet. USCon is closely aligned with LiFE (Lifestyle for Environment), a mission adopted by India to achieve SDG and Net Zero targets, which is possible through sustainable practices and behaviours at the individual and collective levels for the creation of a sustainable future. For USCon across diverse cross-cultural populations, the DISEMBARC framework has been proposed which includes components like Diversity acceptance (D), Inclusivity (I), Sustainability Behaviour (S), Ethics and Values (E), Mindfulness (M), Benevolence and wellbeing (B), Adaptability (A), Resilience (R) and Creativity (C). A conceptual framework of DISEMBARC through analysis of transdisciplinary sustainability literature can form a basis for the development of a universal assessment tool to obtain comparable component-vice and composite LiFE Quotient (LQ). An extensive multi-country study is proposed in partnership with various sustainability centres, to obtain LQ and use it in fostering differential sustainability practices as per the respective socio-cultural milieu of different countries.

Keywords Lifestyle for Environment (LiFE) · Sustainability · Consciousness · Mindfulness · Climate change

---

P. Rishi (✉)
Faculty of Human Resource Management, Indian Institute of Forest Management, Bhopal, India
e-mail: parul@iifm.ac.in

R. Mudaliar
Amity Institute of Psychological Sciences (AIPS), Amity University, Noida, India

W. Leal Filho et al. (eds.), *The Contribution of Universities Towards Education for Sustainable Development*, World Sustainability Series,
https://doi.org/10.1007/978-3-031-49853-4_29

1 Introduction

Humanity is currently facing major environmental challenges. In the last century, human activities have dramatically modified natural processes while significantly affecting social-ecological systems, leading to the current environmental crisis (Ruiz-Mallén and Heras 2020). These environmental challenges warrant the adoption of sustainable practices to tackle these issues. Failure to translate intentions into actual behaviour is one reason that is known in many areas of human action (Torma et al. 2018). In this context, the concept of sustainability has gained widespread attention as individuals, communities and institutions strive to balance economic growth with environmental protection and social responsibility. Various factors have been linked to sustainable consumption in different research studies. A person's internal factors, such as environmental attitudes, values, personal norms, and perceived responsibility have a positive direct effect on engagement with sustainable consumption (Čapienė et al. 2022). People's 'inner worlds' which include their emotions, thoughts, identities and beliefs, lie at the heart of actions for sustainability and have powerful transformative capacity for system change (Ives et al. 2020). While human values play a pertinent role as antecedents of consumers' sustainability consciousness (Balderjahn and Hüttel 2019), developing Sustainability Consciousness (SCon) is seen as a pertinent key in unlocking answers to pressing global environmental issues especially climate change and similar concerns that negatively impact our planet. While Gericke et al. (2019) refer to the concept of sustainability consciousness as an individual's experience and awareness of sustainable development, SCon demands a deep understanding of the complex systems involved in environmental and socio-ecological aspects and pertains to an individual's comprehension and consideration of the same for the long-term health and prosperity of the planet and its ecosystems.

To have a SCon, we must be mindful of our consumption patterns. Geiger et al. (2019) emphasize that mindfulness may foster sustainable consumption behaviour through the reduction of the attitude–behaviour gap, or by weakening material values while increasing subjective well-being. SCon also develops when individuals actively work to reduce/manage waste including E-Waste, adopting sustainable food systems, using eco-friendly products, avoiding single-use plastic and making all these practices part of our lifestyle and culture as envisioned in Mission LiFE.[1] Furthermore, it also involves taking actions to support sustainability, such as endorsing renewable energy, saving water, reducing carbon emissions, and preserving natural resources. Besides, various other factors like mindful decision-making and the use of transdisciplinary, cross-national and co-creational approaches together with additional factors like behavioural, social, cultural, philosophical, ecological, technological, and economic aspects add to making it Unified Sustainability Consciousness (USCon).

The paper aims to decode the concept of Unified Sustainability Consciousness (USCon) through highlighting the interplay of various factors like psychological, philosophical, social, economic and environmental. Further, it proposes a

[1] http://missionlife-moefcc.nic.in/.

framework called DISEMBARC (Diversity Acceptance, Inclusivity, Sustainability Behaviour, Ethics and Values, Mindfulness, Benevolence, Adaptability, Resilience and Creativity) which conceptually explains the possible constituents of USCon. The paper also narrates how this framework is being practised in India's Mission Lifestyle for Environment (LiFE) and in Dayalbagh Educational Institute of India through various sustainability initiatives, collectively intended to meet the targets of SDGs. Being a theory building book chapter, a conceptual model has been proposed based on the review of literature and doctrinal analysis, live examples and case studies. By incorporating these factors and approaches and adopting a sustainable mind set, we can positively impact the planet and work towards creating a more sustainable future for ourselves and for the generations to come.

2 Concept of Unified Sustainability Consciousness (USCon)

The term USCon refers to a shared understanding and appreciation for environmental sustainability that is recognized by individuals and communities worldwide. It involves acknowledging the interconnectedness of environmental, social, and economic issues, as well as the impact our actions have on the planet and future generations. USCon also recognizes the interdependence between humanity and nature taking a holistic approach to sustainability. It goes beyond mere environmental protection understanding that human behaviour is influenced by various factors which can be understood through integrating diverse fields of understanding like psychology, philosophy, sociology and economics and related others. Human factors aligned to SCon differ across cultures and are strongly influenced by socio-cultural environments. USCon prioritizes the importance of making deliberate choices in our day-to-day lives which includes taking steps to decrease our carbon emissions, conserving water and saving energy/adopting alternative modes of energy, minimizing waste and encouraging sustainable practices both individually and within institutions. Additionally, USCon emphasizes the understanding that social ecological and economic systems are interconnected and that it is vital to make decisions with long-term consequences in mind.

USCon recognizes that environmental sustainability is a worldwide concern, necessitating collaboration and cooperation among individuals, communities, institutions, and governments to advocate for sustainable practices and safeguard the environment. An environmentally conscious lifestyle means responsible behaviour in using natural resources to preserve them and includes simple but effective everyday actions such as saving water and electricity, recycling, buying ecological goods, using environmentally friendly means of transportation, as well as vehicle-pooling (Maranzano et al. 2021). Piscitelli and D'Uggento (2022) emphasize that in the first years of life, the family and the educational system play a crucial role in the formation of environmental sensitivity. They can influence the personal attitude of an individual in choosing which behaviours to adopt towards environmental problems. Achieving

USCon requires a shift in mindset, values and behaviours towards a more sustainable and environmentally conscious way of living.

In the next part, we discuss how USCon evolves as a result of the interaction of different contributing factors.

3 The Interplay of Different Factors in USCon

Environmental problems have their origin in human behaviour, and as a result, any solution to environmental issues will require changes in behaviour. USCon encompasses multiple factors that influence our behaviour and decision-making towards a sustainable future. While many disciplines in the social and behavioural sciences offer important perspectives on the behaviours linked with environmental problems, the study of the individual brings a focus on cognitive, social, and motivational processes that provide insights into effective ways to promote change (Schultz 2014).

3.1 Psychological Factors

Several ***psychological factors*** play an important role in shaping and impacting USCon. These psychological factors also shape our beliefs, values, and attitudes towards the environment. Understanding the psychology behind environmental or sustainable behaviours is central to a sustainable future and widespread behaviour change. By understanding these factors, it is possible to develop interventions and strategies to promote sustainable practices and behaviours. However, despite its obvious importance and the substantive impact of research in this domain, knowledge of sustainable behaviour and decision-making is scanty and significantly lags behind other areas in consumer psychology (Trudel 2018). In their attempt to link psychological dimensions to sustainability, Corral-Verdugo et al. (2010) state that since most of the current environmental crisis has been caused by human behaviour, knowledge regarding what predisposes people to behave sustainably is required. Further, they mention that sustainable behaviour is the set of actions resulting in the conservation of natural and social resources, and psychology and other behavioural sciences are committed to studying what incites people to engage in conservationist actions. People are more likely to engage in sustainable behaviour when they feel that their actions can make a difference. To make sustainable decisions and actions in daily life, individuals need to know about sustainability, be conscious of their actions, and be active citizens (Triantafyllidou and Zabaniotou 2022). Behavioural science can also help us to understand the factors that motivate people to change their behaviour towards sustainability. When individuals become knowledgeable about the environmental impact stemming from their choices and recognize the advantages of embracing eco-friendly practices, they become more inclined to adopt such behaviours. For instance, research has shown that providing people with feedback about their environmental impact can be a powerful motivator for behaviour change. Strategies such as prompts, commitments, feedback, social norms, incentives, and convenience have all been shown to effectively promote pro-environmental behaviour—at least in some contexts, for some behaviours, and for some individuals (Schultz 2014).

3.2 *Philosophical Factors*

There are different world philosophies such as Buddhism, Hinduism and Taoism, which highlight the importance of living harmoniously with nature and realizing how all living beings are interconnected. By embracing the wisdom offered by these philosophies, it becomes feasible for us to develop sustainable practices that benefit both human well-being and the natural world. Ravikanth (2021) carried out a thorough inspection on the influence of Hinduism and Jainism, the oldest living religions under the Indian philosophical tradition, over environmental ethics. Central concepts such as *Anekāntavāda, dharma* (non-absolutism*), satyāgraha* (fasting*)*, and *Vasudhaiva Kutumbakam* (one earth, one family, one future) were discussed in both religions and it was shown how these concepts foster nature conservation and environmental awareness. Philosophical factors can shape sustainability consciousness in various ways. Our ethical beliefs and values deeply impact our approach while dealing with sustainability issues. For instance, we might prioritize sustainable practices if we genuinely believe that protecting the environment is our moral duty towards future generations. If we perceive ourselves as being integrated into a wider ecosystem it becomes more likely for us to prioritize sustainable practices. Philosophical beliefs concerning social justice also play an integral part in shaping sustainability consciousness because most sustainability problems have a greater impact on marginalized communities. Furthermore, having philosophical beliefs about holism can enhance comprehension of the interconnectedness between environmental, social and economic systems.

3.3 *Economic Factors*

Further, economic factors can act as both the barriers and promoters of USCon. It is possible that if sustainable products cost more than non-sustainable products, individuals may be less inclined to prioritize sustainability due to financial constraints. Although sustainable products and practices may come at a higher price, they offer long-term advantages such as decreased resource depletion and environmental degradation making them financially viable. When making choices about sustainable habits and behaviours, individuals often consider the cost–benefit analysis. For example, if individuals save money on their electricity bills by using energy-efficient light bulbs, or solar energy, they are more likely to prioritize this practice. Polasky et al. (2019) emphasize that economics, combined with other social and behavioural sciences, is crucial for understanding how it might be possible to shift human behaviour toward achieving sustainable development.

3.4 Social Factors

Additionally, Social factors, such as cultural norms and values, can also influence USCon. For instance, communities that prioritize sustainable practices, such as recycling and conservation, can create social norms (Perry et al. 2021) that promote USCon. Taking into consideration our surroundings, if people in our vicinity prioritize sustainable actions, there is a higher possibility that we will also follow the same. However, observation regarding the neglect of sustainable practices can promote a similar mindset. Education initiatives and awareness-raising campaigns serve as valuable contributions towards understanding how human behaviour affects environmental sustainability on a broader scale. Additionally, social networks hold great potential in disseminating information and providing support for sustainability-related actions. Friends and family members can easily share tips on adopting green practices while motivating each other along the way. Nonetheless, economic inequality poses a major impediment when attempting to cultivate consciousness surrounding sustainability matters; with limited resources available due to financial struggles in meeting basic needs.

3.5 Environmental Factors

Promotion of sustainability consciousness is greatly influenced by ***Environmental factors** too,* creating awareness and shaping individuals' attitudes and behaviours towards sustainable practices. Environmental degradation, such as deforestation, pollution, and biodiversity loss, showcases the negative consequences of unsustainable practices. When individuals witness the detrimental effects of human actions on ecosystems and wildlife, they are likely to be motivated to make more sustainable choices and advocate for environmental conservation. The impacts of climate change also play a crucial role in promoting sustainability consciousness. Extreme weather events, rising temperatures, and changing climatic patterns serve as visible reminders of the urgency to embrace sustainability. The soaring temperatures in various countries of Europe and America and flash floods in Himachal Pradesh and Uttarakhand states of India in 2023 are recent examples of that. These environmental changes increase awareness about how human activities are interconnected with and impact the environment thereby fostering a sense of responsibility to adopt sustainable practices. Access to green spaces like parks, gardens, and natural landscapes also contributes to sustainability consciousness. Being in nature fosters a sense of appreciation and responsibility towards the natural world which may lead to increased sustainability consciousness and a desire to protect these spaces.

The above analysis makes it evident that sustainability consciousness is shaped by a diverse range of factors: psychological, philosophical, economic, social, and environmental aspects. By understanding and addressing these diverse factors we can actively cultivate sustainability consciousness and drive positive change towards a more sustainable future. Besides the above factors, a framework known as DISEMBARC is being proposed which encompasses the essence of USCon as described below.

4 The DISEMBARC Framework

For USCon across diverse cross-cultural populations, the proposed DISEMBARC framework includes the components as depicted in the Fig. 1.

4.1 Diversity Acceptance (D)

The SCon necessitates a comprehensive understanding and utmost reverence for how human systems are intricately intertwined with natural systems across distinct social, cultural and economic circumstances. These multifaceted contexts significantly influence both our association with the environment and the far-reaching implications our actions bear upon diverse communities. The concept of social-ecological resilience

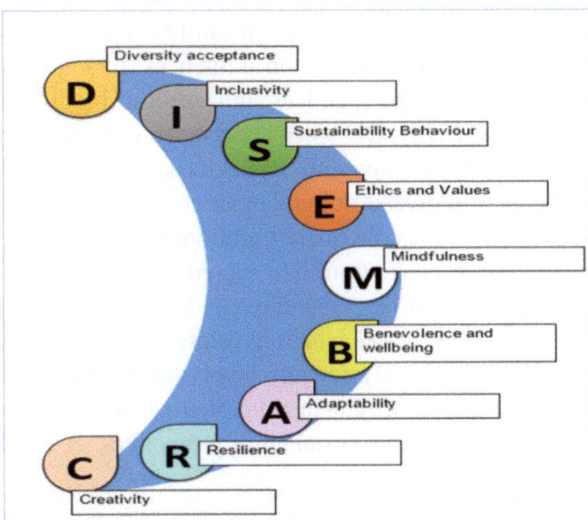

Fig. 1 DISEMBARC framework of sustainability behaviour

strongly emphasizes humanity's inherent place in the biosphere, thus highlighting the profound impact our decisions, values, and choices wield on the planet. Key issues such as democracy, health, poverty, and justice depend on the resilience of the biosphere, which is crucial for human well-being and dignity (Folke et al. 2016).

Diversity acceptance is crucial for promoting sustainability as it allows for the recognition and respect of the pluralistic perspectives, knowledge systems, beliefs, values and cultural practices that are essential for effective and equitable decision-making and help in promoting effective and culturally appropriate policies and practices.[2] Let us take the example of indigenous peoples. They have developed unique and effective methods for managing natural resources sustainably over generations, based on their indigenous knowledge of local ecosystems and their interdependent relationship with the environment. Going deeper into indigenous cultures, their diverse forms of traditional knowledge systems can be used to promote the sustainable development agenda (Magni 2017). It can result in effective and culturally appropriate pro-planet practices leading to better outcomes for both people and the environment. This approach has been successfully implemented in various countries, such as the incorporation of traditional ecological knowledge into forest/land management practices in the US, and Canada and the use of indigenous land management practices in the restoration of degraded landscapes in Australia (Souther et al. 2023; Hosen et al. 2020). One key driver for improving sustainability lies in fostering acceptance towards diversity within our society. This strategic move leads us towards novel paths that enable a panoramic view of daunting challenges surrounding sustainability issues while searching for holistic resolutions. The inclusion of individuals with varying backgrounds allows us access to diverse perspectives that enlighten our understanding of multifaceted problems faced today while shaping robust strategies for a stable future. In addition, innovative solutions achieved through cooperative efforts among heterogeneous teams, bear fruits taking into account social implications alongside economic ramifications and ecological aspects.

The role of bio-cultural diversity in nature conservation and sustainability includes the interconnectedness between cultural and biological diversity, their reciprocal effects, temporal and location influences, intangible cultural heritage, and the link between traditional practices and specific biodiversity aspects (Bridgewater and Rotherham 2019). Diversity acceptance can go a long way in promoting cultural preservation, an important aspect of sustainability, that recognizes the value and importance of diverse cultural practices and traditions. For example, in many indigenous communities, there are traditional practices that have been passed down from generation to generation, like sacred groves at many places in India (Sharma and Kumar 2021), Africa (Ogundayo and Adekunle 2019; Imarhiagbe and Ogwu 2022) and some other places which have proven to be effective in preserving biodiversity and sustainability. By actively involving members of these communities in sustainability initiatives, we can tap into their wealth of traditional knowledge and integrate it into contemporary sustainability practices. A notable illustration of this can be

[2] https://www.ubqmaterials.com/blog-post/why-are-diversity-and-inclusion-important-for-sustainability/.

seen in the engagement of the *Maasai* community in Kenya in conservation endeavours aimed at safeguarding wildlife and their natural habitats. Their traditional practices of pasture management have been found to promote biodiversity and sustainably manage grazing lands, and have been incorporated into modern conservation programs (Wishitemi et al. 2015). The implementation of inclusive and culturally aware conservation initiatives has yielded significant results, positively impacting both human well-being and the environment. These efforts have demonstrated their effectiveness in promoting sustainability. By nurturing an environment that values and embraces diversity, we can cultivate the creation of inclusive and equitable strategies that duly recognize and value the unique perspectives and contributions of all individuals and communities. This approach ensures that sustainability practices are fair and inclusive, taking into account the diverse needs and aspirations of people from various backgrounds and cultures while also addressing the preservation of our natural surroundings.

4.2 Inclusivity (I)

Sustainability doesn't only mean protecting the natural environment or developing smart cities. If we want to achieve holistic sustainability, we must prioritize inclusivity by ensuring that all voices of the voiceless are heard and included in decision-making processes. It contributes towards promoting social integration, embracing fairness and recognizing value within the diversity of individuals from diverse communities. In addition, it is equally important that the social and economic inequalities harming our environment and preventing community growth are also duly addressed. Been et al. (2016) proposed an inclusivity framework for building environmentally sustainable communities which can lead to more equitable and just solutions that address the needs and concerns of all stakeholders, including marginalized and underrepresented communities. Sustainable Development needs to be revisited in line with the inclusivity framework to have a widespread impact on the masses (Tandon and Garg 2009).

Taking an example of the development of renewable energy projects, when stakeholders from diverse backgrounds, especially, marginalized communities are included in decision-making processes, their unique perspectives and concerns, like economic benefits and job opportunities, can be instrumental to ensure the acceptability, effectiveness of projects in the long term along with building trust and support for sustainability initiatives. Similar concerns are required to be addressed while contextualizing smart cities for sustainability and inclusivity (Allam 2018).

4.3 Sustainability Behaviour (S)

To foster sustainable development and shield the environment from harm caused by human activities, sustainability behaviour is crucial as it exemplifies the actions undertaken by individuals, organizations, and communities alike (Rishi et al. 2023). By embodying sustainable practices and behaviours into our lives we possess the capacity to diminish our ecological impact while contributing to an enduring future—our shared responsibility as custodians of this planet. In the Indian context, noteworthy examples substantiating this claim exist like adopting sustainable practices yielded a reduction in greenhouse gas emissions while conserving vital natural resources.

Remarkable leaps forward were made specifically in championing renewable energy sources. Exemplifying efforts via numerous solar power initiatives across India's landscape were made, thus, setting ambitious targets around generation capacity at 175 GW dedicated solely towards renewables by 2022. This is inclusive of an astonishing 100 GW allocation specifically for solar power energy generation besides wind and bioenergy. Though the actual installed renewable energy capacity was 120.90 GW which is about 69% of the overall target by 2022, it set the stage for pioneering ways to replace conventional fossil fuels with cleaner and greener sources of power thus not just reducing greenhouse gas emissions but also epitomizing a future that is healthier and more sustainable.[3]

India has achieved significant strides in the promotion of sustainable green buildings designed to protect natural resources and decrease greenhouse gas emissions. The government has implemented diverse initiatives like the "Energy Conservation Building Code[4]" and the "Green Rating for Integrated Habitat Assessment" program[5] to ensure both citizens' well-being and the preservation of natural resources. Encouraging sustainable behaviours is crucial for fostering positive attitudes and values towards living responsibly and adopting more sustainable lifestyles (Kasser 2009). Diverse approaches contribute towards this change in perspective. Practicing recycling and waste reduction instils greater awareness regarding personal waste generation and its environmental consequences, and encourages altered consumption patterns like opting for minimally packaged products or reusable items. Additionally, transitioning towards more sustainable modes of transportation aids in recognizing the advantages of reducing one's carbon footprint, saving money and enjoying the health benefits associated with regular exercise. Similarly, embracing sustainable practices like using energy-efficient appliances, reducing air conditioning usage and incorporating renewable energy sources fuels a change in attitudes towards energy

[3] https://economictimes.indiatimes.com/industry/renewables/india-misses-re-capacity-target-due-to-low-solar-rooftop-wind-energy-project-installations-parliamentary-panel/articleshow/98870462.cms?from=mdr.

[4] https://beeindia.gov.in/en/energy-conservation-building-code-ecbc#:~:text=The%20Energy%20Conservation%20Building%20Code%20(ECBC)%20sets%20minimum%20energy%20standards,design%20strategies%20%26%20day%20light%20Integration.

[5] https://www.grihaindia.org/about-griha.

consumption. Increasing awareness about personal environmental impact enables individuals to make conscious decisions to reduce their carbon footprint. Millets have been an integral part of Indian food for centuries due to their health benefits, besides being good for the environment with low water and input requirements. Choosing locally produced organic foods like millets promotes an appreciation for sustainable agriculture while advocating for reductions in food waste and endorsing healthier eating habits as propagated in this International Year of millets-2023 declared by FAO of United Nations[6] on the behest of India to create awareness and increase the production and consumption of millet. Such initiatives empower individuals to consciously reshape their lifestyles in alignment with environmental concerns while inspiring others by cultivating a culture rooted in sustainability.

4.4 Ethics and Values (E)

When considering matters related to sustainability, discourse should encompass ethics—investigating conscientiously not only the social but also environmental repercussions together with economics resulting from individual or communal actions. Moreover, this discussion must account for the prominent role played by personal morals, and cherished beliefs intertwined with principles guiding choices about conservation. Ethical and value-based approaches to sustainability encompass a range of practices and policies that prioritize equitable access, environmental protection, and transparency. For instance, promoting equal access to clean energy through renewable technologies like solar panels and wind turbines involves providing financing options, subsidies, and incentives. Similarly, sustainable transportation options such as walking, biking, and public transit are prioritized through investments in affordable alternatives and safe infrastructure. Urban farming initiatives contribute to sustainability by offering fresh food, reducing waste, and fostering community engagement. This is achieved through investments in community gardens, resource allocation to underserved communities, and support for food justice initiatives. Green jobs training programs, including solar panel installation and sustainable agriculture, promote economic growth while reducing carbon emissions, with a focus on equity through training opportunities, apprenticeships, fair wages, and benefits. Precautionary measures are implemented to protect the environment and human health, such as banning untested harmful chemicals and regulating emerging technologies like GMOs. Conservation of biodiversity follows the precautionary principle, protecting endangered species and critical habitats despite uncertainties. Reporting and disclosure requirements enhance transparency by publicly sharing sustainability metrics like greenhouse gas emissions, water usage, and waste management. Independent monitoring and evaluation ensure comprehensive assessments by engaging auditors or third-party evaluators to report on performance. By embracing these

[6] https://www.fao.org/millets-2023/.

ethical and value-based approaches, sustainability is advanced by prioritizing equity, environmental protection, and accountability.

4.5 Mindfulness (M)

The consciousness of environment-friendly lifestyles and the need for the promotion of sustainable behaviour is increasing globally. Consumption-centric mindset depletes natural resources and has negative socio-economic impacts. Mindfulness is the practice of paying attention to the present moment with a non-judgmental and curious attitude (Kabat-Zinn 2005) while understanding our thoughts, feelings, and actions, and how they impact ourselves and others. It can help us develop a deeper connection with the natural world and cultivate a sense of gratitude for the environment. Mindfulness-based approaches are the means to shift consumption patterns towards sustainability by targeting core values and ethical principles. However, empirical evidence on the effects of mindfulness-based approaches in promoting sustainable lifestyles is limited (Geiger et al. 2019). Further, a link between mindfulness practice, sustainability-related behaviours, and various positive outcomes such as compassion, connectedness to nature, reduced materialism, improved well-being, and health-related behaviours was observed in various studies (Ericson et al. 2014).

Mindfulness enhances subjective well-being, empathy, and the clarity of personal values. These factors can potentially foster more sustainable behaviours thus, making advocacy for mindfulness, a policy that not only contributes to sustainability but also promotes overall well-being. However, there is a lack of direct causal evidence supporting the impact of mindfulness on sustainable consumption behaviours. Meta-analytic evidence indicates a negative correlation between mindfulness and impulsive buying tendencies (Geiger et al. 2019). Kabat-Zinn (2013) explains how consciousness can provide mental training towards flexibility and adaptability with the support of neuroscientific research about 'the transformation of consciousness through mindfulness and contemplative practice'. Therefore, there is a likelihood for mindfulness to contribute reasonably towards the development of USCon.

To respectfully contribute towards sustainability, one can explore several engaging strategies. A particularly captivating approach is embarking on Mindful Nature Walks—a practice that encourages individuals to leisurely appreciate and immerse themselves in the beauty of their surroundings. Not only does this activity prompt attention towards visual cues but also necessitates attentive listening to nature's symphony as well as observing its myriad of smells. Partnering these practices with meaningful connections with various flora and fauna enriches the experience significantly. Another effective way of contributing lies within mindful eating. It facilitates a greater awareness regarding the origin of our sustenance while also considering its ecological impact. The process further cultivates a deep appreciation for taste, texture, and scent—engendering a profound sense of thankfulness towards nature. Mindful eating becomes instrumental in promoting sustainability by averting food wastage, enriching meals with locally sourced ingredients like millet and pursuing healthier living habits. Further, engaging in ***conservation activities*** such as planting trees or participating in cleaning our surroundings can help us conscientiously connect with the environment in a meaningful way, while consciously appreciating the beauty of the natural world.

Closely connected to this are ***eco-friendly practices*** such as reducing the usage of single-use plastic, conserving energy and using sustainable products to allow us to become more conscientious and mindful of our impact on the environment. Mission LiFE of India emphatically promotes the same across diverse populations as mentioned in the statement by Narendra Modi, PM of India at COP-26 at Glasgow. "*Mission LiFE can become a mass movement of environmentally conscious Lifestyle. What is needed today is Mindful and Deliberate Utilization, instead of Mindless and Destructive Consumption.*"[7] The application of the DISEMBARC framework of USCon through Mission LiFE will be explained later.

Sustainability has long been ingrained in our traditions, culture, and values. India's per capita carbon footprint is 60% below the global average, attributed to the continuation of sustainable traditional practices in our lifestyle and mindful behaviour. ***Mindful meditation*** is yet another powerful practice that can help us develop a sense of inner peace and calmness and also appreciate the beauty of the natural world more deeply and motivates us to adopt Lifestyle for Environment (LiFE),' an India-led global mass movement to nudge individual and community action to protect and preserve the environment'. Mindfulness plays a vital role in helping us recognize our consumption habits, minimize waste and adopt more sustainable practices in our everyday lives. Moreover, practising mindfulness can inspire individuals to actively participate in group efforts and support sustainability endeavours across various scales, be it local, national, or global.

4.6 Benevolence and Well-Being (B)

Benevolence and well-being are important concepts linked to sustainability as they involve considerations of social and environmental justice, equity, and quality of life. Benevolence is a specific type of altruism that encompasses values such as loyalty, responsibility, honesty, forgiveness, the pursuit of meaning in life, genuine friendship, and spirituality (Gärling 1999). Finding inspiration to safeguard and enhance the welfare of individuals whom we regularly associate with, often known as our "in-group," is paramount. In exploring sustainability, benevolence involves contemplating how our actions affect both people and their surroundings, and then taking active measures to advance social and environmental fairness. Taking further the Indian concept of *'Vasudhaiv Kutumbakam'* and 'One earth-one family future', benevolence proudly encompasses the whole planet as an 'in-group'. Wellbeing is the state of being healthy, happy, and fulfilled, where individuals and communities have access to resources and opportunities inclusively.

[7] http://missionlife-moefcc.nic.in/.

To enhance the likelihood of transitioning towards sustainability, it is crucial to enhance the comprehension of well-being through a relational framework acknowledging the interdependence between human well-being and the overall health of ecosystems, thereby internalizing this connection (Helne and Hirvilammi 2015). Further, Kang et al. (2021) explore the relationship between minimalistic lifestyle, sustainability and emotional wellbeing and concluded that incorporating minimalism into their lifestyle can be valuable for consumers as it can inspire manufacturers and producers to adopt more sustainable modes of production aligned with minimalist principles. One such example is the Dayalbagh Educational Institute (DEI)[8] where students naturally embrace frugal living and minimalism while valuing well-being. The DEI system of holistic education equips students with the conviction and commitment to its core values for social responsibility, sustainable development, and selfless service, emphasizing inclusive growth and empowering marginalized communities. It also integrates innovation, water and air quality, education and healthcare, agriculture and dairy practices, and human values for sustainable development into its educational system thereby succeeding in mapping 15 out of 17 SDGs in its educational system and campus activities. Further, it promotes the principle of achieving more with fewer resources through research and frugal innovation through a spirit of self-sacrifice and respect for the dignity of labour in serving humanity. Dua et al. (2023) suggest that assimilating human 'consciousness and conscientiousness' which are closely aligned to 'values, beliefs, attitudes and spiritual consciousness' can better contribute to SDGs. Therefore, this 'inner dimension of sustainability has the potential to potentiate external dimensions (environmental, social and economic), and thus contribute to holistic Sustainable development.[9] These findings are based on their recent case study of Dayalbagh eco-city, an epitome of sustainability, in India.

Both, benevolence and wellbeing are crucial components that can help promote sustainability by prioritizing the well-being of others and considering the environment and sustainability at the forefront while making decisions. Community farming/gardening is one such activity that promotes social cohesion, improves mental health, and encourages sustainable food production which is regularly practised in Dayalbagh, the eco-village in India.[10]

Institutions that value the welfare of their employees, students and the environment usually create sustainability strategies aimed at reducing their environmental footprint, encouraging sustainable behaviour and contributing to the community. Similarly, individuals who prioritize their well-being often adopt sustainable practices that benefit both themselves and the environment. For example, practising self-care in terms of mental and physical health can foster a positive mindset towards the environment thus resulting in a strong inclination to safeguard and conserve it. That is how USCon gets ingrained in the minds of people for widespread positive impacts on planet Earth.

[8] https://www.dei.ac.in/dei/.

[9] https://uwspace.uwaterloo.ca/handle/10012/19181.

[10] http://sigmasixq.blogspot.com/2019/04/dayalbagh-eco-village.html.

4.7 *Adaptability (A)*

The role of adaptability cannot be undermined when it comes to achieving sustainability amid fluctuating circumstances and conditions. As defined by Reeves et al. (2012), Sustainability entails possessing the capacity for adaptation across ecological, social, and economic domains. Therefore, we must adopt resilient practices as well as robust policies capable of withstanding climate change impact along with other environmental challenges during our comprehensive planning processes covering short- as well as long-term periods. For sustainability initiatives to be successful, individuals must demonstrate adaptability by accepting pro-planet technologies and incorporating them into their lifestyles. Without adaptability, concepts like USCon will remain just ideas without any practical implementation.

To mitigate climate change consequences and decrease reliance on fossil fuels, the energy sector implements sustainable practices such as wind, hydroelectric and solar power. This not only positively impacts our planet but also fosters job opportunities within the renewable energy industry. However, if individuals are disinclined to adapt to these innovative initiatives all endeavours will ultimately prove futile. Another example where adaptability plays a vital role is in implementing sustainable agricultural practices. These practices aim at increasing productivity while reducing the use of harmful chemicals. They also promote soil health, biodiversity conservation, and sustainable food production. Agro-ecology is gaining global recognition as it addresses various aspects including food security, environmental sustainability, biodiversity conservation, and socio-economic equity. Its promotion at technical, institutional and political levels further emphasizes its importance. It is driven by societal demands for healthier food and environments (Tittonell 2020). However, the adaptability principle plays a crucial role in this context. The spiritual community of Dayalbagh eco-village of Agra city is well adapted to Agro-ecological precepts, practised in a symbiotic relationship amongst the plants, animals, human beings and environment for more than 100 years through community volunteering practices. The practices have been further strengthened through the establishment of "*Anupam Upvan*" in the vicinity of river *Yamuna*—a widespread initiative for biodiversity conservation along with precision farming at Dayalbagh, India and the International Centre for Agro-ecology (ICA) in the USA. The goal is to develop innovative solutions and guidelines for present and future sustainability issues.[11]

Additionally, it is important to consider investing in green infrastructure, including urban parks, green roofs, and permeable pavement. These investments have the potential to positively impact air and water quality, decrease urban heat islands and contribute to increased biodiversity. It is crucial to note that the success of these initiatives relies on them being acceptable and adaptable for individuals and communities. Furthermore, a shift towards a circular economy is necessary for minimizing waste and maximizing the reuse and recycling of resources. By doing so we can significantly reduce the environmental consequences associated with production and consumption. This approach also presents economic opportunities that benefit both people and the planet. Embracing a more sustainable and resilient economy should be our focus. In addition, encouraging the adoption of sustainable modes of transportation such as cycling, utilizing public transport and walking is crucial in reducing greenhouse gas emissions. These endeavours also have a positive impact on air quality and simultaneously promote physical activity and healthier lifestyles. We must prioritize these modes of transportation for a greener future.

Adaptability plays a crucial role in promoting sustainability consciousness. One of the ways to promote sustainability consciousness and adaptability is through *resilience planning* which involves identifying and addressing the risks and vulnerabilities of a community or organization to the impacts of climate change and other environmental challenges. It offers a platform to address socio-economic and environmental policies that may enhance climate change impacts (Woodruff et al. 2022). It is further explained in the subsequent section.

4.8 Resilience (R)

Resilience refers to the ability of individuals, communities, and systems to withstand and adapt to change and disruptions, including environmental and social challenges. It is an ability 'to deal with adversity, withstand shocks, and continuously adapt and accelerate as disruptions and crises arise over time'.[12] By exhibiting resilience amidst various environmental as well as social hurdles such as natural calamities, economic

[11] https://www.thestatesman.com/business/international-centre-for-agroecology-to-be-launched-in-new-jersey-usa-1503023935.html.

[12] https://www.mckinsey.com/capabilities/risk-and-resilience/our-insights/resilience-for-sustainable-inclusive-growth.

downturns, or pandemics, individuals along with communities enhance their ability to respond effectively while also recovering from these difficulties. This contributes towards the preservation of socio-economic stability alongside fostering communal responsibility as well as cooperation. Resilience is the capacity of a society to maintain 'cohesion and cultural continuity' during environmentally and socially challenging times, but it involves a built-in cost, which varies across 'time and cultural milieu'(Haldon et al. 2020). Resilience plays a crucial role in promoting sustainability consciousness. It allows us to recognize the impact of human activities on the environment and appreciate the importance of sustainable practices. By adopting these practices individuals and communities can contribute to both sustainability and resilience.

Natural disasters like floods or hurricanes serve as wake-up calls for communities. They highlight the significance of protecting natural habitats and reducing carbon emissions to mitigate climate change impacts. Such events also make us more aware of the importance of waste reduction and resource conservation for promoting sustainability and resilience. Adapting to evolving environmental and social conditions allows individuals and communities to cultivate novel technologies and practices that support sustainability and resilience. For instance, climate-smart agricultural techniques including crop diversification, conservation tillage and agroforestry assist farmers in withstanding challenges like droughts, floods, and extreme weather events while also fostering sustainable land utilization. In addition to this, urban planners and designers hold a significant responsibility in contributing to these efforts as well. Implementing green infrastructure like green roofs, rain gardens, and bios-wales contributes to managing stormwater effectively while mitigating urban heat island effects. This enhances liveability in urban areas while bolstering their resilience.

Adopting renewable energy technologies like wind, geothermal or solar power is an important measure in the fight against greenhouse gas emissions. By doing so, we not only prioritize sustainability but also guarantee energy independence in regions susceptible to power outages or other disturbances. Furthermore, promoting sustainable forms of transportation like public transit, biking, or walking not only leads to reduced greenhouse gas emissions but also enhances air quality. Additionally, it supports public health by encouraging physical activity. Lastly, resilience stimulates innovation and creativity in developing sustainable practices. These innovative solutions play a pivotal role in furthering both sustainability goals along enhancing resilience efforts.

4.9 Creativity and Innovation (C)

Creativity goes hand in hand with consciousness about sustainability—two closely intertwined concepts with the potential for driving environmental progress on multiple fronts such as ecological conservation, social inclusion and economic prosperity. Possessing creative capabilities enables us to conjure up fresh insights, perspectives and problem-solving methods when facing diverse challenges. Unleashing this creative power on matters related to sustainability has resulted in ground-breaking innovations aimed at mitigating pressing societal problems.

Understanding how our environment, limited natural resources and marginalized communities are irrevocably interconnected allows us not only to circumvent present-day challenges by crafting imaginative solutions but ensures their continuity over time. Enthusiasm —coupled with creative thinking— inspires individuals and communities alike to devise novel technologies, practices, and strategies that prioritize ecological harmony, social equity, and prospering economic conditions. Abiding by this vein of thought, creativity has permitted us to bridge gaps in sustainable design and architecture thereby resulting in energy-efficient buildings that incorporate eco-friendly materials and innovative technologies. Consequently, this has paved the way for creating novel transportation solutions that effectively dampen greenhouse gas emissions while bolstering modes of travel rooted in sustainability. Moreover, the impact of creativity far surpasses practical innovations. It beautifully transcends boundaries using artistry and storytelling as vehicles to spread awareness about environmental preservation discuss social dynamics and ultimately incite action towards positive change. Telling stories intricately woven with complexity allows us as participants to fully comprehend and engage with the issues at hand. By fostering a climate characterized by creativity and innovation within the context of sustainability objectives a new breed of solutions emerges —ones capable of addressing highly complex environmental and social predicaments. Beyond its ability to spur change on a global scale, creativity also plays a vital role in aiding public awareness efforts, fostering adaptability and ensuring widespread acceptance of substantial innovative advancements redefining what society is in terms of progress informed by sustainability.

The above analysis of DISEMBARC makes us realize that USCon is not a philosophy or a construct hanging loose in isolation. It is a comprehensive umbrella term with inbuilt components and the judicious interplay of these factors can lead to the world of sustainability. To bring some of its components into action, the very premise of India's Mission LiFE is a welcome beginning towards a sustainable future.

5 USCon Linked to Mission LiFE

During the UN Climate Change Conference at Glasgow on 1st November 2021 (COP 26), Indian Prime Minister Narendra Modi introduced Mission LiFE (Lifestyle for Environment) campaign which is India's contribution to tackle the menace of global climate crisis. On 5th June 2022, on the occasion of World Environment Day, India expanded the vision of LiFE by introducing the LiFE as a Global Movement. This campaign aims to prioritize individual behaviours in global climate action and is intended to be taken forward as a public movement to mobilize individuals to become 'Pro-Planet People'. Thus, Mission LiFE aims to promote simple behavioural changes. Behaviours of individuals and households have major and cumulative impacts on the ecology and sustainable development (Verplanken 2018). Various studies have focused on the need for behavioural change for the benefit of the planet. Huttunen et al. (2021) have emphasized the need for large-scale transitions towards sustainable societies which require people to change the ways they conduct their daily lives in bringing about the needed changes at different levels of society.

Some highlights of the LiFE campaign are given in Fig. 2.

The LiFE campaign is focused on shifting away from a society that mindlessly consumes and disregards the consequences towards a circular economy that promotes mindful consumption. The choices we make when it comes to what we buy, how much we buy, how much we consume and how we dispose of our goods, all have a direct impact on the environment and future generations. Therefore, it's important to understand that each individual's consumption habits accumulate and contribute to the devastation we see today (Trudel 2018). In addition to promoting sustainable consumption, the mission of LiFE encourages individuals to take small actions in their everyday lives that can have a significant impact on fighting climate change worldwide. One of the ways it plans to do this is by utilizing social networks to influence climate-friendly social norms. Mission LiFE also envisions establishing a global community called 'Pro Planet People' (P3) made up of environmentally conscious individuals who will actively promote and adopt eco-friendly lifestyles. This community will create an ecosystem where environmentally friendly behaviours

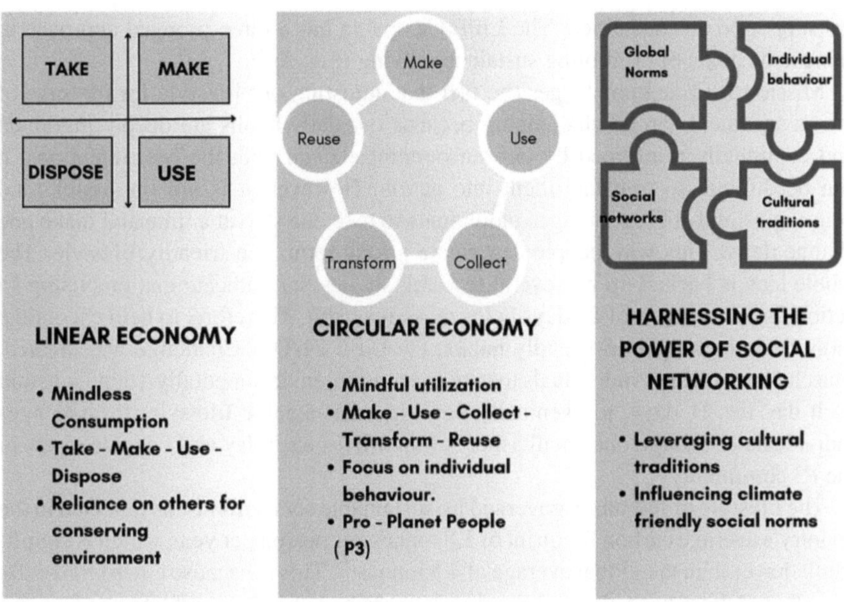

Fig. 2 Highlights of mission LiFE of India

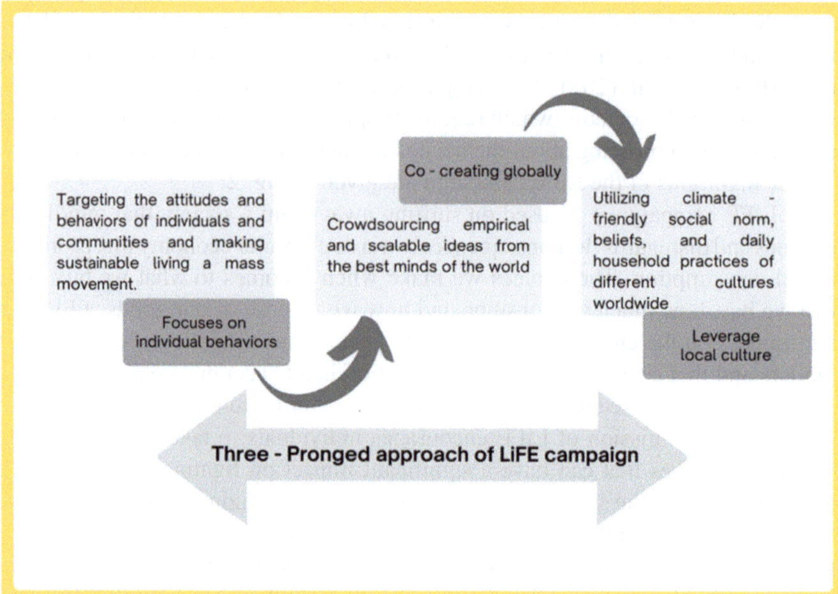

Fig. 3 Mission LiFE's Three-Pronged Approach

are supported and sustained. The LiFE Campaign has a three-pronged approach to achieve its goals of promoting sustainable living (Fig. 3).[13]

Mission LiFE acknowledges the fact that adjusting our lifestyle for the sake of the environment can be challenging because our daily habits are deeply ingrained and continually reinforced by our surroundings. Even with the best intentions, it can be difficult to translate them into action. However, it is not impossible and hence mission LiFE encourages individuals to take one step at a time and make one change daily. This way people can create a long-term, eco-friendly lifestyle. The whole idea is backed up by several research studies that indicate that practising an action for a minimum of 21 days helps make it a habit. Therefore, to help encourage individuals to adopt eco-friendly habits, the LiFE 21-Day Challenge was created. The challenge allows individuals to take one simple, environmentally-friendly action each day for 21 days, to eventually develop a sustainable lifestyle. It challenges individuals to change one small aspect of their lives each day and become a part of the P3 community.

The lifestyle of Indians is governed by sustainable social norms, as reflected in the country's average carbon footprint of 1.8 tonnes per person per year, which is significantly lower than the global average of 4.5 tonnes.[14] This is because environmentally friendly practices have always been deeply ingrained in Indian traditions and culture

[13] http://missionlife-moefcc.nic.in/assets/pdf/LIFE-Brochure-20102022.pdf.

[14] World Bank CO2 emission data. https://data.worldbank.org/indicator/EN.ATM.CO2E.PC?locations=IN-1W.

for ages. Indigenous communities in India have developed knowledge systems that focus on adapting to natural ecosystems. This approach has helped agrarian communities to effectively adapt to the changing environment. For example, practices of planting trees like *Azadirachta indica* (Neem), *Ficus benghalensis* (Banyan) and *Ficus religiosa* (Peepal) alongside built structures to rejuvenate the land in the spring. Additionally, there are many water harvesting techniques used across India, such as Step-Wells in Gujarat and Rajasthan, underground tanks *(Tankaa)* in Tamil Nadu, check dams *(Johads)* in Rajasthan, and the *Zabo* System in Nagaland.[15] These techniques deposit water in pond-like structures on terraced hillsides, which are contextual to local conditions. These practices demonstrate how people in India have traditionally used natural resources judiciously. Indian lifestyle also reflects a mindful relationship with the environment at an individual and household level. For instance, clay utensils are commonly used for cooking and serving food. Street and public food establishments serve food in plant-based biodegradable utensils made from tree leaves, and tea is served in clay pots called *kulhad*.[16] Several elements of the Indian lifestyle, such as water-efficient utensils and clothes washing/drying practices, dietary preferences for plant-based foods, and adaptive architectural forms that minimize electricity consumption, can serve as foundations for Lifestyle for the Environment (LiFE).

Mission LiFE is designed as a worldwide movement that draws inspiration from the eco-friendly culture and heritage practices in India while operating in the present and focusing on the future. It aims to embrace and promote sustainable living practices globally by combining the best of traditional and modern approaches to environmental conservation. As part of Mission LiFE's three-pronged approach, an attempt to leverage Indian culture for reviving age-old traditional practices prevalent in different parts of the country is made possible through a document titled '*Parampara*'.[17] This document aims to make people aware of the numerous sustainable practices that have been ingrained and prevalent in their culture for ages and is a reminder of 'getting back to the roots' in whatever possible way when they try to make lifestyle and behavioural changes.

In commemoration of India's landmark achievement—completing 75 years since independence from the colonial era, Mission LiFE was initiated. As part of this endeavour, a comprehensive roster comprising an impressive array of 75 distinct activities has been curated across seven core domains viz. Transportation, Electricity, Food and Solid Waste Management, Agriculture, Water Conservation Practices, Healthy Lifestyles, Reuse and Recycling. The inclusion criteria sought out highly precise measures that lend themselves well to monitoring and evaluation. By design, these activities are readily adopted by individuals, communities and institutions alike, requiring minimal external support. Furthermore, their implementation can coexist harmoniously with current economic processes, and potentially even amplify economic productivity in the foreseeable future.

---

[15] http://missionlife-moefcc.nic.in/assets/pdf/LIFE-Brochure-20102022.pdf.

[16] http://missionlife-moefcc.nic.in/assets/pdf/LIFE-Brochure-20102022.pdf.

[17] http://missionlife-moefcc.nic.in/assets/pdf/PARAMPARA.pdf.

Mission LiFE aims to translate the vision of LiFE into tangible results by mobilizing at least one billion Indians and other global citizens to take individual and collective action towards protecting and conserving the environment from 2022–23 to 2027–28. The mission sets a goal of making at least 80% of all villages and urban local bodies in India environmentally sustainable by 2028.[18] By promoting sustainable behaviours and encouraging individuals and organizations to take action to reduce their environmental impact, the LiFE campaign helps to create a culture of USCon. Encouraging people and organizations to adopt sustainable lifestyles and make small changes in their daily routines can have a big impact on the environment as a whole. Several studies have proposed the existence of a series of predispositions which allow the appreciation of diversity and the interdependence of person–environment relationships, making it possible to adopt lifestyles that can guarantee the sustainability of socio-ecological systems for present and future generations (Pérez Ibarra et al. 2020).

It is worth highlighting that Mission LiFE has a significant impact on almost all of the Sustainable Development Goals (SDGs) that the global community aims to achieve by 2030. This contribution can be both direct and indirect. Additionally, due to the large population of India, any large-scale behaviour change implemented in the country can have an immediate and direct contribution to the SDGs at a global level. For instance, the Clean India Mission, which aimed to achieve Clean Water and Sanitation (SDG 6), led to a substantial behaviour change in the lives of 550 million Indians. This change reduced 60% of the one billion open defecators in the world, making a significant contribution towards achieving SDG 6.

USCon is the need of the hour and the LiFE campaign emphasizes that we need to make a transition from an existing throwaway culture to a circular economy. The throwaway culture backed by wasteful consumerism is seen as a major culprit in deepening the climate crisis and making climate change mitigation a huge challenge. By encouraging individuals and organizations to take a proactive approach to sustainability, the LiFE campaign helps to create a sense of collective responsibility for the environment and promotes USCon. To create awareness of Mission LiFE among students, youth and the general public in India, various awareness programs, plantation drives, green talks etc. are being organized throughout the country, both in rural and urban areas. Students and educational institutions are especially encouraged to adopt a sustainable lifestyle and green habits along the lines of Mission LiFE to mitigate climate change.

The "Mission LiFE" is especially being incorporated into school curriculums, starting from the basic level, to educate and instil its principles in children. Under the National Education Policy (NEP), the Department of School Education has taken a significant step to engage the younger generation in PM's Mission LiFE and efforts are already underway to incorporate Mission LiFE initiatives at various levels in schools.[19] Also, as part of Mission LiFE, the National Institution of Transforming

[18] http://missionlife-moefcc.nic.in/assets/pdf/LIFE-Brochure-20102022.pdf.

[19] https://thenationalbulletin.in/pm-modis-mission-life-will-be-part-of-the-school-curriculum-children-will-be-given-tips-for-an-eco-friendly-lifestyle.

India, the planning commission (*Neeti Aayog*) called for suggestions and ideas from various academicians, universities, and research institutions in December 2022 for adopting a sustainable lifestyle.

To commemorate the first anniversary of Mission LiFE since its launch as a global movement on June 5, 2022, under the direction of the Ministry of Environment, Forest and Climate Change (MoEF&CC), Indian Institute of Forest Management (IIFM), Bhopal conducted over 1000 events in educational institutes and public places during May 2023 revolving around the seven themes to promote awareness and outreach among students, faculties, staff and civil society. All the events were simple enough to differentially cater to the intellectual levels of participants from all walks of life. The institute also shared resource materials to promote the mission of LiFE in different institutes further. The national data of such events and participants is available on the website '*Merilife*'[20] which is also accessible through an app available on the Google Play Store.

Likewise, various Indian educational institutions are working towards the attainment of SDGs thereby making their contribution in the area of sustainability. Initiatives by educational institutes go a long way in promoting sustainability consciousness among students and people in general. Chakraborty et al. (2021) have emphasized the need for Universities to incorporate SD goals into their research and their operations. Their study investigated the sustainability model of Ashoka University (India) and approaches its commitment through sustainability-focused courses, research, and operations. Mohammadalizadehkorde and Weaver (2018) have emphasized three primary justifications for universities' involvement in sustainability. Firstly, educational and research institutions play a crucial role in generating and sharing knowledge related to sustainability. Secondly, as semi-autonomous communities, these institutions can implement research-based practices and assess their effectiveness. Lastly, universities bear a social responsibility to contribute to the betterment of the communities in their vicinity.

However, some of the previous studies (Bantanur et al. 2015) have found that the concept of sustainability in the Indian higher education system is still in its early stages and lacks extensive research. While the current focus primarily revolves around addressing environmental concerns, Indian universities tend to prioritize aspects such as curriculum, faculty development, intake numbers, and infrastructure, while not giving sufficient consideration to campus sustainability. With the introduction of Mission LiFE, it is hoped that sustainability initiatives will receive a major impetus at Indian Higher Education Institutions (HEIs) because they do have a significant potential to expand their sustainability endeavours. It is imperative to prioritize research and studies aimed at establishing monitoring and assessment systems. These systems will help evaluate the progress of Indian HEIs in terms of sustainability and identify specific areas that require improvement (Priyadarshini and Abhilash 2020; Parvez and Agrawal n.d). Mission LiFE is hoped to bring focused attempts on the front of sustainability by HEIs which will not only strengthen the campaign but will also transform HEIs into torchbearers of USCon.

[20] https://merilife.org/.

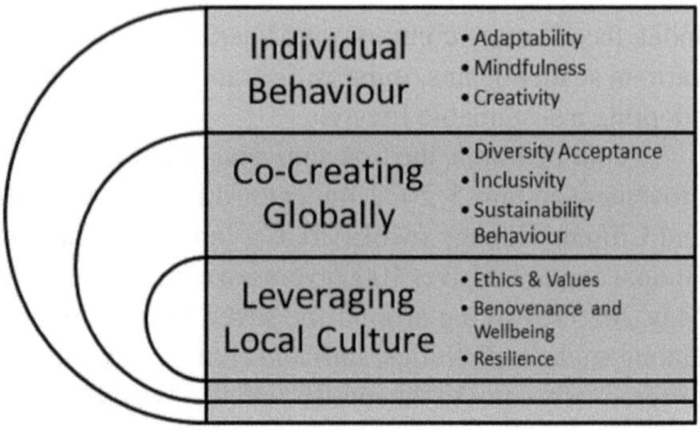

Fig. 4 DISEMBARC as aligned to mission LiFE

6 Final Words—Aligning DISEMBARC with LiFE

The concept of LiFE, which is in close alignment with the conceptual framework—DISEMBARC, is a broad-spectrum plan to develop USCon across the globe. DISEMBARC framework is built in Mission LiFE where the three-pronged approach relates to all the components as given in the Fig. 4.

What is needed is to reach across the globe to the Last, the Least, the Lowest and the Lost[21] (L4), accepting the socio-cultural and economic diversity and leveraging it to create Pro-Planet-People (P3). While focusing on Individual Behaviour, mindful behavioural choices, adaptability to innovative sustainability solutions and trying to reach out to people across regions and nations, particularly children and youth and L4 adopting creative means like sustainability Art, Music, literature, poetry, pictorials, graphics etc., is needed. Co-creating globally calls for accumulating intellectual and creative resources across the globe by reaching out to diverse groups, following the diversity and inclusivity principle and strategically using them to promote sustainability behaviour, one of the major objectives of Mission LiFE. Furthermore, these efforts towards sustainability will be fruitful only when we can leverage the local culture and traditions, giving importance to ethics and values for benevolence and well-being while promoting resilience among individuals and communities. With its abundance of traditional wisdom and environmentally friendly practices, India is well-positioned to take the lead in inspiring behavioural changes towards climate action. By developing innovative, individual-centric nudge mechanisms, that revolve around our existing social norms, and by leveraging technology, the climate change movement can receive a global impetus through LiFE. The quantifiable target of mission LiFE is to reach out to one billion people and this massive challenge will be met by reaching out to more than 80% of villages and urban local bodies by 2028. With the integration of the DISEMBARC framework in Mission LiFE, the promotion of USCon is possible in the times to come. However, it is just a theoretical framework to decode USCon which needs to be tested through a cross cultural study across different regions. Hence, at this stage, without empirical testing of the model, the

[21] https://www.universityworldnews.com/post.php?story=20191007113224304.

proposed DISEMBARC framework is only conceptual in nature till proved through empirical study.

7 Ways Ahead

Although the conceptual alignment is visible and supplemented by a review of literature, an empirical research study across different countries is required to see the relative contribution of different components of DISEMBARC for generating USCon and the way it is instrumental in enhancing the 'LiFE Quotient' (LQ), a possible indicator of Pro-Planet-People (P3). Higher education institutions (HEIs) are facing greater levels of responsibility since adopting and committing to the Agenda 2030 for Sustainable Development (SD) and its 17 Sustainable Development Goals (SDGs). Dedicated need-based capacity-building programmes to promote SDG 13—Climate Action and other sustainability goals can be creatively designed in line with the levels of LQ for diverse population groups to have the desired impact. HEIs are expected to guide various stakeholders on this matter, and also to implement this agenda and the SDGs in their institution (Omazic and Zunk 2021). Ruiz-Mallén and Heras (2020) consider the paramount importance of raising public awareness and reflection on the causes and consequences of the environmental crisis, as well as building capacities to responsibly and creatively deal with related challenges for which Universities and other HEIs can play a crucial role in promoting sustainability through societal engagement. Indian Institute of Forest Management, Bhopal and Dayalbagh Educational Institute, Agra, of India are presenting themselves as leaders in sustainability education, practices and initiatives to propagate to the world that a collective effort of all HEIs in the domain of sustainability research is needed to steer this globally sought after agenda ahead.

References

Allam Z (2018) Contextualising the smart city for sustainability and inclusivity. New Des Ideas 2:124–127

Balderjahn I, Hüttel A (2019) Why consumers buy sustainably: the role of personal values. Mark: ZFP – J Res Manag 41(1):24–38

Bantanur S, Mukherjee M, Shankar R (2015) Sustainability perceptions in a technological institution of higher education in India. Curr Sci 109(12):2198–2203

Been V, Cunningham MK, Ellen IG, Parilla J, Turner MA, Whitney SV, Zimmerman K, Gordon A, Yowell A (2016) Building environmentally sustainable communities: a framework for inclusivity. https://policycommons.net/artifacts/634326/building-environmentally-sustainable-communities/1615672/

Bridgewater P, Rotherham ID (2019) A critical perspective on the concept of biocultural diversity and its emerging role in nature and heritage conservation. People Nat 1(3):291–304. https://doi.org/10.1002/pan3.10040

Čapienė A, Rūtelionė A, Krukowski K (2022) Engaging in sustainable consumption: exploring the influence of environmental attitudes, values, personal norms, and perceived responsibility. Sustainability 14(16), Article 16. https://doi.org/10.3390/su141610290

Chakraborty A, Kumar S, Shashidhara LS, Taneja A (2021) Building sustainable societies through purpose-driven universities: a case study from Ashoka University (India). Sustainability 13(13), Article 13. https://doi.org/10.3390/su13137423

Corral-Verdugo V, Frias-Armenta M, Garcia-Cadena CH (2010) Introduction to the psychological dimensions of sustainability. In: Psychological approaches to sustainability: current trends in theory, research and applications

Dua P, Narayan A, Allamraju A, Dhir A (2023) A holistic approach to achieving SDGs: a case study of Dayalbagh, India. https://uwspace.uwaterloo.ca/handle/10012/19181

Ericson T, Kjønstad BG, Barstad A (2014) Mindfulness and sustainability. Ecol Econ 104:73–79. https://doi.org/10.1016/j.ecolecon.2014.04.007

Folke C, Biggs R, Norström AV, Reyers B, Rockström J (2016) Social-ecological resilience and biosphere-based sustainability science. Ecol Soc 21(3), art41. https://doi.org/10.5751/ES-08748-210341

Gärling T (1999) Value priorities, social value orientations and cooperation in social dilemmas. Br J Soc Psychol 38(4):397–408. https://doi.org/10.1348/014466699164239

Geiger SM, Grossman P, Schrader U (2019) Mindfulness and sustainability: correlation or causation? Curr Opin Psychol 28:23–27. https://doi.org/10.1016/j.copsyc.2018.09.010

Gericke N, Boeve-de Pauw J, Berglund T, Olsson D (2019) The sustainability consciousness questionnaire: the theoretical development and empirical validation of an evaluation instrument for stakeholders working with sustainable development. Sustain Dev 27(1):35–49. https://doi.org/10.1002/sd.1859

Haldon J, Eisenberg M, Mordechai L, Izdebski A, White S (2020) Lessons from the past, policies for the future: resilience and sustainability in past crises. Environ Syst Decis 40(2):287–297. https://doi.org/10.1007/s10669-020-09778-9

Helne T, Hirvilammi T (2015) Wellbeing and sustainability: a relational approach. Sustain Dev 23(3):167–175. https://doi.org/10.1002/sd.1581

Huttunen S, Kaljonen M, Lonkila A, Rantala S, Rekola A, Paloniemi R (2021) Pluralising agency to understand behaviour change in sustainability transitions. Energy Res Soc Sci 76:102067. https://doi.org/10.1016/j.erss.2021.102067

Ives CD, Freeth R, Fischer J (2020) Inside-out sustainability: the neglect of inner worlds. Ambio 49(1):208–217. https://doi.org/10.1007/s13280-019-01187-w

Kabat-Zinn J (2005) Coming to our senses: healing ourselves and the world through mindfulness. Hachette UK.

Kabat-Zinn J (2013) Full catastrophe living, revised edition: how to cope with stress, pain and illness using mindfulness meditation. Hachette UK.

Kang J, Martinez CMJ, Johnson C (2021) Minimalism as a sustainable lifestyle: Its behavioural representations and contributions to emotional well-being. Sustain Prod Consum 27:802–813. https://doi.org/10.1016/j.spc.2021.02.001

Kasser T (2009) Psychological need satisfaction, personal well-being, and ecological sustainability. Ecopsychology 1(4):175–180. https://doi.org/10.1089/eco.2009.0025

Magni G (2017) Indigenous knowledge and implications for the sustainable development agenda. Eur J Educ 52(4):437–447. https://doi.org/10.1111/ejed.12238

Maranzano P, Ascari R, Chiodini PM, Manzi G (2021) Analysis of sustainability propensity of bike-sharing customers using partially ordered sets methodology. Soc Indic Res 157(1):123–138. https://doi.org/10.1007/s11205-020-02333-8

Mohammadalizadehkorde M, Weaver R (2018) Universities as models of sustainable energy-consuming communities? Review of selected literature. Sustainability 10(9), Article 9. https://doi.org/10.3390/su10093250

Ogundayo 'BJ, Adekunle JO (2019) African sacred spaces: culture, history, and change. Rowman & Littlefield

Omazic A, Zunk BM (2021) Semi-systematic literature review on sustainability and sustainable development in higher education institutions. Sustainability 13(14), Article 14. https://doi.org/10.3390/su13147683

Pérez Ibarra RE, Tapia-Fonllem CO, Fraijo-Sing BS, Nieblas Soto N, Poggio L (2020) Psychosocial predispositions towards sustainability and their relationship with environmental identity. Sustainability 12(17), Article 17. https://doi.org/10.3390/su12177195

Perry GLW, Richardson SJ, Harré N, Hodges D, Lyver PO, Maseyk FJF, Taylor R, Todd JH, Tylianakis JM, Yletyinen J, Brower A (2021) Evaluating the role of social norms in fostering pro-environmental behaviours. Front Environ Sci 9. https://www.frontiersin.org/articles/10.3389/fenvs.2021.620125

Piscitelli A, D'Uggento AM (2022) Do young people really engage in sustainable behaviours in their lifestyles? Soc Indic Res 163(3):1467–1485. https://doi.org/10.1007/s11205-022-02955-0

Polasky S, Kling CL, Levin SA, Carpenter SR, Daily GC, Ehrlich PR, Heal GM, Lubchenco J (2019) Role of economics in analyzing the environment and sustainable development. Proc Natl Acad Sci 116(12):5233–5238. https://doi.org/10.1073/pnas.1901616116

Priyadarshini P, Abhilash PC (2020) From piecemeal to holistic: introducing sustainability science in Indian Universities to attain UN-Sustainable Development Goals. J Clean Prod 247:119133. https://doi.org/10.1016/j.jclepro.2019.119133

Ravikanth G (2021) Indian philosophy and environmental ethics. GNOSI: Interdiscip J Hum Theory Praxis 4(1), Article 1

Reeves M, Haanaes K, Love C, Levin S (2012) Sustainability as adaptability. J Appl Corp Financ 24(2):14–22. https://doi.org/10.1111/j.1745-6622.2012.00373.x

Rishi P, Pavan BK, Gupta S, Sinha S (2023) Mindfulness and diversity acceptance as indicators of frugality-linked sustainability behaviour during COVID-19: mediating role of happiness. J Hum Values 09716858231154397. https://doi.org/10.1177/09716858231154397

Ruiz-Mallén I, Heras M (2020) What sustainability? Higher education institutions' pathways to reach the agenda 2030 goals. Sustainability 12(4), Article 4. https://doi.org/10.3390/su12041290

Schultz PW (2014) Strategies for promoting pro-environmental behaviour. Eur Psychol 19(2):107–117. https://doi.org/10.1027/1016-9040/a000163

Sharma S, Kumar R (2021) Sacred groves of India: Repositories of a rich heritage and tools for biodiversity conservation. J Forest Res 32(3):899–916. https://doi.org/10.1007/s11676-020-01183-x

Souther S, Colombo S, Lyndon N (2023) Integrating traditional ecological knowledge into US public land management: knowledge gaps and research priorities. Front Ecol Evol 11:988126. https://doi.org/10.3389/fevo.2023.988126

Tandon BB, Garg A (2009) Inclusivity and sustainable development. Indian J Public Adm 55(4):882–891. https://doi.org/10.1177/0019556120090406

Tittonell P (2020) Assessing resilience and adaptability in agroecological transitions. Agric Syst 184:102862. https://doi.org/10.1016/j.agsy.2020.102862

Torma G, Aschemann-Witzel J, Thøgersen J (2018) I nudge myself: exploring 'self-nudging' strategies to drive sustainable consumption behaviour. Int J Consum Stud 42(1):141–154. https://doi.org/10.1111/ijcs.12404

Triantafyllidou E, Zabaniotou A (2022) From theory to praxis: 'go sustainable living' survey for exploring individual's consciousness level of decision-making and action-taking in daily life towards a green citizenship. Circ Econ Sustain 2(1):113–139. https://doi.org/10.1007/s43615-021-00046-9

Trudel R (2018) Sustainable consumer behaviour. Consum Psychol Rev arcp.1045. https://doi.org/10.1002/arcp.1045

Verplanken B (2018) Promoting sustainability: towards a segmentation model of individual and household behaviour and behaviour change. Sustain Dev 26(3):193–205. https://doi.org/10.1002/sd.1694

Wishitemi BEL, Momanyi SO, Ombati BG, Okello MM (2015) The link between poverty, environment and ecotourism development in areas adjacent to Maasai Mara and Amboseli protected areas, Kenya. Tour Manag Perspect 16:306–317. https://doi.org/10.1016/j.tmp.2015.07.003
Woodruff SC, Meerow S, Stults M, Wilkins C (2022) Adaptation to resilience planning: alternative pathways to prepare for climate change. J Plan Educ Res 42(1):64–75. https://doi.org/10.1177/0739456X18801057

Disciplines Shaping the Higher Education for Sustainable Development Competency Scholarship: Results from a Direct Citation Analysis

Jessica Ostrow Michel, Sarah R. Collins, Peter Siciliano, and Michaela Zint

Abstract A substantial number of higher education for sustainable development (HESD) scholars have and continue to focus on determining what competencies are critical for students to master to support the transition toward sustainability [i.e., to achieve the Sustainable Development Goals (SDGs)]. Given the influence of HESD competency scholarship on the preparation of future leaders, it is important to learn more about the disciplines shaping this HESD sub-field. This study reports findings, including from a direct citation analysis, to answer the following questions: 1) What is the disciplinary background of scholars shaping the HESD competency scholarship?, 2) What are the most prominent journals publishing HESD competency scholarship?, and 3) What, if any, patterns are apparent in how articles on HESD competency cite each other? Results suggest that the scholars shaping the HESD competency sub-field do not come from sufficiently diverse disciplines. In addition, articles on HESD competencies are published mainly in sustainability-focused journals (i.e., *International Journal of Sustainability in Higher Education, Journal of Cleaner Production, Sustainability, Environmental Education Research*, and *Sustainability Science*), with far fewer published in education or other disciplinary journals. Lastly, HESD competency scholars tend to mainly cite each other, particularly across sustainability-focused journals. For a robust set of HESD competencies that will help to ensure that all higher education students will be prepared to contribute to the achievement of the SDGs, it will be necessary for scholars from more

J. O. Michel (✉) · S. R. Collins · P. Siciliano · M. Zint
School for Environment and Sustainability, University of Michigan, 440 Church Street, Ann Arbor, MI 48109-1041, USA
e-mail: michjess@umich.edu

S. R. Collins
e-mail: srsc@umich.edu

P. Siciliano
e-mail: psicili@umich.edu

M. Zint
e-mail: zintmich@umich.edu

W. Leal Filho et al. (eds.), *The Contribution of Universities Towards Education for Sustainable Development*, World Sustainability Series,
https://doi.org/10.1007/978-3-031-49853-4_30

disciplines and non-Western geographies to be engaged in competency research and to publish on the topic in a greater variety of (inter)disciplinary journals.

Keywords Higher education · Sustainability education · Disciplines · Interdisciplinary · Competencies · Direct citation analysis · Bibliometric analysis

1 Introduction

> A sustainable world will require more than just a cadre of sustainable development specialists, and a world in which everyone was first and foremost an expert on sustainable development would hardly be sustainable. It is the unique potential contribution of higher education to prepare engineers, doctors, teachers, managers of all kinds, policy-makers, shipping agents, financial managers, journalists and film directors, whose contribution to the world, in their particular professional capacity, will be one that makes it more rather than less sustainable (Gough and Scott 2008, p. 159).

As suggested by Gough and Scott (2008), our world is in urgent need of citizens working collaboratively toward environmental, economic, and social goals in ways that do not diminish future generations' ability to thrive as well. Most recently, this vision has been reinforced by the United Nations through the 17 Sustainable Development Goals (SDGs), to be achieved by 2030 (UN 2015). Moreover, United Nations Educational, Scientific and Cultural Organization (UNESCO) has challenged the global education community to support achievement of the SDGs through Education for Sustainable Development (ESD) (UNESCO 2020). Education, including higher education, has been identified as one of the key mechanisms to achieve sustainable development, as ESD seeks to ensure future generations will be able to address the challenges the SDGs aim to mitigate through formal curricula, along with non-formal and informal learning experiences (Zint et al. 2024).

The interdisciplinary nature of the challenges the SDGs were developed to address requires that all higher education students, beyond those selecting sustainability-related majors or minors, learn about sustainability (Michel 2020a). Developing a cadre of interdisciplinary leaders [i.e., ones able to engage a variety of disciplines to synthesize new scientific insights (Boone et al. 2020)] willing and able to forge a pathway toward a sustainable future will require, as Gough and Scott (2008) propose, that higher education focus on both sustainable workforce and civic development (Henderson et al. 2022). Higher education has the ability to cultivate engaged citizens through both workforce and civic development by ensuring all disciplines prepare students for understanding how their actions influence our common future.

Higher education institutions can help ensure that such sustainability workforce and civic development will be successful by articulating and adopting higher education for sustainable development (HESD) competencies all students should achieve Competencies are defined by Crick (2008) as "a complex combination of knowledge, skills, understanding, values, attitudes and desire which lead to effective, embodied human action in the world, in a particular domain" (p. 313). Specific disciplinary fields such as business (Abraham and Karns 2009), engineering (Lattuca et al. 2017), and medicine (Koh et al. 2008) have developed competencies that have helped to ensure respective students' success. Competencies are particularly critical in the development of an interdisciplinary field such as sustainability without accreditation standards, to provide guidance on learning outcomes, pedagogies, and assessments (Tigelaar et al. 2004). Johnstone and Soares (2014) found that competency-based education benefits student learning, program assessment and evaluation, and results in synergies between academic preparation and employer needs. In response to the rise of increased environmental, economic, and social crises, much of HESD scholarship has focused on identifying the competencies students need to master to address these intertwined wicked problems[1] and to support the global transformation toward a sustainable, low-carbon world (Blok et al. 2015; Brundiers et al. 2021; Michel et al. 2023).

The body of HESD competency scholarship has and continues to develop rapidly (Brundiers et al. 2021; Hallinger and Chatpinyakoop 2019; Michel et al. 2023). However, we know little about the publishing patterns of this influential body of scholarship, including the disciplines this scholarship is informed by. Given the need to ensure that all higher education students are prepared to contribute to achieving the SDGs, it is important to address this gap. For the purpose of this study, the authors therefore sought to learn more about the scholars engaged in competency scholarship and conducted bibliometric analysis (specifically direct citation analysis) to learn about the peer reviewed journals in which their articles are published. Bibliometric analysis is a popular tool for examining patterns within a particular body of scholarship (Ellegaard and Wallin 2015); in this case, on HESD competencies. More specifically, the aim of this study was to learn about the disciplines and thus, perspectives shaping the influential scholarship on HESD competencies through descriptive and direct citation analyses. To do so, the authors sought to answer the following three research questions:

(1) What is the disciplinary background of the scholars shaping the HESD competency scholarship?
(2) What are the most prominent journals publishing HESD competency scholarship?
(3) What, if any, patterns are apparent in how articles on HESD competency cite each other?

A literature review provides further context for this study and the research questions above, followed by an overview of the methods. Results to answer each of the research questions are presented next, followed by a discussion of findings, identification of limitations, and conclusions for advancing HESD competency scholarship.

[1] *Wicked problems* is a term developed by social policy scholars and adopted by the sustainability field to describe an issue that includes (a) system complexity, (b) goal conflict, and (c) informational uncertainty (Termeer et al. 2019).

2 Literature Review

2.1 Defining Higher Education for Sustainable Development (HESD)

For the purpose of this study, the authors adopted the same definition of sustainability shared in the Brundtland Commission's (1987) report, *Our Common Future,* as "meeting the needs of the present without compromising the ability of future generations to meet their own needs" (p. 1). This vision and the associated SDGs will not be achieved without a transition toward a low-carbon world, to halt the worst of a changing climate (IPCC 2018). Sustainability requires the integration of environmental, economic, and equity considerations through inter- and transdisciplinary approaches by drawing on a range of disciplines including, but not limited to, the natural and social sciences, engineering, the arts and humanities (Kates 2011). Accordingly, the authors employ the term HESD to encompass education at the university level surrounding sustainability and use this term to signify "the process of developing students' sustainability knowledge, attitudes, and behaviors in favor of the environment, and its economic and social implications (Besong and Holland 2015; Cotton and Winter 2010; Leal Filho and Pace 2016; Palmer 2002)" (Michel 2020b, p. 25). This definition is apt as it is intentionally broad enough to encompass learning about sustainability in all its forms across higher education disciplines and settings.

2.2 A Brief Overview of Higher Education Sustainable Development Competency Scholarship

The earliest scholarship on HESD competencies was conducted by Sterling and Thomas (2006), who suggested that HESD competencies should develop students' moral responsibilities. Next, Barth et al. (2007) explored HESD competencies within both curricular and co-curricular learning contexts and found that HESD competencies have both cognitive and non-cognitive dimensions. In turn, Barth and colleagues encouraged HESD competency learning opportunities both in (formally) and out (informally) of the classroom. (Wiek et al. (2011) subsequently wrote one of the most influential and cited publications on HESD competencies (Brundiers et al. 2021; Grosseck et al. 2019; Hallinger and Chatpinyakoop 2019; Michel et al. 2023). Based on a comprehensive literature review, Wiek and colleagues developed a framework that includes five key competencies: systems thinking, anticipatory/futures-thinking, normative/values-thinking, strategic-thinking, and interpersonal/collaboration thinking. Wiek et al. (2016) added another competence, meta-competence. Most recently, an group of international HESD experts came together to reach consensus on a sustainability competency framework, building on Wiek et al.

(2011). This group, as published by Brundiers et al. (2021) added integrated problem solving and self-awareness to the aforementioned sustainability competency framework. The group also suggested that Wiek et al.'s (2011) linear model be changed into a hierarchical one and that specific learning goals be created for those interested in pursuing sustainability careers. Nonetheless, critics have suggested that environmental justice (Clark and Miles 2021) as well as Indigenous and racially diverse scholars/communities' voices and perspectives are missing in this HESD sub-field (Garibay et al. 2016; Garibay and Vincent 2018; Michel et al. 2023; Parsons et al. 2021).

2.3 *Insights from Assessments of Higher Education for Sustainable Development Scholarship*

Recent reviews of the HESD scholarship have found it to have surged in the late twentieth century (Aikens et al. 2016; Leal Filho et al. 2021; Michel et al. 2023). Since the 1970s, the average number of research publications on HESD increased from 1 per year from 1970–1989 to 13.2 per year during 2005–2013 (Aikens et al. 2016). The majority of HESD scholarship has focused on higher education policy and management, with less attention to education (Aikens et al. 2016). Not surprisingly, therefore, one mixed-methods study of HESD scholarship between 2005 and 2014 found that articles on the topic were published most often in science versus education journals (Ávila et al. 2018). At the same time, there is some more evidence, based on a bibliometric study, that the HESD field is becoming more diverse, particularly with regard to the topics studied (Leal Filho et al. 2021).

Scholars have used this same type of bibliometric method used in the latter study to examine the trajectory of research in select HESD sub-fields. For instance, a bibliometric analysis by Cullen (2017) revealed that scholarship on educating business students about sustainability focused on the need for including HESD in business schools and on how business school faculty have integrated sustainability into their teaching. To the best of the authors' knowledge, however, few have assessed the level of engagement by scholars from different disciplines in HESD scholarship. Weiss and Barth (2019) focused their review on HESD implementation and found that while associated scholarship was published across a range of journals, most studies were conducted in the 'Western World'. Hallinger and Chatpinyakoop (2019), who employed a bibliometric analysis to examine HESD scholarship published between 1998 and 2018, similarly found that researchers contributing to this body of literature were mostly from developed countries. Moreover, they identified the top four journals publishing HESD studies: *International Journal of Sustainability in Higher Education (IJSHE), Journal of Cleaner Production (JCP), Sustainability,* and *Environmental Education Research (EER)*. Among the 20 journals they identified as having published on HESD competencies were also a few higher education journals: *Higher*

Education Policy (rank 8), *Higher Education* (rank 11), and *Studies in Higher Education* (rank 16). Although Hallinger and Chatpinyakoop (2019) identified competencies as a key theme in HESD scholarship, they did not study this important sub-field in greater detail, serving as a catalyst for the present study. As suggested by the present study's research questions, the authors sought to examine HESD competency scholarship, in part to compare it with the broader HESD literature in terms of its strengths and limitations regarding disciplinary influences.

3 Methods

3.1 Descriptive Analysis

The present study is an outgrowth of of the authors' prior research assessing the growth of HESD competency research (Michel et al. 2023). To examine the disciplinary background of the scholars shaping HESD competency research, the authors began by exploring the disciplinary backgrounds of the most highly cited and co-cited scholars publishing on the topic through a descriptive analysis.

3.1.1 Data Collection

The departure points for this study were Hallinger and Chatpinyakoop's (2019) list of the (1) 20 most highly *cited* scholars and (2) 20 most highly *co-cited* scholars publishing on HESD between 1998 and 2018 as identified by Scopus. Scholars on the two lists were not mutually exclusive. After removing duplicated scholars, the authors reviewed the publications of the remaining scholars and removed those who published on HESD broadly, but not specifically on HESD competencies. After removal of the aforementioned scholars, a sample of 21 scholars (hereafter referred to as 'primary scholars') remained. Next, the authors conducted an internet search to investigate the disciplinary background of these 21 primary scholars. The authors searched for each scholar' most up-to-date profile via Scopus, Google Scholar, Google, and LinkedIn, to identify their current employer and subsequently current disciplinary focus. The authors also examined their academic websites, reviewed their curriculum vitae, and scanned their scholarship including biographies published in journal articles to verify these initial assessments.

3.1.2 Data Analysis

To determine the disciplinary backgrounds of the 21 primary scholars who published on HESD competencies and answer the first research question, the authors analyzed the foci of their dissertations (as a proxy for the discipline they were trained in) and current disciplines based on their department and title (if there was a specific discipline listed within their title). The authors also reviewed the countries in which the scholars completed their doctoral work and conducted their current research. The authors employed qualitative open coding to group similar labels together into common categorizations (Creswell et al. 2007; Williams and Moser 2019).

By examining scholars' dissertations, the authors were able to determine which disciplines initially influenced scholars' HESD competency work. Dissertations were categorized into broad topics such as the natural sciences which included disciplines like chemistry and biology, for example. Furthermore, scholars' current[2] disciplinary focus was determined by the scholar's home department, job titles (e.g., being a "professor of organizational sustainability") and titles of recent publications. When it was difficult to determine scholars' past or present disciplinary focus due to unavailable or unclear information, the titles and abstracts of recent publications were used to determine a scholar's current research interests to place them in a disciplinary category. For instance, Moore's (2004) dissertation was in urban planning but focused on sustainable communities. As such, Moore's disciplinary focus was categorized into sustainability. Geographic location was categorized by continent because of the sample's limited size.

After completing a direct citation analysis (described further below), the authors engaged in another descriptive analysis of the disciplines of the journals that have published on HESD competencies. The authors determined a journal's discipline by reviewing its aim and scope and allocated it into a broad category. Sustainability and environmental journals, for example, were categorized into one category. Next, a more fine-grained assessment was conducted to determine a journal's sub-discipline. For example, education was assigned to be the *International Journal of Sustainability in Higher Education*'s primary discipline and higher education sustainability its sub-discipline. Similarly, education was assigned to be the *European Journal of Engineering Education's* primary discipline and engineering its sub-discipline.

3.2 Direction Citation Analysis

Consistent with prior studies investigating the trajectory of the growing HESD scholarship (e.g., Aikens et al. 2016; Ávila et al. 2018) including through bibliometric analysis (e.g., Hallinger and Chatpinyakoop 2019; Leal Filho et al. 2021), this study drew on this particular method to assess the 21 primary authors' HESD competency research. As stated by Hallinger and Chatpinyakoop (2019), "[bibliometric analyses] value extends from the capability to document and synthesize broad trends [including disciplinary prominence] that describe the landscape, composition, and intellectual structure of a knowledge base. Thus, the science mapping offers insights into patterns of knowledge that would be difficult to "see" using traditional methods of research

[2] When the authors report on "current" findings, the authors are referring to "currently during data collection," which was in 2021.

review" (pp. 2–3). Specifically, the authors employed direct citation analysis, a form of bibliometric mapping that generates a visual network map to depict relationships between in-text citations (e.g., one article directly citing another) (Boyack and Klavans 2010). This method, like many bibliometric analysis techniques, can take several forms and can be used to analyze different units of measurement within a bibliometric study, such as citation connections between scholars, publications, and journals (Yang and Wang 2015). For the purpose of this study, direct citation analysis between scholarly journals was conducted because the authors sought to learn about the disciplinary landscape of HESD scholarship. Direct citation mapping of journals allowed the authors to understand, (1) where HESD competency scholarship has been published, and (2) to what extent articles published in these journals have cited each other. Results allowed the authors to identify the disciplinary nature of the journals in which HESD competency scholarship has been published and thus, what gaps may exist.

3.2.1 Data Retrieval

Consistent with the procedures for the descriptive analyses, the authors used Hallinger and Chatpinyakoop's (2019) two lists on the most highly cited scholars publishing on HESD as the springboard for data retrieval (i.e., journal article identification). Also consistent with Hallinger and Chatpinyakoop (2019), the authors used Scopus to identify HESD competency scholarship. Scopus includes publications from a broad array of disciplines and the database allows users to directly download bibliographic information. The authors searched Scopus for articles by the 21 primary scholars that have published on HESD competencies. Only publications that had a clear and substantive focus on HESD *competencies*,[3] were published in a peer-reviewed journal between 2006 and 2020.[4] and were published in English were included in this study.

The initial search resulted in 1127 publications by the 21 primary scholars. By using the abovementioned criteria, the sample was reduced to 146 articles (13% of 1127). These 146 articles were written not only by the 21 primary scholars but 249 co-scholars (hereafter termed 'secondary scholars'). Next, the authors replicated the same search process, this time extracting articles written by the 249 secondary

[3] While competency has emerged as the predominant term within sustainability education research and practice, there are a "sea of labels" (Sterling et al. 2017, p. 153) and inconsistencies in referring to associated concepts. Often, interchangeable labels are used (Lester 2014) to refer to a similar concept such as "environmental literacy" (Winter and Cotton 2012) or "learning outcomes" (Buckley and Michel 2020), attributes (Shephard et al. 2011), or capabilities (Thomas and Depasquale 2016). While "competencies" has gained the most acceptance within sustainability education research, the difference between it and other terms tends to be semantic (Brundiers et al. 2021; Lester 2014). As such, the authors included publications on competencies as a concept broadly.

[4] The search was bound to 2006–2020. 2006 was selected as the starting point because this year has been identified as the beginning of HESD scholarship (e.g., Sterling and Thomas 2006). Publications available through Scopus online in 2020 were included, even when their 'print' dates were 2021 (e.g., Brundiers et al. 2021, Leal Filho et al. 2021, and Redman et al. 2021).

scholars. The resulting 5477 articles were again reviewed based on the four aforementioned criteria, resulting in an abridged sample of 121 articles (2% of 5477). Cumulatively across both Scopus searches, 267 peer-reviewed articles from the initial 6604 (4%) identified publications were included in the sample for the direct citation analysis. Including both primary and secondary scholars, these articles were authored by 275 individual scholars.

3.2.2 Data Analysis

The authors used direct citation analysis of journals to create a bibliometric map of the publications publishing articles on HESD competencies. To create this network visualization map, the authors utilized VOSViewer, a commonly used software platform for bibliometric analysis (van Eck and Waltman 2017). To identify the bibliometric patterns present in the sample, first the *publishing* trends in the data were examined, followed by *citation* trends. This approach allowed the authors to learn which journals HESD competency scholarship is typically published in and which journals scholars are drawing on for their sources.

In accordance with the second research question, the authors examined the most prominent journals (as a proxy to understanding scholars' disciplinary focus) publishing articles on HESD competencies, to reveal trends and patterns among journals in which this scholarship is most typically published. Then, in accordance with the third research question, the authors explored the relationships between the articles published by these journals, to determine citation trends (i.e., if scholars cited articles typically published in certain journals over others). This aspect of the analysis looked directly at which data nodes (i.e., journals) had direct citations to each other. By analyzing the connections between journals, the authors sought to find meta level trends in where scholars obtained supporting knowledge for their work on HESD competencies. Because the unit of analysis was not individual articles but rather the journals these articles were published in, pairing journal prominence with journal citation trends yielded helpful insights.

4 Results

4.1 Disciplinary Background of Scholars Shaping HESD Competency Scholarship

To answer the first research question, the authors explored the disciplinary background of the 21 primary scholars who have published on HESD competencies. An overview of results can be found in Table 1. Five scholars were trained in each the natural sciences[5] and sustainability education, four in education, three in sustainability, two in the social sciences, and one in each business and environmental science. Currently, ten scholars are working in sustainability, seven in sustainability education, and one in business, engineering, and the natural sciences, respectively. The remaining scholar's (n = 1) current discipline could not be identified. Overall, these results indicate that most HESD competency scholars began their work in a range of disciplines but over time transitioned to working primarily in sustainability (48%) and sustainability education (33%), representing 81% of the total sample.

The geographic settings in which these scholars conducted their work were also explored. Sixteen completed their doctoral studies in Europe, four in North America, and one in Oceania. Fourteen are currently working in Europe, five in North America, two in Oceania. The authors recognize that the geographic context of a scholar's training and current employment does not represent the scholar's complex identity and background. Nonetheless, it is clear from these data that the primary HESD competency scholars were trained in "western[6]" academia, and thus, likely to hold white, majoritized, and colonized perspectives.

4.2 *Disciplinary Focus of Journals Publishing Articles on Higher Education Sustainability Education Competencies*

Seventy-two academic journals have published articles focused on HESD competencies. Journals with an education focus have most frequently published on HESD competencies (n = 38, followed by those with a focus on sustainability (n = 24), business (n = 6), society (n = 2), and both engineering and human development (n = 1) (Table 2). Among the journals with education as their primary disciplinary focus, the majority focused on higher education (n = 8), some on environmental education (n = 5), and a few on engineering education (n = 2). The remaining education-focused journals covered a variety of sub-disciplines, with one article each. In contrast to the primary scholars who had both social and natural science backgrounds, only one journal had a focus on the natural sciences (in this case, chemistry) and three on science in general.

[5] Although sustainability education could be and is often grouped with sustainability or education, for the purpose of this study, the authors wanted to see the explicit focus on sustainability education separate from the other two categories. The same can be said for sustainability and social science, and environmental science and natural science.

[6] "Western" in this manuscript refers to education and educational contexts that are derived from characteristics of dominant culture, high-income countries with a history of colonial practices.

Table 1 Most highly cited HESD scholars (as per Hallinger and Chatpinyakoop 2019, Tables 2 and 3) with current disciplinary role and field, original work

Scholars	Doctoral country	Disciplinary focus of dissertation	Current country	Current disciplinary focus
Barth	Germany	Education	Germany	Sustainability education
Ceulemans	Belgium	Business	France	Business
Ferrer-Balas	Spain	Natural science	Spain	[a]
Godemann	Germany	Sustainability	Germany	Natural science
Huisingh	USA	Natural science	USA	Sustainability
Keeler (Withycombe)	USA	Sustainability	USA	Sustainability
Lambrechts	Belgium	Sustainability education	Netherlands	Sustainability
Leal Filho[b]	United Kingdom	Sustainability education	Germany	Sustainability
Lozano	Wales	Sustainability	Sweden	Sustainability
Lozano-García	United Kingdom	Natural science	Mexico	Engineering
Lukman	Slovenia	Natural science	Slovenia	Sustainability
Redman	USA	Social science	USA	Sustainability
Rieckmann	Germany	Education	Germany	Sustainability education
Shephard	United Kingdom	Education	New Zealand	Sustainability education
Sterling	United Kingdom	Sustainability education	United Kingdom	Sustainability education
Stoltenberg	Germany	Social science	Germany	Sustainability education
Svanström	Sweden	Natural science	Sweden	Sustainability
Thomas	Australia	Education	Australia	Sustainability education
Tilbury	United Kingdom	Sustainability education	Gibraltar	Sustainability
Wals	USA	Sustainability education	Netherlands	Sustainability education
Wiek	Switzerland	Environmental science	USA	Sustainability

[a]indicates unclear information or no information was found

[b]This scholar holds multiple doctoral degrees; given that this study focuses on sustainability education, the authors analyzed this scholar's sustainability education-focused degree

Table 2 Publication disciplines publishing on HESD competencies, original work

Publication title	# Articles cited	Year of 1st publication	Year of most recent publication	Publication discipline	Publication sub-discipline
International Journal of Sustainability in Higher Education	50	2006	2020	Education	Higher education sustainability
Journal of Cleaner Production	48	2010	2021	Sustainability	Clean production
Sustainability (Switzerland)	35	2013	2020	Sustainability	General
Environmental Education Research	27	2006	2020	Education	Environmental
Sustainability Science	14	2009	2021	Sustainability	Sustainability science
GAIA	5	2007	2018	Sustainability	Environment; solutions
Sustainability (United States)	4	2008	2019	Sustainability	General
European Journal of Engineering Education	4	2008	2019	Education	Engineering
Journal of Teacher Education for Sustainability	4	2010	2017	Education	Sustainability; teacher education
Innovative Higher Education	3	2008	2020	Education	Higher education
Journal of Environmental Education	3	2007	2014	Education	Environmental
Current Opinion in Environmental Sustainability	2	2015	2015	Sustainability	Environmental sustainability
Environment, Development and Sustainability	2	2015	2015	Sustainability	Development; environment
International Journal of Innovation and Sustainable Development	2	2006	2007	Sustainability	Development; technology
Sustainability	2	2010	2012	Sustainability	General

(continued)

Table 2 (continued)

Publication title	# Articles cited	Year of 1st publication	Year of most recent publication	Publication discipline	Publication sub-discipline
Education Sciences	2	2017	2017	Education	General
Journal of Geography in Higher Education	2	2010	2016	Education	Higher education; geography
Journal of Professional Issues in Engineering Education and Practice	2	2013	2017	Education	Engineering
Journal of Transformative Education	2	2009	2018	Education	Transformative
Teaching in Higher Education	2	2020	2020	Education	Higher education
Ambio	1	2019	2019	Sustainability	Environment; society
Environment	1	2011	2011	Sustainability	Environmental—general
Environment and Behavior	1	2020	2020	Sustainability	Environmental behavior
Environmental Science and Policy	1	2019	2019	Sustainability	Environmental policy
International Forestry Review	1	2017	2017	Sustainability	Forestry
International Journal of Environmental Research and Public Health	1	2020	2020	Sustainability	Environmental; public health
International Journal of Sustainable Development and World Ecology	1	2020	2020	Sustainability	Sustainable development
Journal of Environmental Studies and Sciences	1	2013	2013	Sustainability	Environmental studies
Journal of Hunger and Environmental Nutrition	1	2018	2018	Sustainability	Hunger; nutrition

(continued)

Table 2 (continued)

Publication title	# Articles cited	Year of 1st publication	Year of most recent publication	Publication discipline	Publication sub-discipline
Local Environment	1	2016	2016	Sustainability	Justice; policy
Nature Sustainability	1	2020	2020	Sustainability	Policy
Renewable and Sustainable Energy Reviews	1	2017	2017	Sustainability	Renewable energy
Smart Innovation, Systems and Technologies	1	2012	2012	Sustainability	Technology
Survey Review	1	2019	2019	Sustainability	Land management
Sustainable Development	1	2019	2019	Sustainability	Sustainable development
Futures	1	2012	2012	Society	Future society
Journal of Critical Realism	1	2019	2019	Society	Future
Emotion Review	1	2010	2010	Human development	Emotion
Science and Engineering Ethics	1	2013	2013	Engineering	Ethics
Applied Environmental Education and Communication	1	2015	2015	Education	Environmental
Assessment and Evaluation in Higher Education	1	2015	2015	Education	Higher education
Australasian Journal of Engineering Education	1	2019	2019	Education	Engineering
Australian Journal of Environmental Education	1	2013	2013	Education	Environmental
Educacion Quimica	1	2013	2013	Education	Chemistry
Ensenanza de las Ciencias	1	2014	2014	Education	Science teaching

(continued)

Table 2 (continued)

Publication title	# Articles cited	Year of 1st publication	Year of most recent publication	Publication discipline	Publication sub-discipline
European Educational Research Journal	1	2011	2011	Education	Research
European Journal of Education	1	2017	2017	Education	Policy
European Journal of Teacher Education	1	2009	2009	Education	Teacher education
Higher Education	1	2018	2018	Education	Higher education
Higher Education Policy	1	2018	2018	Education	Higher education
Higher Education Research and Development	1	2020	2020	Education	Higher education
Higher Education, Skills and Work-Based Learning	1	2018	2018	Education	Higher education
International Journal of Science Education	1	2017	2017	Education	Science education
Journal of Agricultural Education and Extension	1	2010	2010	Education	Agriculture
Journal of Baltic Science Education	1	2010	2010	Education	Science
Journal of Experiential Education	1	2015	2015	Education	Experiential
Journal of Geoscience Education	1	2017	2017	Education	Environmental
Learning Environments Research	1	2016	2016	Education	Informal learning

(continued)

Table 2 (continued)

Publication title	# Articles cited	Year of 1st publication	Year of most recent publication	Publication discipline	Publication sub-discipline
New Directions for Teaching and Learning	1	2020	2020	Education	Higher education
New Educational Review	1	2014	2014	Education	Research
Policy Futures in Education	1	2013	2013	Education	Public health
Research in Post-Compulsory Education	1	2019	2019	Education	Work-force
Research in Science and Technological Education	1	2020	2020	Education	Technology; science
South African Journal of Education	1	2008	2008	Education	Transformative; sustainability
Teaching and Teacher Education	1	2010	2010	Education	Teacher education
Teoria de la Educacion	1	2017	2017	Education	Pedagogy
International Journal of Entrepreneurship and Innovation	1	2018	2018	Business	Entrepreneurship
International Journal of Management Education	1	2017	2017	Business	Management
Journal of Knowledge Management	1	2019	2019	Business	Knowledge management
Journal of Management and Organization	1	2011	2011	Business	Management; organization
Journal of Small Business and Entrepreneurship	1	2019	2019	Business	Entrepreneurship; small business
Social Responsibility Journal	1	2019	2019	Business	Social responsibility

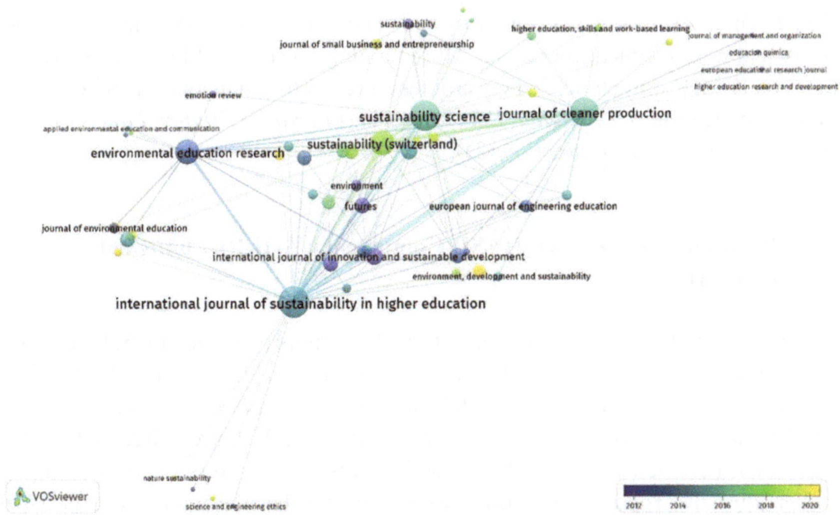

Fig. 1 Publication Prominence of HESD Competency Published Articles

4.2.1 Direct Citation Analysis of Journals

A direct citation analysis map (Fig. 1) shows which of the 72 journals have been the most prominent when it comes to publishing on HESD competencies and the relationships between the journal articles' respective citations. For this map, the authors directed VOSViewer to represent each journal as a separate data node. The node's color denotes the average year of the articles on HESD competencies published by the respective journal (with lighter/warmer colors indicating more recent publications, darker/cooler colors indicating older publications). The node's size is a function of connections to other journals (i.e., the more a journal article cites and/or is cited by articles in other journals, the larger its representative node). Journal prominence (i.e., journals with the largest nodes) in this map is a function of the number of publications each journal has published on HESD competencies, the average year the journal published on the topic, and the number of the journal's articles cited by other journals' articles on the topic.

4.2.2 Prominence of Journals Publishing on HESD Competencies

With regard to the second research question, the journals that have published the highest number of articles on HESD competencies were: *International Journal of Sustainability in Higher Education (IJSHE)* (n = 50), *Journal of Cleaner Production (JCP)* (n = 48), *Sustainability (n = 35), Environmental Education Research (EER)* (n = 27) and *Sustainability Science* (n = 14). The top journals publishing on this topic, with the exception of *IJSHE,* which focuses on sustainability in higher education, and *EER* which focuses on sustainability and education, focus on sustainability more broadly and not on education per se. All remaining journals published less than 10 articles on HESD competencies. Few articles were published in journals with a focus on higher education (i.e., *Higher Education* (1), *Innovative Higher Education* (3), and *Teaching in Higher Education* (2)).

4.2.3 Relationships Between HESD Competency Articles Published by Journals

To answer the study's third research question, the authors examined the relationships of the citations between the HESD articles published by the 72 journals. *IJSHE*'s node was linked to 35 other journals, followed by *Sustainability Science* (34 links), *JCP* (29 links), and *Sustainability* and *EER* (each with 22 links). High link scores suggest that the articles published in these journals cited or were cited by a variety of other journals. The strongest links were between the five main journals listed above, suggesting a high degree of connected scholarship between them, due to the number of times articles used citations published between them. For instance, HESD articles published in *IJSHE* cited and were cited by articles published by many other journals in the sample. *EER*, which published articles on the topic early in comparison to other journals, had articles that were cited by more recent HESD articles in other journals. With regard to journals focused on higher education, there were relatively few citation relationships: *Assessment and Evaluation in Higher Education* (8 links), *Higher Education* (3 links), *Innovative Higher Education* (9 links), and *Teaching in Higher Education* (6 links). In other words, the majority of HESD scholarship is both contained within and linked to journals with a focus on sustainability. While HESD competencies are by nature a topic focused on higher education, they have yet to be published in journals beyond those focused on sustainability.

5 Discussion

The aim of this study was to learn about the disciplines and thus perspectives shaping the scholarship on HESD competencies through descriptive and direct citation analysis. Findings suggest that to date, this body of scholarship was influenced by a small, collaborative set of scholars from closely related disciplines. It is therefore necessary to engage a broader set of scholars to ensure that the HESD competencies that drive curriculum development and assessment have greater potential to prepare students to contribute to the achievement of the SDGs.

5.1 Engaging Scholars Across Disciplines

Based on the present study, the authors found that the HESD competency scholarship has been influenced by scholars representing a limited set of disciplines and that the majority of this scholarship has been published in sustainability-focused journals. This finding is consistent with those by Hallinger and Chatpinyakoop (2019) as well as Leal Filho et al. (2021), who focused on HESD scholarship more broadly. Because scholars typically publish in their primary disciplines' journals, it appears that the majority of HESD competency research, like that in the broader HESD literature, has been conducted by scholars focusing on sustainability. What is particularly surprising, given the nature of HESD competency scholarship, is that involvement by higher education disciplines has been limited.

The SDGs are meant to be a comprehensive set of targets that address all major systems and processes of the contemporary world's economic, environmental, and social landscapes. In this light, interdisciplinary teaching and learning are essential for higher education students to develop multiple frames and lenses through which to address complex sustainability challenges (Pollard et al. 2023). The inadequate contributions from a wider breadth of disciplines, including higher education, is likely limiting HESD competency scholarship, and thus practice, from reaching its potential including with regard to preparing students for contributing to SDG achievements. This is because subject matter experts on sustainability are not experts across all disciplines connected to the SDGs' targets. However, in academia, the responsibility of collaboration tends to fall onto individual scholars, requiring a great amount of effort, given the traditional structures within and across higher education institutions. Due to the inherent siloing of disciplines, interdisciplinarity—though often highly spoken of—remains an often-impractical reality for scholars due to a variety of institutional and funding-related barriers (Peek and Guikema 2021). Scholars thus need support from their organizations to engage in interdisciplinary scholarship.

Higher education institutions can contribute to such interdisciplinary scholarship, for example, by making available funding for interdisciplinary collaborations, supporting the co-teaching of courses (Pollard et al. 2023), and holding lecture series engaging researchers from a range of disciplines. Because the problems that HESD seeks to address are inherently ones that need interdisciplinary solutions, the aforementioned approaches could greatly benefit scholarship on HESD competencies that engages scholars from a broader set of disciplines.

Higher education scholars, in particular, have much to offer HESD competency research including within the context of the SDGs based on their expertise in curriculum development. Curriculum development, "along with the concepts and principles to enhance and assess learning, is one of the fundamental domains of higher education as a field of study" (Freeman et al. 2013, p. 237). While the subject matter expertise from sustainability experts is invaluable, it is insufficient without being coupled with curricular expertise that higher education scholars can contribute. This is along with the subject matter expertise of the individual disciplines that must integrate HESD competencies into their respective fields, if students are to be prepared for contributing to the achievement of the various SDGs. In sum, the authors call for greater collaborations across disciplines, utilizing the strengths of each, to advance the body of HESD competency scholarship and to elucidate both the competencies and curriculum changes needed to support progress toward the SDGs. The authors also wonder about the distinct lack of voices from the real-world application that higher education institutions are preparing students to work in throughout their careers. Are employers, front-line bureaucrats, and primary education settings experiencing the benefit of HESD and sustainability leadership training? Additionally, if the majority of this scholarship is completed by scholars working in sustainability or sustainability education, what might be missed?

5.2 Engaging Scholars from Outside of Western Academia

This study demonstrates that the majority of HESD competency scholarship, like HESD scholarship more broadly, is being conducted but researchers educated by and continuing to work in Western academia. HESD competency scholarship could therefore greatly benefit from the engagement of researchers trained and working in contexts other than the Western world, to help ensure greater progress toward the SDGs including ESD for 2030. Such engagement by a more diverse sense of scholars can help ensure that diversity, equity, and inclusion receive greater attention; along with a stronger focus on justice and innovative solutions to the wicked challenges faced by those living in the geographies facing the brunt of climate change impacts. Knowledge developed within western academia maybe be too far removed from day-to-day life where the enactment of sustainability policies and practices is most needed.

5.3 Mapping to the Sustainable Development Goals

Academic journals are critical to the dissemination and thus progress in scholarly research (Norris 2021). The findings of the present indicated that the majority of HESD competency scholarship to date was published in sustainability journals and that only a few journals focused on other disciplines have done so. The latter publications provide a platform for scholars from diverse disciplines to share their findings, insights, and best practices, including related to HESD competencies in higher education (Norris 2021). By publishing such work, journals can contribute to the advancement of knowledge and understanding in this field, fostering greater awareness and engagement among more diverse scholars (Morton et al. 2022). Moreover, these journals can serve as a catalyst for collaboration and interdisciplinary exchange, encouraging researchers from various disciplines to explore the intersectionality of the SDGs in education. If the SDGs were addressed in articles by a greater number of journals, HESD competencies could become far more robust.

Through disseminating scholarship on HESD competencies, journals not only facilitate the sharing of innovative ideas and approaches, but also inspire educational institutions to integrate sustainability principles into their curricula and practices. Ultimately, the publication of articles on HESD competencies in academic journals can significantly contribute to the realization of the SDGs by promoting transformative learning, shaping policy agendas, and fostering a more sustainable future for generations to come.

6 Limitations

One of the limitations of this study is that it draws on HESD competency scholarship published up to 2020, due to us drawing from and building on the comprehensive literature review by Hallinger and Chatpinyakoop (2019). Nonetheless, given the relatively limited time that has passed since, the authors suspect that these findings would not be significantly different and/or they would not be outweighed by what the authors learned about the status of the majority of publications on the topic. It is the sincere hope of the authors that such scholarship is already underway and/or that this chapter will help to prompt such work. The authors also want to recognize that sustainability is not a novel concept developed by Western academics. Cultural practices from around the globe may not use the term sustainability as a key component of their traditions and lifestyles, rather the tenants of sustainability may be intricately intertwined into every aspect of life. Therefore, the authors want to highlight that this analysis of HESD competency scholarship is limited to scholarship published by Western scholars in English and in peer-reviewed publications. Future research ought to include scholarship published in other languages and examine relevant oral histories.

7 Conclusion

Competencies play an important role in ensuring that higher education institutions develop the transformational leaders society needs to transition toward a sustainable future. All students will need to have the knowledge, skills and dispositions to contribute to the achievement of the SDGs in ways that are environmentally and socially just, which requires them to collaborate in inter- and trans- disciplinary ways. As such, it is important that the continued development and revisions of HESD competencies be informed by scholars from a broad range of disciplines and from outside of western academic contexts. Doing so will not only ensure a diversity of voices, but will also ultimately broaden students' sustainable perspectives.

To date, HESD competency work has already been highly collaborative between sustainability and sustainability education scholars. This foundation has allowed the HESD sub-field to capture and synthesize perspectives from influential scholars, scholarship, and seminal work in the field. The time has come, however, to extend these collaborations to a more diverse set of scholars and to publish in a greater variety of journals including those focused on higher education. Scholars who have traditionally published in journals such as *IJSHE*, *Sustainability Science*, *JCP*, *Sustainability*, and *EER* are encouraged to collaborate with scholars from a greater range of disciplines and non-Western geographies, and to publish in a greater variety of (inter)disciplinary journals. In particular, HESD competency scholarship needs to be published in journals beyond those chiefly focused on sustainability. HESD competency scholarship needs to go beyond being a 'sustainability' topic, to one that benefits higher education students of all backgrounds, regardless of their disciplines and future professions.

It is the authors' hope that this study will contribute toward these goals and thus, support even more robust HESD competency scholarship than has already been completed to date. Existing patterns in HESD competency scholarship reflect traditional methods of building upon landmark citations, but this may mean important ideas are lost when not closely tied to other relevant work. The authors challenge scholars and journal editors alike to expand their traditional approaches to knowledge production, to help ensure that higher education students are better prepared for achieving the SDGs and thus, a future where we may all thrive.

References

Abraham SE, Karns LA (2009) Do business schools value the competencies that businesses value? J Educ Bus 84(6):350–356

Aikens K, McKenzie M, Vaughter P (2016) Environmental and sustainability education policy research: a systematic review of methodological and thematic trends. Environ Educ Res 22(3):333–359

Ávila L, Rossato Facco AL, Bento MHDS, Arigony MM, Obregon SL, Trevisan M (2018) Sustainability and education for sustainability: an analysis of publications from the last decade. Environ Qual Manage 27(3):107–118

Barth M, Godemann J, Rieckmann M, Stoltenberg U (2007) Developing key competencies for sustainable development in higher education. Int J Sustain High Educ 8(4):416–430

Besong F, Holland C (2015) The dispositions, abilities and behaviours (DAB) framework for profiling learners' sustainability competencies in higher education. J Teach Educ Sustain 17(1):5–22

Blok V, Gremmen B, Wesselink R (2015) Dealing with the wicked problem of sustainability: the role of individual virtuous competence. Bus Prof Ethics J 34(3):297–327

Boone CG, Pickett ST, Bammer G, Bawa K, Dunne JA, Gordon IJ, ... Mallee H (2020) Preparing interdisciplinary leadership for a sustainable future. Sustain Sci 15:1723-1733

Boyack KW, Klavans R (2010) Co-citation analysis, bibliographic coupling, and direct citation: which citation approach represents the research front most accurately? J Am Soc Inform Sci Technol 61(12):2389–2404

Brundiers K, Barth M, Cebrián G, Cohen M, Diaz L, Doucette-Remington BS, Dripps W, Habron G, Harré N, Jarchow M, Losch K, Michel J, Mochizuki Y, Rieckmann M, Parnell R, Walker P, Zint M (2021) Key competencies in sustainability in higher education—toward an agreed-upon reference framework. Sustain Sci 16(1):13–29

Brundtland G (1987) Report of the world commission on environment and development: our common future. United Nations General Assembly Document A/42/427
Buckley JB, Michel JO (2020) An examination of higher education institutional level learning outcomes related to sustainability. Innov High Educ 45(3):201–217
Clark SS, Miles ML (2021) Assessing the integration of environmental justice and sustainability in practice: a review of the literature. Sustainability 13(20):11238
Cotton D, Winter J (2010) 'It's not just bits of paper and light bulbs': A review of sustainability pedagogies and their potential for use in higher education. In Jones P, Selby D, Sterling S (eds) *Sustainability education: perspectives and practice across higher education.* Earthscan, New York, NY, pp 39–54
Creswell JW, Hanson WE, Clark Plano VL, Morales A (2007) Qualitative research designs: selection and implementation. Couns Psychol 35(2):236–264
Crick RD (2008) Key competencies for education in a European context: narratives of accountability or care. Eur Educ Res J 7(3):311–318
Cullen JG (2017) Educating business students about sustainability: a bibliometric review of current trends and research needs. J Bus Ethics 145(2):429–439
Ellegaard O, Wallin JA (2015) The bibliometric analysis of scholarly production: how great is the impact? Scientometrics 105(3):1809–1831
Freeman S, Hagedorn LS, Goodchild L, Wright D (2013) Advancing higher education as a field of study: In: Quest of doctoral degree guidelines—commemorating 120 years of excellence. Stylus Publishing, Herndon, VA
Garibay JC, Ong P, Vincent S (2016) Program and institutional predictors of environmental justice inclusion in US post-secondary environmental and sustainability curricula. Environ Educ Res 22(7):919–942
Garibay JC, Vincent S (2018) Racially inclusive climates within degree programs and increasing student of color enrollment: an examination of environmental/sustainability programs. J Divers High Educ 11(2):201–220
Gough S, Scott W (2008) Higher education and sustainable development: paradox and possibility. Routledge
Grosseck G, Țîru LG, Bran RA (2019) Education for sustainable development: evolution and perspectives: a bibliometric review of research, 1992–2018. Sustainability 11(21):6136
Hallinger P, Chatpinyakoop C (2019) A bibliometric review of research on higher education for sustainable development, 1998–2018. Sustainability 11(8):2401
Henderson TS, Michel JO, Bryan A, Canosa E, Gamalski C, Jones K, Moghtader J (2022) An exploration of the relationship between sustainability-related involvement and learning in higher education. Sustainability 14(9):5506
IPCC (2018) Summary for policymakers. In: Masson-Delmotte V, Zhai P, Pörtner H-O, Roberts D, Skea J, Shukla PR, Pirani A, Moufouma-Okia W, Péan C, Pidcock R, Connors S, Matthews JBR, Chen Y, Zhou X, Gomis MI, Lonnoy E, Maycock T, Tignor M, Waterfield T (eds) Global warming of 1.5 °C. An IPCC special report on the impacts of global warming of 1.5 °C above pre-industrial levels and related global greenhouse gas emission pathways, in the context of strengthening the global response to the threat of climate change, sustainable development, and efforts to eradicate poverty. Cambridge University Press, Cambridge, UK and New York, NY, USA, pp 3–24
Johnstone SM, Soares L (2014) Principles for developing competency-based education programs. Change Mag High Learn 46(2):12–19
Kates RW (2011) What kind of a science is sustainability science? Proc Natl Acad Sci 108(49):19449
Koh GCH, Khoo HE, Wong ML, Koh D (2008) The effects of problem-based learning during medical school on physician competency: a systematic review. CMAJ Can Med Assoc J 178(1):34–41
Lattuca LR, Knight DB, Ro HK, Novoselich BJ (2017) Supporting the development of engineers' interdisciplinary competence. J Eng Educ 106(1):71–97

Leal Filho W, Pace P (2016) Teaching education for sustainable development: implications on learning programmes at higher education. In: Leal Filho W, Pace P (eds) Teaching education for sustainable development at university level. World sustainability series. Springer, Cham

Leal Filho W, Will M, Shiel C, Paço A, Farinha CS, Lovren VO, Alvila LV, Platje J, Sharifi A, Vasconcelos C, Fritzen Gomes B, Salvia A, Anholon R, Rampasso I, Quelhas L, Skouloudis A (2021) Towards a common future: revising the evolution of university-based sustainability research literature. Int J Sustain Dev World Ecol 28(6):503–517

Lester S (2014) Professional competence standards and frameworks in the United Kingdom. Assess Eval High Educ 39(1):38–52

Michel JO (2020a) Mapping out students' opportunity to learn about sustainability across the higher education curriculum. Innov High Educ 45(5):355–371

Michel JO (2020b) Toward conceptualizing education for sustainability in higher education. New Dir Teach Learn 2020(161):23–33

Michel JO, Siciliano P, Zint M, Collins S (2023) Toward diversifying higher education sustainability competency scholarship: findings and implications from a bibliometric analysis. Int J Sustain High Educ

Moore JL (2004) Recreating the university from within: sustainability and transformation in higher education (T). University of British Columbia. Retrieved from https://open.library.ubc.ca/collections/ubctheses/831/items/1.0055214

Morton B, Vercueil A, Masekela R, Heinz E, Reimer L, Saleh S, Kalinga C, Seekles M, Biccard B, Chakaya J, Abimbola S (2022) Consensus statement on measures to promote equitable authorship in the publication of research from international partnerships. Anaesthesia 77(3):264–276

Norris T (2021) Journalology: the study of the dissemination of academic knowledge. Brock Educ J 30(2):1–5. https://journals.library.brocku.ca/brocked/index.php/home/article/view/921/410

Palmer J (2002) Environmental education in the 21st century: theory, practice, progress and promise. Routledge, London, England

Parsons M, Fisher K, Crease RP (2021) Environmental justice and indigenous environmental justice. In: Decolonising blue spaces in the anthropocene, pp 39–73

Peek L, Guikema S (2021) Interdisciplinary theory, methods, and approaches for hazards and disaster research: an introduction to the special issue. Risk Anal 41(7):1047–1058.

Pollard JR, Michel JO, Simon AC, Shriberg M (2023) Team-teaching as a promising pathway toward interdisciplinary sustainability competency development. Sustainability 15(15):11534

Redman A, Wiek A, Barth M (2021) Current practice of assessing students' sustainability competencies: a review of tools. Sustain Sci 16(1):117–135

Shephard K, Smith N, Deaker L, Harraway J, Broughton-Ansin F, Mann S (2011) Comparing different measures of affective attributes relating to sustainability. Environ Educ Res 17(3):329–340

Sterling S, Thomas I (2006) Education for sustainability: the role of capabilities in guiding university curricula. Int J Innov Sustain Dev 1(4):349–370

Sterling S, Glasser H, Rieckmann M, Warwick P (2017) Chapter 10 "More than scaling up": a critical and practical inquiry into operationalizing sustainability competencies. In: Envisioning futures for environmental and sustainability education. Wageningen Academic Publishers, pp 681–700

Termeer CJ, Dewulf A, Biesbroek R (2019) A critical assessment of the wicked problem concept: relevance and usefulness for policy science and practice. Policy Soc 38(2):167–179

Thomas I, Depasquale J (2016) Connecting curriculum, capabilities and careers. Int J Sustain High Educ 17(6):738–755

Tigelaar DE, Dolmans DH, Wolfhagen IH, Van Der Vleuten CP (2004) The development and validation of a framework for teaching competencies in higher education. High Educ 48(2):253–268

United Nations (UN) (2015) Transforming our world: the 2030 agenda for sustainable development. Retrieved May 23, 2023, from https://sdgs.un.org/publications/transforming-our-world-2030-agenda-sustainable-development-17981

United Nations Educational, Scientific, and Cultural Organization (UNESCO) (2020) Education for sustainable development a roadmap. Retrieved May 23, 2023, from https://unesdoc.unesco.org/ark:/48223/pf0000374802

van Eck NJ, Waltman L (2017) Citation-based clustering of publications using CitNetExplorer and VOSviewer. Scientometrics 111(2):1053–1070

Weiss M, Barth M (2019) Global research landscape of sustainability curricula implementation in higher education. Int J Sustain High Educ 20(4):570–589

Wiek A, Withycombe L, Redman CL (2011) Key competencies in sustainability: a reference framework for academic program development. Sustain Sci 6(2):203–218

Wiek A, Bernstein M, Foley R, Cohen M, Forrest N, Kuzdas C, Kay B, Withycombe Keeler L (2016) Operationalising competencies in higher education for sustainable development. In: Barth M, Michelsen G, Rieckmann M, Thomas I (eds) Handbook of higher education for sustainable development. Routledge, London, pp 241–260

Williams M, Moser T (2019) The art of coding and thematic exploration in qualitative research. Int Manag Rev 15(1):45–55

Winter J, Cotton D (2012) Making the hidden curriculum visible: sustainability literacy in higher education. Environ Educ Res 18(6):783–796

Yang S, Wang F (2015) Visualizing information science: author direct citation analysis in China and around the world. J Informet 9(1):208–225

Zint M, Michel JO, Collins S, Gallay E, Sloan E, Pineiro ME, Balotin L, Correa V, Dabrowski M, Davoodi B, Dierdorf J, Fernandez D, Miller J, Nakisher I, Roessler C, Schurr P, Siciliano P, Thomson J (2024) Environmental and sustainability education in the USA: pathways to achieving the SDGs. In: Rieckmann M, Thomas R (eds) World review: environmental and sustainability education in the context of the sustainable development goals. Taylor & Francis Group.

ESD in Bosnia and Herzegovina: Implementing Innovative Methodologies in a University Elective Course

Aliaksandr Novikau

Abstract Education is essential for progress toward the United Nations Sustainable Development Goals (SDGs), and universities are pivotal in advancing education for sustainable development (ESD). While many countries have a long history of ESD in higher education, some, including those on the European continent, have only recently begun implementing it. In addition, despite being located in the same geographic region or country, universities often differ significantly in their approaches to ESD. This chapter examines the history and practice of *Social Responsibility and Sustainable Development,* a university elective course primarily targeting first-year students from diverse academic programs at the International University of Sarajevo in Bosnia and Herzegovina. Specifically, the chapter focuses on the implementation of the 17 Rooms methodology during the Spring 2023 iteration of the course. The 17 Rooms is an innovative methodology developed by a joint team from the Brookings Institution and the Rockefeller Foundation in 2018 to stimulate new forms of action for the SDGs. The chapter discusses the experience and pedagogical applicability of the methodology to a tertiary classroom context. The analysis emphasizes the need for more comprehensive and practical approaches to education for sustainable development in higher education, particularly in countries with nascent ESD implementation.

Keywords Bosnia and Herzegovina · Western Balkans · Education for sustainable development · Sustainable development goals · 17 Rooms · International University of Sarajevo

A. Novikau (✉)
Department of Political Science and International Relations, International University of Sarajevo, Hrasnička Cesta 15, 71210 Ilidža, Bosnia and Herzegovina
e-mail: anovikau@ius.edu.ba

W. Leal Filho et al. (eds.), *The Contribution of Universities Towards Education for Sustainable Development*, World Sustainability Series,
https://doi.org/10.1007/978-3-031-49853-4_31

1 Introduction

Universities worldwide play an essential role in creating a sustainable future for all. They contribute to achieving United Nations (UN) Sustainable Development Goals (SDGs) in different ways—beginning from the 1970s, universities have incorporated the ideas and principles of sustainable development into their educational and research programs, campus operations, and local communities (Blasco et al. 2020; Leal Filho et al. 2019; Lozano et al. 2013, 2015).

Although the functions of universities are very diverse, by definition, education is usually viewed as the primary mission of universities. Not surprisingly, much attention is paid to education for sustainable development in higher education (Leal Filho 2010). By design, universities are suitable to catalyze the emergence of novel approaches and avenues that converge toward implementing the SDGs. There is a rising trend among students to seek curricula that adequately equip them to participate in sustainable development. It includes the desire for expanded availability of sustainability-focused courses, the integration of sustainability topics into existing courses that previously lacked such content, the imposition of mandatory sustainability courses within academic programs where they were previously optional, and the establishment of a requirement for all undergraduate students to complete at least one course in the field of sustainable development (Novy et al. 2021). Thus, universities are central in teaching students to think critically about sustainability and develop real-life practices in their local communities (Leal Filho et al. 2019).

There has been inevitable progress in incorporating sustainability into university curricula around the globe (O'Byrne et al. 2015). Yet, although some universities succeed in developing curricula for sustainable development, there are also many universities where actions to address sustainable development are still very limited (Leal Filho et al. 2015).

Even though sustainable development courses are introduced in almost all countries, the progress in ESD differs between geographical regions. It is not surprising because sustainable development depends on national factors such as available financial resources, existing technologies and infrastructure, reporting and accountability mechanisms, and political support (Ávila et al. 2017). For instance, significant progress has been made in the European Union (Lozano et al. 2015). Since 2015, the European Union (EU) has been supporting the sustainability policies in its member states by determining the structure within the union on implementation of Agenda 2030. Not surprisingly, most EU member states have demonstrated significant achievements in ESD. In many other regions, however, the progress in education for sustainable development is often insufficient. Although numerous examples exist of how universities in such countries can still implement education for sustainable development initiatives, the financial, institutional, and political limitations are slowing down the progress in that area.

While the national context undoubtedly holds importance, individual universities' pivotal role in implementing sustainable education cannot be understated. The variations among universities within the same country in the implementation of ESD are often explained by differences at the institutional level. Among the primary barriers encountered on the institutional level are the absence of robust support from university administrations, the dearth of suitable technological infrastructure, and the lack of formal structures such as sustainable development committees (Ávila et al. 2017; Leal Filho et al. 2017). In many universities, endeavors aimed at sustainable development often face obstacles from an inadequate institutional framework. These challenges encompass the absence of established norms, guidelines, accreditation mechanisms, and effective avenues for disseminating and exchanging best practices (Leal Filho et al. 2023).

The dearth of research on ESD within the post-communist countries of Central and Eastern Europe remains evident (Dlouhá et al. 2017a, b; Novikau 2017). However, after joining the European Union, many Central and Eastern European countries, including education, increased their progress toward ESD.

The Western Balkans, a region of Europe that includes Albania and countries that emerged after the break-up of Yugoslavia but not are member of the EU, and is characterized by relatively slow progress in education for sustainable development compared to other European countries. A prominent shift towards ESD that happened on a global scale during the 1990s and early 2000s was almost absent in the Western Balkan countries. This lack of progress can be explained by the region's prevalent political and economic circumstances, marked by instability and conflicts (Uvalic 2001). Without exception, all the Western Balkan states presently experience similar ineffective and underdeveloped political, economic, and administrative capabilities. These internal deficiencies are accompanied by numerous unresolved bilateral concerns and conflicting external actor interests (Bartlett 2021; Kulo and Novikau 2023; Petrovic and Wilson 2021).

Despite the official commitment to sustainable development principles and practices, especially concerning environmental protection, there has been a noticeable absence of significant debates on ESD initiated by the academic community or policymakers in Western Balkans countries. Furthermore, educational governing bodies have displayed considerable ambivalence towards sustainable development, failing to implement ESD into education (Milutinović and Nikolić 2014; Nikolić et al. 2015, Nikolic et al. 2017).

Bosnia and Herzegovina represents a typical example of one of the Western Balkans countries, which, regardless of the granted EU candidate status, demonstrates ineffective and underdeveloped political, economic, and administrative capabilities in political, economic, social, and energy areas (Šiljak and Nielsen 2022). In such an environment, it would be hard to expect that education for sustainable development in the country would be an exception.

In 2015, Bosnia, Herzegovina, and other United Nations Member States committed to implementing Agenda 2030 for Sustainable Development. However, the progress in implementing education for sustainable development is slow. Although the state acknowledges sustainable development as its strategic goal, it does not adequately emphasize the significance of education in fostering sustainable development both at the national and local levels. No dedicated legal provisions have specifically addressed ESD, and, therefore, sustainable development is not formally incorporated as an integral component of the educational curriculum (Višnjić-Jevtić et al. 2022).

Scholarly publications focusing on ESD in higher education in Bosnia and Herzegovina are conspicuously sporadic, even when compared to other Central and Eastern European countries. The existing body of literature notably lacks a comprehensive analysis of ESD at national and especially institutional levels in that country.

To bridge the evident gap in the academic literature on education for sustainable development in Bosnia and Herzegovina, this paper presents a case study detailing the implementation of a specialized SDG-focused elective course at the International University of Sarajevo (IUS).

The objective of this chapter is to describe the experience and lessons learned during designing and teaching a novel SDG-focused course at the IUS. This case study seeks to elucidate the challenges and opportunities that emerged during the introduction and instruction of a course focused on SDGs in Bosnia and Herzegovina, a nation situated in the Western Balkans—a region where ESD confronts multifarious challenges on both national and institutional levels.

The study's limitations are typical to the case study research design—limited generalizability, and possible subjectivity and bias. Yet by illuminating the complexities of implementing an SDG-focused course in a complex educational landscape, this research can still contribute to the study of ESD in Bosnia and Herzegovina, resonating with the broader challenges faced by the entire Western Balkans region. It provides insights for educators who aspire to initiate or improve courses emphasizing sustainable development. It also gives university administrators and policymakers a discerning analysis of the impediments to integrating sustainable development principles into curricula. Thus, the study holds the potential to positively impact the trajectory of ESD in Bosnia and Herzegovina and the broader Western Balkans region.

The chapter is structured as follows. The next section describes institutional contexts for teaching ESD in the IUS. The following section describes introducing and teaching the course *Social Responsibility and Sustainable Development* at the university. The subsequent section discusses the challenges and opportunities identified during the course's introduction and teaching. The final section concludes and offers potential directions for future research.

2 Institutional Context

The International University of Sarajevo was established in 2004 by the Foundation for Education Development Sarajevo in Bosnia and Herzegovina as an international university with all teaching and internal communication conducted in English (Kaminski 2020). Around 2000 students are currently studying in all 3 cycles of studies, coming from more than 45 countries (IUS 2023).

As explained in the previous section, the national context presents a rather unsupportive environment for embedding ESD within higher education institutions. Consequently, the impetus to integrate ESD into curricula originated only at the institutional level.

Intensive efforts to implement the SDGs at the university began in 2018 when it applied for the prestigious world ranking—Times Higher Education Impact Ranking. Times Higher Education Impact Ranking of a few global rankings assess universities against the UN SDGs. At that time, IUS was the only university from Bosnia and Herzegovina to appear on this renowned list. At the beginning of the implementation of the SDG at IUS in 2018, only 4–5 sustainable development goals were covered by university activities, such as Quality Education, Climate Action, Partnership for the Goals, and Reduced Inequalities (IUS 2022). The rating's requirement to include SDGs in teaching, research, and everyday operations was one of the factors that increased the university administration's attention to sustainable development. Thus, the university's strategic plan includes SDG delivery through education and research projects.

2.1 SDG Committee

In 2022, the university established the SDG Executive Committee, consisting of faculty members from different disciplines and a few administrative staff members. Although the committee has received no funding, it still has been able to play a role in the implementation of SDG at the university. Noteworthy endeavors have been undertaken to include the SDGs in university curricula. Analogous to their counterparts in other educational institutions, these efforts have centered around two distinct avenues. First, SDG-related topics have been incorporated into courses that delve into subjects while not explicitly labeled SDG-focused. Although this approach facilitates the dissemination of SDG-related knowledge among a broad spectrum of students, it inherently possesses limitations regarding the depth and extent to which SDGs can be comprehensively covered within such contexts (Bradley 2019). A second option suggests the implementation of specialized courses that explicitly concentrate on sustainable development's multifaceted aspects, which will be discussed in the next section.

2.2 ESSSR and IUSDRP

Membership in international sustainability networks has been highly beneficial for developing SDGs at the university. In October 2020, the university joined two internationally recognized research networks in the area of SDG—the Inter-University Sustainable Development Research Program (IUSDRP) and the European School of Sustainability Science and Research (ESSSR) (IUS 2020).

These organizations have offered curriculum guidelines and recommendations that can assist the university in structuring a comprehensive and rigorous Sustainable Development course that aligns with European sustainability standards (Leal Filho et al. 2020). Collaboration with ESSSR on research projects or initiatives has provided scholars practical learning experiences and opportunities to contribute to ongoing sustainability efforts.

2.2.1 AASHE

The Association for the Advancement of Sustainability in Higher Education (AASHE) is a North American organization that promotes sustainability practices and initiatives within higher education institutions. Although the IUS is not formally a member of the organization, AASHE has been helping develop and implement a course on Sustainable Development at a university. It provided a wealth of resources, including curriculum guides, case studies, and best practices, which aided in designing a comprehensive and practical Sustainable Development course. AASHE's resources can assist in developing a curriculum that aligns with industry standards and addresses the most relevant topics within Sustainable Development. AASHE offers opportunities for networking and collaboration among educators, allowing the university to connect with experienced professionals and other institutions that have successfully implemented similar courses.

3 Course Design

In 2021, a new course titled *Social Responsibility and Sustainable Development* was introduced in the university via an initiative from individual professors. The prime objective of this course has been to provide first-year students with a holistic initiation into sustainable development (Table 1).

The course is structured for 3 ECTS credits, teaching 2 academic hours per week, spanning 15 weeks. This course has been among 10 university elective courses, a stipulation where most academic programs necessitate students to opt for one or two of these electives, usually during the initial year of their academic journey. Consequently, the student cohort enrolled in this course mirrors a diverse mix of

Table 1 Learning outcomes of the course social responsibility and sustainable development

1. Demonstrate an understanding of the importance and relevance of Sustainable Development Goals (SDGs);
2. Enhance both global and local social, environmental, and economic awareness;
3. Understand the relationship between SDGs;
4. Effectively work in teams to accomplish common goals

academic programs, encompassing humanities and social sciences, engineering, and the natural sciences.

In 2021, when the course was introduced, 20 students enrolled. From that moment, the courses started to be offered annually.

In Spring 2023, the course was significantly redesigned to address several main challenges in teaching sustainable development.

First, the course must be tailored to the comprehension level of freshmen students from diverse academic majors, including social sciences and humanities.

Second, although the principles of sustainable development are widely acknowledged, students often struggle to grasp the intricacies and interdisciplinary nature of the concept. Several decades ago, sustainable development was primarily related to environmental issues, natural resource management, and economic development. At the same time, the social and political dimensions of sustainable development have been highlighted only recently. In addition, academicians from environmental sciences or economics usually teach sustainable development courses, who unintentionally focus their courses on environmental or economic dimensions of sustainable development. As a result, students majoring in social sciences and humanities are often unaware of how to connect SDGs with their current studies and future professions. For instance, Psychology or Political Science students who expressed willingness to engage in developing and implementing SDGs within their campuses and communities often faced challenges due to the limited diversity of goals they have been aware of.

Third, one of the pressing inquiries universities grapple with today is the practical translation of education for sustainable development into real-world impact, both locally and globally. This question echoes the broader societal need for actionable strategies to effect positive change (Venkataraman 2009). Unfortunately, the theoretical knowledge obtained by the students in sustainable development classes is often purely theoretical and disconnected from the actual life application of sustainable development. The students' most critical challenge has been to identify an essential sustainable development issue, bring stakeholders, and estimate and attract financing in real life.

In the context of the challenges, a novel approach was undertaken to integrate the principles of the 17 Rooms methodology into the framework of a course.

4 17 Rooms

17 Rooms Initiative was launched in 2018 through a collaboration between the Brookings Institution and the Rockefeller Foundation. It has been designed to galvanize practical initiatives aligned with the United Nations SDGs. The objective of 17 Rooms is to visually illustrate the interconnectedness among these goals while demonstrating their impacts on social, environmental, and economic systems. The methodology involves assembling stakeholders from diverse domains into distinct "Rooms," each aligned with a specific SDG—hence the name—17 Rooms. These

Rooms collaborate to identify and strategize feasible actions over a 12 to 18-month span, encompassing policy adaptations and the creation of practical tools. Guided by principles of inclusivity, gradual progress, and collaborative discourse, the initiative strives to generate concrete outcomes, novel insights, and a shared sense of mission within a diverse community dedicated to advancing the SDGs (Brookings Institution 2023).

Since its inception, the 17 Rooms Initiative has been embraced and customized by various entities, including universities, nongovernmental organizations, municipalities, national bodies, and cross-border regions. These entities have adapted the 17 Rooms approach in distinctive ways to drive progress toward the SDGs. In this vein, the innovative 17 Rooms methodology introduces a unique perspective that offers the potential to address the multifaceted challenges inherent to the SDGs within the context of global higher education. Emerging as a promising paradigm, diverse initiatives have been launched to integrate the 17 Rooms approach into transformative processes across academic campuses.

To orchestrate a 17 Rooms exercise within an educational setting, a systematic approach comprising five distinct phases:

1. **Preparation**. Set objectives, design exercises, and assemble an organizing team.
2. **Room Setup**. Assemble Rooms aligned with specific SDGs, designate room co-leads and extend invitations to participants.
3. **Room Convening**. Arrange a 17 Rooms summit, facilitating within-room discussions, inter-room exchange, and subsequent reporting.
4. **Insight and Action Compilation**. Synthesize insights generated within the Rooms and consolidate actionable proposals into a concise report.
5. **Room Follow-Up and Learning**. Sustain engagement by maintaining contact with the Rooms to identify support opportunities and progress updates within a 12 to 18-month timeframe.

Notably, the nature of 17 Rooms exercises can vary, tailored to local objectives and stakeholders' needs. For instance, the following permutations of 17 Rooms have been proposed:

1. **One-day "flash" events**. Rapidly organized one-day 17 Rooms events can materialize within a short span of six weeks.
2. **A "deeper dive"**. More extensive initiatives might extend over several months, requiring heightened coordination and organization.
3. **"Open-Tent" exercises**. Some 17 Rooms exercises adopt an open-door policy to encourage broad participation.
4. **Highly curated exercises**. Alternatively, closely curated 17 Rooms exercises involve the hand-selection of participants by the organizing team.

In the course, implementation was distinctively characterized by adopting the curated 17 Room exercises approach, wherein the course instructor deliberately selected participants—the deliberate selection aimed to align participants with specific learning outcomes and the course's overarching goals. Driven by time constraints, the course focal point primarily revolved around the first two phases of the 17 Rooms methodology. It encompassed essential aspects such as formulating clear objectives, identifying stakeholders, estimating budgetary considerations, and pinpointing existing challenges.

Importantly, it must be acknowledged that the original intention behind the 17 Rooms Initiative was not to facilitate the teaching of sustainability in a tertial classroom context. Nevertheless, the practical approach to SDG implementation held considerable promise for instructional use. Integrating the 17 Rooms methodology into an educational framework required distilling its complex organizational process into a format suitable for pedagogical purposes.

Consequently, the introduction of the 17 Rooms methodology into the course syllabus was underpinned by dual objectives. Initially, its integration sought to facilitate students' knowledge acquisition in sustainable development through the immersive and action-oriented lens of the 17 Rooms approach. This practical pedagogical strategy enabled students to grasp the intricacies of sustainable development within a tangible context, enhancing their engagement and understanding.

Finally, the incorporation of the 17 Rooms methodology into the course design aimed to expose students to the methodology itself, fostering familiarity and proficiency. It was envisaged that students would be equipped with the skills and insights required to apply the 17 Rooms approach within their academic campuses and local communities upon completing the course. This ambition stemmed from the belief that the methodology's participatory and collaborative nature could serve as a catalyst for impactful initiatives beyond the classroom setting.

Although there were only several prior attempts to include the 17 Rooms methodology in the pedagogical process, the 17 Rooms team from the Brooking Institutions provided valuable resources, advice, and networking with other universities that helped to redesign the course.

This endeavor culminated in comprehensive learning objectives (Table 2) constructed to nurture a holistic comprehension of sustainable development. These objectives are strategically designed to equip students to contribute effectively to a more sustainable world, recognizing the intricate interplay between local and global dynamics.

The course was informally partitioned into roughly equal parts (Table 3). In the first phase, students were introduced to sustainable development, the previous Millennium Goals, and the present-day UN SDGs. Subsequently, each of the 17 SDGs was explored in depth. Given the need to cover all 17 SDGs within 6 weeks, each class session focused on explaining 2–3 SDGs, lasting for two academic hours. It meant dedicating approximately 30–50 min of class time to each SDG. For each goal, the following key aspects were addressed:

1. A United Nations-endorsed definition of the specific goal.
2. An identification of the problem(s) that the goal seeks to address, substantiated with supporting evidence of the problem's existence.
3. An examination of how the problem(s) impacts people universally.
4. A description of the targets and indicators associated with the goal, according to the UN.

Table 2 Course's objectives as per spring 2023 syllabus

This course aims to provide students with a comprehensive introduction to the concepts of sustainable development. The course will utilize the 17 Rooms Initiative methodology as a framework for exploring the 17 Sustainable Development Goals (SDGs) adopted by the United Nations. The 17 Rooms Initiative visually represents the SDGs and highlights the interconnections between various goals and their impact on social, environmental, and economic systems
Throughout the course, students will engage in interactive activities, group discussions, and case studies to develop a deeper understanding of sustainable development and the importance of social responsibility in promoting a sustainable future. The course will also cover the role of individuals, communities, and organizations in promoting sustainable development, providing students with a holistic perspective on the topic
By the end of the course, students will have gained a comprehensive understanding of the concepts of sustainable development and will be equipped with the knowledge and skills to evaluate initiatives to promote a sustainable future critically

Table 3 Class outline (schedule) of the course

Week	Topic(s)
1	Introduction. History of sustainable development. UN sustainable development goals
2	SDGs 1–2
3	SDGs 3–5
4	SDGs 6–8
5	SDGs 9–11
6	SDGs 12–14
7	SDGs 15–17
8	Midterm exam
9	17 rooms initiative. Writing a project. Topic selection, group formation
10	Proposal presentations, discussions, feedback
11	Project: analysis, requirements, challenges
12	No class
13	Student presentations
14	Student presentations
15	Student presentations. Conclusion

5. An exploration of the primary strategies proposed to achieve the goal.

During the eighth week, a two-hour closed-book exam was conducted in class. Students were presented with 20 short-answer open-ended questions, reflecting the content taught over seven weeks—roughly 3 questions each week. Consequently, the first set of 3 questions covered general sustainability concepts, common pool resources, and the Environmental Kuznets Curve. Subsequently, the remaining 17 questions corresponded to the 17 individual SDGs (1 question per each SDG). The nature of these questions varied, ranging from requests for evidence, reasons for the importance of the problem, and identification of specific indicators or targets. Notably, none of the questions demanded precise definitions or numerical values.

The second part of the course unfolded during weeks 9–12, when students were introduced to the 17 Rooms methodology and were apprised of the prerequisites for the term project—a pivotal assignment contributing to 50% of the overall grade. Given the nature of the course, practical implementation of SDG-related projects was not very feasible. Instead, students were tasked with drafting proposals for their own SDG-related projects. These projects were required to be applicable within the university campus or local communities, with the university assuming a central stakeholder role. Hence, the primary aim of the term project was to bridge theoretical knowledge with practical application, anticipating that the proposed projects would later transition into real-world initiatives on campus or within local communities.

The following requirements were made for the project:

1. The term project consists of a paper and presentation.
2. It is a group project. A group of 2 students is required. Each student in the group will receive the same grade.
3. For this assignment, each team writes a project proposal related to contributing to UN SDG goals (no actual activities related!) The team can choose any SDG goal to work on. However,

    a. The proposal should be a well-researched, feasible, and practical plan to address a problem related to the IUS and contribute to the SDGs. The problem should be related to the IUS and could be realistically conducted within the IUS campus or the IUS acting as a major partner.
    b. The proposal should highlight the need for collaboration to achieve sustainable and long-lasting impact.
    c. The topic should be approved in advance by the course instructor.

The following structure of the term project paper was required:

1. **Background of the problem**. Identify the problem related to IUS and explain why it is a problem and for whom.
2. **Objectives**. Explain how your project will contribute to solving the problem above. Identify your targets and explain what exactly you want to change.
3. **Complementarity with SDGs**. Explain how the problem is related to the SDGs. If the problem is related to several SDGs, identify the major one.
4. **Actions**. Explain what exactly you are going to do to achieve your objectives. Be as specific as possible.
5. **Resources**. Identify what you need to achieve your objectives. Create a short list of items/services you need and their approximate costs. Explain where you will get these resources from and how they will be used.
6. **Project team and stakeholders**. Explain who should be involved in the project to achieve your objectives and why. Explain the division of work—who will do what. Show how you will collaborate with other people to ensure the success of your project.

7. **Timeline**. Provide a timeline for your project, indicating when each task will be completed. Be realistic in your timeline.
8. **Indicators**. Explain how you are going to measure if you achieve your objectives. Identify key indicators that will help you track progress toward your objectives. Be as specific as possible.
9. **Sustainability and continuation**. Describe the follow-up of the project after it ends. Provide recommendations for future projects that may build on the successes of your current project.

A similar structure was anticipated for the 10 to 15-min in-class presentations that were part of the assignment.

While stringent anti-plagiarism policies were upheld, students were actively encouraged to study successful instances of SDG-focused projects carried out by other universities.

To assess a comprehensive grasp of the SDGs and their connections with the proposed projects, a cumulative closed-book final exam was conducted in class, constituting 30% of the total grade. This exam covered all course content from the first week, including theoretical insights into sustainable development goals and their practical implementation on campus or within local communities. Thus, the exam questions addressed a broad comprehension of SDGs and the particulars of the group projects.

5 Teaching Process

In the Spring 2023 semester, the course attracted 47, predominantly first-year students. These students came from 11 distinct academic programs, encompassing humanities, social sciences, and engineering. A predominant student contingent was from the Software Engineering program (16 students), Psychology (7 students), and Turkish Language and Literature (6 students) programs.

Initial lectures witnessed moderate student engagement in discussions. Limited interaction was partly attributed to the lecture hall's auditorium-style setup, characterized by tiered seating and fixed furniture, which inadvertently restricted avenues for active participation.

While the midterm examination indicated a solid grasp of sustainable development concepts and most SDGs, it also unearthed an unforeseen challenge concerning not sustainable development per se but rather terms like "target" or "indicator." Instances emerged where questions like "Provide an indicator used to measure progress toward SDG XX's targets" received inaccurate responses due to students' unclear comprehension of the term "indicator." A review session during the first week following the exam addressed these challenges. Primary stumbling blocks were elucidated during this session, and foundational concepts were elaborated upon.

Regarding the term project, student teams formed relatively effortlessly and embarked on developing their project proposals. In total, 22 topics were proposed (Table 4). Two categories of SDGs were predominantly chosen for the term projects. First, students opted for SDGs closely linked to their academic programs. For example, Psychology students frequently chose SDG 3: Good Health and Well-being, concentrating particularly on Target 3.4, which states, "By 2030, reduce by one third premature mortality from non-communicable diseases through prevention and treatment and promote *mental health and well-being*" (emphasis added). Second, goals associated with environmental sustainability and natural resource management, including waste reduction, were also commonly selected.

Although student presentations varied in quality and clarity, all teams successfully presented their proposals.

The final exam demonstrated a significant increase in students' understanding of SDGs and, most importantly, their connections with real life, such as university and community needs.

After completing the course, the best proposals were presented to the university SDG committee, which found them quite promising. In the future, committee representatives could attend student presentations to understand students' views on sustainable development better and establish contacts with the most active teams.

Table 4 Titles of the group project proposed by the students

1.	Preserving local biodiversity on the IUS campus
2.	Improving gender equality at IUS
3.	Healthy lifestyle at IUS
4.	Good health and well-being: mental health among IUS students
5.	The encouragement of better mental health support on the IUS campus
6.	Responsible consumption and production at IUS
7.	How sensor water taps can save water in IUS
8.	Climate change and IUS
9.	Recycling at the International University of Sarajevo
10.	Organic garden: Sustainable agriculture for the university community
11.	Reducing plastic waste on the IUS campus
12.	Healthy food options at IUS
13.	Creating a book exchange space for students at IUS
14.	Roma population in Bosnia and Herzegovina
15.	Empowering rural communities in Bosnia and Herzegovina
16.	Mental health: Raising awareness of students' mental health condition at IUS
17.	Recreational sports clubs at IUS
18.	Spreading awareness about the importance of mental health at IUS
19.	Industrial waste at IUS
20.	Zero waste for a better future for IUS
21.	Implementing a campus-wide recycling program at IUS
22.	Mini hydropower plants in Bosnia and Herzegovina

6 Discussion

The introduction of the course, in general, supports the previous literature that even in the absence of national support for EDS, institutions of higher education can provide ESD and thus contribute to SDGs. The primary role plays the interest of individual professors willing to introduce and teach such courses. The other factors are the university administration and formal and informal institutional structures, such as sustainable development committees and clubs, which can provide additional motivation and support for the professors.

In addition, in the absence of support on the national level and general underdevelopment of ESD in the country, international initiatives such as 17 Rooms and international sustainable development networks such as ESSSR, IUSDRP, and AASHE can play a crucial role by providing ideas and best practices. Moreover, membership in such networks provides additional legitimization for teaching sustainable development at both institutional and national levels. At the same time, the best methodologies and practices from other countries and educational institutions cannot always be easily transferred because of different legal, social, and educational contexts. Consequently, adjustments, sometimes significant, are often necessary.

Teaching SDG-focused courses for first-year students from diverse academic programs is challenging because of insufficient prior knowledge and different academic interests. Nonetheless, as the teaching of the course has demonstrated, students from all academic programs can identify and connect SDGs to their majors. Therefore, introducing novel approaches with a more personalized transfer of knowledge on sustainable development rather than "one-size-fits-all" is crucial for ESD.

7 Conclusion

The introduction of the novel *Social Responsibility and Sustainable Development* course at the International University of Sarajevo marks a substantial stride in promoting education for sustainable development in Bosnia and Herzegovina. Despite the challenging national and institutional contexts, the course exemplifies how universities can proactively address the call for sustainability education. While national policies and resources may be limited, universities can lead by example and create platforms for students to engage with sustainability concepts and actively contribute to real-world challenges.

The integration of the 17 Rooms methodology into the curriculum presents an innovative approach, adapting a real-world SDG-focused strategy for pedagogical application. By engaging students in problem identification and stakeholder selection, the course fosters practical insights into sustainable development within academic timeframes. This approach enhances students' comprehension of sustainable development concepts and equips them with skills applicable beyond the classroom.

The diversity of student participants across various academic programs underscores the widespread interest in sustainability issues among students, encompassing both environmental and societal dimensions. The choice of project topics reflects a proactive approach to addressing local challenges, ranging from waste management to mental health awareness. This diversity of topics highlights universities' potential to contribute to a wide array of sustainable development goals.

Given the limitations of this chapter's case study research design, opportunities for further research emerge. Expanding the investigation to encompass a broader array of institutions and diverse geopolitical contexts could yield a more comprehensive understanding of the challenges and successes in implementing education for sustainable development. Additionally, delving into the long-term impact of such courses on students' careers and their engagement with sustainability concerns post-graduation could provide invaluable insights. Finally, exploring alternative pedagogical strategies that bridge the gap between real-world methodologies and educational settings could offer innovative approaches to teaching sustainable development.

References

Ávila LV, Leal Filho W, Brandli L et al (2017) Barriers to innovation and sustainability at universities around the world. J Clean Prod 164:1268–1278. https://doi.org/10.1016/j.jclepro.2017.07.025

Bartlett W (2021) International assistance, donor interests, and state capture in the Western Balkans. J Contemp Eur Stud 29:184–200. https://doi.org/10.1080/14782804.2020.1794801

Blasco N, Brusca I, Labrador M (2020) Drivers for universities' contribution to the sustainable development goals: an analysis of Spanish public universities. Sustainability 13:89. https://doi.org/10.3390/su13010089

Bradley P (2019) Integrating sustainable development into economics curriculum: a case study analysis and sector wide survey of barriers. J Clean Prod 209:333–352. https://doi.org/10.1016/j.jclepro.2018.10.184

Brookings Institution (2023) 17 rooms. https://www.brookings.edu/projects/17-rooms/. Accessed 8 Aug 2023

Dlouhá J, Glavič P, Barton A (2017a) Higher education in Central European countries—critical factors for sustainability transition. J Clean Prod 151:670–684. https://doi.org/10.1016/j.jclepro.2016.08.022

Dlouhá J, Vintar Mally K, Dlouhý J (2017b) ESD principles in higher education from a perspective of Central and Eastern European countries. Int J Sustain High Educ 18:822–840. https://doi.org/10.1108/IJSHE-03-2016-0045

IUS (International University of Sarajevo) (2020) IUS becomes a member of the world's largest SDG research networks. https://els.ius.edu.ba/en/blog/ius-becomes-member-worlds-largest-sdg-research-networks. Accessed 20 Aug 2023

IUS (International University of Sarajevo) (2022) Implementation of Sustainable Development Goals (SDG) at the International University of Sarajevo (IUS) 2018–2022. https://qao.ius.edu.ba/en/implementation-sustainable-development-goals-sdg-international-university-sarajevo-ius-2018-2022. Accessed 18 Aug 2023

IUS (International University of Sarajevo) (2023) About the university. https://www.ius.edu.ba/en/about-university. Accessed 19 Aug 2023
Kaminski JJ (2020) Succeeding as a Western academic working at a newer foundation university in a developing country. PS Polit Sci Polit 53:532–536. https://doi.org/10.1017/S1049096520000360
Kulo E, Novikau A (2023) The Open Balkan initiative: a step forward towards European integration or running on empty? J Contem Eur Stud 1–13.https://doi.org/10.1080/14782804.2023.2204422
Leal Filho W (2010) Teaching sustainable development at university level: current trends and future needs. J Balt Sci Educ 9:273–284
Leal Filho W (2020) Living labs for sustainable development: the role of the European School of sustainability sciences and research. In: Leal Filho W, Salvia AL, Pretorius RW et al (eds) Universities as living labs for sustainable development. Springer International Publishing, Cham, pp 3–9
Leal Filho W, Abubakar IR, Mifsud MC et al (2023) Governance in the implementation of the UN sustainable development goals in higher education: global trends. Environ Dev Sustain. https://doi.org/10.1007/s10668-023-03278-x
Leal Filho W, Shiel C, do Paço A (2015) Integrative approaches to environmental sustainability at universities: an overview of challenges and priorities. J Integr Environ Sci 12:1–14. https://doi.org/10.1080/1943815X.2014.988273
Leal Filho W, Wu Y-CJ, Brandli LL et al (2017) Identifying and overcoming obstacles to the implementation of sustainable development at universities. J Integr Environ Sci 14:93–108. https://doi.org/10.1080/1943815X.2017.1362007
Leal Filho W, Vargas VR, Salvia AL et al (2019) The role of higher education institutions in sustainability initiatives at the local level. J Clean Prod 233:1004–1015. https://doi.org/10.1016/j.jclepro.2019.06.059
Lozano R, Lozano FJ, Mulder K et al (2013) Advancing higher education for sustainable development: international insights and critical reflections. J Clean Prod 48:3–9. https://doi.org/10.1016/j.jclepro.2013.03.034
Lozano R, Ceulemans K, Alonso-Almeida M et al (2015) A review of commitment and implementation of sustainable development in higher education: results from a worldwide survey. J Clean Prod 108:1–18. https://doi.org/10.1016/j.jclepro.2014.09.048
Milutinović S, Nikolić V (2014) Rethinking higher education for sustainable development in Serbia: an assessment of Copernicus charter principles in current higher education practices. J Clean Prod 62:107–113. https://doi.org/10.1016/j.jclepro.2013.05.028
Nikolić V, Ranitović J, Milutinović S (2015) Greening of higher education in the Republic of Serbia. Envigogika 10. https://doi.org/10.14712/18023061.453
Nikolic V, Milutinovic S, Nedanovski P, Mrnjaus K (2017) ESD professional development of university educators in Serbia, Croatia and Macedonia: a comparative analysis. IJSHE 18:923–938. https://doi.org/10.1108/IJSHE-02-2016-0036
Novikau A (2017) Education for sustainable development and Belarusian higher education: when the state dominates. In: Leal Filho W, Azeiteiro UM, Alves F, Molthan-Hill P (eds) Handbook of theory and practice of sustainable development in higher education. Springer International Publishing, Cham, pp 355–365
Novy JW, Banerjee B, Matson P (2021) A core curriculum for sustainability leadership. Sustainability 13:10557. https://doi.org/10.3390/su131910557
O'Byrne D, Dripps W, Nicholas KA (2015) Teaching and learning sustainability: an assessment of the curriculum content and structure of sustainability degree programs in higher education. Sustain Sci 10:43–59. https://doi.org/10.1007/s11625-014-0251-y
Petrovic M, Wilson G (2021) Bilateral relations in the Western Balkans as a challenge for EU accession. J Contemp Eur Stud 29:201–218. https://doi.org/10.1080/14782804.2020.1865884
Šiljak D, Nielsen KL (eds) (2022) Bosnia and Herzegovina and European integration: obstacles and challenges. International University of Sarajevo, Sarajevo, Bosnia and Herzegovina

Uvalic M (2001) Regional co-operation in Southeast Europe. Southeast Eur Black Sea Stud 1:55–75. https://doi.org/10.1080/14683850108454621

Venkataraman B (2009) Education for sustainable development. Environ Sci Policy Sustain Dev 51:8–10. https://doi.org/10.3200/ENVT.51.2.08-10

Višnjić-Jevtić A, Lepičnik Vodopivec J, Pribišev Beleslin T, Šindić A (2022) Unmasking sustainability in early childhood education: teachers' voices from Bosnia and Herzegovina, Croatia, and Slovenia. Int J Early Child 54:119–137. https://doi.org/10.1007/s13158-022-00321-2

Bridging the Theory–Practice Gap Through Work Integrated Learning-Educating Students as Change Agents Advancing the SDGs: A Case Study of WIL in Higher Education at a Master's Program in Sweden

Karin Alm, Caroline A. Pontoppidan, and Daniela Argento

Abstract This chapter has two key objectives. The first is to explore how Work Integrated Learning (WIL) as a pedagogy can advance education for sustainability in higher education to ensure a sustainable future. The second is to facilitate educators escape dualism between theory and practice while shaping employable graduates able to serve society as change agents. This mixed method study has a qualitative case study design, with an empirical focus on student' engagement throughout WIL in a Swedish Higher Education Institution (HEI) context. The findings show how WIL addresses unanswered questions in the HEIs collaborative practices that can support the sustainable development considering inherent influences of local and national contexts. Extrapolating from this, this chapter contributes by identifying

Lecturer in Business Administration, and Academic leader for the Work Integrated Learning at the master's program at Faculty of Business, Kristianstad University, Sweden and the Co-ordinator and, Academic leader for the Principles for Responsible Management Education (PRME) at Kristianstad University, Sweden.

Associate Professor in Accounting and Auditing, and Academic Director at the office of Responsible Management Education (PRME) at Copenhagen Business School, Denmark and Associate Professor in Accounting and Auditing at the Faculty of Business, Kristianstad University, Sweden.

Associate Professor in Management Control and Accounting at the Faculty of Business, Kristianstad University Sweden.

K. Alm (✉) · C. A. Pontoppidan · D. Argento
Faculty of Business, Kristianstad University, Kristianstad, Sweden
e-mail: Karin.Alm@hkr.se

C. A. Pontoppidan
e-mail: cap.acc@cbs.dk

D. Argento
e-mail: Daniela.argento@hkr.se

C. A. Pontoppidan
Copenhagen Business School, Frederiksberg, Denmark

W. Leal Filho et al. (eds.), *The Contribution of Universities Towards Education for Sustainable Development*, World Sustainability Series,
https://doi.org/10.1007/978-3-031-49853-4_32

how WIL can enable students as future change agents through engaging with industry and communities while applying theory on real-life challenges of implementing the Sustainable Development Goals (SDGs).

Keywords Work-integrated learning (WIL) · Sustainable development goals (SDGs) · Change agents · Sustainability literacy · Dualism of theory practice

1 Introduction

In contemporary society there is an urgent need to advance transformations towards sustainability. In 2020, UNESCO released 'Education for Sustainable Development (ESD)—A Roadmap'. This roadmap is a framework supporting the global advancement of SDGs and serves for implementation of ESD. A call for transformative competencies is echoed by international organisations (OECD 2023; UNESCO 2022) and seen as an increasingly important theme in higher education research (Argento et al. 2023). Society calls for graduates equipped with a new set of competencies enabling them to act as change agents (Redman and Wiek 2021). Aware of this need, Higher Education Institutions (HEIs) programs including Work-Integrated Learning (WIL) can improve graduates' competencies by providing a bridge between theory and practice (Björck 2021). WIL relies on the development of relationships between the university, industry, and community organizations (Fleming et al. 2018) aimed to escape dualism between theory and practice (Björck and Johansson 2019).

In Sweden, Work Integrated Education (Kristianstad University 2023a) is known as Verksamhetsintegrerat Lärande (VIL) (Högskolan Kristianstad 2023), a pedagogy that fosters application of theory to develop students as life-long learners. In this chapter we extend our argument on how WIL can lead to students' transformative competencies (see Argento et al. 2023), exploring further how WIL supports students in becoming future sustainable development change agents. HEIs are in a unique position to advance the mobilization of knowledge capital (Steiner and Scherr 2013) required to transform society to meet the UN Sustainable Development Goals (SDGs) Agenda 2030 (Unite Nations n.d.). Since 2018 Kristianstad University works actively within the UN based initiative PRME (The Principles for Responsible Management Education) and are signed up as PRME Champions for the third cycle 2022/2023 (Kristianstad University 2023b).

HEI educators are challenged to consider, not only the SDGs of their core disciplinary interest, but also other corresponding SDGs (Lozano et al. 2019). Adopting the SDGs more explicitly, requires an integration of a broad range of topics spanning across social, economic, and environmental challenges (Leal Filho et al. 2023). Previous research in higher education shows an accelerating as well as recognized interest in research on the role and methods to integrating the SDGs into educational programs (Alm et al. 2021a, b; Argento et al. 2023; Cavalcanti-Bandos et al. 2021; Chankseliani and McCowan 2021; Leal Filho et al. 2019).

Hence, HEIs have an important role in advancing teaching methods that bridge students' knowledge, skills and mindset to become change agents driving the transition for a sustainable society (Hesselbarth and Schaltegger 2014). Sustainability literacy refers to the knowledge, skills, and mindset necessary to make informed decisions for the benefit of oneself, others, and the planet (Cavalcanti-Bandos et al. 2021). The concept is multidimensional and encompasses economic, social, and environmental aspects in line with Elkington's (1994) theory of Triple bottom Line, often referred to as people, planet, prosperity. Education plays a crucial role in promoting sustainability literacy. If designed well it could be expected to provide a platform for students to learn about sustainability issues, engage in critical thinking, and develop problem-solving skills (Cavalcanti-Bandos et al. 2021).

WIL as a concept emerged more than a decade ago (Patrick et al. 2008) and has been advanced. It was defined as an educational approach that combines relevant work-based experiences to allow students to integrate theory with practice of work as an intentional component of the curriculum. This educational approach requires the engagement of three stakeholders, namely the student, the HEI, and the workplace. WIL has since been embraced by researchers to advance understanding of the relationship between theory and practice (Björck 2020, 2021; Björck and Johansson 2019). WIL is designed so that it consists of moments of on-campus theoretical lectures as well as practice lectures combined with placement-based off-campus training. Students engage in so-called real-world-like activities on campus to acquire both technical and soft skills and corroborate them with the activities conducted off-campus. This approach supports graduate employability because it is founded on a theory–practice terminology in which conceptual pairings such as theory–practice, academia, real world, and study-work are used as opposite terms to define graduate employability (Björck 2021).

HEIs however, face challenges while working with WIL. These challenges can, for example, be traced to the development of trust in relationships between academia, industry and often community organizations (Fleming et al. 2018). Still, WIL is a valuable pedagogical method because of its potential for providing a bridge between theory and practice that nurtures employable graduates (Björck 2021). WIL is considered as a pedagogical approach to ascribe meaning to graduate employability by reducing the creation of the so-called theory–practice gap for students. Past research has acknowledged that great care is required in the design of WIL to avoid aggregation of dualism (Björck and Johansson 2019). That is, WIL goals and activities need to be planned and conducted in order to integrate theoretical and practical knowledge. Björck and Johansson (2019) explicitly argue that "a non-dualistic terminology must use neither this nor any other distinctions which disconnect theory from practice but instead signify that they are inseparable" (p. 10). The authors suggest that a non-dualistic terminology could be useful in two ways: (1) to avoid polarizing theory and practice for students and (2) to broaden students' understanding of the intrinsic link between theory and practice. Therefore, a non-dualistic understanding of learning advances the idea that theory and practice always coexist in learning. Hence, theories (principles, ideas, etc.) are always learnt in and through different forms of (social) practices. While some interesting findings have been reported in the literature, researchers have asked for further explorations on what such promising WIL environments look like (Björck 2021).

The importance of HEIs in advancing sustainable development has been emphasized in contemporary research. University Social Responsibility (USR), as emphasized by Bokhari (2017), is an aspect of moving forward toward implementation of the Agenda 2030 in HEIs, given thc potential role of HEIs in the transition to a more sustainable society. Additionally, looking at the SDGs from an academic perspective, it can be noticed that USR encourages universities to incorporate environmental, social values and ethical principles into their research, teaching, and community engagement (Alm et al. 2021b). We consider that the educator's effort must prevent silo-oriented teaching, having an open mindset regarding multidisciplinary knowledge from different disciplines (Alm et al. 2021a; Levi and Rothstein 2018). Advancing challenge-based learning may create and lead to a more mature knowledge associated with integrating the SDGs in a pedagogical setting.

This chapter has a twofold purpose. It aims to explore (1) how WIL as a pedagogical learning environment can advance students as change agents for a sustainable future and, (2) how WIL can promote a non-dualistic terminology to ultimately address discourses aiming to bridge theory and practice.

2 Literature Review

An urgent need for higher education, across disciplines, to embrace sustainability is critical to ensure continuation of life on the planet (see for example Jim and Shen 2016). Educating future change agents involves teaching that embrace global issues such as, for example, climate change, biodiversity loss, and socio-economic inequalities, and requires ambitious and whole-scale transformations of societies worldwide (Redman and Wiek 2021; Scoones et al. 2020). The importance of climate change education in schools has been increasingly recognized, with researchers and experts calling for more research in the climate-health nexus. Hindley (2022) who explored the gap between ambitions and climate change education in the context of HEI argue that even where formal commitments and institutional incentives exist to encourage action there seem to be a need for speeding up a culture around climate change education reassuring student's climate literacy. Moreover, a broader educational agenda at all levels of the education system has been proposed as a crucial component of the global response to climate change. However, while there is an increasing recognition of the importance of climate change education, there are still gaps and challenges in its delivery (Hindley 2022). Further research is needed to explore these gaps and develop strategies for effective integration of climate change education into school curricula which requires a bottom-up understanding of what students know about sustainability (Kuehl et al. 2023).

In higher education, teaching seeks to provide students with ethical, cognitive, personal, and generic competences connected to complex knowledge, skills, and mindset (Gale et al. 2021). As highlighted by Gale et al. (2021) teaching students as future change agents includes educating students on essential skills for innovative and critical thinking, ability to engage in collaborative partnerships, capacity to work across boundaries, and resilience. Moreover, in terms of mindset, students need courage, enthusiasm, commitment and optimism to develop an open mindedness, tenacity, and adaptability.

Lozano et al. (2019) suggest several pedagogical methods for actively engaging students in the learning process. These include for example case studies, dialogues, teamwork brainstorming, solving real problems from society, projects, debates, peer assessment and videos. All these approaches require that students participate actively, promoting active learning. It can be suggested though that WIL projects asking students to collaborate with a partner organization and write a case study can produce desirable learning outcomes that reduce (if not close) the debated theory–practice gap. WIL can more explicitly serve as a platform for research-informed teaching which is of value to academic staff and their institutions (Xia et al. 2015).

Previous literature addressed critical gaps in educating students key-competencies: system thinking competence, anticipatory competence, strategic competence, normative competence, and interpersonal competence for sustainability (Redman and Wiek 2021; Wieck et al. 2011). Researchers have attended to pedagogies that can support HEIs to develop these competencies, for example, by focusing on bringing in real-world problems into the classroom (Brundiers et al. 2010). WIL as a pedagogy adds to previous research on how letting students work with real-world problems (Alm et al. 2021b; Brundiers et al. 2010) allowing students to engage and re-engaging between theory and practice while simultaneously analysing real-life sustainability challenges. WIL thus has the potential to strengthen sustainability competencies required for students' future work-life. Research has explored the capability HEIs have in advancing change in society, focusing on how higher education can facilitate change in its own institutions (Stephens et al. 2008). HEIs, thus have a responsibility following Agenda 2030 to adopt pedagogies that can advance sustainability competencies and students that graduate with the capabilities of acting as change agents for a sustainable future. These competencies enable graduates to envision sustainable futures, based on the SDGs, and develop effective and efficient strategies to realize them (Giangrande et al. 2019).

This chapter focusses on WIL as a pedagogy that can promote students as change agents for sustainability. Change agents can be defined as societal actors that promote change that disrupts the status quo (Hsu and Chien 2022). Hesselbarth and Schaltegger (2014) conducted a study that advances a competency matrix for students to graduate as corporate change agents for sustainability. Their study highlights that change agents develop sustainability as a factor of success in their working environment. By connecting the literature on WIL with the notion of change agents this chapter advances knowledge on the importance of reducing the role of dualism to nurture the future change agents (i.e., the graduating students).

3 Method

A mixed-methods study was undertaken to address the research questions. The study participants were 30 master students attending a one-year master program in Business Administration, specializing in Auditing and Control or International Business and Marketing, enrolled at Kristianstad University, Sweden.

Student data was collected through an online survey and their WIL project group assignments which were in the format of a reflective group report. The **following nine survey questions** were designed to understand how students assess their own learning achieved through the WIL project, their contribution to the WIL organization and the wider impact of the WIL track:

1. WIL is a way to connect you as students with work life in your learning process. Please elaborate on your learning outcome working with the SDGs throughout the project.
2. Working with WIL benefits the local businesses with theoretical ideas students bring to the table. In what ways do you feel you contributed to businesses working with WIL?
3. Addressing the SDGs through WIL, do you see yourselves as possible future change agents in society and different organizations?
4. What is your takeaway on the SDGs framework, based on your WIL track experiences?
5. In what ways could universities benefit from implementing WIL in their course syllabus and programs?
6. Discuss two–three different corporate social responsibility/sustainability issues that you learned during thew WIL project. Why do you think they are the most essential?
7. What recommendations did your group made to the WIL organization to focus on sustainability? What were the main reason you recommended them?
8. What aspects of sustainability/CSR knowledge gained during the WIL project can you potentially apply in your future workplace? Give at least two examples and motivate why and how they are helpful.
9. Further reflections and comments?

The student survey, namely written reflective answers, aimed to evaluate students own understanding of their WIL work and related learnings. The survey was administered to the 37 students attending the 2021/2022 academic year and 30 students completed the survey (response rate: 81%). Table 1 provides an overview of the survey responses.

A content analysis was conducted (Miles and Huberman 1994), where the responses from the survey and the written reflective group reports were carefully read, categorised, and analysed by the researchers. The results from the survey were categorized by the nine survey questions and students' responses anonymised through an id-number, while the reflective group reports were categorised as follows: Academic year, WIL partner organisation, title of the report, SDGs goals, purpose of the project.

Table 1 Survey responses

Major	Enrolled	Participated	Response rate
Auditing and control (AC)	13	8	61.5
International business and marketing (IBM)	24	22	91.6
Total	37	30	81

Source Prepared by the author

4 Results

The pedagogical setting of WIL involves five blocks (Fig. 1) which are progressively developed and built to create a learning progression for students' life-long learning. The pedagogical setting for the WIL model learning environment is that WIL is given throughout the fall semester on all four master's courses at the master program in business administration. Each master course involves a WIL module of 1,5 credits and WIL gives in total 6 credits.

WIL and the advancement of SDGs in Block 5 which takes place in the first master course in the spring we arrange The Sustainability Day, a full day arrangement taking place in the university auditorium where students present their finalized WIL projects. The academic reports all include managerial guidelines and recommendations for their WIL organisation and to be further discussed at the event. All partner organisations are invited to the event together with faculty members, and teachers. The idea is that this event supports the objectives of that the WIL model creates a feedback loop between students, educators and WIL organisations enabling an interactive dialogue supporting graduates as life-long learners for a sustainable development. The event ends with the Diploma Certificate Ceremony celebrating students' contributions serving society as change agents.

A year later we did an interview with one of our graduates from our master's program a year after graduation this student proved how **WIL educates future change agents**. The story told by one of our graduates, who grew up in Gambia in West Africa, show how WIL through change of mindsets can support future change agents for sustainable development. The student in this story had a bachelor's degree in business administration with a focus on auditing from the University of Gambia, and a few years of work experience and wanted to study a master's degree in business administration at Kristianstad University in Sweden. We asked the alumni to share his story, this is his story:

> Above all, I appreciate that you get the opportunity to immerse yourself in one subject at a time. In one of the stages, we students would collaborate with companies outside the university to evaluate their sustainability plans and strategies and come up with recommendations. It really was an aha experience, he remembers.

In Gambia the agriculture employs a large part of the country's population. On small family farms, food is grown for the table, but also for sale in the nearby markets. As an alumnus of Kristianstad University, the student has continued to stay

BLOCK	CONTENT	AIM	OUTCOME
Block 1	- Kick-off with WIL partners - Students read scientific article about SDGs in HEIs and analyze a real-life organization's communication related to sustainability and SDGs	- Identify practical issues with SDGs - Understand the role of theory in explaining organizational communication and work with sustainability and SDGs	Unveil the gap between practice and theory
Block 2	- Student groups contact and visit the WIL partners - Student groups prepare the WIL research proposal	- Establish a relationship with the WIL partner organizations - Develop an academic approach to SDGs	Build the bridge between theory and practice
Block 3	- Student groups conduct intense field work and gain knowledge from the WIL partner	- Awareness of WIL partners' challenges with working with sustainability and SDGs	Experience real-life settings
Block 4	- Student groups concentrate on theoretically informed analysis of their experience and draw conclusions	- Application of theories to make sense of practice, i.e., to abstract the learning from the real-life setting - Change agent mindset	Bridge the dualism between practice and theory
Block 5	The Sustainability Day - Public presentation - Diploma Certificate Ceremony - Survey	- Self-assessment of learning achieved through the WIL project, contribution to the WIL partner and the wider impact of the WIL track	Escape dualism supporting graduates serving society as change agents

Fig. 1 The pedagogical learning environment for the WIL model. *Source* Author's own

in contact with the faculty and shared his story on how he after graduation engaged in an entrepreneurial sustainable project using solar panels to give agricultural products a longer life in a climate-friendly warehouses to be built in his home country Gambia.

> Through the WIL project I have received training in developing my curiosity, and in believing in myself. At the same time, this is not just about me. We can all have so much more impact if we think and act sustainably. I will also carry this with me in the future.

4.1 Survey Results

Students' reflections show how WIL works aiming to bridge the duality between theory and practice to create learning progression for graduates' lifelong learning.

> A very interesting project that taught me a lot, not only about sustainability but also about teamwork and leadership. It was refreshing to connect academia with practice and see some "real-world" results. (Stud.IBM.1).

These results show how working across different positions and industries provide students with new insights that may help fill the gaps between theory and practice. Furthermore, students said that while combining theory with practice they learned how education is always a key pillar for sustainable development as people that are skilled and knowledgeable can believe in making a change themselves.

> As we have learned, education is always a key pillar to sustainable developments, because people that are aware and knowledgeable can make a change themselves. This then affects all other important factors such as in our case SDG 7 and SDG 12, because people will likely be more motivated to act on them if they are aware and experience the difference first-hand. Getting people to a level of welfare where they can start thinking about sustainable issues instead of just their own needs and survival (Maslow's Hierarchy) is therefore crucial. (Stud.IBM.2).

This then could affect the advancement of implementing the SDGs because people will likely be more motivated to act on them if they feel experienced. Getting people to a level of welfare where they can start thinking about sustainable issues instead of just their own needs and survival one student compared to "Maslow's hierarchy of needs" is therefore crucial.

> I don't know to which extent this experience will relate to future activities, but since I am personally motivated to make an impact, the knowledge gained here will help me. It created new perspectives for me in terms of hopes for reality and working with the possibilities given. (Stud. IBM 7).

Another student shared:

> We all can make a change. We just need to do it little by little. Collaboration is what proceed us move forward. (Stud. IBM 5).

This study seeks to contribute with a better understanding of how WIL can serve to advance students understanding of climate change in higher education. As one student suggested:

> Use of green energy; easy implementable and huge impact of the climate and water recycling; easy implementable also has a huge impact. (Stud AC 4).

4.2 Coding the WIL Project Reports

Through the coding of the reflective reports, we captured the purpose and research questions used by the students and the SDGs covered in the student projects. It needs to be recognized that the SDGs covered can reflect that the companies are based in Sweden. The group written reports show a variety of partner organizations involved in the WIL projects. To mention a few, the sample of organisations include manufacturing, industrial and services companies. Table 2 provides a draft synthesis of key findings of four WIL-reports.

Table 2 WIL report coding scheme outcome

Academic year	WIL partner organisation	Title of the report	SDG Goals	Purpose of the project
2019	Nederman AB is a Swedish environnmental technology company operating on a global market, The Clean Air tech company: We shape the future for clean air	Communicating sustainability: an analysis of Nederman's internal sustainability communication	SDG 8 SDG 9 SDG 13	The purpose of this study is to understand and explore how sustainability is communicated throughout the organization of Nederman
2022	Malmberg AB is a Swedish clean tech company with complete solutions in the fields of water treatment. Biogas. Geothermal energy, and drilling. To overcome the biggest challenge of our generation we must aim for nothing: Generation Zero	Market positioning through green marketing	SDG 3 SDG 6 SDG 7 SDG 12	The purpose of this study is to explore how an organization may position themselves through green marketing, in accordance with United Nations Sustainable Development Goals
2022	Lyckeby is a Swedish starch company in the food industry, the business concept is founded on potatoes, processing and utilizing the raw material	Meeting market demands. A study on Lyckeby	SDG 3 SDG 9 SDG 12 SDG 13	The purpose of the study is to understand how Lyckeby works with meeting market demands for plant-based food products
2022	IMS Murtra Nonwowens AB is manufacture non-woven fabrics and distribute them for use in multiple sectors: automotive, medical, liquid filtration, and agriculture. Guided by the SDGs the company aim to play a relevant role in the change and reduce impact on the planet	Murtra Nonwowens	SDG 12 SDG 13	The purpose of this study is geared towards providing insight regarding what we can learn about what has been done in the implementation of SDG goal 12 Responsible production and consumption and make recommendations that will boost its sustainability goals, strengthen is CSR portfolio without jeopardizing the company's competitive strengths

Source Prepared by the author

5 Discussion and Reflections

The results show how the proposed WIL model, if carefully implemented, voice students' literacy on sustainability arguing that using real-life cases creates a pedagogical setting that supports the development of students' sustainability competencies (Chankseliani and McCowan 2021). We herewith consider that through WIL students get profound knowledge about the actual goals as they are. Students then learn how to critically reflect upon and analyse how SDGs can have an impact on and contribute to the organizations' sustainability work. They see firsthand how organizations have the freedom and possibility to choose different goals to work with and how they work in real life with implementing these goals. Combining theory and practice built around these goals and how organizations implement them and the work towards achieving the SDGs affects their everyday operations and the strategic decisions made. This stimulates and strengthens students' capacity to think in terms of change and change agency required for sustainable development. The pedagogical setting for our developed WIL model, as shown in Fig. 1, is suggested to provide practical insights into how educators can help students bridging the gap between theory and practice (Björk 2021). Moreover, the content analysis revealed that even though the difference in purpose of their WIL projects reports the groups were especially interested in learning about how the partner organization is working with the achievement of the chosen SDGs and if the organisation focused their attention on a specific goal and in that case why. The analysis of the survey answers reveals that students felt that universities could benefit from implementing WIL, and how WIL as a pedagogical model was bridging the gap between students theoretical understanding and real-world settings, may foster students as change agents. Students expressed that WIL created a better understanding of sustainability, and not just referring to the environmental dimension that is typically ascribed to sustainability, and thereby clarify some misconceptions. The results further suggests that HEI can benefit from an increased collaboration between universities, business and the society.

Given the design of the WIL model students interact across groups and act as discussants and critical friends of each other's reports on various occasions, and through self-assessment learning students widen their understanding of sustainable development in sharing how organisations strategy, SDGs and steering mechanism can differ substantially. Pointing at that, we found that the designed WIL model addressed in this chapter serves well as a pedagogical learning environment to reduce dualism between theory and practice (Fleming et al. 2018). The student's writing process, analysing and peer-reviewing we believe promotes critical thinking and sustainability literacy. By peer-reviewing other group's projects, the students get a broadened perspective of the subject as well a deeper insight into how different companies work in various ways with the SDGs and the impact it may have on a company's organization, work process, and daily activities. We found that WIL, if carefully designed, creates a feedback-loop between students and educators, practice

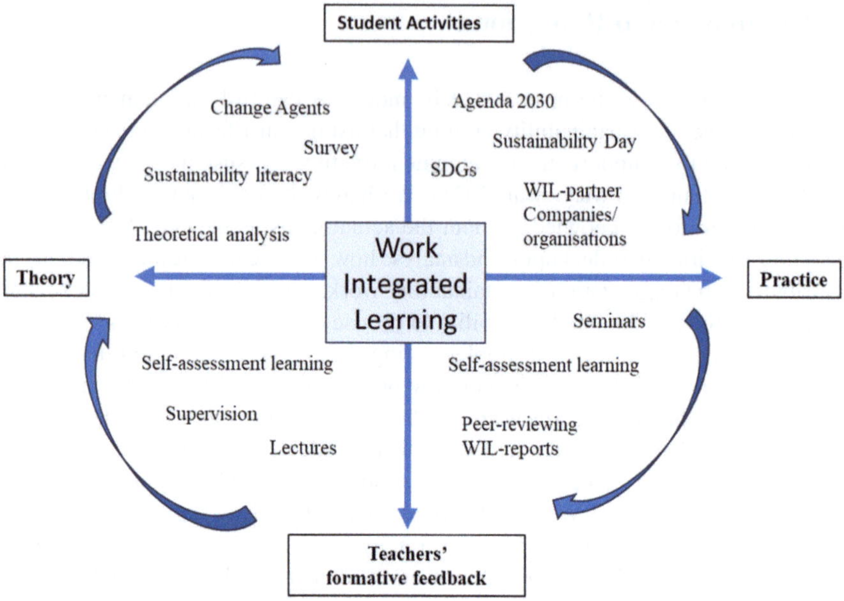

Fig. 2 Work integrated learning pedagogical learning environment through formative teaching and learning. *Source* Author's own

and theory and students' self-assessment learning and multi-faceted real-life experience. Figure 2 demonstrates the efforts in all dimensions composing the feed-back loop.

Moreover, the outcomes of WIL reports demonstrate how students' change in mindset benefits and profits this work by interpreting and analysing a situation from several important perspectives, with climate change being one (Hindley 2022). This chapter suggests that WIL may foster both students' sustainability literacy and climate literacy through education for sustainable development and of learning how organizations work and collaborate with society (Hindley 2022; Cavalcanti-Bandos et al. 2021). Especially since several WIL groups in their project reports chose the Triple Bottom Line as a part of their theoretical framework, which indicate that students consider organizations to focus equally on people, planet, and prosperity. Hence, through education for sustainable development, students are expected to obtain capabilities in planning, executing sustainability research (Cavalcanti-Bandos et al. 2021). We argue that the results show that students gain knowledge, skills and notably change of mindset through applying and solving of real problems (Gale et al. 2021). Solutions that are not only core to WIL partner organizations but also to graduates serving society as future change agents for sustainable development.

6 Conclusions

Related to WIL as a pedagogical learning environment we argue if carefully designed teaches students through complexity of real-life settings and fostering students as future change agents and by bridging theory and practice sustainability agency.

This chapter propose that students through WIL get profound knowledge about the Agenda 2030 and knowledge about challenges implementing and applying the SDGs in a real-life setting. Through WIL students see firsthand how organizations have the freedom and possibility to choose different goals and how they work in real-life situations with implementing these goals. WIL seeks an iterative process through engaging theory and the real-life cases. Students then learn how they, themselves in their future work-life, can have an impact on the organizations' sustainability work. Through WIL we argue students not only gain knowledge, skills and change in mindset, but also develop their strategic thinking preparing them for their future work-life (Redman and Wiek 2021).

This chapter further argues that WIL as a learning environment gives access to contemporary challenges within different organisations and teaches students to "think through complexity, learn through dialogue and communication, engage in deep reflection, develop worldview and values sensitivity, and assess when activities support or detract from achieving the SDGs" (Leal Filho et al. 2019, p. 286). Moreover, this provides HEI educators with continuous access to real and current sustainability challenges that allows for a continuous feedback-loop between students and educators. The continuous feedback-loop also ensures that HEI educators get empirical insights into the challenges of organisations practice (Fig. 2). This then stimulates and strengthens educators' capacity to think in terms of change and change agency required for education for sustainability. We propose that students' survey reflections show that WIL, as proposed, contributes to bridge the duality between theory and practice to create learning progression for graduates' lifelong learning. Hence, this chapter contributes by showing how WIL does not only support students' transformative competencies for a sustainable future (Argento et al. 2023). WIL also by regular and responsive formative teaching through a continuous feedback-loop between students' self-assessment, educators, and partner organisations contributes to support reduction of dualism to create learning progression for graduates' life-long learning. This we propose ultimately enable students as future change agents for a sustainable future.

For continued future research we conclude that previous research has not engaged the full potential of using and recognizing WIL as an important pedagogical model for bridging theory and practice in education for sustainable development and as a learning environment contributing to the needed global transformations by educating graduates as future change agents.

References

Alm K, Melén M, Aggestam-Pontoppidan C (2021a) Advancing SDG competencies in higher education: exploring an interdisciplinary pedagogical approach. Int J Sustain High Educ 22(6):1450–1466. https://doi.org/10.1108/IJSHE-10-2020-0417

Alm K, Beery TH, Eiblemeier D, Fahmy T (2021b) Students' learning sustainability—implicit, explicit or non-existent: a case study approach on students' key competencies addressing the SDGs in HEI program. Int J Sustain High Educ 22(8). https://doi.org/10.1108/IJSHE-12-2020-0484

Argento D, Alm K, Pontoppidan AC (2023) Transformative competencies for a sustainable future. Work Integrated Learning (WIL) at a master's program in business administration. Kristianstad University Press

Björck V (2020) The idea of academia and the real world and its ironic role in the discourse on work-integrated learning. Stud Contin Educ 42(1):1–16. https://doi.org/10.1080/0158037X.2018.1520210

Björck V (2021) Taking issue with how the work-integrated learning discourse ascribes a dualistic meaning to graduate employability. High Educ 82(2):307–322. https://doi.org/10.1007/s10734-020-00650-y

Björck V, Johansson K (2019) Problematising the theory–practice terminology: a discourse analysis of students' statements on work-integrated learning. J Furth High Educ 43(10):1363–1375. https://doi.org/10.1080/0309877X.2018.1483016

Bokhari AA (2017) Universities social responsibility (USR) and sustainable development: a conceptual framework. Int J Econ Manage Stud 4(12):8–16. https://doi.org/10.14445/23939125/ijems-v4i12p102

Brundiers K, Wiek A, Redman CL (2010) Real-world learning opportunities in sustainability: from classroom into the real world. Int J Sustain High Educ 11(4):308–324. https://doi.org/10.1108/14676371011077540

Cavalcanti-Bandos MF, Quispe-Prieto S, Paucar-Caceres A, Burrowes-Cromwell T (2021) Provision of education for sustainability development and sustainability literacy in business programs in three higher education institutions in Brazil, Colombia and Peru. Int J Sustain High Educ 22(5):1467–6370. https://doi.org/10.1108/IJSHE-07-2020-0247

Chankseliani M, McCowan T (2021) Higher education and the sustainable development goals. High Educ 81(1):1–8. https://doi.org/10.1007/s10734-020-00652-w

Elkington J (1994) Towards the sustainable corporation: win-win-win business strategies for sustainable development. Calif Manage Rev 36:90–100. https://doi.org/10.2307/41165746

Fleming J, McLachlan K, Pretti TJ (2018) Successful work-integrated learning relationships: a framework for sustainability. Int J Work-Integr Learn 19(4):321–335, 322

Gale J, Alemdar M, Cappelli C, Morris D (2021) A mixed methods study of self-efficacy, the sources of self-efficacy, and teaching experience. Front Educ 6.https://doi.org/10.3389/feduc.2021.750599

Giangrande N, White RM, East M, Jackson R, Clarke T, Saloff Coste M, Penha-Lopes G (2019) A competency framework to assess and activate education for sustainable development: addressing the UN sustainable development goals 4.7 challenge. Sustainability 11(10):2832. https://doi.org/10.3390/su11102832

Hindley A (2022) Understanding the gap between university ambitions to teach and deliver climate change education. Sustainability 2022(14):13823. https://doi.org/10.3390/su142113823

Hesselbarth C, Schaltegger S (2014) Educating change agents for sustainability—learnings from the first sustainability management master of business administration. J Clean Prod 62:24–36. https://doi.org/10.1016/j.jclepro.2013.03.042

Hsu C-C, Chien F (2022) The impact of high economic growth and technology advancement on extensive energy production in China: evidence using NARDL model. Environ Sci Pollut Res. Scopus. https://doi.org/10.1007/s11356-022-22205-7

Högskolan Kristianstad (2023) Verksamhetsförlagd utbildning (VFU). https://www.hkr.se/utbildningar/student-pahkr/studierna/verksamhetsforlagd-utbildning-vfu/

Jim W, Shen J-P (2016) Higher education for sustainable development: a systematic review. Int J Sustain High Educ 17(5):633–651. https://doi.org/10.1108/IJSHE-01-2015-0004

Kristianstad University (2023a) Work integrated education. https://www.hkr.se/en/study-at-hkr/during-your-studies/work-integrated-education/

Kristianstad University (2023b) PRME report 2023. Sharing information on progress: responsibility, ethics and sustainability. Kristianstad. https://www.hkr.se/contentassets/5441d85ea1e74604bcbb7fc6ba6e72f4/hkr-prme-report-2023.pdf

Kuehl C, Sparks AC, Hodges H, Smith ERAN (2023) Exploring sustainability literacy: developing and assessing a bottom-up measure of what students know about sustainability. Front Sustain. https://doi.org/10.3389/frsus.2023.1167041

Leal Filho W, Shiel C, Paço A, Mifsud M, Ávila LV, Brandli LL, Molthan-Hill P, Pace P, Azeiteiro UM, Vargas VR, Caeiro S (2019) Sustainable development goals and sustainability teaching at universities: falling behind or getting ahead of the pack? J Clean Prod 232:285–294. https://doi.org/10.1016/j.jclepro.2019.05.309

Leal Filho W, Viera Trevisan L, Simon Rampasso I, Anholon R, Pimenta Dinis MA, Londero Brandli L, Sierra J, Lange Salvia A, Pretorius R, Nicolau M, Paulino Pires Eustachio JH, Mazutti J (2023) When the alarm bells ring: why the UN sustainable development goals may not be achieved by 2030. J Clean Prod 407:137108. https://doi.org/10.1016/j.jclepro.2023.137108

Lozano R, Barreiro-Gen M, Lozano FJ, Sammalisto K (2019) Teaching sustainability in European higher education institutions: assessing the connections between competences and pedagogical approaches. Sustainability 11(6):1602

Levi L, Rothstein B (2018) Universities must lead on sustainable development goals. University World News, The Global Window on Higher Education

Miles MB, Huberman AM (1994) Qualitative data analysis, 2nd edn. Sage Publications, London

OECD Future of Education and Skills (2023) Conceptual learning framework. transformative competencies for 2030. OECD. https://www.oecd.org/education/2030-project/teaching-and-learning/learning/transformative/competencies/Transformative_Competencies_for_2030_concept_note.pdf

Patrick C-J, Peach D, Pocknee C, Webb F, Fletcher DM, Pretto G (2008) The WIL (work integrated learning) report: a national scoping study. Australian Learning and Teaching Council, Canberra

Redman A, Wiek A (2021) Competencies for advancing transformations towards sustainability. Front Educ 6:785163. https://doi.org/10.3389/feduc.2021.785163

Scoones I, Stirling A, Abrol D, Atela J, Charli-Joseph L, Eakin H, Ely A, Olsson P, Pereira L, Priya R, van Zwanenberg P, Yang L (2020) Transformations to sustainability: combining structural, systemic and enabling approaches. Curr Opin Environ Sustain 42:65–75. https://doi.org/10.1016/j.cosust.2019.12.004

Steiner G, Scherr J (2013) Higher education for complex real-world problems and innovation: a tribute to Heufler's industrial design approach. Creat Educ 04(07):130–136. https://doi.org/10.4236/ce.2013.47A2016

Stephens JC, Hernandez ME, Román M, Graham AC, Scholz RW (2008) Higher education as a change agent for sustainability in different cultures and contexts. Int J Sustain High Educ 9(3):317–338. https://doi.org/10.1108/14676370810885916

UNESCO (2022). Knowledge-driven actions: transforming higher education for global sustainability. https://doi.org/10.54675/YBTV1653

United Nations (n.d.) The sustainable development agenda. https://www.un.org/sustainabledevelopment/development-agenda/

Wieck A, Withycombe L, Redman CL (2011) Key competencies in sustainability: a reference framework for academic program development. Sustain Sci 6(2):203–218

Xia J, Caulfield C, Ferns S (2015) Work-integrated learning: linking research and teaching for a win-win situation. Stud High Educ 40(9):1560–1572. https://doi.org/10.1080/03075079.2014.882302

MIX
Papier aus verantwortungsvollen Quellen
Paper from responsible sources
FSC® C105338

If you have any concerns about our products, you can contact us on
ProductSafety@springernature.com

In case Publisher is established outside the EU, the EU authorized representative is:
Springer Nature Customer Service Center GmbH
Europaplatz 3, 69115 Heidelberg, Germany

Printed by Libri Plureos GmbH
in Hamburg, Germany